Physical Principles of Sedimentary Basin Analysis

Presenting a rigorous treatment of the physical and mechanical basis for the modeling of sedimentary basins, this book supplies geoscientists with practical tools for creating their own models. It begins with an introduction to the properties of porous media, linear elasticity, continuum mechanics and rock compressibility – providing a thorough grounding for their use later in the text. A chapter on the modeling of burial histories is then followed by a series of chapters on heat flow, subsidence, rheology, flexure and gravity, which consider sedimentary basins in the broader context of the Earth's lithosphere. Later chapters then cover the topics of pore space cementation, compaction and fluid flow.

This volume introduces basic, state-of-the-art models and demonstrates how results can be easily reproduced with simple tools such as MATLAB and Octave (codes are available online at www.cambridge.org/9780521761253). Throughout the book the main equations are derived from first principles, and their basic solutions are obtained and then applied. More technical details are supplied in notes, and the text is illustrated with real-world examples, applications and test exercises. This book is therefore a key resource for graduate students, academic researchers and oil industry professionals looking for an accessible introduction to quantitative modeling of sedimentary basins.

MAGNUS WANGEN has worked in the field of sedimentary basin-modeling since the late 1980s – conducting research on a wide range of topics. He obtained a Dr. Scient. degree in applied mathematics from the University of Oslo in 1993 with a thesis on the modeling of heat and fluid flow in sedimentary basins. Since the early 1990s he has developed two complementary basin simulators used by the oil industry. The first simulator deals with heat flow on a lithospheric scale, fluid flow, compaction and overpressure in sedimentary basins through the geohistory, while the second simulates hydrocarbon generation and migration. He is currently a research scientist at the Institute for Energy Technology in Norway. This book is based on a course in basin analysis that Dr. Wangen taught for a number of years while an assistant professor at UNIK (an affiliate of the University of Oslo at Kjeller).

PHYSICAL PRINCIPLES
OF SEDIMENTARY BASIN
ANALYSIS

Magnus Wangen

Institute for Energy Technology, Norway

CAMBRIDGE
UNIVERSITY PRESS

CAMBRIDGE
UNIVERSITY PRESS

University Printing House, Cambridge CB2 8BS, United Kingdom

One Liberty Plaza, 20th Floor, New York, NY 10006, USA

477 Williamstown Road, Port Melbourne, VIC 3207, Australia

4843/24, 2nd Floor, Ansari Road, Daryaganj, Delhi - 110002, India

79 Anson Road, #06-04/06, Singapore 079906

Cambridge University Press is part of the University of Cambridge.

It furthers the University's mission by disseminating knowledge in the pursuit of education, learning and research at the highest international levels of excellence.

www.cambridge.org
Information on this title: www.cambridge.org/9781108446969

© Magnus Wangen 2010

First published 2010
First paperback edition 2017

A catalogue record for this publication is available from the British Library

ISBN 978-0-521-76125-3 Hardback
ISBN 978-1-108-44696-9 Paperback

To Ingeborg and Reidar

We often fail to realize how little we know about a thing until we
attempt to simulate it on a computer.
Donald Knuth, The Art of Computer Programming
(vol. 1, 3rd edn., p. 298)

Contents

Preface

God is in the details.
The devil is in the details.
Ludwig Mies van der Rohe (among others)

This book is based on lecture notes about the physical processes that govern sedimentary basins. The notes were the basis for a one-semester seminar named "Heat and fluid flow in sedimentary basins" offered by UNIK, an affiliate of the University of Oslo at Kjeller. As the title suggests, this book is about the physical principles of processes in sedimentary basins, for instance, heat and fluid flow. The subject is approached by deriving the basic equations from fundamental principles such as mass and energy conservation. The equations are then solved for simple problems that give insight into the processes.

It should be possible to reproduce most of the solutions, calculations and plots presented in the book with a modest effort and basic computer facilities. Reproduction of the results is the only way to ensure that the results are correct.

The book is written primarily for students who want to study heat and fluid flow in sedimentary basins from a physical point of view, and need to do their own modeling. The book requires some background in mathematics, and knowledge of continuum mechanics is an advantage. The reader should be familiar with calculus and linear algebra. It would be advantageous to be familiar with partial differential equations like the heat equation, Fourier series and complex numbers. As long as the reader is familiar with differentiation and integration, and has a sufficient interest in mathematics, she or he should be able to follow the derivations. Several linear (partial) differential equations are solved, but all details are provided. The aim has been to make the book as self-contained as possible by deriving all results that are presented. Details are necessary in this respect, in order to make the text complete and self-contained.

The book is meant as an introduction to and a primer for modeling, and it therefore covers the basic (state-of-the-art) models. It is not meant to cover the latest developments in the various fields. It does not attempt to cover the historical development of the various subjects either. This is reflected by the reference list, which easily could have been expanded 10 or 100 times. Each chapter has a last section with a few references that may serve as a starting point for further reading. A problem with writing such a book is to decide what should be included and what should be left out. Important topics such as sedimentology, seismics, diagenesis and models for hydrocarbon generation and migration are left out.

Examples and applications of the models are shown. But geological processes are often very complex and specific examples often have several more aspects than those captured by the actual model. These other aspects are mentioned, but a discussion often leads far beyond this book and into special disciplines like for instance sedimentology, structural geology, geochemistry or petrology. The purpose of the examples is to show how the models work with real data, and the setting is therefore chosen to be as simple as possible.

This book can used in different ways depending on the goals, the students' background and the amount of lecturing per week. It is, for example, possible to take two main routes through the book, one with respect to subsidence, rheology, flexure and gravity (Chapters 7, 8, 9 and 10) and another with respect to fluid flow (Chapters 11, 12, 13, 14 and 15). Common for both routes are the following chapters: properties of porous media, continuum mechanics, burial histories and heat flow (Chapters 2, 3, 5 and 6 respectively). Details of derivations are provided in notes that follow the sections, which may be left to the students to go through.

Inevitably, some errors will remain in this book, and in order to correct them I ask that any that are noticed are reported. It would also be great to know if there are alternative and simpler derivations than the ones presented or if there are better examples. Any suggestions that could improve the book will be greatly appreciated.

It is my hope that this book will be useful for anyone interested in a quantitative modeling of processes related to sedimentary basins and Earth science.

Acknowledgments

This text is based on a seminar held at UNIK over six years, a seminar that benefited greatly from the support of Idar Åsen and Kristin Scheen from the staff.

Much of what I know about the modeling of sedimentary basins and of geo-processes in general has been learned through cooperation with others – especially the following friends and colleagues from IFE: Leif Kristian Alm, Bjørg Andresen, Børre Antonsen, Egil Brensdal, Bjørn Fossum, Jan Kihle, Erik Løw, Gotskalk Halvorsen, Olaf Huseby Kjersti Iden, Harald Johansen, Ingar Johansen, Thormod Johansen, Pål Tore Mørkved, Ingrid Anne Munz, Jiri Müller, Harald Hancke Olsen, Tom Pedersen, Jan Sagen, Antoine Saucier, Torfinn Skardhamar, Bent Barman Skaare, Kjell Solberg, Jan Søreng, Torbjørn Throndsen and Åse Unander.

Finally, this project would never have happened without the support and the love from my wife Mona and our children Mia-Sofie, Fredrik, Daniel, Anna-Lousie and Lars Olav.

1

Preliminaries

1.1 Notation

Most problems in this book are solved in 1D along the vertical axis. It is natural to let the surface be at $z = 0$ and to have the z-axis pointing downwards, with positive z-coordinates for the subsurface. An advantage with this choice is that the acceleration of gravity is positive. A potential problem with the z-axis pointing downwards is that Fourier's law gives negative heat flow – heat that flows in the opposite direction to the positive z-axis. A simple solution to this problem is to drop the minus-sign in Fourier's law when the heat flow is computed in practical problems. There is a similar problem with Darcy's law, with the same simple solution. But Fourier's law and Darcy's law retain their minus signs when equations are derived. The full xyz-axis system is right-handed as shown in Figure 1.1b.

Vectors are written with lower case bold letters, as for instance, $\mathbf{v}$, $\mathbf{n}$ or as $\mathbf{n}^T = (n_1, \ldots n_2)$, where T denotes the transpose. Matrices are written with upper case bold letters, for instance like $\mathbf{A}$ and $\mathbf{R}$. The matrix elements are A_{ij} or R_{ij}, where the indices may be x, y and z for the respective spatial directions. Another example of a matrix is

$$K = \begin{pmatrix} k_{xx} & k_{xy} & k_{xz} \\ k_{yx} & k_{yy} & k_{yz} \\ k_{zx} & k_{zy} & k_{zz} \end{pmatrix}. \tag{1.1}$$

Scalar products can be written in several different ways depending on what is most convenient. Here are some examples:

$$\mathbf{x} \cdot \mathbf{y} = \mathbf{x}^T \mathbf{y} = x_1 y_1 + x_2 y_2 + x_3 y_3 = \sum_{i=1}^{3} x_i y_i. \tag{1.2}$$

The second example shows the scalar product as a matrix product, where the vectors are written as row and column matrices. It is often convenient to write summations using what is called Einstein's summation convention, which says that summation is understood for every pair of equal indices. Here is an example: the scalar product

$$\mathbf{x} \cdot \mathbf{y} = \sum_{i=1}^{3} x_i y_i \tag{1.3}$$

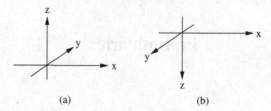

Figure 1.1. *(a) A right-handed coordinate system with the z-axis pointing upwards. (b) A right-handed coordinate system with the z-axis pointing downwards.*

is simply written as

$$\mathbf{x} \cdot \mathbf{y} = x_i y_i \tag{1.4}$$

when using Einstein's summation convention. Here is another example:

$$\sum_{j=1}^{3} \sigma_{ij} n_j = \sigma_{ij} n_j \tag{1.5}$$

which shows the summation over a pair of equal indices. The summation convention is often very useful, but it may lead to confusion. For instance, it implies that $K_{ii} = K_{11} + K_{22} + K_{33}$, which is the sum over the diagonal elements. If we want K_{ii} to denote one (single) diagonal element we have to state that explicitly. One pair of equal indices may be replaced by another pair of equal indices because there is a summation over them – for example $K_{ii} = K_{jj}$. There is never summation over x, y and z when they are used as indices. It is always possible to use these indices as numbers, where $x = 1$, $y = 2$ and $z = 3$. We therefore have that

$$\mathbf{n} = (n_x, n_y, n_z)^T \quad \text{is the same as} \quad \mathbf{n} = (n_1, n_2, n_3)^T. \tag{1.6}$$

An important point is the notation for dimensionless quantities. When depth z is scaled with a characteristic depth h it is written $\hat{z} = z/h$. A hat above a symbol denotes a dimensionless quantity. For example, dimensionless spatial coordinates, time and temperature are $\hat{x}$, $\hat{y}$, $\hat{z}$, $\hat{t}$ and $\hat{T}$.

1.2 Further reading

Riley *et al.* (1998) and Kreyszig (2006) are two comprehensive guides to mathematical methods for physics and engineering.

2

Properties of porous media

2.1 Porosity

Sediments and sedimentary rocks are porous media, and a porous medium is a solid with holes in it. The holes (pores) are normally connected and a fluid may flow through the pore space. The passage from one pore to another is through a pore throat, although there is not always a clear distinction between a pore and a pore throat. The way in which the pores are connected and the size of the pore throats control how permeable a porous medium is for fluid flow. The volume of the pore space controls its capacity to store fluid. Figure 2.1a shows an illustration of a porous medium made of a regular arrangement of spherical grains of equal size. It is a simple idealization of sediments and sedimentary rocks. A real rock has a much more complex pore space than the regular packing of spheres, as seen from the thin section in Figure 2.1b. It consists of grains of a variety of sizes, shapes and minerals. The pore space in rocks is also the result of a complex interplay of mechanical and chemical processes. The *porosity* is the volume fraction of void space of a porous medium, expressed as

$$\phi = \frac{V_p}{V_t} \tag{2.1}$$

where V_p is the volume of the void space and V_t is the total volume (of both solid and void) of the sample. An alternative way to measure the void space is to relate it to the solid volume of the rock rather than the total volume. This property, called the void ratio, is

$$e = \frac{V_p}{V_t - V_p} = \frac{\phi}{1 - \phi}. \tag{2.2}$$

The solid volume of the rock is the difference between the total volume V_t and the void volume V_p. Equation (2.2) can also be inverted to give an expression of the porosity as a function of the void ratio,

$$\phi = \frac{e}{e + 1}. \tag{2.3}$$

As we will see later it is often more convenient to work with the void ratio than the porosity.

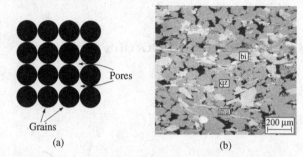

Pores

Grains

(a)   (b)

Figure 2.1. *(a) A regular porous medium made of grains of equal size. (b) A SEM image of a sandstone where the pore space is black and the quartz grains are gray. (qz = quartz, mu = muscovite and bi = biotite)*

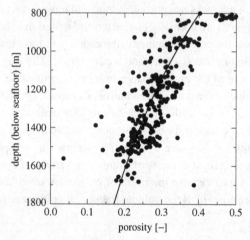

Figure 2.2. *Porosity of clays and silts as a function of depth. Data is from ODP site 1276, leg 210, see Sawyer and Fackler (2007). The porosity–depth trend is fitted with the function $\phi(z) = 0.79 \exp(-z/1180)$, where z is the depth below seafloor in meters.*

It is not possible to obtain a meaningful porosity unless the bulk volume V_t contains a large number of grains. The porous medium is said to be *homogeneous* if the porosity is (almost) constant regardless of where in the medium the volume V_t is taken, and V_t is then called a representative elementary volume, REV. There are two types of porosity – connected and unconnected. It is only the volume of the connected pores that is normally included in the porosity. The term *effective porosity* is used to underline that only connected pores are included.

Sediments and rocks rarely are homogeneous. A characteristic feature of sediments and sedimentary rocks is their layered structure caused by deposition processes. Sedimentary rocks are therefore often strongly heterogeneous in the direction normal to the bedding plane. Figure 2.2 shows an example of clay and silt porosity in a 1000 m depth interval. This is a typical example of the large scatter often seen in sediment porosity, where there

Table 2.1. *Porosity–depth data for lithologies in the North Sea from Sclater and Christie (1980).*

Lithology	ϕ_0	z_0 [m]
Shale	0.63	1960
Sand	0.49	3703
Chalk	0.70	1408
Shaly sandstone	0.56	2464

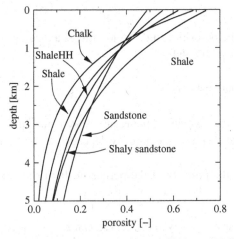

Figure 2.3. *Porosity–depth trends from Sclater and Christie (1980) and Helland-Hansen et al. (1988) (denoted HH).*

are considerable jumps in the porosity over short depth intervals. The porosity–depth trend in Figure 2.2 is fitted with the function

$$\phi(z) = \phi_0 \exp(-z/z_0) \tag{2.4}$$

where ϕ_0 is the surface porosity and z_0 is a depth that characterizes the compaction. The depth z is measured from the sediment surface. This porosity function was first applied by Athy (1930) to the porosity of sedimentary basins and has later been named the Athy function. Figure 2.3 shows the Athy function fitted against data for the lithologies shale, sandstone, chalk and shaly-sandstone from the North Sea. The parameters are obtained by Sclater and Christie (1980) and are listed in Table 2.1. Remember that these curves are smooth trends fitted against observations with a large scatter in the porosities. Another point is that the porosity varies from basin to basin depending on the deposition history and the temperature history. Fortunately, we rarely need to know the detailed porosity when dealing with compaction, subsidence or overpressure build-up on a basin

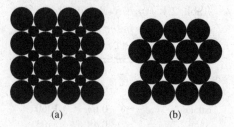

(a)   (b)

Figure 2.4. *Regular porous media. (a) Two grain sizes. (b) Rhombohedral packing.*

scale. We will meet the Athy porosity function later since it is a convenient function to work with.

Exercise 2.1 What is the porosity and the void ratio of a medium with exactly the same amount of void space as solid volume?

Exercise 2.2 Derive relationship (2.3).

Exercise 2.3 Calculate the porosity of the regular porous medium shown in Figure 2.1a, where all grains have equal size. Is the porosity dependent on the grain size (the radius)?

Exercise 2.4 Calculate the porosity of the porous medium shown in Figure 2.4. Notice that grains of different sizes allow for denser packing.

Exercise 2.5 The porosity of a dry rock sample can be measured from the increase in weight by filling the pore space by a wetting fluid. What is the porosity of a sample if its mass increases with Δm, when it is filled with a fluid of density ϱ, and it has the total volume V_t?

2.2 The correlation function and specific surface

A porous medium can specified by the *characteristic* function f defined as

$$f(\mathbf{x}) = \begin{cases} 0, & \text{when } \mathbf{x} \text{ is in a grain} \\ 1, & \text{when } \mathbf{x} \text{ is in the pore space} \end{cases} \tag{2.5}$$

where the porosity of the volume V is seen to be

$$\phi = \frac{1}{V} \int_V f(\mathbf{x}) \, dV. \tag{2.6}$$

The characteristic function can be used to define the two-point correlation function

$$C(\mathbf{r}_1, \mathbf{r}_2) = \frac{1}{V} \int_V f(\mathbf{x} + \mathbf{r}_1) f(\mathbf{x} + \mathbf{r}_2) \, dV \tag{2.7}$$

which expresses how likely it is that the porous medium is void at position $\mathbf{r}_2$, when it is void at position $\mathbf{r}_1$. The medium is *statistically homogeneous* when the correlation function depends only on the distance $r = |\mathbf{r}_2 - \mathbf{r}_1|$ as

$$C(\mathbf{r}_1, \mathbf{r}_2) = C(\mathbf{r}_2 - \mathbf{r}_1) = C(r). \tag{2.8}$$

From the definition (2.7) it follows that the statistically homogeneous correlation function has the following properties:

$$C(0) = \phi \quad \text{and} \quad \lim_{r \to \infty} C(r) = \phi^2. \tag{2.9}$$

The latter relation assumes that the pore space is uncorrelated between any two positions separated by a "large" distance. An important reason for introducing the two-point correlation function is that it gives the surface area of the pore space per unit volume – *the specific surface area*. It is obtained from the two-point correlation function by the following simple relation (Berryman, 1987):

$$S = -4 \left. \frac{dC(r)}{dr} \right|_{r=0}. \tag{2.10}$$

This relation also holds for anisotropic porous media as shown in Note 2.1. The correlation function can be found experimentally for real porous media or exactly for simple models, and once it is obtained we will have the porosity from relation (2.9) and the specific surface area from relation (2.10).

Note 2.1 The derivation of the expression (2.10) follows Berryman (1987). We first introduce the angular average of the correlation function

$$
\begin{aligned}
C_a(r) &= \frac{1}{4\pi} \int C\left(r \mathbf{n}_r(\theta, \varphi)\right) \sin\theta \, d\theta \, d\varphi \\
&= \frac{1}{4\pi V} \int \int_V f(\mathbf{x}) f(\mathbf{x} + r\mathbf{n}_r) \, dV \, \sin\theta \, d\theta \, d\varphi \\
&= \frac{1}{4\pi V} \int \int_{V_p} f(\mathbf{x} + r\mathbf{n}_r) \, dV \, \sin\theta \, d\theta \, d\varphi \tag{2.11}
\end{aligned}
$$

where $\mathbf{n}_r(\theta, \varphi)$ is the unit vector in the direction of r. The integration of φ is from 0 to 2π, and the integration of θ is from 0 to π, see Figure 2.5. The last equality holds because

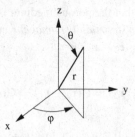

Figure 2.5. *A position in space is given by the spherical coordinates* (r, θ, φ). *The angles, φ from 0 to 2π and θ from 0 to π, parameterize the surface of sphere with radius r.*

$f(\mathbf{x}) = 1$ for $\mathbf{x}$ in the pore space V_p and otherwise 0. The derivative of the angular average of the correlation function is

$$
\begin{aligned}
\frac{dC_a(r)}{dr} &= \frac{1}{4\pi V} \iint \int_{V_p} \frac{\partial f(\mathbf{x} + r\mathbf{n}_r)}{\partial r} \, dV \, \sin\theta \, d\theta \, d\varphi \\
&= \frac{1}{4\pi V} \iint \int_{V_p} \mathbf{n}_r \cdot \nabla f(\mathbf{x} + r\mathbf{n}_r) \, dV \, \sin\theta \, d\theta \, d\varphi \\
&= \frac{1}{4\pi V} \iint \int_{V_p} \nabla \cdot \left(\mathbf{n}_r \, f(\mathbf{x} + r\mathbf{n}_r) \right) dV \, \sin\theta \, d\theta \, d\varphi \\
&= \frac{1}{4\pi V} \iint \int_{A_p} \mathbf{n} \cdot \mathbf{n}_r \, f(\mathbf{x} + r\mathbf{n}_r) \, dA \, \sin\theta \, d\theta \, d\varphi.
\end{aligned}
\tag{2.12}
$$

The volume integral is converted to a surface integral by means of the divergence theorem, and $\mathbf{n}$ is the outward unit vector of the surface A_p of the pore space. The coordinate system is now centered at $\mathbf{x}$ with $\mathbf{n} = \mathbf{n}_z$, which gives that $\mathbf{n} \cdot \mathbf{n}_r = \cos\theta$. The outward normal vector $\mathbf{n}$ of the surface of the pore space is pointing upwards, which means that the surface is locally in the xy-plane around $\mathbf{x}$. The solid is locally above the xy-plane and the pore space is locally below the xy-plane. The integral

$$
\begin{aligned}
I &= \int_0^{2\pi} \int_0^{\pi} \sin\theta \, \mathbf{n} \cdot \mathbf{n}_r \, f(\mathbf{x} + r\mathbf{n}_r) \, d\theta \, d\varphi \\
&= 2\pi \int_0^{\pi} \sin\theta \, \cos\theta \, f(\mathbf{x} + r\mathbf{n}_r) \, d\theta \\
&= 2\pi \int_{\pi/2}^{\pi} \sin\theta \, \cos\theta \, d\theta \\
&= -\pi
\end{aligned}
\tag{2.13}
$$

in the limit $r \to 0$. The function $f(\mathbf{x} + r\mathbf{n}_r)$ is then 0 for $\mathbf{n}_r(\theta, \varphi)$ pointing into the solid, with angles θ from 0 to $\pi/2$. Inserting the integral I into expression (2.12) gives that $dC/dr = -A_p/4V$, where A_p is the surface area of the pore space and the ratio A_p/V is the specific surface.

2.3 The penetrable grain model

The porosity and the characteristic function are not exactly known for other than some simple porous media. One example of such a porous medium is N randomly placed spherical grains of equal radius in a volume V. This model is called the penetrable grain model because the grains are allowed to overlap. A porous medium of penetrable spheres is shown in Figure 2.6a. The inverse of the penetrable sphere model, where solid and void are interchanged, is shown to the right. The inverse model is sometimes called a "Swiss cheese" model, because the pores are now overlapping spheres.

The porosity of the penetrable grain model is equal to the probability that a given point inside V is not overlapped by any of the N grains of volume V_g,

$$\phi = \left(1 - \frac{V_g}{V}\right)^N. \tag{2.14}$$

The probability that a point in V is overlapped by a single grain is V_g/V, when it is assumed that the grains are uniformly distributed. We can replace the volume V by the grain density $\varrho = N/V$, which is the number of grains per unit volume. The porosity is then

$$\phi = \left(1 - \frac{\varrho V_g}{N}\right)^N \tag{2.15}$$

which becomes

$$\phi = \exp(-\varrho V_g) \tag{2.16}$$

in the limit $N \to \infty$. (We have that $(1 + x/N)^N \to e^x$ when $N \to \infty$.) A porous medium of penetrable spheres of radius a has

$$V_g = (4/3)\pi a^3 \tag{2.17}$$

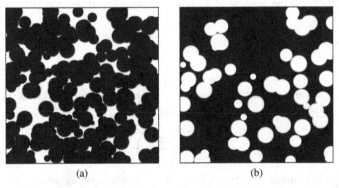

(a)   (b)

Figure 2.6. *(a) Porous medium formed by overlapping spheres. (b) The inverse porous medium of the overlapping sphere model where the solid and the void are interchanged.*

and the porosity is

$$\phi = \exp\left(-\frac{4}{3}\varrho\pi a^3\right). \tag{2.18}$$

The pair correlation function $C(r)$ can be obtained in a similar way as the porosity. It is equal to the probability that two points separated by a distance r inside the volume V is not occupied by a pair of spheres. Using the number density ϱ and going to the limit of a large number of spheres gives

$$C(r) = \exp\left(-\varrho V_2(r)\right) \tag{2.19}$$

where V_2 is the volume of two overlapping spheres separated by a distance r. The volume V_2 is

$$V_2 = \frac{4\pi a^3}{3}\left[1 + \frac{3}{4}\left(\frac{r}{a}\right) - \frac{1}{16}\left(\frac{r}{a}\right)^3\right], \quad r < 2a \tag{2.20}$$

where the sphere radius is a. When $r \geq 2a$ the spheres do not overlap and we have that V_2 is the sum of the volumes of the two spheres,

$$V_2 = \frac{8\pi a^3}{3}, \quad r \geq 2a. \tag{2.21}$$

Relation (2.10) for the specific surface of the porous medium gives

$$S = -4\left.\frac{dC(r)}{dr}\right|_{r=0} = 4\pi a^2 \varrho\phi \tag{2.22}$$

where ϕ is the porosity given by (2.18).

We will later need an expression for the specific surface as a function of the porosity. This is for applications where the porosity is lost due to precipitation of minerals, and where the precipitation process is controlled by the available specific surface. Before we derive an expression for the specific surface as a function of the porosity, we will first find an expression for the volume of a grain as a function of the porosity. We have from (2.16) that

$$V_g(\phi) = V_0\frac{\ln\phi}{\ln\phi_0} \tag{2.23}$$

where the initial porosity ϕ_0 and the initial grain volume V_0 are used to eliminate the number density. From the volume of a grain (2.17) we obtain the following expression for the radius of a sphere as a function of the porosity:

$$a(\phi) = a_0\left(\frac{\ln\phi}{\ln\phi_0}\right)^{1/3}. \tag{2.24}$$

Finally, the specific surface (2.22) as a function of the porosity becomes

$$S(\phi) = -\frac{3\phi\ln\phi_0}{a_0}\left(\frac{\ln\phi}{\ln\phi_0}\right)^{2/3}. \tag{2.25}$$

The expressions (2.23) and (2.24) apply for the penetrable sphere model. These expressions can easily be converted to the inverse model, the "Swiss cheese model," by replacing ϕ by $1 - \phi$.

Exercise 2.6 Find the number density as a function of the porosity (2.18), and use this relation for the number density to show that the pair correlation function (2.19) becomes

$$
C(r) = \begin{cases} \exp\left[\ln\phi \left(1 + \frac{3}{4}\left(\frac{r}{a}\right) - \frac{1}{16}\left(\frac{r}{a}\right)^3\right)\right], & r < 2a \\[3mm] \phi^2, & r \geq 2a \end{cases}
\tag{2.26}
$$

when (2.20) and (2.21) give the volume V_2. Check that the correlation function fulfills both relations (2.9).

Exercise 2.7 Show that volume of two overlapping spheres separated by a distance $r < 2a$ is given by expression (2.20). Use that the volume of the carlot of a sphere is $V_k = (\pi h/6)(3r_k^2 + h^2)$, where h is the carlot height and r_k is the carlot radius, see Figure 2.7b.
Solution: We have that $V_2 = 2(V_g - V_k)$ where V_g is the volume of a sphere and V_k is the volume of one of the two carlots formed by the overlapping spheres, see Figure 2.7a. The height of the carlot is $h = a - r/2$, and the carlot radius is $r_k^2 = a^2 - (a-h)^2 = a^2 - r^2/4$. The volume of the carlot is then

$$
V_k = \frac{\pi}{6}\left(a - \frac{r}{2}\right)\left[3a^2 - \frac{3r^2}{4} + \left(a - \frac{r}{2}\right)^2\right] = \frac{\pi}{6}\left(4a^2 - 3a^2 r + \frac{r^3}{2}\right).
\tag{2.27}
$$

We check that $V_k(r = 0) = 4\pi a^3/6$, which is a half sphere. Another check is that two spheres only touching each other have $V_k(r = 2a) = 0$. The volume of the two overlapping spheres $V_2 = 2(V_g - V_k)$ is then found by inserting the volume of a sphere $V_g = (4/3)\pi a^3$ and the carlot volume V_k from (2.27).

Exercise 2.8 (a) Derive expression (2.23). (b) Derive expression (2.24). (c) Derive expression (2.25).

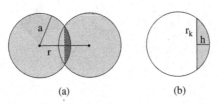

(a)   (b)

Figure 2.7. (a) Two overlapping spheres of radius a that are separated by a distance r. (b) The volume of the carlot is given by the height h and the carlot radius r_k.

2.4 Darcy's law

Darcy's law is an expression for the flux of fluid that is flowing through a porous medium in response to a pressure difference. Figure 2.8 shows a cylindrical core with length l where the fluid is driven through by a pressure difference $\Delta p_f = p_2 - p_1$. The flux v_D of the fluid is measured as volume per cross-section area and per time, and it is expressed as follows by Darcy's law:

$$v_D = -\frac{k}{\mu}\frac{\Delta p}{\Delta l}. \tag{2.28}$$

The minus sign is introduced because the flow is positive in the direction of the pressure drop. The Darcy flux is, for a given core and a given fluid, proportional to the pressure difference across the core and inversely proportional to the length of the core. There are two parameters in Darcy's law (2.28) and they are the *permeability* k and the fluid *viscosity* μ. The permeability is a rock property characterizing the rocks' ability to conduct fluid, while the viscosity is a fluid property expressing the fluids' resistance to deformations. The unit often used for permeability is the darcy (D) or millidarcy (mD). A permeability of 1 D gives the flux 1 cm/s for a fluid with viscosity 1 cP (centipoise) when the pressure gradient is 1 atm/cm. One darcy becomes

$$1\,D = \frac{(1\text{ cm/s})(1\text{ cP})(1\text{ cm})}{1\text{ atm}} = \frac{(10^{-2}\text{ m/s})(10^{-3}\text{ Pa s})(10^{-2}\text{ m})}{1.01325 \cdot 10^5\text{ Pa}}$$
$$= 0.986923 \cdot 10^{-12}\text{ m}^2 \tag{2.29}$$

where we have that $1\text{ atm} = 1.01325 \cdot 10^5$ Pa, and that $1\text{ cP} = 10^{-3}$ Pa s. SI units for permeability are m^2, and without much loss of accuracy we let $1\text{ D} = 10^{-12}$ m^2 and $1\text{ mD} = 10^{-15}$ in applications.

The permeability has been measured for a large range of sediments and rocks, and it is a property that spans several orders of magnitude. Figure 2.9 shows the permeability range for some common rocks. Notice how difficult it is to constrain the permeability within one rock type, for instance sandstone.

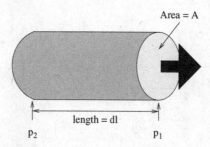

Figure 2.8. *The Darcy flow through a core plug is proportional to the pressure difference* $\Delta p = p_2 - p_1$ *and inversely proportional to the length of the core* Δl.

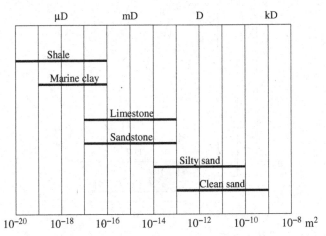

Figure 2.9. *The permeability for common sediments and sedimentary rocks in units* $\log_{10}(m^2)$. *Notice how unconstrained the permeability really is. Even the permeability of one rock type will normally span several orders of magnitude.*

The Darcy flux is related to the average fluid velocity v_f by multiplication with the porosity ϕ

$$v_D = \phi v_f \tag{2.30}$$

assuming that the fluid is only flowing through the fraction ϕ of the area of a cross-section. An important point concerning compacting and deforming porous media is that the Darcy flux is measured relative to the solid skeleton. The Darcy flux in terms of velocities is therefore

$$\mathbf{v}_D = \phi \cdot (\mathbf{v}_f - \mathbf{v}_s) \tag{2.31}$$

where $\mathbf{v}_s$ is the velocity of the solid skeleton.

Darcy's law (2.28) can also be rewritten as the force balance

$$\text{sum forces} = \Delta p A + \frac{\mu}{k} v_D A \, dl = 0 \tag{2.32}$$

where the first term is the pressure force and the second term is the viscous friction force. Newton's second law of motion gives that the momentum is conserved when the forces add up to zero, and Darcy's law can therefore be viewed as an expression for conservation of momentum.

Exercise 2.9 How large a pressure difference is needed across a core plug of length 30 cm, cross-section 100 cm^2 in order for 1 liter per hour to flow through? The core plug has the permeability 1 mD and the viscosity of the fluid is 10^{-3} Pa s. (Answer: 8.33 MPa)

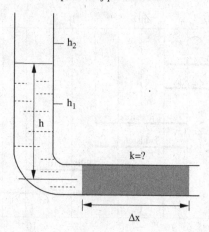

Figure 2.10. *See Exercise 2.10.*

Exercise 2.10 Fluid is driven through a porous core (or plug) by the weight of the fluid in a cylinder, see Figure 2.10. Show that the permeability of the plug is

$$k = \frac{\mu \Delta x \ln (h_2 / h_1)}{\varrho_f g t} \tag{2.33}$$

when the time t is needed for the fluid to drop from height h_2 to h_1. Assume that the area of the cross-section is the same for the cylinder as for the core.

2.5 Potential flow and gravity

We have seen that a fluid pressure gradient drives fluid through a porous medium. But, in the vertical direction we have to subtract the effect of gravity, and Darcy's law becomes

$$v_D = -\frac{k}{\mu} \left(\frac{\partial p_f}{\partial z} - \varrho_f g \right) \tag{2.34}$$

where p_f is the fluid pressure and ϱ_f is the fluid density. The z-axis is pointing downwards, which implies that the weight of a fluid column is increasing with increasing z-coordinate. We see that a fluid pressure equal to the weight of the fluid column, $p_f = \varrho_f g z$, leads to zero Darcy flux. This explains why we had to subtract $\varrho_f g$ in the vertical direction in Darcy's law. In three dimensions Darcy's law can be written as

$$\mathbf{v}_D = -\frac{k}{\mu} \left(\nabla p_f - \varrho_f g \mathbf{n}_z \right) \tag{2.35}$$

where $\mathbf{n}_z$ is the unit vector in the vertical direction. It is often convenient to express Darcy's law in terms of a pseudo-potential Φ defined as the fluid pressure minus a hydrostatic pressure $p_{h,0}$ relative to a reference level as

$$\Phi = p_f - p_{h,0} \tag{2.36}$$

where the hydrostatic pressure

$$p_{h,0} = \int_{z_0}^{z} \varrho_f g \, dz \tag{2.37}$$

is relative to the reference level z_0. A natural choice for a reference level could for instance be the sea level. The reason for introducing the pseudo-potential Φ is that it becomes a true potential when the fluid density is constant. In that case we can write

$$\mathbf{v}_D = -\frac{k}{\mu} \nabla \Phi \tag{2.38}$$

and the Darcy flow becomes proportional to the gradient of Φ, where Φ is the potential. Flow that is proportional to the gradient of a quantity is potential flow, and the quantity is denoted the potential. The unit for the potential Φ is the same as for pressure (Pa in SI units).

It is common to express Darcy's law in terms of the gradient of the *hydraulic head* instead of the fluid flow potential. The hydraulic head h is the fluid flow potential Φ divided by $\varrho_f g$, and Darcy's law becomes

$$\mathbf{v}_D = -\frac{k \varrho_f g}{\mu} \nabla h. \tag{2.39}$$

The hydraulic head measures the fluid pressure in terms of the corresponding height of a fluid column. The factor $k \varrho_f g / \mu$ is called the hydraulic conductivity, and it has SI units m/s. The hydraulic conductivity becomes the Darcy flux driven by a gradient of the hydraulic head equal to 1.

2.6 Permeability as a function of porosity

Fluid flow on a pore scale is from one pore through a pore throat to a neighbor pore. The permeability of the rock is determined by two properties – the size of the pore throats and how well the pores are connected. We will now look at a simple model that accounts for the size of the pore throats, which is a porous medium made of parallel tubes. It is possible to obtain a relationship between the permeability and the porosity for this model, and it leads to the often used Kozeny–Carman relationship for permeability. It turns out that this permeability model applies for porous media of well sorted grains, like for instance well sorted sands and sandstones. Equal grain size corresponds to equal pore throats and a constant tube radius. The starting point for Darcy flow in such a medium is laminar flow in a single tube, which is written as

$$\bar{u} = -\frac{r^2}{8\mu} \frac{dp}{dx} \tag{2.40}$$

where $\bar{u}$ is the average velocity of the fluid in the pipe, r is the radius of the pipe and μ is the viscosity. The Darcy flux v_D, which is the volume rate of fluid per unit time out of a cross-section, is then

$$v_D = -\frac{N\pi r^2}{A}\bar{u} = -\frac{N\pi r^4}{8A\mu}\frac{dp}{dx} \qquad (2.41)$$

for a cross-section A with N pipes, when it is assumed that the cross-section A is normal to the tubes. The permeability is then identified as

$$k = \frac{N\pi r^4}{8A}. \qquad (2.42)$$

The porosity of the medium of pipes is

$$\phi = \frac{N\pi r^2}{A} \qquad (2.43)$$

which is used to replace the pore radius in the permeability (2.42). Using the porosity (2.43) in the Darcy flux (2.41) gives that $v_D = \phi\bar{u}$, which is the general relationship (2.30) between the average fluid velocity in the pores and the Darcy flux. The permeability of the medium becomes

$$k(\phi) = \frac{A}{8\pi N}\phi^2 \qquad (2.44)$$

when the porosity (2.43) replaces the tube radius in expression (2.42). The permeability decreases as $\sim\phi^2$ with decreasing porosity, and it is proportional to the area of the cross-section (A), and inversely proportional to the number of pipes (N). The permeability can also be written as

$$k(\phi) = k_0 \cdot \left(\frac{\phi}{\phi_0}\right)^2 \quad \text{with} \quad k_0 = \frac{\phi_0^2 A}{8\pi N} \qquad (2.45)$$

where $k = k_0$ is the permeability at the reference porosity $\phi = \phi_0$. Notice that permeability has units of length squared (m^2). We have already seen that the specific surface of the pore space has units of inverse length, (m^2/m^3). It is therefore possible to express the permeability k with the inversely of the specific surface squared. A porous medium of tubes has the specific surface area

$$S = \frac{2\pi Nr}{A} \quad \text{or} \quad S^2 = \frac{4N\pi\phi}{A} \qquad (2.46)$$

when the tube radius is replaced by the porosity. The permeability as a function of porosity can then be written

$$k(\phi) = \frac{\phi^3}{2S^2}. \qquad (2.47)$$

This form of permeability function becomes the Kozeny–Carman relationship

$$k(\phi) = \frac{\phi^3}{C\,S_s^2\,(1-\phi)^2} \qquad (2.48)$$

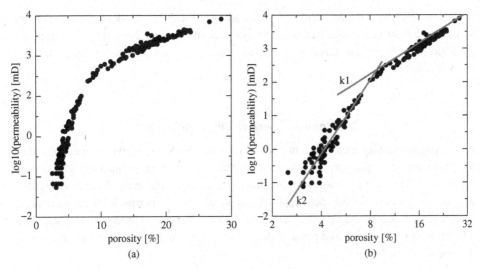

Figure 2.11. *The permeability of Fountainbleau sandstone taken from Bourbie and Zinszner (1985). (a) Linear x-axis for the porosity. (b) Log-scale x-axis for the porosity.*

when S is expressed as $S = (1 - \phi)S_s$, where S_s is the specific surface with respect to the volume of solid instead of the total volume and C is a factor. The permeability (2.48) was proposed by Kozeny (1927) and used by Carman (1937) with $C = 5$ instead of 2. The Kozeny–Carman function does not always fit observations, and empirical relationships are used instead. One common empirical porosity–permeability expression is simply

$$k(\phi) = k_c \, \phi^m \tag{2.49}$$

where the exponent m are fitted with numbers in the range from 2 to 7, and where k_c is constant. Figure 2.11a shows a data set for the Fountainbleau sandstone taken from Bourbie and Zinszner (1985), where the permeability is plotted as a function of porosity. The sandstone is well sorted, and it has a grain size around 250 μm. We see from Figure 2.11a that the porosity is in the range from 2% to 30%, while the permeability range is large – it spans six orders of magnitude. This well sorted sandstone is probably rather unusual because it is clean (100% quartz) with a large span in both porosity and permeability. Figure 2.11b shows a log–log plot of the permeability as function of porosity, and it appears to be a cross-over from one type of permeability model to another at $\sim$8% porosity. Bourbie and Zinszner (1985) fitted the two permeability functions, $k_1(\phi) = 0.303 \, \phi^{3.05}$ and $k_2(\phi) = 2.75 \cdot 10^{-5} \, \phi^{7.33}$ (see Figure 2.11b), where ϕ is a percentage. These functions are of the form (2.49), with $m = 3.05$ and $m = 7.33$, respectively. The Kozeny–Carman permeability, which is $\sim\phi^3$, fits the permeability measurements for $\phi > 8$. The permeability in this regime is controlled by the size of pore throats. The other part of the curve, where $\phi < 8$, is where quartz cementation has filled the original pore space sufficiently for pore throats to become closed. The reduction in permeability in this part is not only due to reduced pore throats, but also to reduced connectivity between the pores.

Nelson (2004b) discusses alternative porosity–permeability models for sedimentary rocks in the light of a number of porosity–permeability studies. Nelson (2005) discusses the importance of the pore throat size in addition to the porosity as a parameter controlling the permeability.

2.7 Empirical permeability relationships

Porosity–permeability measurements often show a large range in the permeability data for nearly the same porosity. This is seen for the two sandstone permeabilities shown in Figure 2.12. This figure shows two more things – there is a trend in the data where increasing porosity gives increasing permeability, and we see that the permeability measurements span several orders of magnitude. The permeability is therefore plotted as $\log_{10}$ values as a function of the porosity, and permeability observations on this form are often fitted with the empirical relationship

$$\log_{10} k = a\phi + b \tag{2.50}$$

where a and b may be parameters in a linear-least-squares fit. The permeability data in Figure 2.12 are fitted with the following two lines: (a) $\log_{10}(k) = 14.9\phi - 1.94$ and (b) $\log_{10}(k) = 19.9\phi - 2.4$. The linear relationship (2.50) can be rewritten as

$$k(\phi) = k_c \exp(a\phi) \tag{2.51}$$

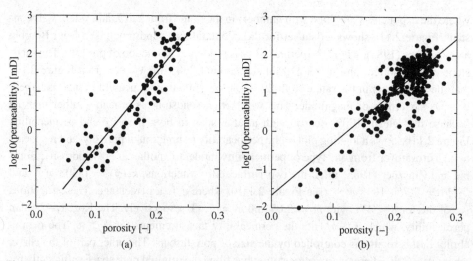

Figure 2.12. *Sandstone permeability in units mD plotted as* $\log_{10}$*-values. The straight lines are linear-least-squares fit of the observations. (a) Data from Bloch et al. (2002) and (b) from Dutton et al. (2003).*

where $k_c = \exp(b)$ is the permeability at zero porosity. Another possibility is to write the permeability as

$$k(\phi) = k_0 \exp\left(a(\phi - \phi_0)\right) \tag{2.52}$$

where $k_0 = \exp(b + a\phi_0)$ is the permeability at a reference porosity ϕ_0. The permeability a controls how strongly the permeability depends on the porosity, and the parameter b is just $\ln(k_c)$. The linear fit to the log-values of the permeability is common with respect to different types of lithologies, clay, shale and carbonate rocks, not just sandstones as in Figure 2.12.

We have so far only looked at permeabilities for sandstones. It is a much studied lithology with respect to petrophysical properties because oil and gas reservoirs are often found in sandstones. Shales on the other hand are much less studied with respect to permeability, although it is commonly the dominating lithology in sedimentary units. Their very low permeability is one reason why there are so few measurements. It is difficult to measure since very small leaks in the testing set-up can easily lead to too large permeabilities. Neuzil (1994) compiled and compared 12 laboratory data sets for clays and shales over a porosity range from almost 0 to 80%. The corresponding span in permeability is nearly nine orders of magnitude. The data compiled by Neuzil (1994) show an even larger scatter in permeability values for the same porosity than the sandstone permeabilities in Figure 2.12. There is also a clear trend in the data sets with a decreasing permeability with decreasing porosity. Neuzil (1994) estimated that the permeability of shales and clays decrease with one order of magnitude for a 13% loss in porosity. This gives that $a = 7.7$ in the log–linear expression (2.50). He also estimated that the permeabilities could be as low as 10^{-23} m^2. Figure 2.13a shows two trends for clays and shales based on the data set of Neuzil (1994),

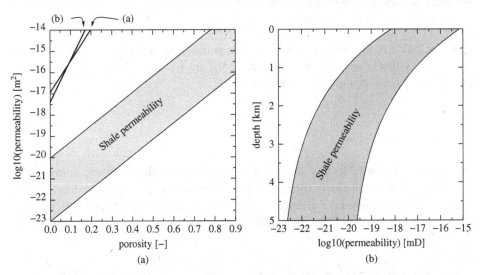

Figure 2.13. (a) Sandstone and shale permeabilities as a function of porosity. (b) Shale permeability as a function of depth.

one where $b = -23$ and another where $b = -20$. We notice that the log–linear relationship is useful both for sandstones, clays and shales. It is often the best porosity–permeability description available unless detailed measurements are made. The sandstone permeabilities from Figure 2.12 are also plotted in Figure 2.13, and they are three orders of magnitude above the upper limit for shale permeability. Figure 2.13b shows how the permeability might decrease with depth for shale. The porosity–permeability area in Figure 2.13a is mapped to permeability depth using Athy's depth–porosity function (2.4) with Sclater and Christie (1980) parameters in Table 2.1 for shale. We see that we may expect shale permeability as low as 10^{-20} m^2 in the depth interval below 1.5 km. It has been claimed that the permeability of shale and clay are dependent on scale because of heterogeneities like for instance fractures. Fractures exist on all length scales and can be important for the average permeability with increasing block size. Neuzil (1994) also compiled permeability estimates from inverse hydrological analyses that cover a much larger length scale than what is possible to measure in the laboratory and found that these permeabilities are consistent with the low permeabilities from the compilation of laboratory measurements.

A third group of rocks that should be mentioned are carbonates, which are also less studied with respect to permeability than the sandstones. The carbonates have a pore space that often contains vugs or small cavities. To what degree the vugs are connected and therefore important for the permeability may be field or site dependent. Nelson (2004b) discusses two data sets for carbonates in addition to several sandstones, and it appears that the permeability of the two carbonates are in a similar porosity–permeability range as the sandstones. Mallon and Swarbrick (2008) compare two chalk data sets from the North Sea and conclude that routine core analyses may overestimate the permeability by orders of magnitude for low permeability cores. Nelson (2004a) and Nelson and Kibler (2003) present a collection of porosity and permeability data sets for core plugs in silicilastic rocks (which may be downloaded via the internet).

Exercise 2.11 Let the permeability be k_1 at a porosity ϕ_1 and k_2 at a different porosity ϕ_2. Show that a and b in the linear fit (2.50) are $b = \log_{10}(k_2/k_1)/(\phi_2 - \phi_1)$ and $a = (\phi_2 \log_{10}(k_1) - \phi_1 \log_{10}(k_2))/(\phi_2 - \phi_1)$.

2.8 The rotation matrix

We will need to rotate vectors and matrices in the work with anisotropic permeability. Rotations are done with a matrix $\mathbf{R}(\theta)$ that transforms a vector from one coordinate system to a coordinate system rotated an angle θ. It is straightforward to derive in 2D. Figure 2.14 shows the transformation of a vector from the unprimed coordinate system to the primed coordinate system, and we have

$$x' = a\cos(\alpha - \theta) = a\cos\alpha\cos\theta + a\sin\alpha\sin\theta = x\cos\theta + z\sin\theta$$
$$z' = a\sin(\alpha - \theta) = a\sin\alpha\cos\theta - a\cos\alpha\sin\theta = -x\sin\theta + z\cos\theta$$

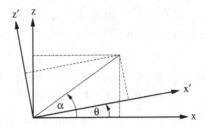

Figure 2.14. *The primed coordinate frame is rotated an angle θ relative to the unprimed coordinate frame.*

because $x = a \cos \alpha$ and $z = a \sin \alpha$, and using matrix notation we can write

$$\begin{pmatrix} x' \\ z' \end{pmatrix} = \begin{pmatrix} \cos\theta & \sin\theta \\ -\sin\theta & \cos\theta \end{pmatrix} \begin{pmatrix} x \\ z \end{pmatrix} = \mathbf{R}(\theta) \begin{pmatrix} x \\ z \end{pmatrix}. \tag{2.53}$$

The rotation matrix can be checked by trying out $\theta = 0$ (no rotation), $\theta = \pi/2$ (rotation by a right angle) and $\theta = \pi$ (rotation 180°). A special case is when the angle θ is small, ($\theta \ll 1$). We can then make the approximations $\cos\theta \approx 1$ and $\sin\theta \approx \theta$, and the rotation matrix becomes

$$\mathbf{R}(\theta) \approx \begin{pmatrix} 1 & \theta \\ -\theta & 1 \end{pmatrix}. \tag{2.54}$$

A nice feature of the rotation matrix is that the inverse of the matrix is the same as the transposed matrix. This property is time-saving in numerical applications where for instance each finite element requires its own rotation matrix and its inverse. The numerical inversion of rotation matrices for a large number of elements could be time consuming. The application of the rotation matrix does not change the length of a vector, and it left as an exercise to show that equation (2.53) preserves the vector length.

The rotation matrix $\mathbf{R}(\theta)$ takes a (fixed) vector from a given coordinate system to a coordinate system rotated an angle θ. It is seen directly from Figure 2.14 that the rotation of the vector an angle θ with respect to a (fixed) coordinate system is the same as a rotation of the coordinate system an angle $-\theta$.

A gradient operator

$$\nabla = \mathbf{n}_x \frac{\partial}{\partial x} + \mathbf{n}_z \frac{\partial}{\partial z} \tag{2.55}$$

rotates to a gradient operator in the primed system just as a vector, and we have that $\nabla' = \mathbf{R}\nabla$.

A simple way to obtain the entries in the rotation matrix is to make the scalar products of the unit vectors along the axes in the two respective coordinate systems. In the 2D example above we see that

$$\mathbf{R} = \begin{pmatrix} \mathbf{n}'_x \cdot \mathbf{n}_x & \mathbf{n}'_x \cdot \mathbf{n}_z \\ \mathbf{n}'_z \cdot \mathbf{n}_x & \mathbf{n}'_z \cdot \mathbf{n}_z \end{pmatrix} \tag{2.56}$$

because we have that $\mathbf{n}'_x = (\cos\theta, \sin\theta)$ and $\mathbf{n}'_z = (-\sin\theta, \cos\theta)$. This form of writing the rotation matrix is easily generalized to 3D, where we have that

$$\mathbf{R} = \begin{pmatrix} \mathbf{n}'_x \cdot \mathbf{n}_x & \mathbf{n}'_x \cdot \mathbf{n}_y & \mathbf{n}'_x \cdot \mathbf{n}_z \\ \mathbf{n}'_y \cdot \mathbf{n}_x & \mathbf{n}'_y \cdot \mathbf{n}_y & \mathbf{n}'_y \cdot \mathbf{n}_z \\ \mathbf{n}'_z \cdot \mathbf{n}_x & \mathbf{n}'_z \cdot \mathbf{n}_y & \mathbf{n}'_z \cdot \mathbf{n}_z \end{pmatrix}. \tag{2.57}$$

We see that the entries in the rotation matrices are simply the elements of the primed unit vectors in the unprimed system. This form of the rotation matrix is useful in numerical applications, where it might be necessary to compute a rotation matrix for each element in a finite element grid. The scalar product of two unit vectors $\mathbf{n}'_i$ and $\mathbf{n}_j$ is also the cosine of the angle between the unit vectors, because $\mathbf{n}'_i \cdot \mathbf{n}_j = \cos(\alpha_{ij})$, where α_{ij} is the angle between $\mathbf{n}'_i$ and $\mathbf{n}_j$. The rotation matrix is therefore sometimes called the *direction cosine matrix*.

The 2D rotation matrix (2.53) rotates round the y-axis. Such rotations in 3D around the x-, y- and the z-axes are

$$\mathbf{R}_x(\alpha) = \begin{pmatrix} 1 & 0 & 0 \\ 0 & \cos\alpha & \sin\alpha \\ 0 & -\sin\alpha & \cos\alpha \end{pmatrix}, \quad \mathbf{R}_y(\beta) = \begin{pmatrix} \cos\beta & 0 & -\sin\beta \\ 0 & 1 & 0 \\ \sin\beta & 0 & \cos\beta \end{pmatrix} \tag{2.58}$$

and

$$\mathbf{R}_z(\gamma) = \begin{pmatrix} \cos\gamma & -\sin\gamma & 0 \\ \sin\gamma & \cos\gamma & 0 \\ 0 & 0 & 1 \end{pmatrix}. \tag{2.59}$$

These rotation matrices can be multiplied together:

$$\mathbf{R}(\alpha, \beta, \gamma) = \mathbf{R}_x(\alpha)\,\mathbf{R}_y(\beta)\,\mathbf{R}_z(\gamma) \tag{2.60}$$

where any possible rotation is specified by the three angles α, β and γ. An important point is that the product (2.60) depends on the order of the multiplication of the rotation matrices.

Exercise 2.12 Show that the length of a vector remains the same after a rotation of the coordinate system by an angle θ.

Exercise 2.13 Show the following equalities:

$$\mathbf{R}(\theta)^{-1} = \mathbf{R}(-\theta) = \mathbf{R}(\theta)^T. \tag{2.61}$$

Solution: The first equality is obvious because $\theta - \theta = 0$. The second equality follows from expression (2.57) for the rotation matrix, where the inverse rotation is obtained by interchanging the primed and unprimed basis vectors.

Exercise 2.14 A coordinate system has as unit vectors

$$\mathbf{n}'_x = \frac{1}{\sqrt{3}}(1, 1, 1)^T, \quad \mathbf{n}'_y = \frac{1}{\sqrt{6}}(1, -2, 1)^T, \quad \mathbf{n}'_z = \frac{1}{\sqrt{2}}(1, 0, -1)^T. \tag{2.62}$$

(a) Check that the unit vectors are orthonormal.
(b) Find the rotation matrix (2.57).
(c) Check that the transpose of the rotation matrix is its inverse.

Exercise 2.15
(a) Show that the rotation of the gradient operator is the same as the rotation of a vector.
(b) Show that the Laplace operator is rotation invariant.
Solution: (a) Equation (2.53) is a linear transformation of a coordinate position in the unprimed to the primed reference system. Let f be a function of x and z. The chain rule of differentiation then gives

$$\frac{\partial f}{\partial x} = \frac{\partial f}{\partial x'}\frac{\partial x'}{\partial x} + \frac{\partial f}{\partial z'}\frac{\partial z'}{\partial x} = \cos\theta\,\frac{\partial f}{\partial x'} - \sin\theta\,\frac{\partial f}{\partial z'} \qquad (2.63)$$

$$\frac{\partial f}{\partial z} = \frac{\partial f}{\partial x'}\frac{\partial x'}{\partial z} + \frac{\partial f}{\partial z'}\frac{\partial z'}{\partial z} = \sin\theta\,\frac{\partial f}{\partial x'} + \cos\theta\,\frac{\partial f}{\partial z'} \qquad (2.64)$$

which is the same as

$$\nabla f = \mathbf{R}^{-T}\nabla' f \qquad (2.65)$$

or

$$\nabla' = \mathbf{R}\nabla. \qquad (2.66)$$

(b) We have that

$$\nabla'^2 = (\nabla')^T\nabla' = (\mathbf{R}\nabla)^T(\mathbf{R}\nabla) = \nabla^T\mathbf{R}^T\mathbf{R}\nabla = \nabla^T\nabla = \nabla^2. \qquad (2.67)$$

2.9 Anisotropic permeability

The permeability of rocks is rarely isotropic. Especially sedimentary rocks often have a layered structure caused by the deposition process. The permeability normal to the layers is often much less than the permeability parallel to the layers because of low permeable sheets. It is therefore necessary to distinguish between the two perpendicular directions. Assume that a rock has a layered structure in the z-direction, and that the permeability in the xy-plane is isotropic. Darcy's law for both perpendicular directions at the same time can be written as the vector

$$\mathbf{v}_D = \begin{pmatrix} v_x \\ v_z \end{pmatrix} = \begin{pmatrix} -\frac{k_x}{\mu}\frac{\partial\Phi}{\partial x} \\ -\frac{k_z}{\mu}\frac{\partial\Phi}{\partial z} \end{pmatrix} = -\frac{1}{\mu}\begin{pmatrix} k_x & 0 \\ 0 & k_z \end{pmatrix}\nabla\Phi \qquad (2.68)$$

where k_x is the permeability in the bedding plane and k_z is the permeability normal to the bedding plane. The anisotropic permeability is a diagonal matrix

$$\mathbf{K}' = \begin{pmatrix} k_x & 0 \\ 0 & k_z \end{pmatrix}. \qquad (2.69)$$

Properties of porous media

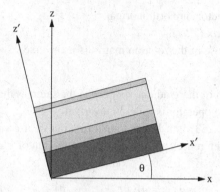

Figure 2.15. *The rotated block has a different permeability in the x'-direction than in the z'-direction because of the layered structure.*

The reference frame where the permeability becomes a diagonal matrix is called the *principal coordinate system*, and the permeability values are the *principal permeabilities*. The principal system is generally not aligned with the coordinate system we are working in, where the x-axis is horizontal and the z-axis is vertical, because sedimentary layers may be folded or tilted as shown in Figure 2.15. The diagonal permeability matrix (tensor) of the tilted principal system must therefore be rotated to the coordinate system we are using. The rotation of the Darcy flux, which is a vector, leads to the rotation of the permeability tensor. The Darcy flux in the principal system is marked with a prime, and the rotation is expressed as

$$\mathbf{v}_D \;=\; \mathbf{R}\mathbf{v}'_D = -\frac{1}{\mu}\mathbf{R}\mathbf{K}'\nabla'\Phi \tag{2.70}$$

where $\mathbf{R} = \mathbf{R}(\theta)$ is the rotation matrix that rotates a vector an angle θ from the primed to the unprimed system (see Figure 2.15). The primed gradient operator is for the primed system. We can now insert the identity matrix written as $\mathbf{R}^{-1}\mathbf{R} = \mathbf{I}$ as follows:

$$\mathbf{v}_D \;=\; -\frac{1}{\mu}\mathbf{R}\mathbf{K}'\,\mathbf{R}^{-1}\,\mathbf{R}\nabla'\Phi \;=\; -\frac{1}{\mu}\mathbf{K}\nabla\Phi \tag{2.71}$$

where we identify

$$\mathbf{K} = \mathbf{R}\mathbf{K}'\mathbf{R}^{-1} \tag{2.72}$$

as the rotated permeability matrix (tensor) and $\nabla = \mathbf{R}\nabla'$ as the rotated gradient operator in the actual (unprimed) coordinate system.

Permeability is in general anisotropic and it is represented by a permeability matrix (tensor)

$$\mathbf{K} = \begin{pmatrix} k_{xx} & k_{xy} & k_{xz} \\ k_{yx} & k_{yy} & k_{yz} \\ k_{zx} & k_{zy} & k_{zz} \end{pmatrix}. \tag{2.73}$$

We will assume that there exists a coordinate system where the permeability tensor becomes diagonal. The permeability tensor in any other reference frame with a different orientation is obtained by rotation of the diagonal permeability tensor. The application of the rotation matrix on a diagonal matrix leads to a symmetric matrix. The existence of a diagonal permeability matrix therefore implies the symmetry property of the permeability tensor. This means that $k_{yx} = k_{xy}$, $k_{zx} = k_{xz}$ and $k_{yz} = k_{zy}$, and that there are only six independent elements in the tensor (2.73).

2.10 Directional permeability

The Darcy flux $\mathbf{v}_D$ is in the case of potential flow

$$\mathbf{v}_D = -\frac{1}{\mu} \mathbf{K} \nabla \Phi \qquad (2.74)$$

but the direction of the Darcy flux ($\mathbf{v}_D$) and the gradient of the potential ($\nabla \Phi$) are not necessarily the same in the case of anisotropic permeability. We will now introduce the *directional permeability*, which is the (scalar) permeability in the direction of the potential gradient. The directional permeability can be thought of as the permeability of a thin cylindrical core taken in the wanted direction of the anisotropic rock. The gradient of the potential can be written

$$\nabla \Phi = |\nabla \Phi| \mathbf{n} \qquad (2.75)$$

where $\mathbf{n}$ is the unit vector in the direction of the gradient. The Darcy flux in the direction of $\mathbf{n}$ is $v_n = \mathbf{n} \cdot \mathbf{v}_D$, and the permeability in the same direction is therefore

$$k_n = \frac{v_n}{\frac{1}{\mu}|\nabla \Phi|} = \mathbf{n}^T \mathbf{K} \mathbf{n}. \qquad (2.76)$$

The directional permeability (2.76) becomes

$$k_n = k_x n_x^2 + k_y n_y^2 + k_z n_z^2 \qquad (2.77)$$

in the principal system, where the permeability tensor is diagonal. The directional permeability (2.77) can also be represented by an ellipsoid, and to see that we let $n_x = x/r$, $n_y = y/r$ and $n_z = z/r$ where $x^2 + y^2 + z^2 = r^2$. In the case $r = 1/\sqrt{k_n}$ we get the general equation for an ellipsoid

$$\frac{x^2}{a^2} + \frac{y^2}{b^2} + \frac{z^2}{c^2} = 1 \qquad (2.78)$$

with $a = 1/\sqrt{k_x}$, $b = 1/\sqrt{k_y}$ and $b = 1/\sqrt{k_z}$. The radius of the ellipsoid is therefore an expression for directional permeability, and the semi-axis represents the principal permeability. The Darcy flow $\mathbf{v}_D$ is actually normal to the tangent plane of the ellipsoid at the point (x, y, z), a property that is shown in Note 2.2 (see Figure 2.16).

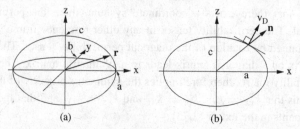

Figure 2.16. *(a) The scalar directional permeability k_n gives the radius of the ellipsoid as $r = 1/\sqrt{k_n}$. (b) The unit vector $\mathbf{n}$ is in the direction of the gradient of the potential, and the Darcy flux vector $\mathbf{v}_D$ is normal to the tangent plane of the ellipsoid.*

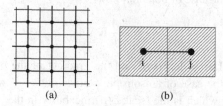

Figure 2.17. *(a) The cell centers in a grid are connected by transmissibilities. (b) The transmissibility connecting cell i and j.*

The direction permeability can be used in numerical applications to compute the *transmissibilities*, which is the conductivity of the bonds that connect the cells in a grid. Figure 2.17a shows a grid where the cell centers are connected by bonds. The fluid flow in Figure 2.17b from cell i to cell j is the transmissibility between the two cells multiplied by the pressure difference. The flow rate (in units of volume fluid per time) from cell i to cell j is

$$q = A v_n = A \frac{k_n}{\mu} \frac{(p_j - p_i)}{l_{ij}} \tag{2.79}$$

where A is the area of the cross-section between the cells, k_n is the directional permeability between the cell centers, $p_j - p_i$ is the pressure difference between the cells and l_{ij} is the distance between the cell centers. The directional permeability is made with the unit vector in the direction of the pressure gradient, which is the direction of the transmissibility. Most of the parameters in the flow rate (2.79) are given together as the *transmissibility*

$$T_{ij} = \frac{A k_n}{\mu l_{ij}}. \tag{2.80}$$

The flow rate q from cell i to cell j is therefore the transmissibility T_{ij} times the pressure difference $p_j - p_i$ between the cells. For a strongly anisotropic rock the directional permeability will be quite nonuniform for connections pointing in different directions.

Note 2.2 The Darcy flux vector $\mathbf{v}_D$ is normal to the tangent plane of the ellipsoid, because it is proportional to the gradient of the ellipsoid function

$$f(x, y, z) = \frac{x^2}{a^2} + \frac{y^2}{b^2} + \frac{z^2}{c^2}. \tag{2.81}$$

We have that $\nabla f = (2xk_x, 2yk_y, 2zk_z) = 2r(k_x n_x, k_y n_y, k_z n_z)$, which is parallel to $\mathbf{v}_D$. Recall that $\nabla \Phi$ is parallel to $\mathbf{n}$ and $\mathbf{v}_D$ is therefore parallel to $\mathbf{K}\,\mathbf{n} = (k_x n_x, k_y n_y, k_z n_z)$.

Exercise 2.16 The permeability in the principal system is

$$\mathbf{K} = \begin{pmatrix} k_x & 0 \\ 0 & k_z \end{pmatrix}. \tag{2.82}$$

(a) Show that the directional permeability is $k_n = k_x \cos^2 \theta + k_z \sin \theta$ in the direction of the unit vector $\mathbf{n} = (\cos\theta, \sin\theta)$.
(b) Let $k_z \ll k_x$ and show that the directional permeability is reduced from its maximum at $\theta = 0$ to approximately half at $\theta = 45°$.

Exercise 2.17 Show how the directional permeability $k_n = \mathbf{n}^T \mathbf{K}\,\mathbf{n}$ for a general anisotropic permeability tensor can be rewritten using a diagonal tensor.
Solution: Let $\mathbf{R}$ be the matrix that rotates the coordinate system into the principal coordinate system, which makes $\mathbf{D} = \mathbf{R}\mathbf{K}\mathbf{R}^T$ diagonal. We can then write

$$k_n = \mathbf{n}^T \mathbf{R}^T \mathbf{R}\mathbf{K}\mathbf{R}^T \mathbf{R}\mathbf{n} = \mathbf{m}^T \mathbf{D}\,\mathbf{m} \tag{2.83}$$

where $\mathbf{m} = \mathbf{R}\mathbf{n}$ is the unit vector in the principal coordinate system.

2.11 Average permeability

Sedimentary rocks are often layered due to the deposition of different lithologies. An example of a block of sediments with a layered structure is shown in Figure 2.18. We will need the average permeability of such a block, when the spatial resolution

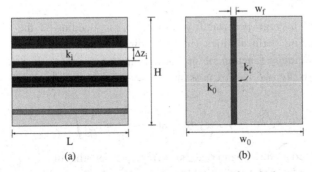

Figure 2.18. *(a) A block of layered sedimentary rock, where layer i has thickness z_i and permeability k_i. (b) A layer of rock with a vertical fracture zone.*

is the size of the block. The average permeability k_x is first found in the x-direction, (parallel to the bedding), where it is the permeability that yields the average Darcy flux. The average Darcy flux v_x is obtained from the total rate of fluid passing through the block

$$v_x \sum_i \Delta z_i = -\sum_i \frac{\Delta z_i k_i}{\mu} \frac{\Phi}{L} \tag{2.84}$$

where k_i is the permeability (in the x-direction) of the ith layer, Δz_i is the layer thickness, L is the width of the block and Φ is the potential difference across the block. Since the average Darcy flux v_x is related to the average permeability k_x, we get

$$v_x = -\frac{k_x}{\mu}\frac{\Phi}{L} \quad \text{where} \quad k_x = \frac{1}{H}\sum_i k_i \Delta z_i \tag{2.85}$$

where $H = \sum_i \Delta z_i$ is the height of the block. The average permeability k_x along the bedding is the *arithmetic* average of the bedding permeability.

Next we find the average permeability normal to the bedding. We note that the Darcy flow normal to the bedding is the same through each layer. The average Darcy flux normal to the bedding is therefore

$$v_z = -\frac{k_i}{\mu}\frac{\Delta \Phi_i}{\Delta z_i} \tag{2.86}$$

where $\Delta \Phi_i$ is the potential difference across the ith layer. The potential difference across the entire block in the z-direction is

$$\Phi = \sum_i \Delta \Phi_i = -v_z \mu \sum_i \frac{\Delta z_i}{k_i}. \tag{2.87}$$

The average permeability in the z-direction is the permeability that gives v_z from Darcy's law when there is a potential difference Φ applied across a block with height $H = \sum_i \Delta z_i$. We then have

$$v_z = -\frac{k_z}{\mu}\frac{\Phi}{H} \quad \text{where} \quad \frac{1}{k_z} = \frac{1}{H}\sum_i \frac{\Delta z_i}{k_i}. \tag{2.88}$$

We see that the average permeability normal to the bedding is the *harmonic* average of the bedding permeabilities (in the z-direction). In case the bedding permeability is a continuous function of the z-coordinate $k = k(z)$, instead of a discrete bedding permeability, we can express the average permeabilities (2.85) and (2.88) as integrals

$$k_x = \frac{1}{H}\int k(z)\,dz \quad \text{and} \quad \frac{1}{k_z} = \frac{1}{H}\int \frac{dz}{k(z)}. \tag{2.89}$$

It is also possible to go further and find the average permeability of more general structures than those that are layered, but the problem is then that the flow field inside the block is no longer 1D.

Exercise 2.19 shows when a vertical fracture dominates the average permeability of a layer. This example is in 2D, while it might be that fracture flow in 3D is more similar to channel flow in thread-like structures in the fracture zone.

Exercise 2.21 studies the anisotropy of a formation made of two layers of different thickness and different permeabilities. The anisotropy ratio of the formation is the ratio of the average parallel and the average normal permeabilities. The exercise shows that the anisotropy ratio is a function of the thickness ratio of the two layers and the ratio of their permeabilities (as we would expect).

Exercise 2.18 Show that the average permeability of a thick layer of sediments is

$$k_z = \frac{n(z_2 - z_1)}{z_0} \cdot \frac{k_1 k_2}{k_1 - k_2} \tag{2.90}$$

where the sediment permeability is a function of the porosity, $k(\phi) = k_0 \cdot (\phi/\phi_0)^n$, and the porosity is a function of the depth, $\phi(z) = \phi_0 \exp(-z/z_0)$. The layer is in the depth interval from z_1 to z_2, and the permeabilities at z_1 and z_2 are k_1 and k_2, respectively.

Exercise 2.19 Let us assume that a layer of rock with later extent w_0 has a vertical fracture zone of width w_f where $w_o \gg w_f$ (see Figure 2.18b). The rock has the permeability k_0 and the fracture zone has the permeability k_f.
(a) Show that the average vertical permeability of the layer can be approximated as

$$k_{av} \approx k_0 + k_f \left(\frac{w_f}{w_0}\right). \tag{2.91}$$

(b) Show that permeability of the layer can be characterized as either rock-dominated or fracture-dominated as follows:

$$k_{av} = \begin{cases} k_0, & k_f w_f \ll k_0 w_0, & \text{rock-dominated} \\ k_f(w_f/w_0), & k_f w_f \gg k_0 w_0, & \text{fracture-dominated.} \end{cases} \tag{2.92}$$

The fracture zone is negligible with respect to vertical fluid flow in the rock-dominated regime. But the layer has an average permeability that is much larger than the rock permeability in the fracture-dominated regime, and most of the vertical fluid flow, therefore, takes place in the fracture zone.
(c) Assume that the fracture zone is a parallel plate fracture, which has the permeability $k_f = w_f^2/12$. What is the condition for a fracture-dominated fracture in terms of the fracture width w_f?
(d) How wide must the fracture be for the layer to be fracture-dominated when $k_0 = 10^{-18}$ m^2 and $w_0 = 100$ m?
Solution:
(a) The average permeability is the linear average in this case

$$k_{av} = \frac{k_0 w_0 + k_f w_f}{w_0 + w_f} \approx k_0 + k_f \left(\frac{w_f}{w_0}\right) \tag{2.93}$$

when $w_0 \gg w_f$.

(b) The average permeability can be written as

$$k_{av} = k_0\left(1 + \frac{k_f w_f}{k_0 w_0}\right) \tag{2.94}$$

which shows that the average permeability is much larger than the rock permeability k_0 when $k_f w_f \gg k_0 w_0$.

(c) The condition for fracture-dominated average permeability becomes $w_f^3/12 \gg k_0 w_0$ or $w_f \gg (12 k_0 w_0)^{1/3}$.

(d) $w \gg 10^{-5}$ m.

Exercise 2.20 Let a thin horizontal sheet of thickness h_1 be inserted into a layer of rock with thickness h_0. The sheet and the layer have the permeabilities k_1 and k_0, respectively. The sheet has a thickness that is much less than the layer ($h_1 \ll h_0$).

(a) Show that the average vertical permeability of the composite layer is

$$k_{av} = \begin{cases} k_0, & h_1/h_0 \ll k_1/k_0, & \text{rock-dominated} \\ k_1(h_0/h_1), & h_1/h_0 \gg k_1/k_0, & \text{sheet-dominated.} \end{cases} \tag{2.95}$$

There are two regimes, one where the sheet does not alter the permeability of the block significantly (rock-dominated), and another regime where the sheet controls the permeability of the block (sheet-dominated).

(b) Assume that there is a pressure difference Δp across the layer in the vertical direction. Show that this pressure difference is almost entirely across the sheet when the average permeability is sheet-dominated.

Solution:

(a) We have

$$k_{av} = \frac{h_0 + h_1}{(h_0/k_0) + (h_1/k_1)} \approx \frac{k_0}{1 + \frac{k_0 h_1}{k_1 h_0}}. \tag{2.96}$$

(b) The vertical Darcy flow through the layer is

$$u = \frac{k_{av}}{\mu}\frac{\Delta p}{(h_0 + h_1)} \approx \frac{k_1}{\mu}\frac{\Delta p}{h_1} \tag{2.97}$$

when $h_0 \gg h_1$.

Exercise 2.21 A rectangular block is composed of two parts – one part of permeability k_1 and thickness h_1 and another part of permeability k_2 and thickness h_2, as shown by Figure 2.19. The permeabilities parallel and normal to the layer are

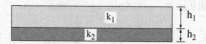

Figure 2.19. *A block is composed of two layers – one with permeability k_1 and thickness h_1 and another of permeability k_2 and thickness h_2. See Exercise 2.21.*

$$k_\| = \frac{k_1 h_1 + k_2 h_2}{h_1 + h_2} \qquad (2.98)$$

$$k_\perp = \frac{h_1 + h_2}{(h_1/k_1) + (h_2/k_2)} \qquad (2.99)$$

respectively. The anisotropy is expressed by the ratio of the permeabilities $R = k_\|/k_\perp$.
(a) Show that R can be written as

$$R = \frac{(h_1 k_1 + h_2 k_2)(h_1 k_2 + h_2 k_1)}{(h_1 + h_2)^2}. \qquad (2.100)$$

(b) Use the result from (a) to rewrite R as

$$R = \frac{(1 + R_h R_k)(1 + R_h/R_k)}{(1 + R_h)^2} \qquad (2.101)$$

where $R_h = h_2/h_1$ and $R_k = k_2/k_1$. The anisotropy ratio R depends only on the thickness ratio h_2/h_1 and permeability ratio k_2/k_1 as we would expect. The anisotropy should be independent of both length scale and permeability scale of the formation.
(c) Verify that $R = 1$ for $R_k = 1$ regardless of R_h.

2.12 Fourier's law and heat conductivity

The heat flux q through a solid material is given by Fourier's law

$$q = -\lambda \frac{\partial T}{\partial x} \qquad (2.102)$$

as the product of the heat conductivity λ of the material and the temperature gradient in the material, $\partial T/\partial x$. The heat flux is measured as energy (J) through a cross-section (m^2) per time (s), and it is therefore given in units $W\,m^{-2}$. The thermal gradient has units °C/m, which implies that units for the thermal conductivity is $W\,m^{-1}\,K^{-1}$.

The heat conductivities of the sediments and rocks show a large variability because of pore fluids and different mineral compositions of the solid skeleton (see Table 2.2). There are also cases of heat conductivity measurements on the same (or nearly the same) rocks by different laboratories that are quite different. For example Midttømme *et al.* (1998) measured lower heat conductivities than Bloomer (1981) for claystones and mudstones from the UK, and Norden and Förster (2006) measured higher heat conductivities than in previous studies for sandstones in the Northeast German Basin.

The heat conductivity of minerals are much better constrained than those for rocks, since they have a specific crystal structure and a chemical formula. Clauser and Huenges (1995) have collected the heat conductivities for a large number of minerals. The heat conductivity of water is considerably lower than for most minerals, which makes the porosity an important controlling factor for the average heat conductivity of fluid saturated rocks and sediments. The average bulk heat conductivity is often related to the heat conductivities of the solid and the fluid using the porosity as a normalizing weight. A commonly used average is the geometric mean

Table 2.2. *A range of heat conductivities for sedimentary rocks at 20°C (Blackwell and Steele, 1989).*

Lithology	Heat conductivity [W m^{-1} K^{-1}]
Claystone and siltstone	0.80–1.25
Shale	1.05–1.45
Sand	1.70–2.50
Sandstone	2.50–4.20

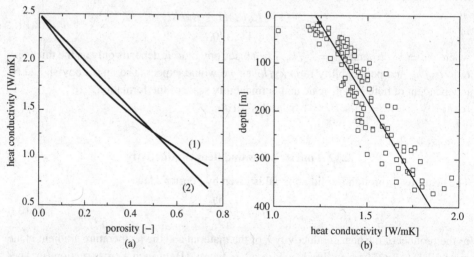

Figure 2.20. *(a) Curve (1) is the average heat conductivity (2.103) when $\lambda_f = 0.64$ W/Km and $\lambda_s = 2.5$ W/Km. The straight line (curve 2) is a linear approximation, which is equal at the two porosities $\phi = 0$ and $\phi = 0.5$. (b) In-situ heat conductivity measurements from ODP Leg 122 Hole 763A (Pribnow et al., 2000). A linear-least-squares fit to the data is also shown.*

$$\lambda(\phi) = \lambda_f^{\phi}\lambda_s^{(1-\phi)} = \lambda_s \left(\frac{\lambda_f}{\lambda_s}\right)^{\phi} \tag{2.103}$$

where λ_f and λ_s are the heat conductivities of the fluid and the rock, respectively. The geometric mean is between the arithmetic and harmonic means, and it does not have a simple physical basis as the other two means. (These three ways of averaging are compared in Exercise 2.25.) The geometric mean of heat conductivities $\lambda_s = 2.5$ W m^{-1} K^{-1} and $\lambda_f = 0.64$ W m^{-1} K^{-1} is shown in Figure 2.20a for ϕ in the interval from 0 to 0.75. A linear approximation that is equal to the average heat conductivity at $\phi = 0$ and $\phi = 0.5$ is also shown. A proper value of λ_s or the bulk heat conductivity are often difficult to constrain unless observations are made. There might even be a large scatter in values for the same lithology inside the same formation, as shown by Norden and Förster (2006) for sedimentary rocks in the Northeast German Basin.

Table 2.3. *Mean values for the heat conductivity of various rock types are taken from Seipold (1998). The temperature range is at least 0 °C to 400 °C.*

Rock types	λ_0 [W m^{-1} K^{-1}]	c_0 [10^{-4}°C^{-1}]
Amphibolites	2.4	4.6
Basalts	2.5	3.6
Granites	3.2	13.0
Granulites	2.7	9.9
Gneiss	3.0	10.4
Pyroxenites	2.9	4.1
Serpentinites	2.2	2:4
Olivine rocks	5.1	16.2

Table 2.4. *The heat conductivities of the minerals are taken from Clauser and Huenges (1995). The temperature range is 0 °C to 400 °C.*

Mineral	λ_0 [W m^{-1} K^{-1}]	c_0 [10^{-4}°C^{-1}]
Salt (NaCl)	6.2	48.5
SiO_2	5.9	23.0
$(Mg,Fe)SiO_3$	3.8	8.5

Figure 2.20b shows in situ measurements of heat conductivity from ODP leg122 hole 763A (Pribnow *et al.*, 2000). There is an increasing heat conductivity with depth, which can be explained by a decreasing porosity with depth. Heat conductivity–depth trends can be made by combining the average heat conductivity (2.103) with a porosity–depth function. Exercise 2.27 shows how it is possible to estimate a porosity–depth function using a data set like the one in Figure 2.20b.

Table 2.3 shows the average values for the heat conductivity of various rock types. These rocks have a heat conductivity in the interval from 2.2 W m^{-1} K^{-1} to 3.2 W m^{-1} K^{-1}, with a mean value 2.7 W m^{-1} K^{-1}, when olivine rocks are excluded. We notice from Table 2.4 that salt has a high heat conductivity at surface conditions compared to common rocks.

The heat conductivity of rocks show normally a slight decrease with increasing temperature. This temperature dependence is often expressed as

$$\lambda(T) = \frac{\lambda_0}{1 + c_0 T} \tag{2.104}$$

where T is in °C and λ_0 is a reference heat conductivity at $T = 0$ °C. The temperature dependence is accounted for by the parameter c_0, and we see that the heat conductivity halves when the temperature increases from $0\,°C$ to $T = 1/c_0$. Table 2.3 shows examples of the coefficient c_0 for various rocks. For example $c_0 = 5 \cdot 10^{-4}°C^{-1}$ gives a reduction of 7% when the temperature increases from $0\,°C$ to $150\,°C$. The temperature dependence is therefore negligible in the basin sediments and also to some degree in the crust, because of the uncertainty with respect to rock types and the value of λ_0.

Increasing pressure leads to increasing heat conductivity for rocks and minerals. The increase is expressed by adding a factor $1 + \alpha_0 p$ to the temperature-dependent heat conductivity. The relationship (2.104) therefore generalizes to

$$\lambda(p, T) = \lambda_0 \left(\frac{1 + \alpha_0 p}{1 + c_0 T} \right). \tag{2.105}$$

Both pressure and temperature increase with depth, and their effect on the heat conductivity are in opposite directions. The coefficient α_0 has been measured in the range 0.05 to 0.2 GPa^{-1} for crystalline rocks by Seipold (1995). If we estimate the pressure at 4 km depth to be 0.1 GPa we get that the heat conductivity increases by 2% by using the upper bound for α_0. The effect is therefore negligible in sedimentary basins.

Most rocks have an anisotropic heat conductivity because of anisotropy in the mineral texture, like for instance foliation. The anisotropy is often measured with the ratio of the heat conductivities parallel to the layering and normal to the layering, when isotropy is assumed in the bedding plane. The anisotropy ratio has been measured to be in the range 1.1 to 1.5 for clay and claystones and mudstones from the UK (Midttømme *et al.*, 1998). Even stronger anisotropy might appear on the formation scale because of interlayering of sandstones and shales.

The heat conductivity is in general a tensor quantity

$$\lambda = \begin{pmatrix} \lambda_{xx} & \lambda_{xy} & \lambda_{xz} \\ \lambda_{yx} & \lambda_{yy} & \lambda_{yz} \\ \lambda_{zx} & \lambda_{zy} & \lambda_{zz} \end{pmatrix} \tag{2.106}$$

which is symmetric, just like the permeability. Exactly the same results apply for the heat conductivity tensor as for the permeability tensor, for instance rotation, principal values, directional heat conductivity and averaging.

The heat conductivity appears quite often in the heat equation as thermal diffusivity, $\kappa = \lambda/\varrho c$, which is the heat conductivity (λ) divided by the density (ϱ) and the specific heat capacity (c). Table 2.5 shows the density, specific heat capacity and the thermal diffusivity for several common minerals. Except for quartz the thermal diffusivity is $\sim 10^{-6}\ m^2\,s^{-1}$. A sediment matrix (skeleton) is a mixture of several minerals and we are normally interested in the average density and the average specific heat capacity for the solid. The average density of a mineral ensemble, where mineral i has mass m_i and volume V_i, is simply

$$\varrho_{\mathrm{av}} = \frac{\sum m_i}{\sum V_i} = \sum \frac{m_i}{V_i} \frac{V_i}{V_{\mathrm{tot}}} \tag{2.107}$$

Table 2.5. *Density, heat conductivity and specific heat capacity for common minerals in sediments. The data are taken from Goto and Matsubayashi (2008).*

Mineral	Density kg/m3	Thermal conductivity $W\,m^{-1}\,K^{-1}$	Specific heat $J\,kg^{-1}\,K^{-1}$	Thermal diffusivity $10^{-6}m^2\,s^{-1}$
Quartz	2648	7.69	741	3.92
Albite	2620	2.20	776	1.08
Anorthite	2760	1.68	745	0.82
Orthoclase	2570	2.32	707	1.28
Muscovite	2831	2.32	796	1.03
Illite	2660	1.85	808	0.86
Smectite	2608	1.88	795	0.91
Chlorite	2800	5.15	818	2.25
Calcite	2710	3.59	820	1.62
Seawater	1024	0.59	3993	0.15
Mud (grain)	2731	3.40	758	1.64

where $V_{tot} = \sum V_i$ is the total volume of the solid. The average density is therefore calculated with respect to the volume fraction V_i/V_{tot} of each mineral. The density of common minerals are ~ 2600 kg m^{-3} according to Table 2.5, which is a fair estimate for the average density of a sediment matrix. The average specific heat capacity is calculated similarly, but with respect to the mass fraction m_i/m_{tot} of each mineral in the bulk solid

$$c_{av} = \frac{\sum c_i m_i}{\sum m_i} = \sum c_i \frac{m_i}{m_{tot}} \quad (2.108)$$

where $m_{tot} = \sum m_i$ is the total mass. Table 2.5 shows that a typical average specific heat capacity is 800 J kg^{-1} K^{-1}.

We notice from Table 2.5 that water has a heat conductivity that is less than half the heat conductivity of minerals, and that it has a specific heat capacity that is about five times higher than for minerals. Water has less than half the density of the minerals and the heat capacity per unit volume for water is therefore approximately double that for minerals. The porosity is therefore an important parameter for a sediment's ability to conduct and store heat.

Exercise 2.22 The temperature at the basin surface and at the depth 2000 m is measured to be 0 °C and 90 °C, respectively. The heat flux through the sedimentary column is $q = 0.035$ W m^{-2}. What is the average heat conductivity of the interval?

Exercise 2.23 How much does the average heat conductivity given by function (2.103) change when the porosity is changed by $\Delta\phi$?

Exercise 2.24 (a) Show that a linear average heat conductivity function

$$\lambda(\phi) = \lambda_s \cdot (1 - c\,\phi) \quad (2.109)$$

which is equal to the average heat conductivity function (2.103) at the two porosities $\phi = 0$ and $\phi = \phi_2$ has c given by

$$c = \frac{1}{\phi_2}\left(1 - \exp\left(\phi_2 \ln(\lambda_f/\lambda_s)\right)\right).$$ (2.110)

(b) What is c when $\lambda_f = 0.64$ W/Km and $\lambda_s = 2.5$ W/Km?

(c) How much does a change $\Delta\phi$ in the porosity change the linear average heat conductivity?

Exercise 2.25 Plot the following three functions of ϕ in the interval from 0 to 1 for $a = 1$ and $b = 0.1$:

$$\begin{array}{lll}
f_a(\phi) = \phi a + (1 - \phi)b & \text{arithmetic} \\
f_g(\phi) = a^\phi b^{(1-\phi)} & \text{geometric} \\
f_h(\phi) = \dfrac{1}{(\phi/a) + ((1 - \phi)/b)} & \text{harmonic.}
\end{array}$$ (2.111)

Which average is largest, in between and least?

Exercise 2.26 The temperature-dependent heat conductivity may be represented as

$$\lambda(T') = \frac{1}{A + BT'}$$ (2.112)

where T' is the temperature in kelvin. An example is the data provided by Seipold (1995, 1998). Show that (2.112) can be rewritten in the form of (2.104) where $\lambda_0 = 1/(A + BT_0)$, $c_0 = B/(A + BT_0)$ and $T_0 = 273°$C.

Exercise 2.27 This exercise shows how in situ measurements of sediment heat conductivities can be used to estimate a porosity–depth function. Let heat conductivity as a function of porosity be linearly approximated as

$$\lambda(\phi) = \lambda_s(1 - c\phi)$$ (2.113)

where $c = 0.988$. (See Exercise 2.24.) Assume that the Athy porosity–depth function also can be linearly approximated,

$$\phi(z) = \phi_0 \exp(-z/z_0) \approx \phi_0 (1 - z/z_0),$$ (2.114)

which is a reasonable approximation for shallow depths compared to z_0.

(a) Show that a linear representation of *in-situ* measurements of heat conductivity as a function of depth

$$\lambda(z) = \lambda_0 + az$$ (2.115)

gives the following parameters in Athy's porosity function:

$$\phi_0 = \frac{1}{c}\left(1 - \frac{\lambda_0}{\lambda_s}\right) \quad \text{and} \quad z_0 = \frac{1}{a}\left(\lambda_s - \lambda_0\right).$$ (2.116)

(b) The linear-least-squares fit to the heat conductivity data in Figure 2.20b is $\lambda_0 = 1.22$ W m^{-1} K^{-1} and $a = 0.0016$ W m^{-2} K^{-1}. What are the parameters ϕ_0 and z_0 in this case?

2.13 Further reading

Guéguen and Palciaukas (1994) give an introduction to rock physics. Dullien (1979) examines the relationship between transport properties and pore structure of porous materials. Monicard (1980) deals with laboratory measurements of rock properties. Turquato (2002) is a broad presentation of construction and reconstruction of porous media and their properties. Clauser and Huenges (1995) present thermal conductivities for rocks and minerals.

3

Linear elasticity and continuum mechanics

3.1 Hooke's law, Young's modulus and Poisson's ratio

Forces that act on a body lead to deformations. A material is said to be linear elastic if the deformations are proportional to the forces. How forces and deformations are related is most easily shown in a basic one-dimensional experiment where a rod is stretched a distance dl by a force F pointing in the same direction as the rod. Figure 3.1 shows such a stretched rod. The relative elongation of the rod is the longitudinal *strain*, $\varepsilon = dl/l_0$, where l_0 is the initial length of the rod. The *stress* acting on the rod is the force per cross-section area, $\sigma = F/A$. The stress and the strain for linear elastic materials are related as

$$E = \frac{\text{stress}}{\text{strain}} = \frac{F}{A}\frac{l_0}{dl} \tag{3.1}$$

where the constant of proportionality, E, is *Young's modulus*. The relation (3.1) is called *Hooke's law*. The radius of a cross-section of the rod becomes reduced by $dr = r - r_0$ when it is stretched, where r and r_0 are the radius of the stretched and unstretched rod, respectively. The relative amount of reduction is measured as the transverse strain, $-dr/r_0$. A minus sign has been added because dr is a negative quantity. It turns out that the ratio of transverse strain and longitudinal strain is a constant for linear elastic materials

$$\nu = \frac{\text{transverse strain}}{\text{longitudinal strain}} = \frac{dr}{dl}\frac{l_0}{r_0} \tag{3.2}$$

where the constant of proportionality is *Poisson's ratio*.

Exercise 3.1 Show that the work needed to stretch a rod a distance s is $W = \frac{1}{2}EV(s/l_0)^2$, where $V = Al_0$ is the volume of the rod. Notice that the energy per volume becomes $\frac{1}{2}E\epsilon^2$ $\frac{1}{2}\sigma\epsilon$, where strain is $\epsilon = s/l_0$ and stress is $\sigma = E\epsilon$.

3.2 Bulk modulus

A body will have its volume reduced if it is subjected to a normal stress across its whole surface. This normal stress could, for instance, be produced by immersing the body in a fluid with pressure p. The volume change caused by the pressure p is denoted dV, and

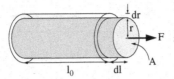

Figure 3.1. *A rod of length l and cross-section area A is stretched by a force F. The rod is dl longer after being stretched and its radius is reduced by dr.*

the relative volume change is the strain, $\varepsilon = dV/V_0$, where V_0 is the initial volume of the body. The stress and the strain are linearly related for linear elastic materials:

$$K = \frac{\text{stress}}{\text{strain}} = -\frac{p\,V_0}{dV} \tag{3.3}$$

where the constant K is the *bulk modulus*. The minus sign is added because dV is negative for a positive (compressive) pressure p. The definition of the bulk modulus is related to the definition of compressibility

$$C = -\frac{1}{V}\frac{dV}{dp} \tag{3.4}$$

which shows that the bulk modulus (3.3) is an inverse čompressibility, $K = 1/C$.

Exercise 3.2 Show that the compressibility can be expressed as

$$C = -\frac{1}{\varrho}\frac{\partial \varrho}{\partial p} \tag{3.5}$$

in terms of the density.

3.3 Shear modulus

A third type of strain is produced by tangential surface forces, as shown in Figure 3.2. The force F acts parallel to the top surface with area A of the rectangular block, and the top surface becomes sheared relative to the lower surface. The strain is in this case measured as the ratio of horizontal displacement to the height of the block, $\varepsilon = dx/h = \tan\theta$. The shear stress is the tangential force divided by the surface area, and the stress–strain relation is

$$G = \frac{\text{stress}}{\text{strain}} = \frac{F\,h}{A\,dx} = \frac{F}{A\tan\theta} \approx \frac{F}{A\theta}. \tag{3.6}$$

For small deformations, when $\tan\theta \approx \theta$, we have that the *shear modulus* G is the shear stress over the deformation angle. Young's modulus, Poisson's ratio and the shear modulus are not independent. It turns out that there are only two independent moduli for linear elastic materials. The shear modulus can be expressed by Young's modulus and Poisson's ratio as

$$G = \frac{E}{2(1+v)} \tag{3.7}$$

which is shown in Exercise 3.18.

Linear elasticity and continuum mechanics

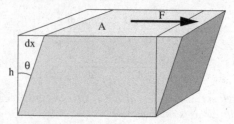

Figure 3.2. *A rectangular block is subjected to a tangential force F over its top surface with area A. The shear force has moved the upper surface a distance dx relative to the lower fixed surface. The height of the block is h.*

Exercise 3.3 Show that the work needed to shear the block in Figure 3.2 by an angle θ is $W = \frac{1}{2}G\theta^2 V$, where $V = Ah$ is the volume of the block.

3.4 Strain

We have seen that the strain of a stretched rod is $\varepsilon = dl/l_0$, when the rod's initial length is l_0 and it is stretched a distance dl. A point x along the initial rod becomes displaced a distance $u(x) = x\, dl/l_0$ by the stretching. Such a displacement function makes it possible to obtain the strain locally by measuring the difference in displacement of two nearby points. The strain at a position x along the rod is the difference in displacement between two points x and $x + dx$ relative to the distance dx separating the points, $\varepsilon = (u(x + dx) - u(x))/dx \approx \partial u/\partial x$ (assuming that dx is small). Notice that it is differences in the displacement field that produce strain. For example, a constant displacement does not result in any strain, only a pure translation. The use of a (scalar) displacement function to obtain the strain can be generalized to three dimensions by use of the displacement vector $u_i = u_i(x_1, x_2, x_3), i = 1, 2, 3$. The difference in displacement between two nearby points separated by (dx_1, dx_2, dx_3) can now be written

$$
\begin{aligned}
du_i &= \frac{\partial u_i}{\partial x_j} dx_j \\
&= \left\{ \frac{1}{2}\left(\frac{\partial u_i}{\partial x_j} + \frac{\partial u_j}{\partial x_i} \right) + \frac{1}{2}\left(\frac{\partial u_i}{\partial x_j} - \frac{\partial u_j}{\partial x_i} \right) \right\} dx_j \\
&= \left(\varepsilon_{ij} + \mathcal{R}_{ij} \right) dx_j.
\end{aligned}
\tag{3.8}
$$

(Notice that the summation convention is used, where there is summation over all pairs of equal indices.) The partial derivative $\partial u_i/\partial x_j$ has been rewritten as a sum of a symmetric part

$$
\varepsilon_{ij} = \frac{1}{2}\left(\frac{\partial u_i}{\partial x_j} + \frac{\partial u_j}{\partial x_i} \right)
\tag{3.9}
$$

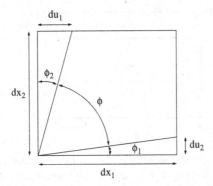

Figure 3.3. *The deformation of a right angle.*

and an antisymmetric part

$$\mathcal{R}_{ij} = \frac{1}{2}\left(\frac{\partial u_i}{\partial x_j} - \frac{\partial u_j}{\partial x_i}\right) \tag{3.10}$$

(ε_{ij} is symmetric because $\varepsilon_{ij} = \varepsilon_{ji}$, and $\mathcal{R}_{ij}$ is antisymmetric because $\mathcal{R}_{ij} = -\mathcal{R}_{ji}$.) The reason for expressing $\partial u_i/\partial x_j$ as these two parts is that the symmetric part ε_{ij} expresses volume and shape changes, while the antisymmetric part $\mathcal{R}_{ij}$ expresses rotations. We will later see that we have to distinguish rotations from volume and shape changes when deformation (strain) is related to forces (stress).

We have already observed that a diagonal element of the symmetric part ε_{ij}, for instance ε_{11}, represents relative length change in the x-direction. (If two points along the x-axis, x_1 and $x_1 + dx_1$, become displaced by $u_1(x_1)$ and $u_1(x_1 + dx_1)$ respectively, then the relative length change between the points is $du_1/dx_1 \approx \partial u_1/\partial x_1 = \varepsilon_{11}$.)

The off-diagonal elements of ε_{ij} are the amount a right-angle becomes deformed. See Figure 3.3, where we have that $\phi_1 \approx du_2/dx_1$ and $\phi_2 \approx du_1/dx_2$, because we can make the approximation $\tan\phi \approx \phi$ for small angles (resulting from small deformations). The right-angle then becomes deformed by an amount

$$\phi_1 + \phi_2 = \frac{du_1}{dx_2} + \frac{du_2}{dx_1} = 2\,\varepsilon_{12}. \tag{3.11}$$

The off-diagonal term ε_{12} is thus one-half the amount a right-angle becomes deformed. The interpretation of the two other off-diagonal elements ε_{13} and ε_{23} is similar.

It is the sum of the diagonal elements of ε_{ij} that expresses local volume changes. This is seen by considering a rectangular box with sides dx_i ($i = 1, 2, 3$) before it is deformed. The sides of the box are $dx_i + du_i$ after the deformation, where du_i is the difference in displacement of the two diagonal opposite corners $\mathbf{x}$ and $\mathbf{dx}$, see Figure 3.4. The difference in volume of the box caused by the deformation is then

$$dV = du_1 dx_2 dx_3 + dx_1 du_2 dx_3 + dx_1 dx_2 du_3$$
$$= \left(\frac{du_1}{dx_1} + \frac{du_2}{dx_2} + \frac{du_3}{dx_3}\right) dx_1\, dx_2\, dx_3 \tag{3.12}$$

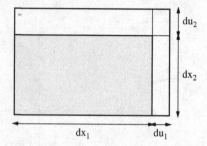

Figure 3.4. *A rectangular box with initial sides dx_1 and dx_2 is displaced by a deformation, and the difference in the displacement of the two parallel sides are du_1 and du_2. The change in volume is $dV = dx_1 du_2 + du_1 dx_2$, when the second-order contribution $du_1 du_2$ to dV is ignored. (It is already assumed that $du_i \ll dx_i$.)*

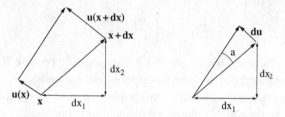

Figure 3.5. *Left: the two points $\mathbf{x}$ and $\mathbf{x} + d\mathbf{x}$ are displaced by the vectors $\mathbf{u}(\mathbf{x})$ and $\mathbf{u}(\mathbf{x} + d\mathbf{x})$, respectively. Right: the two points are separated by $d\mathbf{x} + d\mathbf{u}$ after the displacement, where $d\mathbf{u} = \mathbf{u}(\mathbf{x} + d\mathbf{x}) - \mathbf{u}(\mathbf{x})$. The distance $d\mathbf{u} + d\mathbf{x}$ between the two points after the displacement is the initial distance vector between the points $d\mathbf{x}$ rotated on an angle a.*

and we get

$$\frac{dV}{V} \approx \frac{\partial u_i}{\partial x_i} = \varepsilon_{ii} \tag{3.13}$$

(where the summation convention is used). The relative volume change, also called the *dilatation*, is simply the sum of the diagonal elements of ε_{ij}.

We will now look at the antisymmetric part $\mathcal{R}_{ij}$ in two dimensions to convince ourselves that it really expresses rotation. The difference in the displacement du_i between two points initially separated by (dx_1, dx_2) becomes

$$\begin{bmatrix} du_1 \\ du_2 \end{bmatrix} = \begin{bmatrix} 0 & -a \\ a & 0 \end{bmatrix} \begin{bmatrix} dx_1 \\ dx_2 \end{bmatrix} \tag{3.14}$$

where $\mathcal{R}_{12} = -\mathcal{R}_{21} = -a$. See Figure 3.5. The distance between the two points in the deformed state is $(dx_1', dx_2') = (dx_1 + du_1, dx_2 + du_2)$, which can be written

$$\begin{bmatrix} dx_1' \\ dx_2' \end{bmatrix} = \begin{bmatrix} 1 & -a \\ a & 1 \end{bmatrix} \begin{bmatrix} dx_1 \\ dx_2 \end{bmatrix}. \tag{3.15}$$

The displacement vector between the two points in the deformed state is a rotation by an angle a of the displacement vector of the initial state. This rotation matrix is identical to the rotation matrix in (2.54), which applies for small rotation angles.

3.5 Stress

Stress is a generalization of pressure, and similar to pressure it is defined as force per unit area of a surface. The pressure in a fluid is a special case of stress where the stress is always normal to the surface. Let the force $\mathbf{F}$ be acting on a (small) plane with surface area A, as shown in Figure 3.6. It is assumed that the plane is so small that the force can be considered constant over the surface. The stress vector acting on the plane is then the force divided by the surface area, $\mathbf{S} = \mathbf{F}/A$. The vector $\mathbf{S}$ can be decomposed into a part normal to the plane, $\mathbf{S}_n$, and a part tangential to the plane, $\mathbf{S}_t$, and we have that $\mathbf{S} = \mathbf{S}_n + \mathbf{S}_t$. Notice that the stress on the opposite side of the plane is $-\mathbf{S}$ because the force on that side of the plane is $-\mathbf{F}$. This follows from Newton's law which says that the sum of all forces is zero on a body at rest.

A stress state can be related to a coordinate system by using the forces acting on the surface planes of a cube oriented parallel to the coordinate axes, (see Figure 3.6). The force acting on the plane that is normal to the x-direction gives the stress vector $\boldsymbol{\sigma}_x$, when it is divided by the surface area. The vector $\boldsymbol{\sigma}_x$ can be written in terms of its vector components as $\boldsymbol{\sigma}_x = (\sigma_{xx}, \sigma_{xy}, \sigma_{xz})$, where σ_{xx} is the normal stress and where σ_{xy} and σ_{xz} are the tangential (or shear) stress decomposed in the y- and z-directions respectively. The stresses on the sides with outward unit vector pointing in the positive y- and z-directions are found in the same way. We then get three stress vectors with nine components:

$$\boldsymbol{\sigma}_x = (\sigma_{xx}, \sigma_{xy}, \sigma_{xz}) \tag{3.16}$$

$$\boldsymbol{\sigma}_y = (\sigma_{yx}, \sigma_{yy}, \sigma_{yz}) \tag{3.17}$$

$$\boldsymbol{\sigma}_z = (\sigma_{zx}, \sigma_{zy}, \sigma_{zz}). \tag{3.18}$$

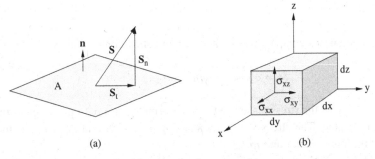

(a)   (b)

Figure 3.6. (a) *The stress acting on a plane can be decomposed into a normal stress component* ($\mathbf{S}_n$) *and a tangential (or shear) component* ($\mathbf{S}_t$). (b) *The stress components on the sides of a cube.*

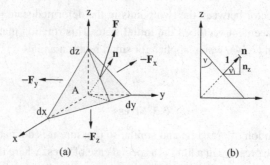

(a)   (b)

Figure 3.7. (a) The force acting on an arbitrary oriented triangle is balanced by the forces on the orthogonal sides of the tetrahedron. (b) A cutting plane through the tetrahedron in which both the z-axis and the normal vector lie. Notice that **n** is first translated to the point on the triangle where it passes through the origin. It is then seen that $\frac{1}{2}dx\,dy$ is equal to $A\,n_z$.

The components of the stress vectors form the matrix

$$\boldsymbol{\sigma} = \begin{pmatrix} \sigma_{xx} & \sigma_{xy} & \sigma_{xz} \\ \sigma_{yx} & \sigma_{yy} & \sigma_{yz} \\ \sigma_{zx} & \sigma_{zy} & \sigma_{zz} \end{pmatrix}. \tag{3.19}$$

We will refer to this matrix as the stress tensor, although tensors are not formally introduced. The knowledge of the stress state in terms of these components allow us to compute the force and the stress state on any (infinitesimal) plane with any orientation. The stress vector **S** on the triangle in Figure 3.7, oriented by an arbitrary unit normal vector **n**, can be computed by first finding the force **F** on the orthogonal sides. We have from Newton's second law that forces on all sides of the tetrahedron have to cancel each other for the tetrahedron to be at rest. The force **F** on the triangle is therefore

$$\mathbf{F} = \mathbf{S}\,A = \frac{1}{2}\left(\boldsymbol{\sigma}_x\,dy\,dz + \boldsymbol{\sigma}_y\,dx\,dz + \boldsymbol{\sigma}_z\,dx\,dy\right) \tag{3.20}$$

where **S** is the stress vector on the triangle, and where dx, dy and dz are the corner positions of the triangle relative to the origin. The areas of the orthogonal sides are related to the area A as $\frac{1}{2}dx\,dy = A\,n_z$, $\frac{1}{2}dx\,dz = A\,n_y$ and $\frac{1}{2}dy\,dz = A\,n_x$. Using these areas, the stress vector S is written in a compact way using the summation convention as

$$S_i = \sigma_{xi}n_x + \sigma_{yi}n_y + \sigma_{zi}n_z = n_j\sigma_{ji} \tag{3.21}$$

Premultiplication of the stress tensor by the unit normal vector (of a plane) gives the stress vector on the plane. The stress in a particular direction with unit vector **m** is the scalar product of the stress vector and **m**,

$$S_m = \mathbf{S} \cdot \mathbf{m} = n_j\sigma_{ji}m_i. \tag{3.22}$$

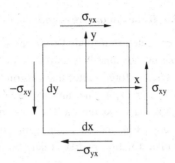

Figure 3.8. *The shear stress that acts on the surface of an infinitesimal square element in the xy-plane. (The box has unit height in the direction normal to the plane.)*

The size of the normal stress on the plane is the scalar product of the stress vector **S** and the normal vector **n**

$$S_n = S_i \, n_i = n_j \sigma_{ji} \, n_i,$$ (3.23)

which yields the normal stress and tangential (or shear) stress on the plane

$$\mathbf{N} = S_n \mathbf{n} \quad \text{and} \quad \mathbf{T} = \mathbf{S} - \mathbf{N},$$ (3.24)

respectively.

The stress tensor turns out to be symmetric, a property that follows from the assumption that any infinitesimal volume element of a continuum is torque-free. Recall that the torque is a vector quantity defined as $\mathbf{M} = \mathbf{r} \times \mathbf{F}$. The torque $\mathbf{M}$ on an infinitesimal small element in the xy-plane, as shown in Figure 3.8, is found by adding the contribution from all four sides, and we get

$$\mathbf{M} = 2\left((\sigma_{xy}dy)\frac{dx}{2} - (\sigma_{yx}dx)\frac{dy}{2} \right)\mathbf{e}_z$$
$$= (\sigma_{xy} - \sigma_{yx}) \, dx \, dy \, \mathbf{e}_z.$$ (3.25)

There is one contribution from each side, where the first term is seen to be the shear force $F_y = \sigma_{xy}dy$ acting on the left side of the element with the distance $dx/2$ to the origin. A torque-free element ($\mathbf{M} = \mathbf{0}$) implies that $\sigma_{xy} = \sigma_{yx}$. The same argument can be applied for the elements in the xz- and yz-planes as well. Therefore, it follows that the stress tensor is symmetric, $\sigma_{ij} = \sigma_{ji}$, and that it has only six independent components. The symmetry of the stress tensor allows the stress vector (3.21) to be written as the matrix–vector product

$$S_i = \sigma_{ij}n_j.$$ (3.26)

A later section shows that it is always possible to find a coordinate system where the stress tensor becomes diagonal. It is often an advantage to work in such a coordinate system because of simplifications.

Exercise 3.4 Show that the stress tensor $\sigma_{ij} = p\delta_{ij}$ gives the stress state in a fluid (at rest). (Hint: The fluid acts on any plane with the normal stress p.)

3.6 Rotation of stress and strain

The stress and strain are often wanted in a rotated coordinate system. The rotation of the stress and strain tensor is done in the same way as rotation of the permeability tensor in Section 2.9. Force is a vector, and the force vector that acts on a plane with area A and outward unit normal vector n_j is $f_i = \sigma_{ij} n_j A$. The same expression in vector–matrix notation is $\mathbf{f} = \boldsymbol{\sigma} \mathbf{n} A$, where $\boldsymbol{\sigma} = (\sigma_{ij})$ is the stress tensor. The force vector $\mathbf{f}$ becomes rotated by multiplication by the rotation matrix $\mathbf{R}$, and the components in the rotated coordinate system are denoted by a prime. (The rotation matrix is derived in Section 2.8.) The rotated force becomes

$$\mathbf{f}' = \mathbf{Rf} \tag{3.27}$$

$$= \mathbf{R}\boldsymbol{\sigma}\mathbf{n}A \tag{3.28}$$

$$= \mathbf{R}\boldsymbol{\sigma}\mathbf{R}^{-1}\mathbf{Rn}A \tag{3.29}$$

$$= \boldsymbol{\sigma}'\mathbf{n}'A \tag{3.30}$$

where we had $\mathbf{R}^{-1}\mathbf{R} = 1$ (the identity matrix). Recall the nice property of the rotation matrix that the inverse matrix is the transpose, $\mathbf{R}^{-1} = \mathbf{R}^T$ (see Exercise 2.13). The stress tensor in the rotated coordinate system is therefore $\boldsymbol{\sigma}' = \mathbf{R}\boldsymbol{\sigma}\mathbf{R}^{-1}$, or

$$\sigma'_{ij} = R_{ik}\,\sigma_{kl}\,R_{lj}^{-1} \tag{3.31}$$

when it is written out with indices. The strain tensor is rotated the same way as the stress tensor. A small line segment represented by the vector $\mathbf{ds} = \boldsymbol{\epsilon}\,\mathbf{dx}$ is rotated into

$$\mathbf{ds}' = \mathbf{R}\,\mathbf{ds}$$

$$= \mathbf{R}\boldsymbol{\epsilon}\,\mathbf{ds}$$

$$= \mathbf{R}\boldsymbol{\epsilon}\mathbf{R}^{-1}\mathbf{R}\,\mathbf{ds}$$

$$= \boldsymbol{\epsilon}'\,\mathbf{ds}' \tag{3.32}$$

which gives that the rotated strain tensor is $\boldsymbol{\epsilon}' = \mathbf{R}\boldsymbol{\epsilon}\mathbf{R}^{-1}$.

Note 3.1 The strain tensor in the rotated coordinate system is

$$\epsilon'_{ij} = \frac{1}{2}\left(\frac{\partial u'_i}{\partial x'_j} + \frac{\partial u'_j}{\partial x'_i}\right) \tag{3.33}$$

where u'_i and x'_j are the displacement field and coordinates of the rotated system, respectively. They are both related to the unrotated displacement field and coordinates by use of the rotation matrix, $u'_i = R_{ik}u_k$ and $x'_l = R_{lm}x_m$, where the inverse relationship is $x_l = R_{lm}^{-1}x'_m$. The rotated strain tensor becomes

$$\epsilon'_{ij} = \frac{1}{2}\left(\frac{\partial u'_i}{\partial x'_j} + \frac{\partial u'_j}{\partial x'_i}\right)$$

$$= \frac{1}{2}\left(R_{ik}\frac{\partial u_k}{\partial x'_j} + R_{jk}\frac{\partial u_k}{\partial x'_i}\right)$$

$$= \frac{1}{2}\left(R_{ik}\frac{\partial u_k}{\partial x_l}\frac{\partial x_l}{\partial x'_j} + R_{jk}\frac{\partial u_k}{\partial x_l}\frac{\partial x_l}{\partial x'_i}\right)$$

$$= \frac{1}{2}\left(R_{ik}\frac{\partial u_k}{\partial x_l}R_{lj}^{-1} + R_{jk}\frac{\partial u_k}{\partial x_l}R_{li}^{-1}\right). \tag{3.34}$$

The last term above can be rewritten using that $R_{li}^{-1} = R_{il}$

$$R_{jk}\frac{\partial u_k}{\partial x_l}R_{li}^{-1} = R_{li}^{-1}\frac{\partial u_k}{\partial x_l}R_{jk} = R_{il}\frac{\partial u_k}{\partial x_l}R_{kj}^{-1} = R_{ik}\frac{\partial u_l}{\partial x_k}R_{lj}^{-1} \tag{3.35}$$

which finally gives that $\epsilon'_{ij} = R_{ik}\epsilon_{kl}R_{lj}^{-1}$. (The steps are, firstly, swapping sides for R_{jk} and R_{li}^{-1}, using that $R_{li}^{-1} = R_{il}$ and then swapping the indices k and l.)

Exercise 3.5

(a) Show that the 2D stress state represented by the stress tensor

$$\sigma = \begin{pmatrix} \sigma_{xx} & \sigma_{xz} \\ \sigma_{xz} & \sigma_{zz} \end{pmatrix} \tag{3.36}$$

becomes rotated into a stress tensor with components

$$\sigma'_{xx} = \sigma_{xx}\cos^2\theta + \sigma_{zz}\sin^2\theta + 2\sigma_{xz}\sin\theta\cos\theta \tag{3.37}$$

$$\sigma'_{zz} = \sigma_{xx}\sin^2\theta + \sigma_{zz}\cos^2\theta - 2\sigma_{xz}\sin\theta\cos\theta \tag{3.38}$$

$$\sigma'_{xz} = (-\sigma_{xx} + \sigma_{zz})\sin\theta\cos\theta + \sigma_{xz}(\cos^2\theta - \sin^2\theta) \tag{3.39}$$

by use of the rotation matrix (2.53). There are only three independent components of the stress tensor σ in 2D, because $\sigma_{xz} = \sigma_{zx}$.

(b) Show that $\sigma'_{xx} + \sigma'_{zz} = \sigma_{xx} + \sigma_{zz}$.

(c) Show that shear stress vanishes when the rotation angle is such that

$$\tan\theta = \frac{2\sigma_{xz}}{\sigma_{xx} - \sigma_{zz}}. \tag{3.40}$$

(Use that $\sin\theta\cos\theta = \frac{1}{2}\sin 2\theta$ and that $\cos 2\theta = \cos^2\theta - \sin^2\theta$.)

3.7 Principal stress

It is often an advantage to work with normal stress, because a plane with normal stress does not have shear stress. The normal stress has to be proportional to the normal vector, and we have

$$\sigma_{ij} n_j = \sigma n_i. \tag{3.41}$$

The unit normal vector **n** and the normal stress σ are seen to be the eigenvector and the eigenvalue, respectively, of the stress tensor. It is shown in Notes 3.2 and 3.3, using linear algebra, that there are three real eigenvalues (σ_1, σ_2 and σ_3) and that the three corresponding eigenvectors ($\mathbf{n}_1$, $\mathbf{n}_2$ and $\mathbf{n}_3$) are orthogonal, because the stress tensor is real and symmetric. Consequently, the eigenvectors give three orthogonal directions with only normal stress. A plane oriented with one of the eigenvectors as the normal vector will have the corresponding eigenvalue as the normal stress. The orthogonal system is the *principal system*, and the corresponding stress is the *principal stress*. By convention, the largest principal stress is numbered σ_1 and the least principal stress is numbered σ_3. The eigenvalues are a solution of

$$\begin{vmatrix} \sigma_{11} - \sigma & \sigma_{12} & \sigma_{13} \\ \sigma_{21} & \sigma_{22} - \sigma & \sigma_{23} \\ \sigma_{31} & \sigma_{32} & \sigma_{33} - \sigma \end{vmatrix} = 0 \tag{3.42}$$

which becomes the polynomial

$$\sigma^3 - I_1 \sigma^2 + I_2 \sigma - I_3 = 0 \tag{3.43}$$

that has three solutions σ_i, $i = 1, 2, 3$. The coefficients of the polynomial are

$$\begin{aligned} I_1 &= \sigma_{11} + \sigma_{22} + \sigma_{33} \\ I_2 &= \sigma_{22}\sigma_{33} - \sigma_{23}\sigma_{32} + \sigma_{33}\sigma_{11} \\ &\quad - \sigma_{31}\sigma_{13} + \sigma_{11}\sigma_{22} - \sigma_{12}\sigma_{21} \\ I_3 &= \sigma_{11}\sigma_{22}\sigma_{33} - \sigma_{11}\sigma_{23}\sigma_{32} + \sigma_{12}\sigma_{23}\sigma_{31} \\ &\quad - \sigma_{12}\sigma_{21}\sigma_{33} + \sigma_{13}\sigma_{21}\sigma_{32} - \sigma_{13}\sigma_{22}\sigma_{31} \end{aligned} \tag{3.44}$$

and they are called invariants because the solutions for the normal stress σ_i have to be independent of the coordinate system used to represent the stress tensor. The eigenvalues σ_i are the same regardless of how the stress tensor σ is rotated. It is shown rigorously in Exercise 3.8 that I_1, I_2 and I_3 really are invariant under rotation. We also have that any combination of the invariants is also an invariant. An important use of the invariants is the formulation of yield criteria. We will encounter the first invariant later because it is proportional to the mean stress $\sigma_m = \frac{1}{3} I_1$. The stress tensor in the principal system is simply

$$\sigma = \begin{pmatrix} \sigma_1 & 0 & 0 \\ 0 & \sigma_2 & 0 \\ 0 & 0 & \sigma_3 \end{pmatrix} \tag{3.45}$$

and the normalized eigenvectors become the unit vectors of the orthonormal system

$$\mathbf{e}_1 = (1, 0, 0)^T, \qquad \mathbf{e}_2 = (0, 1, 0)^T, \qquad \mathbf{e}_3 = (0, 0, 1)^T. \tag{3.46}$$

Working with the stress tensor is considerably simplified in the principal system. The stress vector on any plane oriented with the normal vector **n** in the principal system is

$$\mathbf{S} = \boldsymbol{\sigma}\mathbf{n} = (\sigma_1 n_1, \sigma_2 n_2, \sigma_3 n_3)^T \tag{3.47}$$

and the normal stress on the plane becomes

$$\sigma = \mathbf{n} \cdot \mathbf{S} = \sigma_1 n_1^2 + \sigma_2 n_2^2 + \sigma_3 n_3^2. \tag{3.48}$$

The shear stress on the plane is most easily given to the power of two:

$$\tau^2 = (\sigma_1 - \sigma_3)^2 \, n_1^2 n_3^2 + (\sigma_1 - \sigma_2)^2 \, n_1^2 n_2^2 + (\sigma_2 - \sigma_3)^2 \, n_3^2 n_2^2 \tag{3.49}$$

as shown in Note 3.4.

Note 3.2 *Real eigenvalues*. We have that $\boldsymbol{\sigma}\mathbf{n}_i = \sigma_i \mathbf{n}_i$ and we want to show that the eigenvalues σ_i are real for a real and symmetric matrix. Some more notation is needed before we can do that. The complex conjugate of a scalar a and a matrix $\mathbf{A}$ are denoted a^* and $\mathbf{A}^*$, respectively. The combined operation of complex conjugation and transposing is denoted $\mathbf{A}^\dagger = (\mathbf{A}^*)^T$. The stress tensor does not change by taking the transpose and the conjugate, $\boldsymbol{\sigma}^\dagger = \boldsymbol{\sigma}$, because it is symmetric and real. The transpose and conjugate of $\boldsymbol{\sigma}\mathbf{n}_i = \sigma_i \mathbf{n}_i$ is therefore $\mathbf{n}_i^\dagger \boldsymbol{\sigma} = \sigma_i^* \mathbf{n}_i^\dagger$ and right multiplication by $\mathbf{n}_i$ gives $\mathbf{n}_i^\dagger \boldsymbol{\sigma}\mathbf{n}_i = \sigma_i^* \mathbf{n}_i^\dagger \cdot \mathbf{n}_i$. Left multiplication of $\boldsymbol{\sigma}\mathbf{n}_i = \sigma_i \mathbf{n}_i$ by $\mathbf{n}_i^\dagger$ gives $\mathbf{n}_i^\dagger \boldsymbol{\sigma}\mathbf{n}_i = \sigma_i \mathbf{n}_i^\dagger \cdot \mathbf{n}_i$. We have now two expressions for $\mathbf{n}_i^\dagger \boldsymbol{\sigma}\mathbf{n}_i$ and their difference is $(\sigma_i^* - \sigma_i)\mathbf{n}_i^\dagger \cdot \mathbf{n}_i = 0$, which implies that $\sigma_i^* = \sigma_i$ because $\mathbf{n}_i^\dagger \cdot \mathbf{n}_i \neq 0$.

Note 3.3 *Orthogonal eigenvectors*. The orthogonality of the eigenvectors of a real and symmetric matrix are shown by a similar reasoning as in Note 3.2. Let $\boldsymbol{\sigma}\mathbf{n}_i = \sigma_i \mathbf{n}_i$ and $\boldsymbol{\sigma}\mathbf{n}_j = \sigma_j \mathbf{n}_j$. Taking the transpose of the first equation before right multiplication by $\mathbf{n}_j$, and then left multiplication of the second equation by $\mathbf{n}_i^T$ gives $\mathbf{n}_i^T \boldsymbol{\sigma}\mathbf{n}_j = \sigma_i \mathbf{n}_i \cdot \mathbf{n}_j = \sigma_j \mathbf{n}_i \cdot \mathbf{n}_j$, and the difference becomes $(\sigma_i - \sigma_j)\mathbf{n}_i \cdot \mathbf{n}_j = 0$. The eigenvectors for different eigenvalues are therefore orthogonal. In the case that the eigenvalues are equal, $\sigma_i = \sigma_j$, it is not possible to conclude that $\mathbf{n}_i \cdot \mathbf{n}_j = 0$. But any linear combination of eigenvectors with the same eigenvalue is a new eigenvector, and it is therefore possible to construct two new eigenvectors that are orthogonal, for instance $\mathbf{m}_i = \mathbf{n}_i$ and $\mathbf{m}_j = \mathbf{n}_j - a\mathbf{n}_i$, where $a = (\mathbf{n}_i \cdot \mathbf{n}_j)/(\mathbf{n}_i \cdot \mathbf{n}_i)$.

Note 3.4 Figure 3.9 shows the shear stress, normal stress and the stress vector, and the size of the shear stress follows from the Pythagorean theorem:

$$\tau^2 = S^2 - \sigma^2 = (\sigma_1 n_1)^2 + (\sigma_2 n_2)^2 + (\sigma_3 n_3)^2 - \left(\sigma_1 n_1^2 + \sigma_2 n_2^2 + \sigma_3 n_3^2\right)^2. \tag{3.50}$$

We expand the last parentheses and regroup the terms to get

$$\tau^2 = \sigma_1^2 n_1^2 (1 - n_1^2) + \sigma_2^2 n_2^2 (1 - n_2^2) + \sigma_3^2 n_3^2 (1 - n_3^2)$$
$$- 2\sigma_1 \sigma_2 n_1^2 n_2^2 - 2\sigma_2 \sigma_3 n_2^2 n_3^2 - 2\sigma_1 \sigma_3 n_1^2 n_3^2. \tag{3.51}$$

Figure 3.9. *The stress vector, normal stress and the shear stress on a plane.*

The next step is to use that the normal vector has unit length, $n_1^2 + n_2^2 + n_3^2 = 1$, and to do the replacements $1 - n_1^2 = n_2^2 + n_3^2$, $1 - n_2^2 = n_1^2 + n_3^2$ and $1 - n_3^2 = n_1^2 + n_2^2$. By collecting terms we get

$$\tau^2 = \left(\sigma_1^2 - 2\sigma_1\sigma_2 + \sigma_2^2\right) n_1^2 n_2^2 + \left(\sigma_2^2 - 2\sigma_2\sigma_3 + \sigma_3^2\right) n_2^2 n_3^2 + \left(\sigma_3^2 - 2\sigma_3\sigma_1 + \sigma_1^2\right) n_3^2 n_1^2$$

(3.52)

which is (3.49) for the shear stress.

Exercise 3.6 Show that the invariants (3.44) expressed with principal stresses are

$$I_1 = \sigma_1 + \sigma_2 + \sigma_3 \tag{3.53}$$

$$I_2 = \frac{1}{2}(\sigma_1^2 + \sigma_2^2 + \sigma_3^2 - I_1^2) \tag{3.54}$$

$$I_3 = \sigma_1\sigma_2\sigma_3. \tag{3.55}$$

Exercise 3.7 Show that the invariants (3.44) for the general stress tensor can be written

$$I_1 = \sigma_{ii} \tag{3.56}$$

$$I_2 = \frac{1}{2}(\sigma_{ij}\sigma_{ij} - I_1^2) \tag{3.57}$$

$$I_3 = |\boldsymbol{\sigma}|. \tag{3.58}$$

Exercise 3.8 Show that the invariants (3.44) really are invariant under rotation of the stress tensor. In other words, if $\boldsymbol{\sigma}' = \mathbf{R}\boldsymbol{\sigma}\mathbf{R}^{-1}$ is a rotated stress tensor, show that the invariants of $\boldsymbol{\sigma}'$ and $\boldsymbol{\sigma}$ are the same.
Solution:
This solution is based on the representation of the identity matrix known as the Kronecker delta δ_{ij}, where δ_{ij} is one whenever $i = j$ and otherwise zero. Since the Kronecker delta δ_{ij} is the identity matrix we have that $R_{ik}^{-1} R_{kj} = \delta_{ij}$. We will also use that the transpose of the rotation matrix is the inverse, $R_{ij}^{-1} = R_{ij}^T = R_{ji}$.
(1) The first invariant of $\boldsymbol{\sigma}'$ is

$$I_1' = \sigma_{ii}' = R_{ij}\sigma_{jk}R_{ki}^{-1} = R_{ki}^{-1}R_{ij}\sigma_{jk} = \delta_{kj}\sigma_{jk} = \sigma_{kk} = I_1. \tag{3.59}$$

(2) Exercise 3.7 shows that is sufficient to show that $J = \sigma_{ij}\sigma_{ij}$ is invariant under rotation of the stress tensor. The invariant of $\boldsymbol{\sigma}'$ is

$$J' = \sigma'_{ij}\sigma'_{ij} = (R_{ik}\sigma_{kl}R_{lj}^{-1})(R_{ip}\sigma_{pq}R_{qj}^{-1})$$
$$= (R_{pi}^{-1}R_{ik})(R_{lj}^{-1}R_{jq})\sigma_{kl}\sigma_{pq}$$
$$= \delta_{pk}\delta_{lq}\sigma_{kl}\sigma_{pq}$$
$$= \sigma_{pq}\sigma_{pq}$$
$$= J. \tag{3.60}$$

(3) The third invariant of σ' is

$$I'_3 = |\sigma'| = |\mathbf{R}\sigma\mathbf{R}^{-1}| = |\mathbf{R}||\sigma||\mathbf{R}^{-1}| = |\sigma| = I_3 \tag{3.61}$$

because $|\mathbf{R}| = |\mathbf{R}^{-1}| = 1$.

3.8 Mohr's circles

The principal stresses can be used to express the normal stress and the shear stress on any given plane. The normal and shear stresses are obtained by adding the forces that act normal to and tangential to a plane. Figure 3.10 shows a plane in 2D with a unit normal vector that makes an angle θ with the direction of the largest principal stress σ_1. The largest principal stress is horizontal and the least principal stress is vertical, and could for example be the lithostatic stress. The horizontal force acting on the vertical side is $F_1 = \sigma_1 A \cos\theta$ and the vertical force on the base is $F_3 = \sigma_3 A \sin\theta$, where A is the area of the plane. These forces are decomposed normal to the plane and tangential to the plane, and the normal force (F_n) and the tangential force (F_t) are

$$F_n = F_1 \cos\theta + F_3 \sin\theta \tag{3.62}$$
$$F_t = -F_1 \sin\theta + F_3 \cos\theta \tag{3.63}$$

which lead to the normal stress $\sigma = F_n/A$ and the shear stress $\tau = F_t/A$

$$\sigma = \sigma_1 \cos^2\theta + \sigma_3 \sin^2\theta \tag{3.64}$$
$$\tau = -(\sigma_1 - \sigma_3)\sin\theta\cos\theta. \tag{3.65}$$

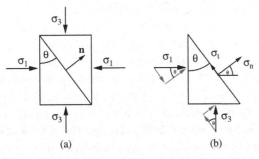

(a)   (b)

Figure 3.10. *(a) A plane has a unit normal vector that makes an angle θ with the largest principal stress. (b) The normal and the tangential stresses on the slanted surface are obtained by adding forces.*

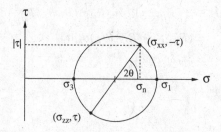

Figure 3.11. *A Mohr's circle gives the normal stress σ and the shear stress τ on a plane with a normal vector that is rotated an angle θ relative to the largest principal stress. The least principal stress is σ_3 and the largest principal stress is σ_1.*

These expressions can be rewritten using the trigonometric relations $\cos^2 \theta = \frac{1}{2}(1 + \cos 2\theta)$, $\sin^2 \theta = \frac{1}{2}(1 - \cos 2\theta)$ and $\sin 2\theta = 2 \sin \theta \cos \theta$, and we get

$$\sigma = \frac{1}{2}(\sigma_1 + \sigma_3) + \frac{1}{2}(\sigma_1 - \sigma_3) \cos 2\theta \qquad (3.66)$$

$$\tau = -\frac{1}{2}(\sigma_1 - \sigma_3) \sin 2\theta. \qquad (3.67)$$

The equations show that the normal stress σ and the shear stress τ on any 2D plane can be represented by a circle as shown in Figure 3.11, where the normal vector makes an angle θ relative to the largest principal stress. Such a representation of the normal and the shear stresses in terms of the principal stresses is called a *Mohr's circle*. Equations (3.66) and (3.67) show that Mohr's circle has the diameter $\sigma_1 - \sigma_3$ and its center is at position $\frac{1}{2}(\sigma_1 + \sigma_3)$ along the σ-axis. An inspection of equation (3.67) shows that the shear stress is zero on planes making an angle $\theta = 0$ or $\theta = \pi/2$ with the largest principal stress. These planes have the corresponding normal stresses $\sigma = \sigma_1$ and $\sigma = \sigma_3$, respectively, which is as expected from Figure 3.11.

The Mohr's circle in Figure 3.11 shows several properties of stress states. Shear stress is zero for an isotropic stress state ($\sigma_1 = \sigma_3$). Zero normal stress is not possible unless σ_1 and σ_3 have opposite signs, because the normal stress is constrained to the interval $[\sigma_3, \sigma_1]$. We also see that the maximum shear stress is $\tau = \frac{1}{2}(\sigma_1 - \sigma_3)$.

Mohr's circles can represent normal stress and shear stress in 3D as well. Equations (3.48) and (3.49) give the normal stress and shear stress, respectively, in 3D. These equations take a unit vector related to the principle system. Equations (3.64) and (3.65), which make a 2D Mohr's circle in the σ_1–σ_3-plane, are recovered from equations (3.48) and (3.49) by setting $n_2 = 0$. In the same way, setting $n_1 = 0$, gives a 2D Mohr's circle in the σ_2–σ_3-plane, and setting $n_3 = 0$ gives a 2D Mohr's circle in the σ_1–σ_2-plane. Figure 3.12a shows these three circles. The general case of a unit normal vector that is not in any of the three coordinate planes gives shear and normal stress in the area between the three Mohr's circles. The stress states between the Mohr's circles can be explored by making the normal vector

$$n_1 = \cos \alpha \cos \beta, \quad n_2 = \cos \alpha \sin \beta, \quad n_3 = \sin \alpha \qquad (3.68)$$

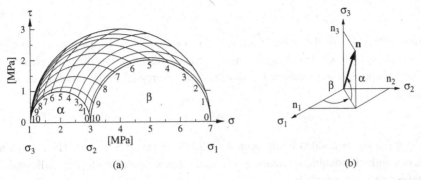

Figure 3.12. (a) Mohr's circles representing the normal stress and the shear stress in 3D. (b) A normal vector.

which is shown in Figure 3.12b. The normal vector can be inserted into the normal stress (3.48) and shear stress (3.49), and both quantities become functions of the angles α and β. Selecting an α-value and tracing out the curve for β from 0 to $\pi/2$ makes one curve in the family of curves shown in Figure 3.12a. These curves start at the σ_1–σ_3-circle for $\beta = 0$, and goes down to the σ_2–σ_3-circle for $\beta = \pi/2$. The other family of curves is produced by fixing a β-value and tracing out a curve for α from 0 to $\pi/2$. Figure 3.12a shows α- and β-curves for the interval from 0 to $\pi/2$ divided into 10 steps. For example α-curve number i has $\alpha = i\pi/20$, and β-curve j has $\beta = j\pi/20$. Knowledge of the unit normal vector $\mathbf{n}$ in terms of angles α and β makes it possible to look up the normal stress and the shear stress by the corresponding i and j indices.

Exercise 3.9 Show that the rotation of the stress tensor in the principal system by an angle θ gives

$$\sigma_{xx} = \frac{1}{2}(\sigma_1 + \sigma_3) + \frac{1}{2}(\sigma_1 - \sigma_3)\cos 2\theta \qquad (3.69)$$

$$\sigma_{zz} = \frac{1}{2}(\sigma_1 + \sigma_3) - \frac{1}{2}(\sigma_1 - \sigma_3)\cos 2\theta \qquad (3.70)$$

$$\tau = -\frac{1}{2}(\sigma_1 - \sigma_3)\sin 2\theta \qquad (3.71)$$

This stress state is plotted in Figure 3.11. Hint: $\boldsymbol{\sigma}' = \mathbf{R}\boldsymbol{\sigma}\mathbf{R}^{-1}$, where

$$\boldsymbol{\sigma}' = \begin{pmatrix} \sigma_{xx} & \tau \\ \tau & \sigma_{zz} \end{pmatrix}, \boldsymbol{\sigma} = \begin{pmatrix} \sigma_1 & 0 \\ 0 & \sigma_3 \end{pmatrix} \quad \text{and} \quad \mathbf{R} = \begin{pmatrix} \cos\theta & \sin\theta \\ -\sin\theta & \cos\theta \end{pmatrix} \qquad (3.72)$$

The stress tensor $\boldsymbol{\sigma}$ gives the normal stresses on two orthogonal sides of a square, where the least principal stress is horizontal and the largest principal stress is vertical.

3.9 Stress ellipsoid

The stress tensor σ_{ij} makes it possible to compute the force on any plane, as shown in Section 3.5. The force vector $\mathbf{f}$ on a plane oriented with a normal vector $\mathbf{n}$ of unit length is simply

$$f_i = \sigma_{ij} n_j A \qquad (3.73)$$

where A is the area of the plane. Section 3.7 shows that it is always possible to choose a coordinate system where the stress tensor becomes diagonal. The diagonal elements σ_{ii} are the principal stresses denoted σ_i, and the force expressed in the principal coordinate system is

$$f_1 = \sigma_1 n_1 A, \qquad f_2 = \sigma_2 n_2 A, \qquad f_3 = \sigma_3 n_3 A, \qquad (3.74)$$

and the force vector divided by the area A is the stress vector $\mathbf{S} = \mathbf{f}/A$. The stress vector on planes with all possible orientations can now be represented by a stress ellipsoid. The normal vector $\mathbf{n}$ has length 1:

$$\mathbf{n} \cdot \mathbf{n} = n_1^2 + n_2^2 + n_3^2 = 1 \qquad (3.75)$$

and the components (3.74) of the force vector give the components of the unit vector,

$$n_1 = \frac{S_1}{\sigma_1}, \qquad n_2 = \frac{S_2}{\sigma_2}, \qquad n_3 = \frac{S_3}{\sigma_3}. \qquad (3.76)$$

This unit vector is inserted into expression (3.75) for unit length

$$\left(\frac{S_1}{\sigma_1}\right)^2 + \left(\frac{S_2}{\sigma_2}\right)^2 + \left(\frac{S_3}{\sigma_3}\right)^2 = 1 \qquad (3.77)$$

which is the equation for an ellipsoid. For a given stress tensor (σ_1, σ_2 and σ_3) the stress vector on planes of all possible orientations trace out an ellipsoid. Figure 3.13 shows an example of a stress ellipsoid in 3D, and a stress ellipse in 2D. The stress ellipsoid should not be confused with the ellipsoid for directional permeability or directional heat conductivity, see Note 3.5. On the other hand, it is straightforward to make a strain ellipsoid in the same way as the stress ellipsoid, which is Exercise 3.10. From geometry we have that the volume of the stress ellipsoid is $V = (4/3)\pi \sigma_1 \sigma_2 \sigma_3$ and that the area of the stress ellipse is $A = \pi \sigma_1 \sigma_3$.

Note 3.5 We have now used ellipsoids to represent two different properties of a tensor. Section 2.10 shows the ellipsoid for the directional permeability – the permeability in the direction of the gradient of the potential. Such an ellipsoid expresses a different type of quantity than the stress ellipsoid. The ellipsoid for directional permeability gives

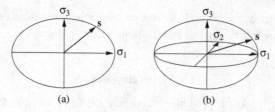

(a)   (b)

Figure 3.13. *The stress vector traces out an ellipse in 2D (a) and an ellipsoid in 3D (b).*

$k_n = \mathbf{n}^T \mathbf{K} \mathbf{n}$, where $\mathbf{n}$ is a unit vector, while the stress ellipsoid expresses all possible stress vectors $\mathbf{S} = \boldsymbol{\sigma} \mathbf{n}$.

Exercise 3.10 Show how the strain tensor can be written as an ellipsoid in an analogous way to the stress tensor.

3.10 Deviatoric stress

The ductile flow of rocks is often related to how far away the stress is from an isotropic state. A measure for this difference is the *deviatoric stress* defined as the stress tensor minus the mean stress

$$\boldsymbol{\sigma}' = \begin{pmatrix} \sigma_{xx} - \sigma_m & \sigma_{xy} & \sigma_{xz} \\ \sigma_{yx} & \sigma_{yy} - \sigma_m & \sigma_{yz} \\ \sigma_{zx} & \sigma_{zy} & \sigma_{zz} - \sigma_m \end{pmatrix} \tag{3.78}$$

where the mean stress is

$$\sigma_m = \frac{1}{3}\sigma_{ii} = \frac{1}{3}\left(\sigma_{xx} + \sigma_{yy} + \sigma_{zz}\right). \tag{3.79}$$

Using the more compact notation we write that

$$\sigma'_{ij} = \sigma_{ij} - \sigma_m \delta_{ij}. \tag{3.80}$$

In the principal system we have

$$\boldsymbol{\sigma}' = \begin{pmatrix} \sigma_1 - \sigma_m & 0 & 0 \\ 0 & \sigma_2 - \sigma_m & 0 \\ 0 & 0 & \sigma_3 - \sigma_m \end{pmatrix} \tag{3.81}$$

where σ_1, σ_2 and σ_3 are the principal stress components. We notice that the component $\sigma'_3 = \sigma_3 - \sigma_m$ is always less than (or equal to) zero. The deviatoric stress tensor therefore contains negative components, regardless of how compressive the stress state might be. The mean stress is one third of the stress invariant I_1, and we therefore have

$$\sigma_m = \frac{1}{3}\left(\sigma_{xx} + \sigma_{yy} + \sigma_{zz}\right) = \frac{1}{3}\left(\sigma_1 + \sigma_2 + \sigma_3\right) = \frac{1}{3}I_1. \tag{3.82}$$

The deviatoric stress measures how far away a stress state is from the isotropic mean stress. An often-used alternative to the deviatoric stress is the *differential stress*, defined as the difference between the largest and the least principal stress components,

$$\sigma_d = \sigma_1 - \sigma_3. \tag{3.83}$$

In the quite common situation with $\sigma_2 = \sigma_3$ the deviatoric stress is proportional to the differential stress

$$\boldsymbol{\sigma}' = \begin{pmatrix} \frac{2}{3} & 0 & 0 \\ 0 & -\frac{1}{3} & 0 \\ 0 & 0 & -\frac{1}{3} \end{pmatrix}(\sigma_1 - \sigma_3), \tag{3.84}$$

a property shown in Exercise 3.11. Nevertheless, deviatoric stress should not be confused with differential stress, because differential stress is just one value, while the deviatoric stress is a tensor with up to six different values.

Exercise 3.11 Show that equation (3.84) gives the deviatoric stress in the principal system when $\sigma_2 = \sigma_3$. Such a stress state is common in triaxial testing where a cylindrical rock specimen is subjected to axial compression with the largest principal stress, and confinement with $\sigma_2 = \sigma_3$. See Figure 8.9.

Solution: Since $\sigma_m = \frac{1}{3}(\sigma_1 + 2\sigma_3)$ we get

$$\sigma_1' = \sigma_1 - \frac{1}{3}(\sigma_1 + 2\sigma_3) = \frac{2}{3}(\sigma_1 - \sigma_3)$$

$$\sigma_3' = \sigma_3 - \frac{1}{3}(\sigma_1 + 2\sigma_3) = -\frac{1}{3}(\sigma_1 - \sigma_3).$$

Exercise 3.12 Show that the first invariant I_1' for a deviatoric stress tensor is 0.

3.11 Linear stress–strain relations

We have seen that stress and strain are related for linear elastic materials, and we have defined three moduli for such materials. It is now time to generalize these relations in terms of the stress and strain tensors. We start by adding the strain contributions in the x-direction, where we have

$$\varepsilon_{xx}^{(1)} = \frac{1}{E}\sigma_{xx} \tag{3.85}$$

$$\varepsilon_{xx}^{(2)} = -\nu\,\varepsilon_{yy} = -\frac{\nu}{E}\,\sigma_{yy} \tag{3.86}$$

$$\varepsilon_{xx}^{(3)} = -\nu\,\varepsilon_{zz} = -\frac{\nu}{E}\,\sigma_{zz}. \tag{3.87}$$

The first part is Hooke's law (3.1), and the next two parts are the ratio of transverse and longitudinal strain (3.2) in the y- and z-directions. The transverse strain (ε_{yy} and ε_{zz}) is related to the transverse stress (σ_{yy} and σ_{zz}) by Hooke's law. The total strain in the x-direction is the sum of all three contributions, because we are dealing with a linear elastic material:

$$\varepsilon_{xx} = \frac{1}{E}\sigma_{xx} - \frac{\nu}{E}\,\sigma_{yy} - \frac{\nu}{E}\,\sigma_{zz}. \tag{3.88}$$

The total strain in the y- and z-directions is obtained in a similar way, and these are respectively

$$\varepsilon_{yy} = -\frac{\nu}{E}\sigma_{xx} + \frac{1}{E}\sigma_{yy} - \frac{\nu}{E}\,\sigma_{zz} \tag{3.89}$$

$$\varepsilon_{zz} = -\frac{\nu}{E}\sigma_{xx} - \frac{\nu}{E}\sigma_{yy} + \frac{1}{E}\sigma_{zz}. \tag{3.90}$$

We have already defined the shear modulus as the ratio of shear stress over strain with equation (3.6). The definition of the shear modulus for plane strain in (3.6) becomes generalized as

$$G = \frac{\text{shear stress}}{\text{deformation angle}} = \frac{\sigma_{xy}}{\partial u_x / \partial y + \partial u_y / \partial x} = \frac{\sigma_{xy}}{2\varepsilon_{xy}}. \tag{3.91}$$

Equation (3.88) shows that the shear strain and the shear stress (deformation angle) are proportional. The relationship between the shear stress and shear strain for the xz- and yz-directions are similar, and all three relationships are

$$\varepsilon_{xy} = \frac{1+v}{E} \sigma_{xy} \tag{3.92}$$

$$\varepsilon_{xz} = \frac{1+v}{E} \sigma_{xz} \tag{3.93}$$

$$\varepsilon_{yz} = \frac{1+v}{E} \sigma_{yz} \tag{3.94}$$

where it is used that

$$G = \frac{E}{2(1+v)}. \tag{3.95}$$

Exercise 3.18 derives equation (3.95) for the shear modulus G in terms of E and v. The all-together six stress–strain relationships (3.88)–(3.90) and (3.92)–(3.94) can be collected into one equation

$$-\varepsilon_{ij} = \frac{1}{E}(1+v)\sigma_{ij} - \frac{v}{E}\sigma_{kk}\delta_{ij} \tag{3.96}$$

where the summation convention applies to the term σ_{kk} (which is $\sigma_{xx} + \sigma_{yy} + \sigma_{zz}$), and where δ_{ij} is the Kronecker delta defined by

$$\delta_{ij} = \begin{cases} 1 & i = j \\ 0 & i \neq j \end{cases}. \tag{3.97}$$

Notice that a *minus sign* has been added to the left-hand side of equation (3.96). The reason for this sign reversal is that compressive stress states are more common in the crust than stress states of tension, and it is therefore conventional in rock mechanics to associate a positive stress with compression (negative strain).

Equation (3.96) for the strain as a linear function of stress can be inverted to an expression for stress as a linear equation of strain (see Exercise 3.14):

$$-\sigma_{ij} = 2G\varepsilon_{ij} + \lambda \varepsilon_{kk}\delta_{ij} \tag{3.98}$$

where

$$\lambda = \frac{vE}{(1+v)(1-2v)}. \tag{3.99}$$

The stress–strain relationship (3.98) is the Lamé equation, and the two parameters G and λ are called the Lamé coefficients. Notice that both Lamé coefficients are expressed by Young's modulus and Poisson's ratio.

Exercise 3.13

(a) The vertical stress into the crust is $\sigma_{zz} = \varrho_b g z$ (where ϱ_b is the rock bulk density and g is the constant of gravity), and assume that the rock behaves as a linear elastic material. Show that the isotropic horizontal stress $\sigma_h = \sigma_{xx} = \sigma_{yy}$ then becomes

$$\sigma_h = \frac{\nu}{1-\nu}\, \varrho_b g z \qquad (3.100)$$

when there is zero strain in the horizontal direction ($\varepsilon_{xx} = \varepsilon_{yy} = 0$). Hint: start with equations (3.89) and (3.90).

(b) Is the horizontal stress less than or larger than the vertical (lithostatic) stress?

Exercise 3.14 Show that the Lamé equation (3.98) is the inverse of equation (3.96). Hint: first, show that $\sigma_{kk} = -\frac{1}{E}(1-2\nu)\varepsilon_{kk}$.

Exercise 3.15 Show that the bulk modulus can be expressed as follows by Young's modulus (E) and Poisson's ratio (ν)

$$K = \frac{E}{3(1-2\nu)}. \qquad (3.101)$$

Hint: use definition (3.13) for the volume strain and let the pressure be defined as the average normal stress, $p = \frac{1}{3}\sigma_{kk}$.

Exercise 3.16 Show that the bulk modulus becomes

$$K = \frac{2}{3}G + \lambda \qquad (3.102)$$

when expressed with the Lamé coefficients.

Exercise 3.17 Show that the Lamé coefficients (3.99) and (3.95) can be inverted to the following expressions for Poisson's ratio and Young's modulus:

$$\nu = \frac{\lambda}{2(G+\lambda)} \quad \text{and} \quad E = \frac{G(2G+3\lambda)}{G+\lambda}. \qquad (3.103)$$

Hint: notice that $\lambda = 2G\nu/(1-2\nu)$.

Exercise 3.18

(a) Let $\sigma_1 = -\sigma_2$ in Figure 3.14, and show that $\sigma_{xy} = -\sigma_1 = \sigma_2$ and that the normal stress is zero.

(b) Show that

$$\epsilon_1 = -\frac{1}{E}(1+\nu)\sigma_1 \quad \text{and} \quad \epsilon_2 = \frac{1}{E}(1+\nu)\sigma_1. \qquad (3.104)$$

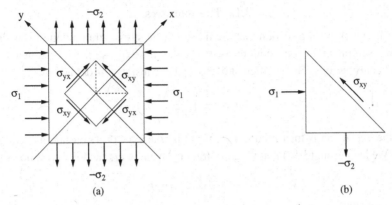

Figure 3.14. *(a) The square is compressed in the horizontal direction and stretched in the vertical direction. (b) The inner (rotated) square has shear stress $\sigma_{xy} = -\sigma_1 = \sigma_2$ and zero normal stress, when $\sigma_1 = -\sigma_2$. This stress condition is called* pure shear *as opposed to the stress condition in Figure 3.2, which is called* simple shear.

(c) Show that

$$\epsilon_{xy} = \epsilon_1 \tag{3.105}$$

which gives that the shear modulus is

$$G = \frac{E}{2(1 + v)}. \tag{3.106}$$

Solution: (a) The force balance on the triangle in Figure 3.14b, in the direction of the shear stress, is

$$\sqrt{2}\sigma_{xy} = -\frac{1}{\sqrt{2}}\sigma_1 + \frac{1}{\sqrt{2}}\sigma_2 \tag{3.107}$$

which gives that $\sigma_{xy} = -\sigma_1 = \sigma_2$, when $\sigma_1 = -\sigma_2$. The force balance in the x-direction shows that the normal stress is zero.

(b) The strain in the x- and z-directions is

$$-\epsilon_1 = \frac{1}{E}\sigma_1 - \frac{v}{E}\sigma_2 = \frac{1}{E}(1 + v)\sigma_1 \tag{3.108}$$

$$-\epsilon_2 = -\frac{v}{E}\sigma_1 + \frac{1}{E}\sigma_2 = -\frac{1}{E}(1 + v)\sigma_1. \tag{3.109}$$

(Recall the sign convention saying that compressive stresses are positive. The sign reversal is also applied to the shear stress–strain relationship, see preceding section.)

(c) Strain is rotated as a matrix in the same way as stress. The normal strain is therefore zero ($\epsilon_{xx} = 0$) and the shear strain is $\epsilon_{xy} = -\epsilon_1 = \epsilon_2$. Shear strain and shear stress are then related as $\epsilon_{xy} = -\epsilon_1 = \frac{1}{E}(1 + v)\sigma_1 = -\frac{1}{E}(1 + v)\sigma_{xy}$, which gives the shear modulus (3.106), when compared with the relation $\sigma_{xy} = -2G\epsilon_{xy}$.

3.12 Thermal stress

We saw in the preceding section that the linear elastic stress–strain relation was obtained by adding several strain contributions. Isotropic thermal expansion can be added to the list of strain contributions. The coefficient of thermal expansion β is defined by

$$\beta = \frac{1}{V}\left(\frac{\partial V}{\partial T}\right)_p \tag{3.110}$$

where the expansion is measured at a constant pressure p (or a constant average normal stress). We have from (3.13) that volume strain is given by ε_{kk}, and it then follows that

$$\varepsilon_{kk} \approx \frac{\Delta V}{V} \approx \beta \Delta T. \tag{3.111}$$

The average strain contribution in each spatial direction is therefore

$$\varepsilon_{ij}^{(4)} = \frac{1}{3}\beta \Delta T \delta_{ij} \tag{3.112}$$

and the stress–strain relationship (3.98) becomes generalized to

$$-\varepsilon_{ij} = \frac{1}{E}(1+v)\,\sigma_{ij} - \frac{v}{E}\sigma_{kk}\,\delta_{ij} - \frac{1}{3}\beta \Delta T \delta_{ij}. \tag{3.113}$$

It is often convenient to have the inverse of equation (3.113), in other words, the stress tensor expressed by the strain tensor, which is

$$-\sigma_{ij} = 2G\varepsilon_{ij} + \lambda\,\varepsilon_{kk}\delta_{ij} - \frac{1}{v}\lambda\,\beta \Delta T \tag{3.114}$$

in terms of the Lamé coefficients.

3.13 Thermal stress compared with lithostatic stress

The horizontal stress down into the crust is calculated in Exercise 3.13 assuming that the rock behaves like a linear elastic material. It is shown that the (isotropic) horizontal stress is proportional to the vertical (lithostatic) stress σ_b as

$$\sigma_h = \frac{v}{1-v}\,\sigma_b \tag{3.115}$$

when the rock is confined in the horizontal direction ($\varepsilon_{xx} = \varepsilon_{yy} = 0$), and the horizontal stress state is isotropic ($\sigma_h = \sigma_{xx} = \sigma_{yy}$). We can do the same calculation with thermal stress too. From equation (3.113) the strain in the horizontal plane is

$$-\varepsilon_{xx} = -\varepsilon_{yy} = \frac{1}{E}(1+v)\sigma_h - \frac{v}{E}(2\sigma_h + \sigma_b) - \frac{1}{3}\beta \Delta T. \tag{3.116}$$

Zero horizontal strain $\varepsilon_{xx} = \varepsilon_{yy} = 0$ gives the horizontal stress

$$\sigma_h = \frac{v}{1-v}\sigma_b + \frac{E\beta \Delta T}{3(1-v)}. \tag{3.117}$$

The two terms in the right-hand side of the equation are the contributions to the horizontal stress from the lithostatic stress and the horizontal thermal stress, respectively.

These contributions can be compared when the horizontal stress (3.117) is rewritten as

$$\sigma_h = \frac{v}{1-v}\left(1 + \frac{E\beta c}{3v\varrho_b g}\right)\varrho_b g z \tag{3.118}$$

where the lithostatic stress is $\sigma_b = \varrho_b g z$ and the temperature is given by a constant thermal gradient c ($\Delta T = c z$). One sees that the number

$$N_T = \frac{E\beta c}{3v\varrho_b g} \tag{3.119}$$

measures the size of the thermal stress relative to the lithostatic stress as a cause for horizontal stress. We get $N_T = 0.96$ using typical values for the parameters like $v = 0.25$, $E = 60$ GPa, $\beta = 1 \cdot 10^{-5}$ K^{-1}, $c = 0.03$ K/m and $\varrho_b g = 2.5 \cdot 10^4$ Pa/m, which means that the thermal expansion of the rock is almost equally important as compression caused by the lithostatic stress.

3.14 Buoyancy and effective stress

We will first examine a body immersed in a fluid, and show the renowned *Archimedes' principle*. It states that the upward vertical buoyancy force is equal to the weight of the fluid displaced by the body. Archimedes' principle is shown for a solid box of bulk density ϱ_b and volume $V = Ah$, where A is the surface area and h is the thickness (see Figure 3.15). The surface forces from the fluid on the vertical sides balance each other, but the lower and upper horizontal surfaces are acted on by the forces $\mathbf{F}_1 = -\varrho_f g A(z+h)\mathbf{e}_z$ and $\mathbf{F}_2 = \varrho_f g A \mathbf{e}_z$, respectively, where z is the depth down to the upper surface. The gravitational force on the body is $\mathbf{F}_3 = \varrho_b V g \mathbf{e}_z$, and the sum of these three forces $\mathbf{F}_i$ ($i = 1, 2, 3$) is

$$\mathbf{F} = (\varrho_b - \varrho_f)V g \mathbf{e}_z. \tag{3.120}$$

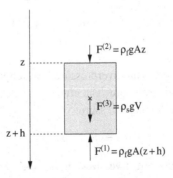

Figure 3.15. *The gravitational force on the body is counteracted by the surface forces from the fluid. (The surface forces on the vertical sides balance each other and they are therefore not shown.)*

We see that the gravitational force on the body is counteracted by the weight of the fluid displaced by the body, which implies that the body will sink (**F** points downwards) when $\varrho_b > \varrho_f$, or otherwise float.

The reason for deriving Archimedes' principle is that sediments are normally immersed in fluid and therefore experience buoyancy. Let us consider a column of sediments of height dz, porosity ϕ, matrix density ϱ_s and a horizontal surface area A. The net gravitational force of the sediments on the base of the column is then

$$F = (\varrho_s - \varrho_f)gA(1 - \phi)dz, \tag{3.121}$$

because the volume of fluid displaced by the solid (sediments) is $A(1 - \phi)\,dz$. The stress on the base of the column becomes

$$d\sigma' = F/A = (\varrho_s - \varrho_f)(1 - \phi)g\,dz \tag{3.122}$$

where the vertical stress σ' is called the *effective stress*. It is written properly as an integral

$$\sigma' = \int_0^z (\varrho_s - \varrho_f)\Big(1 - \phi(z)\Big)g\,dz \tag{3.123}$$

because the porosity is a function of depth. The effective stress plays an important role in controlling the compaction of the sediments. The porosity (or the void) ratio of loose (unlithified) sediments is shown with laboratory experiments to be a function of the effective stress.

We will now introduce two more vertical stress definitions. The first one is the *hydrostatic fluid pressure*,

$$p_h = \int_0^z \varrho_f g\,dz \tag{3.124}$$

which is simply the stress from the weight of the water column. The second one is the vertical *lithostatic stress*

$$p_b = \int_0^z \Big(\phi\varrho_f + (1 - \phi)\varrho_s\Big)g\,dz \tag{3.125}$$

which is the stress from the weight of both the fluid and the solid sediment matrix. The difference between the lithostatic stress and the hydrostatic fluid pressure is the effective vertical stress, $\sigma' = p_b - p_h$.

It often turns out that sediments are overpressured, which means that the fluid pressure p_f is larger than the hydrostatic pressure. The effective stress is therefore generally stated as the lithostatic stress minus the fluid pressure,

$$\sigma' = p_b - p_f. \tag{3.126}$$

It is important to note that the effective stress is not the same as the average stress on grain–grain contacts. The stress at grain contacts may be quite large because the weight of a sedimentary column can be distributed over small contact areas. The vertical effective stress is simply the weight of the sedimentary column minus the buoyancy on a macroscopic xy-surface. The pores can be thought of as so small that they become an unimportant perforation of planes that carry stress or force. The effective stress concept dates back to Terzaghi (Terzaghi, 1925) and his studies of compaction of soils in civil engineering.

Exercise 3.19 Let the porosity by related to depth by Athy's function (2.4). Show that the lithostatic pressure becomes

$$p_b(z) = \varrho_s g z + (\varrho_s - \varrho_w) g \phi_0 z_0 (1 - e^{-z/z_0}) \tag{3.127}$$

as a function of depth z.

3.15 Effective stress in 3D

We have so far only considered effective stress in the vertical direction. It has already been mentioned that the vertical effective stress controls how loose sediments compact. The generalization to 3D of vertical effective stress (3.126) is the effective stress tensor

$$\sigma'_{ij} = \sigma_{ij} - \alpha \, p_f \delta_{ij} \tag{3.128}$$

where α is the *Biot coefficient*. The Biot coefficient is 1 in the vertical effective stress (3.126), and it is therefore often approximated by 1. The effective stress tensor controls deformations (strain) of porous media just like the effective vertical stress controls deformation (compaction) of porous sediments. In the case of linear elasticity the Lamé equation becomes

$$-\sigma'_{ij} = -(\sigma_{ij} - \alpha \, p_f \delta_{ij}) = 2G\varepsilon_{ij} + \lambda \, \varepsilon_{kk} \delta_{ij}. \tag{3.129}$$

Our sign convention implies that a positive fluid pressure leads to positive strain, which causes expansion of the porous media. The inverse of the Lamé equation is just like equation (3.96) except that strain is now produced by effective stress instead of stress:

$$-\varepsilon_{ij} = \frac{1}{E}(1+v) \sigma'_{ij} - \frac{v}{E} \sigma'_{kk} \delta_{ij}. \tag{3.130}$$

One way to get some feeling of the effective stress is to compute the volume strain ε_{kk} from equation (3.130), which gives

$$\frac{\Delta V}{V} = \frac{(\alpha p - \bar{\sigma})}{K} \tag{3.131}$$

where $\bar{\sigma} = \frac{1}{3}\sigma_{kk}$ is the average stress in the solid porous matrix. (See Exercise 3.15 and Section 3.2.) The volume strain (3.131) shows that

$$\Delta V > 0, \quad \text{when } \alpha p > \bar{\sigma} \tag{3.132}$$

$$\Delta V < 0, \quad \text{when } \alpha p < \bar{\sigma}. \tag{3.133}$$

Our sign convention now tells us that the porous matrix expands when the pore fluid pressure is larger than the average normal stress. We assume for simplicity that $\alpha = 1$. The porous matrix is thus compressed by the normal forces when the pore fluid pressure is less than the average normal stress. Finally, the pore fluid pressure and the average normal stress balance each other when they are equally large.

Exercise 3.20 The fluid pressure acting on any (infinitesimal) surface with area dS and orientation $\mathbf{n}$ (given by the outward unit vector $\mathbf{n}$) is $\mathbf{p} = -\varrho_f g z \mathbf{n}$. (The minus sign is needed because the force from the hydrostatic pressure acts in the direction opposite to the outward unit normal vector.) Use the divergence theorem

$$\int_S \mathbf{a} \cdot \mathbf{n} \, dS = \int_V \nabla \cdot \mathbf{a} \, dV \tag{3.134}$$

to show that the total hydrostatic force acting on the surface of a body with volume V and arbitrary shape is $\mathbf{F} = -\varrho_f g V \mathbf{e}_z$.
Solution:

$$F_i = -\int_S \varrho_f g z \delta_{ij} n_j \, dS = -\int_V \frac{\partial}{\partial x_j}(\varrho_f g z \delta_{ij}) \, dV = -\varrho_f g V \delta_{zi}. \tag{3.135}$$

The divergence theorem is applied to each vector component of the force separately. We used that $\varrho_f g z n_i$ can be written as a scalar product of $a_j = \varrho_f g z \sigma_{ij}$ and n_j, that $\partial z / \partial x_j = \delta_{zj}$ and that $\delta_{zj} \delta_{ij} = \delta_{zi}$.

3.16 Euler and Lagrange coordinates

A moving or deforming continuum like the stretching of the crust or the subsidence and compaction of the sediments in a basin can be modeled in two alternative reference frames (coordinate systems). One option is to relate the motion of the "particles" constituting the continuum to a fixed coordinate system. Such a fixed reference frame is called an *Eulerian* coordinate system. The alternative coordinate system is one that follows the motion of the "particles." In other words, the particles are at rest in this reference frame, which is called a *Lagrangian* coordinate system. The Euler and the Lagrange coordinates are related. The Lagrange coordinate $\mathbf{a}$ of a particle is now taken to be its position at time $t = 0$. The Euler position of this particle is

$$\mathbf{x} = \mathbf{x}(\mathbf{a}, t) \tag{3.136}$$

where the $\mathbf{a}$-vector is its unique label, which therefore remains constant through time. (See Figure 3.16.) We will assume that the inverse of equation (3.136) exists,

$$\mathbf{a} = \mathbf{a}(\mathbf{x}, t), \tag{3.137}$$

which tells where the particle in position $\mathbf{x}$ was at time $t = 0$. The velocity of a particle following the path $\mathbf{x} = \mathbf{x}(\mathbf{a}, t)$ is

$$\mathbf{V}(\mathbf{a}, t) = \frac{\partial \mathbf{x}(\mathbf{a}, t)}{\partial t} \tag{3.138}$$

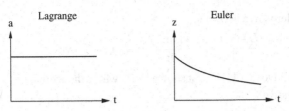

Figure 3.16. *(Left) A Lagrange coordinate system follows the movement of the "particles" of the continuum. The particles are therefore at rest in the Lagrange coordinate system. (Right) An Euler coordinate system is a fixed reference frame in which the "particles" move.*

and it is normally expressed with reference to an arbitrary position $\mathbf{x}$ as

$$\mathbf{v} = \mathbf{v}(\mathbf{x}, t) = \mathbf{V}\Big(\mathbf{a}(\mathbf{x}, t), t\Big). \tag{3.139}$$

As already suggested, we will use an upper-case letter for a property calculated by Lagrangian coordinates and the corresponding lower-case letter for the same property calculated by Eulerian coordinates. As an example, we have

$$f(\mathbf{x}, t) = F\Big(\mathbf{a}(\mathbf{x}, t), t\Big) \quad \text{and} \quad F(\mathbf{a}, t) = f\Big(\mathbf{x}(\mathbf{a}, t), t\Big) \tag{3.140}$$

where f and F represent the same property of the continuum, like for instance velocity, temperature or density. It is important to distinguish between the velocity (or any other property) calculated by the Lagrangian coordinates and the Eulerian coordinates, because the functions like $\mathbf{V}$ and $\mathbf{v}$ are different. Both functions are expressions for the same velocity, but they take different arguments. The time derivative of a property is different in the two coordinate systems. We see from equation (3.140) that

$$\frac{\partial F(\mathbf{a}, t)}{\partial t} = \frac{\partial f(\mathbf{x}, t)}{\partial t} + \frac{\partial f(\mathbf{x}, t)}{\partial x_i} \frac{\partial x_i(\mathbf{a}, t)}{\partial t} \tag{3.141}$$

because the position $\mathbf{x}$ is not a constant of time. The right-hand side of equation (3.141) accounts for the fact that we have to move along with the particle in order to measure the time rate of change of a property of a specific particle. This derivative is called the *material derivative* or *convective derivative* and it is often denoted

$$\frac{Df(\mathbf{x}, t)}{dt} = \frac{\partial f(\mathbf{x}, t)}{\partial t} + \mathbf{v}(\mathbf{x}, t) \cdot \nabla f(\mathbf{x}, t). \tag{3.142}$$

Notice from equation (3.141) that the material derivative in Eulerian coordinates is the partial derivative in Lagrangian coordinates. In the next section we will need the material derivative of the Jacobian connecting the Eulerian and Lagrangian coordinate systems. The *Jacobian* is the following determinant:

$$J(\mathbf{a}, t) = \begin{vmatrix} \frac{\partial x_1}{\partial a_1} & \cdots & \frac{\partial x_1}{\partial a_3} \\ & \cdots & \\ \frac{\partial x_3}{\partial a_1} & \cdots & \frac{\partial x_3}{\partial a_3} \end{vmatrix} \tag{3.143}$$

and the material derivative becomes

$$\frac{DJ(\mathbf{x}, t)}{dt} = \frac{\partial J(\mathbf{a}, t)}{\partial t} = J(\mathbf{a}, t) \, \nabla \cdot \mathbf{v}(\mathbf{x}, t). \tag{3.144}$$

The material derivative (3.144) is first shown in 1D where the Jacobian is $J(a, t) = \partial x / \partial a$, and the time derivative is

$$\begin{aligned}
\frac{\partial J}{\partial t}(a, t) &= \frac{\partial}{\partial t} \left(\frac{\partial x}{\partial a}(a, t) \right) = \frac{\partial}{\partial a} \left(\frac{\partial x}{\partial t}(a, t) \right) = \frac{\partial v}{\partial a}(a, t) \\
&= \frac{\partial v}{\partial a}(x(a, t), t) \\
&= \frac{\partial v}{\partial x}(x, t) \frac{\partial x}{\partial a}(a, t) \\
&= \frac{\partial v}{\partial x}(x, t) \, J(a, t). \tag{3.145}
\end{aligned}$$

Note 3.6 A little more work is needed to show expression (3.144) for the general case of any number of spatial dimensions. First we need to recall the definition of the determinant function before we can undertake this task. A determinant is defined as a scalar function $d = d(\mathbf{b}_1, \ldots, \mathbf{b}_n)$ of n vectors with the following properties:

(1) It is linear in each slot:

$$d \left(\ldots, c_1 \mathbf{b}_k^{(1)} + c_2 \mathbf{b}_k^{(2)}, \ldots \right) = c_1 \, d \left(\ldots, \mathbf{b}_k^{(1)}, \ldots \right) + c_2 \, d \left(\ldots, \mathbf{b}_k^{(2)}, \ldots \right)$$

(2) The determinant vanishes if any two vectors are equal:

$$d(\mathbf{b}_1, \ldots, \mathbf{b}_n) = 0, \text{ if } \mathbf{b}_i = \mathbf{b}_j \text{ for } i \neq j$$

(3) It is unity for the identity matrix:

$$d(\mathbf{e}_1, \ldots, \mathbf{e}_n) = 1, \text{ where } \mathbf{e}_i \text{ is the unit vector in direction } i.$$

Equation (3.144) is now shown for 3D, but the same approach applies equally well for a different number of spatial dimensions. Using the definition of a determinant the Jacobian in 3D is

$$J = d(\mathbf{b}_1, \mathbf{b}_2, \mathbf{b}_3) \tag{3.146}$$

where the vector $\mathbf{b}_i$ is

$$\mathbf{b}_i = \left(\frac{\partial x_i}{\partial a_1}, \frac{\partial x_i}{\partial a_2}, \frac{\partial x_i}{\partial a_3} \right). \tag{3.147}$$

We then get

$$\frac{\partial J}{\partial t} = d \left(\frac{\partial \mathbf{b}_1}{\partial t}, \mathbf{b}_2, \mathbf{b}_3 \right) + d \left(\mathbf{b}_1, \frac{\partial \mathbf{b}_2}{\partial t}, \mathbf{b}_3 \right) + d \left(\mathbf{b}_1, \mathbf{b}_2, \frac{\partial \mathbf{b}_3}{\partial t} \right) \tag{3.148}$$

where the components of $\partial \mathbf{b}_i / \partial t$ are

$$\frac{\partial}{\partial t} \left(\frac{\partial x_i}{\partial a_j} \right) = \frac{\partial}{\partial a_j} \left(\frac{\partial x_i}{\partial t} \right) = \frac{\partial}{\partial a_j} \left(v_i(\mathbf{x}(\mathbf{a}, t), t) \right) = \frac{\partial v_i}{\partial x_s} \frac{\partial x_s}{\partial a_j}. \tag{3.149}$$

(Notice that the summation convention tells us that there is a summation over all pairs of the same index, in this case s.) We therefore have

$$\frac{\partial \mathbf{b}_i}{\partial t} = \left(\frac{\partial v_i}{\partial x_s} \frac{\partial x_s}{\partial a_1}, \frac{\partial v_i}{\partial x_s} \frac{\partial x_s}{\partial a_2}, \frac{\partial v_i}{\partial x_s} \frac{\partial x_s}{\partial a_3} \right) = \frac{\partial v_i}{\partial x_s} \mathbf{b}_s. \qquad (3.150)$$

Properties (1) and (2) from the definition of a determinant give that

$$\frac{\partial J}{\partial t} = \frac{\partial v_1}{\partial x_s} d(\mathbf{b}_s, \mathbf{b}_2, \mathbf{b}_3) + \frac{\partial v_2}{\partial x_s} d(\mathbf{b}_1, \mathbf{b}_s, \mathbf{b}_3) + \frac{\partial v_3}{\partial x_s} d(\mathbf{b}_1, \mathbf{b}_2, \mathbf{b}_s) = (\nabla \cdot \mathbf{v}) J \quad (3.151)$$

which is what we wanted to show.

Exercise 3.21 Show that the acceleration in Eulerian coordinates is Dv/dt.

3.17 An important Lagrange coordinate

We will in later chapters deal with subsiding sedimentary basins where the sediments compact and loose porosity during deposition and burial. Either an Euler or a Lagrange coordinate system can be used to describe the compaction of a sedimentary basin. The Euler coordinate system has the advantage that it can in principle be any coordinate system, but the disadvantage is that the material (time) derivative makes the equations more difficult to solve. A Lagrange description avoids the material derivative, but it might be difficult to find a convenient Lagrangian coordinate system. Fortunately, it turns out that the net (porosity-free) amount of sediments, when it is measured as the height above the base of the basin, is a convenient choice for a vertical Lagrangian coordinate. We will use ζ as notation for this Lagrangian coordinate, and the ζ-position of a sedimentary "grain" is constant, given that the movement of the sediments is restricted to the vertical direction. The relationship between the ζ and the real z-coordinate is given by

$$d\zeta = (1 - \phi)dz \qquad (3.152)$$

which clearly shows that $d\zeta$ is the net volume of solid in the interval dz. The net volume of solid is a constant given that the density of the solid is constant, an assumption that is normally justified. The ζ-coordinate is measured from the base of the basin (or a sedimentary layer) while the z-coordinate is measured from the basin or water surface with the z-axis pointing downwards. The z-coordinate for a given ζ-coordinate is therefore

$$z = \int_{\zeta}^{\zeta^*} \frac{d\zeta}{1 - \phi(\zeta)} \qquad (3.153)$$

where the porosity ϕ is a function of ζ, and the ζ-coordinate of the basin surface is denoted ζ^*. Figure 3.17 gives an example of what a 2D basin looks like in a xz-coordinate system and in the corresponding $x\zeta$-coordinate system. The Lagrangian ζ-coordinates in the next chapters are used in several applications that involve sediment compaction (decreasing porosity with depth).

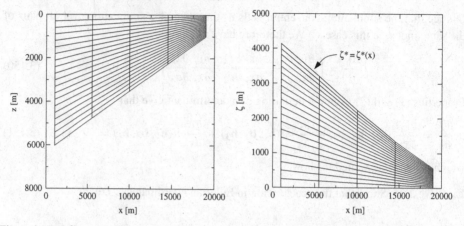

Figure 3.17. *Left: A 2D sedimentary basin is shown in the xz-coordinate system. Right: The same basin as to the left is shown as net (porosity-free) thicknesses measured from the basement. The total net thickness of the basin along the x-axis is denoted ζ^*. The position of a grid node in the $x\zeta$-coordinate system sticks to the sedimentary grains, and is there constant regardless of deposition of more sediments, compaction or eventual water depth changes.*

3.18 Conservation laws in 1D

Conservation of quantities like mass, energy or momentum are the most fundamental laws of nature. We will start by showing how conservation of mass is expressed in 1D by considering the flow of mass through a box located from x to $x + \Delta x$ with a cross-section area A (see Figure 3.18). The mass flowing into the box (from the left side) during a time step Δt, and the mass flowing out of the box (at the right side) during the same time step are

$$m_{\text{in}} = v(x, t)\, \Delta t\, A\, \varrho(x, t) \tag{3.154}$$

$$m_{\text{out}} = v(x + \Delta x, t)\, \Delta t\, A\, \varrho(x + \Delta x, t) \tag{3.155}$$

respectively, where ϱ is the density. The mass gain (or loss) becomes

$$\Delta m = m_{\text{in}} - m_{\text{out}} \approx -\frac{\partial}{\partial x}\left(v\varrho\right)\Delta V\, \Delta t \tag{3.156}$$

where $\Delta V = A\, \Delta x$ is the volume of the box. The increase (or decrease) of mass during the time step Δt is the difference between the mass m_2 in the box at time $t + \Delta t$ and the mass m_1 in the box at time t. These masses are

$$m_1 = \varrho(x_c, t)\, V \tag{3.157}$$

$$m_2 = \varrho(x_c, t + \Delta t)\, V \tag{3.158}$$

respectively, where x_c is the coordinate of the box center, and the difference becomes

$$\Delta m = m_2 - m_1 \approx \frac{\partial \varrho}{\partial t}\Delta V\, \Delta t. \tag{3.159}$$

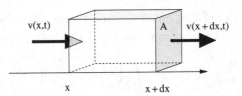

Figure 3.18. *The figure shows mass flowing through a box. The velocity of the mass entering the box (at the left) is $v(x, t)$ and the velocity of the mass leaving the box (at the right) is $v(x + \Delta x, t)$.*

Since mass is a conserved property we have

$$m_{\text{in}} - m_{\text{out}} = m_2 - m_1 \tag{3.160}$$

which gives

$$\frac{\partial \varrho}{\partial t} + \frac{\partial (v\varrho)}{\partial x} = 0 \tag{3.161}$$

in the limit where both Δx and Δt approach zero. This is the equation for mass conservation, which is also called the *continuity equation.*

Exercise 3.22 Show that equation (3.161) can be rewritten as

$$\frac{D\varrho}{dt} + \varrho\frac{\partial v}{\partial x} = 0. \tag{3.162}$$

Exercise 3.23 Show that a constant density ϱ in 1D implies that the velocity v has to be constant too.

Exercise 3.24 Show that equation (3.161) becomes

$$\frac{\partial \varrho}{\partial t} + \frac{\partial (v\varrho)}{\partial x} = S(x, t) \tag{3.163}$$

when $S(x, t)$ is the rate of mass production per unit volume and per time in the position x at time t.

3.19 Mass conservation

The equation for mass conservation can be derived by considering a volume $V(t)$ that always contains the same "particles" and therefore the same mass. We then have

$$\frac{d}{dt}\int_{V(t)} \varrho \, dV = 0 \tag{3.164}$$

where $dV = dx\,dy\,dz$. The time differentiation of the integral is not straightforward to carry out because the volume $V(t)$ is a function of time. The integral is therefore rewritten in Lagrangian coordinates as

$$\frac{d}{dt}\int_{V_0'} \varrho J \, dV' = 0 \tag{3.165}$$

using the Jacobian J (3.143), where $dV' = da_1 da_2 da_3$. The volume V_0' is a constant volume in the Lagrangian coordinate system, because the particles are at rest there:

$$\frac{d}{dt} \int_{V_0'} \varrho\, J\, dV' = \int_{V_0'} \left(\frac{\partial \varrho}{\partial t} J + \varrho \frac{\partial J}{\partial t} \right) dV'$$

$$= \int_{V_0'} \left(\frac{\partial \varrho}{\partial t} + \varrho \nabla \cdot \mathbf{v} \right) J\, dV'$$

$$= \int_{V(t)} \left(\frac{D\varrho}{dt} + \varrho \nabla \cdot \mathbf{v} \right) dV. \qquad (3.166)$$

Equation (3.166) uses that $\partial J / \partial t = J \nabla \cdot \mathbf{v}$ in Lagrange coordinates, and that the partial time derivative in Lagrange coordinates becomes the material derivative in Euler coordinates. Since equation (3.166) is zero we get the expression for mass conservation

$$\frac{D\varrho}{dt} + \varrho \nabla \cdot \mathbf{v} = 0. \qquad (3.167)$$

This is the generalization to any number of spatial dimensions of equation (3.161) for 1D mass conservation. (See Exercise 3.24.) Mass conservation (3.167) can be used to derive the following useful relationship:

$$\frac{d}{dt} \int_{V(t)} \varrho\, f\, dV = \int_{V(t)} \varrho \frac{Df}{dt}\, dV \qquad (3.168)$$

where f is any scalar function, see exercise 3.25.

Exercise 3.25 Verify equation (3.168).

Exercise 3.26 Show that mass conservation for an incompressible medium ($\varrho = $ constant) is given by $\nabla \cdot \mathbf{v} = 0$.

Exercise 3.27 Derive Reynolds transport theorem

$$\frac{d}{dt} \int_{V(t)} f\, dV = \int_{V(t)} \frac{\partial f}{\partial t}\, dV + \int_{\partial V(t)} f \mathbf{v} \cdot \mathbf{n}\, dA. \qquad (3.169)$$

Hint: switch to Lagrangian coordinates, use equation (3.144) and then Gauss's theorem (3.171).

3.20 Momentum balance (Newton's second law)

We will now derive an expression for Newton's second law which says that *the time rate of change of momentum for a body is equal to the sum of all forces acting on it.* This law is written as

$$\frac{d}{dt} \int_{V(t)} \varrho\, v_i\, dV = \int_{V(t)} \varrho\, g\, \delta_{i,z}\, dV + \int_{\partial V(t)} \sigma_{ij} n_j\, dA \qquad (3.170)$$

where the "body" is defined as all particles contained by the volume $V(t)$. The volume may change shape and size, but it always contains the same particles. The left-hand side of equation (3.170) is the rate of change of momentum, and the right-hand side has as the first term the body force (gravity) and as the second term the surface force acting on the body. The Kronecker delta $\delta_{i,z}$ makes sure that gravity acts only in the vertical direction. We have seen that the force acting on a surface with area dA and outward unit normal vector n_j is $\sigma_{ij} n_j dA$. Notice that $\partial V(t)$ is used as a notation for the entire surface of the volume $V(t)$. The integral over the surface is converted to a volume integral by means of Gauss's theorem,

$$\int_{\partial V(t)} f_j n_j \, dA = \int_{V(t)} \frac{\partial f_j}{\partial x_j} dV. \tag{3.171}$$

Furthermore, the rate of change of momentum in direction i is rewritten as

$$\frac{d}{dt} \int_{V(t)} \varrho \, v_i \, dV = \int_{V(t)} \varrho \, \frac{Dv_i}{dt} \, dV \tag{3.172}$$

by means of equation (3.168) when $f = v_i$. Newton's second law as expressed by equation (3.170) is therefore

$$\varrho \, \frac{Dv_i}{dt} = \varrho \, g \, \delta_{i,z} + \frac{\partial \sigma_{ij}}{\partial x_j}. \tag{3.173}$$

The acceleration Dv_i/dt is often small compared with the constant of gravity, and the left-hand side of equation (3.173) can be approximated by zero. Newton's second law then reduces to a force balance

$$\frac{\partial \sigma_{ij}}{\partial x_j} = -\varrho \, g \, \delta_{i,z} \tag{3.174}$$

and it is in this last form that we will be using Newton's second law. But we have to change the sign of the right-hand side before we can use equation (3.174) with a z-axis pointing downwards. A 1D model along the z-axis gives after sign reversal

$$\frac{\partial \sigma_{zz}}{\partial z} = \varrho \, g \tag{3.175}$$

and an integration then gives $\sigma_{zz} = \varrho g z$ as wanted. The force balance (3.174) is now written once more, adapted for a z-axis pointing downwards,

$$\frac{\partial \sigma_{ij}}{\partial x_j} = \varrho \, g \, \delta_{i,z}. \tag{3.176}$$

We have already seen that effective stress is the cause for deformations in a fluid-saturated porous medium. The fluid pressure acts to expand the medium while (positive) normal stress will compress it. The force balance (3.176) becomes

$$\frac{\partial \sigma'_{ij}}{\partial x_j} + \alpha \frac{\partial p_f}{\partial x_i} = \varrho \, g \, \delta_{i,z} \tag{3.177}$$

when it is rewritten in terms of the effective stress (3.128). We see that in order to solve for effective stress we must know the fluid pressure (or we must solve for both effective stress and fluid pressure simultaneously).

Exercise 3.28 Solve equation (3.176) for the displacement in the vertical direction caused by gravity in a 1D crust, assuming linear elasticity and zero horizontal strain. The bulk density ϱ is constant. There is no displacement at the top of the column ($z = 0$), and the top of the column has zero (vertical) stress.

Solution: A 1D crust with no lateral strain ($\varepsilon_{xx} = \varepsilon_{yy} = 0$) is the starting point. The linear elastic stress–strain relationship (3.98) is then

$$\sigma_{zz} = -(2G + \lambda)\varepsilon_{zz} = -(2G + \lambda)\frac{du}{dz} \tag{3.178}$$

where u is the displacement in the vertical direction. We get

$$\frac{d\sigma_{zz}}{dz} = -(2G + \lambda)\frac{d^2u}{dz^2} = \varrho\,g \tag{3.179}$$

when the stress–strain relationship (3.178) is inserted into equation (3.176) for the force balance. Equation (3.179) is integrated twice, and gives the vertical displacement

$$u(z) = -\frac{\varrho\,g}{(2G + \lambda)}\left(\frac{1}{2}z^2 + Az + B\right). \tag{3.180}$$

It follows from $\sigma_{zz}(z=0) = 0$ that $du/dz(z=0) = 0$, which implies that $A = 0$. Zero displacement at the surface gives that $B = 0$. The solution for the displacement is therefore

$$u(z) = -\frac{\varrho\,g}{2(2G + \lambda)}z^2 \tag{3.181}$$

or alternatively

$$u(z) = -u_0\left(\frac{z}{h}\right)^2 \tag{3.182}$$

where the coefficient $u_0 = \frac{1}{2}\varrho\,g\,h^2/(2G + \lambda)$ is the maximum displacement at the base ($z = h$) of the column. Notice that the displacement is upwards. The vertical strain is

$$\varepsilon_{zz} = \frac{du}{dz} = -\varepsilon_0\frac{z}{h} \tag{3.183}$$

where the maximum strain is $\varepsilon_0 = \varrho\,g\,h/(2G + \lambda)$. The stress in the vertical direction is found from the displacement, equation (3.178), and it is

$$\sigma_{zz} = -(2G + \lambda)\varepsilon_{zz} = \sigma_0\frac{z}{h} \tag{3.184}$$

where the maximum stress σ_0 found at the base of the column is $\sigma_0 = \varrho\,g\,h$. This is precisely the stress caused by the weight of the column. What is the displacement at the depth $h = 1000$ m when $E = 50$ GPa and $\nu = 0.25$?

Exercise 3.29 Show that equation (3.177) for effective stress becomes

$$\frac{\partial \sigma'_{ij}}{\partial x_j} = (\varrho - \varrho_f)\, g\, \delta_{i,z} \tag{3.185}$$

when the fluid pressure is hydrostatic $p_f = \varrho_f\, g\, z$ and the Biot coefficient is $\alpha = 1$.

Solution: We have

$$\frac{\partial p_f}{\partial x_i} = \frac{\partial(\varrho_f g z)}{\partial x_i} = \varrho_f g\frac{\partial z}{\partial x_i} = \varrho_f g \delta_{i,z} \tag{3.186}$$

(where it is assumed that both the fluid density ϱ_f and the acceleration of gravity are constants).

Exercise 3.30 Find the vertical displacement caused by gravity in a 1D fluid-saturated linear-elastic sedimentary column. Assume that the fluid pressure is hydrostatic $p_f = \varrho_f g z$, the horizontal strain is zero, the displacement at the top of column $(z = 0)$ is zero, the top of the column $(z = 0)$ is unstressed, and the Biot coefficient is $\alpha = 1$.

Solution: It is the effective stress that causes deformations in a saturated porous medium, and the equilibrium equation in the case of a hydrostatic fluid is

$$\frac{\partial \sigma'_{ij}}{\partial x_j} = (\varrho - \varrho_f)\, g\, \delta_{i,z} \tag{3.187}$$

as seen from Exercise 3.29. Using the Lamé equations we get

$$\frac{d\sigma'_{zz}}{dz} = -(2G + \lambda)\frac{d^2 u}{dz^2} = (\varrho - \varrho_f)\, g. \tag{3.188}$$

The solution of equation (3.188) is therefore the same as in Exercise 3.28, except that the solution yields effective stress instead of stress, and that the density that enters the solution is $\varrho - \varrho_f$. The maximum effective stress is $\sigma'_0 = (\varrho - \varrho_f)\, g\, h$ at the base of the column $(z = h)$, and we get the stress at the same depth by adding the fluid pressure $p_h = \varrho_f g h$, which is then $\sigma_0 = \varrho_f g h$. We see that there will be no deformation of the sedimentary column when the fluid density is equal to the bulk density, $(\varrho - \varrho_f = 0)$.

Exercise 3.31 Show that the fluid density has to be equal to the matrix (solid) density if the bulk density and the fluid density are equal.

Solution: The bulk density is $\varrho = (1 - \phi)\varrho_s + \phi\varrho_f$ where ϕ is the porosity, ϱ_s and ϱ_f are the matrix and fluid densities, respectively. We then get

$$\varrho - \varrho_f = (1 - \phi)(\varrho_s - \varrho_f) \tag{3.189}$$

which implies that if $\varrho_s = \varrho_f$ then $\varrho = \varrho_f$. (It is assumed that $\phi < 1$.)

Exercise 3.32

(a) Show that the equilibrium equations (3.177) become

$$\frac{d\sigma'_{zz}}{dz} = (\varrho - \alpha\varrho_f)g - \alpha\frac{dp_e}{dz} \tag{3.190}$$

along the z-axis in a 1D model, when the overpressure p_e is used instead of the fluid pressure.

(b) Let the overpressure $p_e(z)$ be known as a function of depth z. Show that the displacement along the z-axis becomes

$$u(z) = -\frac{(\varrho - \alpha\varrho_f)g}{2(2G + \lambda)} z^2 + \frac{\alpha}{(2G + \lambda)} \int_0^z p_e(z')dz' \qquad (3.191)$$

when the bulk density ϱ is constant.

Exercise 3.33 Show that the displacement $u(z)$ along the z-axis is

$$u(z) = -\frac{(\varrho_s - \varrho_w)}{(2G + \lambda)} \left(\frac{1}{2}z^2 + \phi_0 z_0(z_0 - z - z_0 e^{-z/z_0})\right) \qquad (3.192)$$

for a hydrostatic basin, when the bulk density is $\varrho = \phi\varrho_w + (1 - \phi)\varrho_s$ and the porosity as a function of depth is $\phi = \phi_0 \exp(-z/z_0)$. The displacement field is zero ($u = 0$) at the surface ($z = 0$) and there is zero stress at the surface, $\sigma_{zz} = -(2G + \lambda)du/dz = 0$ for $z = 0$.

3.21 Particle paths and streamlines

The *particle paths* are the traces made by flowing particles. They are always tangential to the flow field, and are therefore the solution of

$$\frac{d\mathbf{r}}{dt} = \mathbf{v}(\mathbf{r}, t) \qquad (3.193)$$

when the vector field is known. The vector equation (3.193) can also be written as

$$\frac{dx}{v_x} = \frac{dy}{v_y} = \frac{dz}{v_z} = dt \qquad (3.194)$$

which later will be a useful representation of the particle paths. The same equation also follows from the vector product

$$\mathbf{v} \times d\mathbf{r} = 0 \qquad (3.195)$$

where $d\mathbf{r} = (dx, dy, dz)$ is a small line segment along the path.

A *streamline* is the path given by the flow field at a specific point in time. The streamlines therefore coincide with the particle paths when the flow field is independent of time. The following flow field illustrates the difference between particle paths and streamlines:

$$\frac{dx}{dt} = v_x = \frac{x}{t_0 + t} \quad \text{and} \quad \frac{dy}{dt} = v_y = \frac{y}{t_0} \qquad (3.196)$$

where t_0 is a parameter. An integration of equations (3.196) gives the particle path

$$x(t) = x_0 \cdot \left(1 + \frac{t}{t_0}\right) \quad \text{and} \quad y(t) = y_0 \, e^{t/t_0} \qquad (3.197)$$

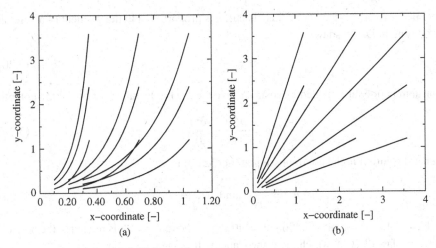

Figure 3.19. (a) Particle paths. (b) Streamlines.

for a particle that is at position (x_0, y_0) at time $t = 0$. The paths that follow the fixed flow field of a given time are the streamlines. The flow field at the time $t = t_1$ gives the following equations for the streamlines:

$$\frac{dx}{ds} = v_x = \frac{x}{t_0 + t_1} \quad \text{and} \quad \frac{dy}{ds} = v_y = \frac{y}{t_0}. \tag{3.198}$$

The streamlines are parameterized by s, because it is not necessarily time. An integration of (3.198) gives the streamlines

$$x(s) = x_0\, e^{s/(t_0+t_1)} \quad \text{and} \quad y(s) = y_0\, e^{s/t_0} \tag{3.199}$$

which go through the point (x_0, y_0) for $s = 0$. The particle paths (3.197) and the streamlines (3.199) are shown in Figure 3.19, when $t_0 = 1$ and $t_1 = 0$.

3.22 Streamlines in 2D

Streamlines for stationary potential flow in 2D are an important special case for several reasons. Firstly, a stream function can be obtained from the flow field, which is constant for each streamline. Secondly, the potential is related to the stream function through a pair of relationships called the Cauchy–Riemann equations. Finally, the streamlines and the iso-potential curves are always normal (see the example in Figure 3.21). To see this we start with mass conservation for an incompressible fluid in 2D,

$$\nabla \cdot \mathbf{u} = \frac{\partial u_x}{\partial x} + \frac{\partial u_y}{\partial y} = 0 \tag{3.200}$$

where $\mathbf{u} = (u_x, u_y)$ is the Darcy flux. The flux is proportional to the gradient of a potential p according to Darcy's law

$$\mathbf{u} = -\frac{k}{\mu}\nabla p. \tag{3.201}$$

A constant permeability k and viscosity μ give the 2D Laplacian equation for the potential

$$\nabla^2 p = \frac{\partial^2 p}{\partial x^2} + \frac{\partial^2 p}{\partial y^2} = 0. \tag{3.202}$$

A general solution to the continuity equation (3.200) is

$$u_x = \frac{\partial \Psi}{\partial y} \quad \text{and} \quad u_y = -\frac{\partial \Psi}{\partial x} \tag{3.203}$$

which is a solution for any function $\Psi = \Psi(x, y)$. The function Ψ is related to the potential by Darcy's law (3.201), which gives the Cauchy–Riemann equations

$$\frac{\partial \Psi}{\partial y} = -\frac{k}{\mu}\frac{\partial p}{\partial x} \quad \text{and} \quad \frac{\partial \Psi}{\partial x} = \frac{k}{\mu}\frac{\partial p}{\partial y}. \tag{3.204}$$

The factor k/μ does not belong to the Cauchy–Riemann equations and is normally left out. If the Laplace equation is solved for the potential, and the Darcy velocities are known, it is possible to obtain the function Ψ by integration, since it follows from (3.203) that

$$\Psi = \int u_x \, dy \quad \text{or} \quad \Psi = -\int u_y \, dx. \tag{3.205}$$

The stream function Ψ is constant along each streamline, which is a property of the stream function that follows from

$$d\Psi = \frac{\partial \Psi}{\partial x}dx + \frac{\partial \Psi}{\partial y}dy = -u_x \, dx + u_y \, dy = 0 \tag{3.206}$$

because for streamlines we have

$$\frac{dx}{u_x} = \frac{dy}{u_y}. \tag{3.207}$$

The streamlines are also orthogonal to the iso-potential curves (in the case when the flow field is given by the gradient of a potential), because the Cauchy–Riemann equations give that

$$\nabla p \cdot \nabla \Psi = 0. \tag{3.208}$$

The gradient ∇p is normal to the iso-potential curves and the gradient $\nabla \Psi$ is normal to the streamlines. The streamlines and the iso-potential curves are therefore orthogonal.

Another direct consequence of the Cauchy–Riemann equations is that the stream function is also a solution of the Laplace equation

$$\frac{\partial^2 \Psi}{\partial x^2} + \frac{\partial^2 \Psi}{\partial y^2} = 0. \tag{3.209}$$

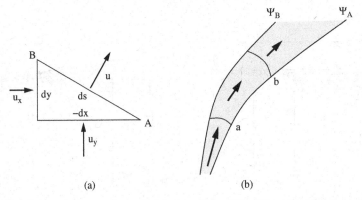

Figure 3.20. *(a) The sum of the volumetric flow rate through the sides of a triangle is zero. (b) The volumetric flow rate crossing the curves a and b or any other curve that connects the two streamlines is $\Psi_B - \Psi_A$, which is the difference between the stream functions for the two streamlines. The area between two streamlines is called a* streamtube.

This is an alternative route to the stream function instead of first solving for the potential, then obtaining the flow field from the potential and finally the stream function from the flow field. A possible problem with the solution of (3.209) is that it may be difficult to specify the value of the stream function along open boundaries.

The differential (3.206) gives an interpretation of the stream function as shown by Figure 3.20. Mass conservation requires that the volumetric flow rate between points A and B is the sum of the volumetric flow rates through the orthogonal sides. We therefore have that $d\Psi = -u_y\,dx + u_x\,dy$ is the volumetric flow flow rate between A and B. The volumetric flow rate between any points A and B is therefore

$$\int_A^B d\Psi = \Psi_B - \Psi_A. \tag{3.210}$$

The difference in value between two stream functions is the volumetric flow rate between the two streamlines. It is the volumetric flow rate crossing any curve that connects the two streamlines, see Figure 3.20. Since the stream function is constant along a streamline there is no flow rate between two points along the same streamline. The area between two streamlines is called a *streamtube*.

The rest of the section gives an example of streamlines and iso-potential curves for meteoric fluid flow driven by water saturated topography with a cosine shape (see Figure 3.21). The potential is a solution of the Laplace equation (3.202), for a rectangular cross-section of the subsurface with width l and height h. The boundary conditions are zero fluid flow through the boundaries, except for the surface. The topography creates a surface potential p_s given by the cosine function

$$p_s(x, z{=}h) = \varrho g h_0 \left(1 + \cos(\omega x)\right) \tag{3.211}$$

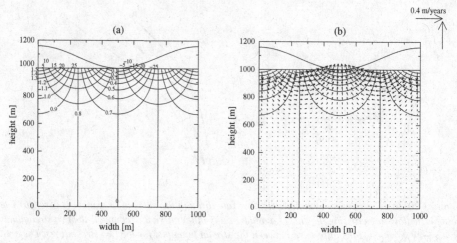

Figure 3.21. *An example of stationary meteoric fluid flow. Iso-potential curves are in units MPa, and the stream function is in units* m^2/years. *(a) The streamlines and the iso-potential curves are orthogonal. (b) The iso-potential curves and the Darcy flux.*

where $\omega = 2\pi/l$ is the wave number, ϱ is the water density and g is the constant of gravity. The maximum difference in height between the lowest and highest positions along the surface is $2h_0$. The solution of the Laplace equation (3.202) with these boundary conditions is

$$p(x, z) = \varrho g h_0 \left(1 + \frac{1}{c} \cos(\omega x) \cosh(\omega z)\right) \qquad (3.212)$$

where $c = \cosh(\omega h_0)$. The maximum potential along the surface is $p_0 = 2\varrho g h_0$. It is straightforward to verify that this is a solution of the Laplace equation by inserting $p(x, y)$. The Darcy flux (3.201) then becomes

$$u_x = \frac{u_0}{c} \sin(\omega x) \cosh(\omega z) \quad \text{and} \quad u_z = -\frac{u_0}{c} \cos(\omega x) \sinh(\omega z) \qquad (3.213)$$

where $u_0 = (k/\mu)\varrho g h_0 \omega$. We see that the Darcy flux u_0 is the maximum value for u_x. The flux u_0 is also an accurate estimate for the maximum of u_z whenever $\sinh(\omega h)/c = \tanh(\omega h) \approx 1$, a condition that may be written $\omega h > 2$. The boundary conditions for the fluid flux are straightforward to verify from the Darcy fluxes (3.213). The stream function is obtained from one of the Darcy flux components, which gives

$$\Psi = \int u_x \, dz = \frac{\Psi_0}{c} \sin(\omega x) \sinh(\omega z) \qquad (3.214)$$

where $\Psi_0 = k\varrho g h_0/\mu$ is approximately the maximum value for the stream function when $\omega h > 2$.

The parameters used in the case shown in Figure 3.21 are $h = l = 1000$ m, $h_0 = 80$ m, $k = 1 \cdot 10^{-15}$ m^2 and $\mu = 1 \cdot 10^{-3}$ Pa s. The maximum potential is $p_0 = 2\varrho g h_0 = 1.6$ MPa, the maximum Darcy flow is $u_0 = (k/\mu)\varrho g h_0 \omega = 0.15$ m/years, and the maximum stream function is $\Psi_0 = k\varrho g h_0/\mu = 25$ m^2/years. (The flux u_0 is the maximum for u_z, and Ψ_0 is the maximum of Ψ, because tanh$(2\pi) = 0.999$.) These maximum values serve as a simple check of the plots in Figure 3.21.

3.23 Further reading

A standard reference to continuum mechanics is Aris (1962). Haug and Langtangen (1997) gives a short review of the basics in continuum mechanics. One of many good text books in this field is Mase (1970) in Schaum's outline series. Another is the classic work of Fung (1965). Davis and Selvadurai (1996, 2002), Jaeger *et al.* (2007) give a comprehensive treatment of stress and strain in geomechanics.

4

Compressibility of rocks and sediments

4.1 Rock compressibility

Volume changes of a porous rock are linearly related to the pore pressure and the bulk pressure changes within certain bounds. For large variations in the pressure a rock volume may behave non-linearly or eventually fracture. The bulk pressure is defined as the mean normal stress acting on the surface of the rock sample, and it is taken to be the same as the *confining pressure* in poroelasticity. It is now possible to define four rock compressibilities in terms of the bulk volume V_b and the pore volume V_p with respect to the bulk pressure p_b and pore pressure p_f, and they are

$$\alpha_{bc} = -\frac{1}{V_b} \left(\frac{\partial V_b}{\partial p_b} \right)_{p_f} \tag{4.1}$$

$$\alpha_{bp} = +\frac{1}{V_b} \left(\frac{\partial V_b}{\partial p_f} \right)_{p_b} \tag{4.2}$$

$$\alpha_{pc} = -\frac{1}{V_p} \left(\frac{\partial V_p}{\partial p_b} \right)_{p_f} \tag{4.3}$$

$$\alpha_{pp} = +\frac{1}{V_p} \left(\frac{\partial V_p}{\partial p_f} \right)_{p_b}. \tag{4.4}$$

These compressibilities are given the following names: α_{bc} is the *drained bulk compressibility*, α_{bp} is the *bulk volume expansion coefficient*, α_{pc} is the *drained pore compressibility* and α_{pp} is the *pore volume expansion coefficient*. A drained compressibility is measured at a constant pore fluid pressure, because the fluid must be allowed to leave or enter (drain) the sample in order to keep the pressure constant. Another way to express these compressibilities is to write down the linear relationships between volume changes and pressure changes:

$$\frac{\Delta V_b}{V_b} = -\alpha_{bc}\Delta p_b + \alpha_{bp}\Delta p_f \tag{4.5}$$

$$\frac{\Delta V_p}{V_p} = -\alpha_{pc}\Delta p_b + \alpha_{pp}\Delta p_f. \tag{4.6}$$

Notice that the compressibilities α_{bp} and α_{pp} with respect to the pore fluid are written with a plus sign. That is because a positive fluid pressure increment is a tensile pressure that expands the rock. On the other hand, a positive bulk pressure increment is a compressive pressure that reduces the rock volume.

These compressibilities are not independent of each other or the compressibility of a (pure) solid matrix material, α_s. The following relations between the compressibilities follow from the work of Zimmerman *et al.* (1986). The bulk volume and the pore volume are both functions of the bulk pressure and the pore fluid pressure. These volumes are therefore written as

$$V_b = V_b(p_b, p_f) \quad \text{and} \quad V_p = V_p(p_b, p_f). \tag{4.7}$$

A change in the pressure state is now given by the pressure pair (dp_b, dp_f). Assuming that the rock behaves linearly with respect to pressure changes we can write that

$$dV_b(dp_b, dp_f) = dV_b(dp_b, 0) + dV_b(0, dp_f). \tag{4.8}$$

The volume change caused by a pressure change (dp_b, dp_f) is decomposed into a volume change from $(dp_b, 0)$ and $(0, dp_f)$, respectively, using the assumption of linearity. Although the assumption about linearity is not in general valid for rocks, it is at least valid for "small" changes to a given pressure state. The next step is the observation that a volume change caused by a uniform stress increment (dp, dp) turns out to be $dV_b = -\alpha_s \, dp$. The reason for this is that a volume change caused by equally large pressure changes in the pore fluid as in the bulk is the same as if the pores were filled with matrix material. Equation (4.8) can therefore be written as

$$(-\alpha_s \, dp) = (-\alpha_{bc} \, dp) + (\alpha_{bp} \, dp) \tag{4.9}$$

where the definition of the compressibilities α_{bc} and α_{bp} from equations (4.1) and (4.2), respectively, are used. This yields the first relation between the compressibilities,

$$\alpha_{bp} = \alpha_{bc} - \alpha_s. \tag{4.10}$$

An analogous argument can now be applied to the pore volume. The pressure change (dp, dp) is applied to V_p, and the resulting pore volume change $dV_p(dp, dp)$ is decomposed into the pore volume change from the bulk pressure $dV_p(dp, 0)$ and the pore volume change from the pore pressure $dV_p(0, dp)$,

$$dV_p(dp, dp) = dV_p(dp, 0) + dV_p(0, dp). \tag{4.11}$$

Using again that the volume change from a uniform pressure change (dp, dp) is the same as if the rock had the pore space filled up by matrix material, we can write equation (4.11) as

$$(-\alpha_s \, dp) = (-\alpha_{pc} \, dp) + (\alpha_{pp} \, dp) \tag{4.12}$$

where the pore space compressibilities, given by definitions (4.3) and (4.4), are used. This leads to the second relationship between the compressibilities:

$$\alpha_{pp} = \alpha_{pc} - \alpha_s. \tag{4.13}$$

A third relation between the compressibilities is based on a theorem in linear elastic mechanics called Betti's reciprocal theorem. It applies when a linearly elastic structure is subjected to two separate force systems, and it states that the work done by the force of the first system during deformations of the second system is equal to the work done by the force of second system during deformations of the first system. (This theorem is shown in Note 4.1.) The pressure dp_b and the corresponding deformation $dV_b(dp_b, 0)$ is one force system and the pressure dp_f and the corresponding deformation $dV_p(dp_b, 0)$ is the second force system. Betti's reciprocal theorem implies that

$$- dV_b(dp_b, 0) \cdot dp_f = dV_p(0, dp_f) \cdot dp_b. \tag{4.14}$$

The minus sign is needed because the pressure increment dp_f is tensile while the increment dp_b is compressive. Equation (4.14) is rewritten as

$$- (-V_b \alpha_{bc} \, dp_b) \, dp_f = (V_p \alpha_{pp} \, dp_f) \, dp_b \tag{4.15}$$

in terms of compressibilities. Then we get the third relation between the compressibilities:

$$\alpha_{bc} = \phi \alpha_{pp} \tag{4.16}$$

where it is used that $V_p = \phi V_b$. The three relationships (4.10), (4.13) and (4.16) can be used to express any three of the four compressibilities α_{bc}, α_{bp}, α_{pc} and α_{pp} in terms of ϕ, α_s and any remaining fourth compressibility. The bulk compressibility α_{bc} is considered to be the "fundamental" compressibility, because it is easier to measure than the other compressibilities. The other three compressibilities are in terms of ϕ, α_s and α_{bc}:

$$\alpha_{bp} = \alpha_{bc} - \alpha_s \tag{4.17}$$

$$\alpha_{pp} = \left(\alpha_{bc} - (1 + \phi)\alpha_s \right)/\phi \tag{4.18}$$

$$\alpha_{pc} = (\alpha_{bc} - \alpha_s)/\phi. \tag{4.19}$$

Figure 4.1 gives an idea of what these compressibilities (4.17)–(4.19) might be. We notice that the compressibilities α_{pp} and α_{pc} depend on the porosity. The numbers used are $\alpha_s = 2.7 \cdot 10^{-11}$ Pa^{-1} for pure quartz and $\alpha_{bc} = 4 \cdot 10^{-11}$ Pa^{-1} for the Fontainebleau sandstone, where these parameters are based on Song and Renner (2008). The compressibilities derived above are constant (for a given porosity) assuming a homogeneous rock without any specific texture or composition. It turns out that most rocks are heterogeneous on all length scales, from the grain scale to the basin scale, with variations in porosity, mineralogy and texture. The compressibilities are also to some degree dependent on effective pressure, as for instance seen in the study by Song and Renner (2008). It is therefore difficult to assign precise numbers for these compressibilities.

Note 4.1 *Betti's reciprocal theorem* is shown in 1D by stretching a bar. We recall that the force needed to keep a bar stretched a distance Δl is $F = A E \, \Delta l / l$, and that the work done stretching the bar a distance Δl is $W = (1/2)A E \, \Delta l^2 / l$, where A is the cross-section of

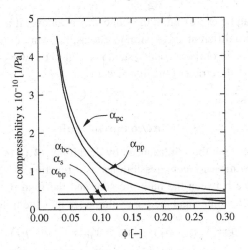

Figure 4.1. *The compressibilities (4.19) as a function of porosity.*

the bar and E is Young's modulus. Let us look at the following expression for work where two different displacements are used:

$$W = \frac{1}{2} A E \frac{\Delta l_1 \Delta l_2}{l}.$$ (4.20)

This expression for work can be rewritten in terms of the force F in two alternative ways:

$$\frac{1}{2} F_1 \Delta l_2 = \frac{1}{2} F_2 \Delta l_1$$ (4.21)

where F_i is the force corresponding to the displacement Δl_i ($i = 1, 2$). Equation (4.21) is Betti's reciprocal theorem: if a linearly elastic structure is subjected to two separate force systems, the work done by the force of the first system during deformations of the second system is equal to the work done by the second system during deformations of the first system.

Betti's reciprocal theorem can be shown generally using tensors. The stress state σ_{ij} in a linear elastic material is related to the strains ε_{kl} by the generalized Hooke's law $\sigma_{ij} = E_{ijkl}\varepsilon_{kl}$. (Einstein's summation convention is used, where there is summation over each pair of equal indices.) A different strain state $\bar{\varepsilon}_{kl}$ will produce a different stress state $\bar{\sigma}_{ij} = E_{ijkl}\bar{\varepsilon}_{kl}$. The following scalar is now written using both strain states:

$$W = \frac{1}{2} E_{ijkl} \bar{\varepsilon}_{ij} \varepsilon_{kl}.$$ (4.22)

This scalar can be rewritten using the stresses corresponding to the two different strain states as follows:

$$W = \frac{1}{2} \bar{\sigma}_{kl} \varepsilon_{kl} = \frac{1}{2} \sigma_{ij} \bar{\varepsilon}_{ij}.$$ (4.23)

The last equality is Betti's reciprocal theorem, and it is only based on the symmetry $E_{ijkl} = E_{klij}$ of the Hooke tensor. The scalar W is an expression for work, and Betti's

reciprocal theorem can be expressed in the same way as above: if a linearly elastic structure is subjected to two different stress–strain systems, the work done by the stress of the
first system with the strain of the second system is equal to the work done by the stress of
the second system with the strain of the first system.

4.2 More compressibilities

More compressibilities than those introduced by definitions (4.1)–(4.4) are measured and
used. It is common to measure compressibilities as a function of the effective pressure,
which is the difference between the bulk pressure and the fluid pressure, instead of the
bulk pressure. The bulk volume and the pore volume are now the following functions:

$$V_b = V_b(p_s, p_f) \quad \text{and} \quad V_p = V_p(p_s, p_f) \tag{4.24}$$

where the effective pressure is

$$p_s = p_b - p_f. \tag{4.25}$$

The compressibilities of the bulk volume and the pore volume with respect to effective
pressure and fluid pressure are

$$\frac{1}{K} = -\frac{1}{V_b}\left(\frac{\partial V_b}{\partial p_s}\right)_{p_f} \tag{4.26}$$

$$\frac{1}{K_s'} = -\frac{1}{V_b}\left(\frac{\partial V_b}{\partial p_f}\right)_{p_s} \tag{4.27}$$

$$\frac{1}{K_p} = -\frac{1}{V_p}\left(\frac{\partial V_p}{\partial p_s}\right)_{p_f} \tag{4.28}$$

$$\frac{1}{K_\phi} = -\frac{1}{V_p}\left(\frac{\partial V_p}{\partial p_f}\right)_{p_s} \tag{4.29}$$

where the notation for the moduli follow Wang (2000). The bulk pressure is here taken to
be the same as the confining pressure in Wang (2000). These compressibilities have the
following names: $1/K$ is the *drained bulk compressibility*, $1/K_s'$ is the *unjacketed bulk
compressibility*, $1/K_p$ is the *drained pore compressibility* and $1/K_\phi$ is the *unjacketed pore
compressibility*. The drained compressibilities are defined by a volume change at constant
(pore) fluid pressure in response to an effective pressure change. The pore fluid pressure
can only be kept constant at changing bulk pressure by allowing fluid to move out of
or into (by draining) the pore space, because the pore volume is changing with changing
bulk pressure. The unjacketed compressibilities are defined by volume changes in response
to changes in the pore fluid pressure, when the effective pressure is constant. The effective pressure is kept constant by letting the bulk pressure follow the pore pressure, which
is achieved in experiments by measuring the volume changes in a porous sample that is
immersed in the fluid. The (confining) fluid pressure on the walls of the sample are then

the same as the pore fluid pressure. Since the rock sample is immersed in the fluid it wears no "jacket," and thus the name *unjacketed*. Changes in the effective pressure and the fluid pressure lead to the following volume changes:

$$\frac{\Delta V_b}{V_b} = -\frac{1}{K}\Delta p_s - \frac{1}{K_s'}\Delta p_f \tag{4.30}$$

$$\frac{\Delta V_p}{V_p} = -\frac{1}{K_p}\Delta p_s - \frac{1}{K_\phi}\Delta p_f. \tag{4.31}$$

The compressibilities defined in terms of effective pressure and fluid pressure are not independent of the compressibilities defined in terms of bulk pressure and fluid pressure. These new volume compressibilities are related to the previous compressibilities by the chain rule of differentiation. The following relationships are derived in Note 4.2:

$$\alpha_{bc} = \frac{1}{K} \tag{4.32}$$

$$\alpha_{bp} = \frac{1}{K} - \frac{1}{K_s'} \tag{4.33}$$

$$\alpha_{pc} = \frac{1}{K_p} \tag{4.34}$$

$$\alpha_{pp} = \frac{1}{K_p} - \frac{1}{K_\phi}. \tag{4.35}$$

The inverse relationships, which are straightforward to obtain, are given for the sake of completeness:

$$\frac{1}{K} = \alpha_{bc} \tag{4.36}$$

$$\frac{1}{K_s'} = \alpha_{bc} - \alpha_{bp} \tag{4.37}$$

$$\frac{1}{K_p} = \alpha_{pc} \tag{4.38}$$

$$\frac{1}{K_\phi} = \alpha_{pc} - \alpha_{pp}. \tag{4.39}$$

Note 4.2 The relationship between the two sets of compressibilities follows from the relationship between the pressures, where $p_s = p_b - p_f$. Let V be either the bulk volume or the pore volume, and let the pair of compressibilities with respect to the pressures p_b and p_f be C_1 and C_2, respectively. A volume change is then related to the pressure change as

$$\begin{aligned}\frac{\Delta V}{V} &= -C_1\Delta p_b + C_2\Delta p_f \\ &= -C_1\Delta p_s + (-C_1 + C_2)\Delta p_f.\end{aligned} \tag{4.40}$$

The compressibilities with respect to the pressures p_s and p_f are $1/K_1$ and $1/K_2$, respectively, and the volume change is written

$$\frac{\Delta V}{V} = -\frac{1}{K_1}\Delta p_s - \frac{1}{K_2}\Delta p_f \tag{4.41}$$

which then gives

$$\frac{1}{K_1} = C_1 \quad \text{and} \quad \frac{1}{K_2} = C_1 - C_2 \tag{4.42}$$

or

$$C_1 = \frac{1}{K_1} \quad \text{and} \quad C_2 = \frac{1}{K_1} - \frac{1}{K_2}. \tag{4.43}$$

These are precisely the relationships (4.32) and (4.33) for the bulk volume, and relationships (4.34) and (4.35) for the pore volume.

Note 4.3 The relationships between the two different set of compressibilities follow from the chain rule of differentiation. Let $V(x, y)$ be a volume with respect to the pressures x and y, and let $V(s, t)$ be the same volume with respect to a new pressure pair s and t. In the case of effective pressure the two alternative pressure pairs are related as $s = x - y$ and $t = y$. The inverse relationship between the pressures are $x = s + t$ and $y = t$. Differentiation then leads to

$$\frac{\partial V}{\partial s} = \frac{\partial V}{\partial x}\frac{\partial x}{\partial s} + \frac{\partial V}{\partial y}\frac{\partial y}{\partial s} = \frac{\partial V}{\partial x} \tag{4.44}$$

$$\frac{\partial V}{\partial t} = \frac{\partial V}{\partial x}\frac{\partial x}{\partial t} + \frac{\partial V}{\partial y}\frac{\partial y}{\partial t} = \frac{\partial V}{\partial x} + \frac{\partial V}{\partial y}. \tag{4.45}$$

These partial derivative can be applied to the definitions of the compressibilities, which then give the relationship between them.

4.3 Compressibility of porosity and the solid volume

This section shows how the porosity and the volume of the solid matrix depend on (confining) bulk pressure and (pore) fluid pressure in a porous rock. Both relationships are based on the compressibilities for the bulk volume and the pore volume of Section 4.2. The porosity is treated first, where a change in the porosity is

$$\Delta\phi = \Delta\left(\frac{V_p}{V_b}\right) = \phi\left(\frac{\Delta V_p}{V_p} - \frac{\Delta V_b}{V_b}\right) \tag{4.46}$$

where the porosity is $\phi = V_p/V_b$ (the pore volume over the bulk volume). It then follows from (4.5) and (4.6) for changes in the bulk and pore volume that

$$\frac{\Delta\phi}{\phi} = (\alpha_{bc} - \alpha_{pc})\Delta p_b + (\alpha_{pp} - \alpha_{bp})\Delta p_f. \tag{4.47}$$

The relative change of the solid volume with respect to changes in the bulk volume and pore volume is

$$\frac{\Delta V_s}{V_s} = \frac{1}{(1-\phi)} \frac{\Delta V_b}{V_b} - \frac{\phi}{(1-\phi)} \frac{\Delta V_p}{V_p}. \tag{4.48}$$

The compressibility of the solid volume and the bulk volume are equal at zero porosity. Inserting relationships (4.5) and (4.6) for the bulk and the pore volume, we get

$$\frac{\Delta V_s}{V_s} = \frac{1}{(1-\phi)} \Big((\phi\alpha_{pc} - \alpha_{bc})\Delta p_b + (\alpha_{bp} - \phi\alpha_{pp})\Delta p_f \Big). \tag{4.49}$$

The porosity and the solid volume could have been used instead of the bulk volume and the pore volume as a basis for the definitions of the four basic compressibilities. How the compressibilities of the porosity and the solid volume are related to the already introduced compressibilities are summarized as follows:

$$\frac{1}{\phi}\left(\frac{\partial \phi}{\partial p_b}\right)_{p_f} = \alpha_{bc} - \alpha_{pc} \tag{4.50}$$

$$\frac{1}{\phi}\left(\frac{\partial \phi}{\partial p_f}\right)_{p_b} = \alpha_{pp} - \alpha_{bp} \tag{4.51}$$

$$\frac{1}{V_s}\left(\frac{\partial V_s}{\partial p_b}\right)_{p_f} = \alpha_{bc} - \phi\alpha_{pc} \tag{4.52}$$

$$\frac{1}{V_s}\left(\frac{\partial V_s}{\partial p_f}\right)_{p_b} = \phi\alpha_{pp} - \alpha_{bp}. \tag{4.53}$$

We have already shown by equation (4.19) that the compressibility of a pure solid is $\alpha_s = \alpha_{bc} - \phi\alpha_{pc}$. It turns out that later in the formulation of pressure equations we need the porosity change (4.47) and solid volume change (4.49).

Exercise 4.1 Derive relationship (4.46) for the change $\Delta\phi$ in the porosity.
Solution: We have $\phi = V_p/V_b$ and therefore

$$\Delta\phi = \Delta\left(\frac{V_p}{V_b}\right) = \frac{-\Delta V_b\, V_p + V_b\,\Delta V_p}{V_b^2} \tag{4.54}$$

$$= -\left(\frac{V_p}{V_b}\right)\left(\frac{\Delta V_b}{V_b}\right) + \left(\frac{V_p}{V_b}\right)\left(\frac{\Delta V_p}{V_p}\right) \tag{4.55}$$

which is (4.46).

Exercise 4.2 Derive relationship (4.48) for the change $\Delta V_s/V_s$ in the solid volume.

Solution: We have

$$\frac{\Delta V_s}{V_s} = \frac{\Delta(V_b - V_p)}{V_b - V_p} = \frac{\Delta V_b}{V_b - V_p} - \frac{\Delta V_p}{V_b - V_p} \tag{4.56}$$

$$= \frac{V_b}{(V_b - V_p)}\left(\frac{\Delta V_b}{V_b}\right) - \frac{V_p}{(V_b - V_p)}\left(\frac{\Delta V_p}{V_p}\right) \tag{4.57}$$

$$= \frac{1}{(1 - \phi)}\frac{\Delta V_b}{V_b} - \left(\frac{\phi}{1 - \phi}\right)\frac{\Delta V_p}{V_p} \tag{4.58}$$

which is relationship (4.48).

4.4 Effective pressure coefficients

The compressibilities were defined by relative changes of the bulk volume and the pore volume in response to changes in the bulk pressure and the pore fluid pressure, as given by equations (4.5) and (4.6). These equations can also be written as

$$\frac{\Delta V_b}{V_b} = -\alpha_{bc}(\Delta p_b - \gamma_b \Delta p_f) \tag{4.59}$$

$$\frac{\Delta V_p}{V_p} = -\alpha_{pc}(\Delta p_b - \gamma_p \Delta p_f) \tag{4.60}$$

in terms of the *effective pressure coefficients*

$$\gamma_b = \frac{\alpha_{bp}}{\alpha_{bc}} = 1 - \frac{K}{K_s'} \tag{4.61}$$

$$\gamma_p = \frac{\alpha_{pp}}{\alpha_{pc}} = 1 - \frac{K_p}{K_\phi}. \tag{4.62}$$

It turns out that the effective pressure coefficients often have values close to 1, which implies that $1/K_s' \gg 1/K$ and $1/K_\phi \gg 1/K_p$. Effective pressure coefficients close to 1 can also be written as $\alpha_{bp} \approx \alpha_{bc}$ and $\alpha_{pp} \approx \alpha_{pc}$, which means that the rock then is equally compressible with respect to the (confining) bulk pressure as to the (pore) fluid pressure. Effective pressure coefficients can also be introduced for the porosity and the solid volume.

4.5 Compaction of sediments

Soils are different from rocks by being loose aggregates of particles. Different types of soil can be grouped with respect to their mineral content and the size of the particles. For instance (clean) sand is comprised of quartz grains of a relatively large size, while clay is comprised of very fine particles. Although there is a great difference between soils like sand and clay all soils are found to compact by loss of void space caused by a rearrangement and crushing of the grains. The compressibility of the mineral grains themselves is considered negligible. A direct way to measure compaction of soils is to use an oedometer test, where

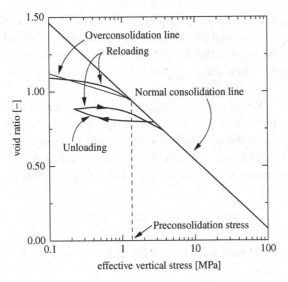

Figure 4.2. *The normal consolidation line and the overconsolidation line. (The normal consolidation line (4.63) has the parameters $e_0 = 1$, $\sigma_0' = 1$ MPa and $C_c = 0.2$.)*

a confined cylindrical test sample is compressed axially. The oedometer test is close to the compaction that often takes place in nature because the test sample is not allowed to expand in the plane normal to the cylinder. The results of a compaction experiment with an oedometer can be plotted in a semi-logarithmic plot like the one shown in Figure 4.2. The void ratio can be approximated by a straight line when plotted as a function of $\ln(\sigma'/\sigma_0')$, where σ' and σ_0' are the effective stress and a reference effective stress, respectively. This straight line is called the *normal consolidation line,* and it is written

$$e = e_0 - C_c \ln\left(\frac{\sigma'}{\sigma_0'}\right) \tag{4.63}$$

where the slope of the line C_c is called the *compression index.* The void ratio e_0 is the surface void ratio obtained at the reference effective stress σ_0'. The reference effective stress σ_0' can be set to 1 kPa, which corresponds to the weight of 10 cm of water. A reference effective stress also serves to make the argument to the ln-function dimensionless. The normal consolidation line is an approximation that applies for effective stresses over a limited range, typically from 10 kPa to 10^4 kPa. We notice that the void ratio of equation (4.63) for the normal consolidation line can be made negative for a sufficiently large effective stress. Such stresses are obviously beyond the interval where the normal consolidation line applies, because negative void ratios do not make sense.

When a soil sample is unloaded the soil does not decompact and recover its original void ratio, like an elastic spring that always recovers its length when unloaded. The path followed during unloading can be approximated by another line called the *overconsolidation line.* When reloaded (after unloading) the soil follows the overconsolidation line until the effective stress exceeds the maximum effective stress the soil has been subjected to. The maximum effective stress the soil has previously experienced is called the *preconsolidation stress*, and the slope of the preconsolidation line is called the *swelling index*. The ratio of the preconsolidation stress σ'_{max} and the current effective stress σ' is called the *overconsolidation ratio*, OCR $= \sigma'_{max}/\sigma'$.

We have seen that Young's modulus for linear elastic materials was defined as stress over strain, where the strain is the relative elongation. The normal consolidation line can be used to compute a similar modulus for soils, where the strain is the relative thickness of a soil sample. The thickness of a sample can be expressed by the porosity (or the void ratio) using the net (porosity-free) thickness ζ of the sample, which is a constant during compression:

$$h = \frac{\zeta}{1-\phi} = (1+e)\zeta. \tag{4.64}$$

A change Δe in the void ratio leads to a change $\Delta h = \zeta \Delta e$ in the height of the sample. The strain then becomes

$$\frac{\Delta h}{h} = \frac{\Delta e}{1+e}. \tag{4.65}$$

The stress over strain, where the strain is caused by this change in the effective stress, is

$$E' = \frac{\Delta \sigma'}{\Delta h/h} = \frac{1+e}{C_v}\sigma'. \tag{4.66}$$

The modulus E' is not constant since it is proportional to σ' and $1+e$. This statement tells us that the soil gets "stiffer" the more it is compressed.

Exercise 4.3 What is the subsidence caused by a weight $\Delta\sigma' = 10$ kPa placed on top of a $h = 2$ m clay layer? The clay layer covers incompressible bedrock and has never been preconsolidated. The normal consolidation line for the clay is $e_0 = 1.5$, $C_v = 0.2$ and $\sigma'_0 = 1$ kPa. It is assumed that $e = e_0$ through the entire thickness of the clay layer.
Solution: The change in the void ratio caused by the load is $\Delta e = C_v \ln(\Delta\sigma'/\sigma'_0) = 0.46$, which leads to the subsidence $\Delta h = h \, \Delta e/(1+e_0) = 0.37$ m.

Exercise 4.4 Show that the void ratio of the normal consolidation line becomes zero for the effective vertical stress $\sigma' = \sigma'_0 \exp(e_0/C_v)$. What is the stress for $e_0 = 1.5$, $C_v = 0.2$ and $\sigma'_0 = 1$ kPa?

Exercise 4.5 Derive equation (4.66).

4.6 Gravitational compaction of a hydrostatic clay layer

The weight of sediments alone will make the sediments compact. How much the porosity is reduced with depth due to gravitational compaction can be computed using the normal consolidation line assuming hydrostatic conditions. The void ratio as a function of the effective stress is then given by equation (4.63). The vertical effective stress under hydrostatic conditions is given by

$$\sigma' = \int_0^z (\varrho - \varrho_f) g (1 - \phi) \, dz \tag{4.67}$$

where the z-axis is pointing downwards and the surface of the layer is $z = 0$. The integral (4.67) is not trivial to carry out in the z-coordinate, because the porosity is not yet known as a function of z. The trick is now to switch to the Lagrangian coordinate defined by the position in the layer measured as net (porosity-free) rock above the base of the layer. This is the so-called ζ-coordinate, and we have that $d\zeta = (1 - \phi) dz$. (The ζ-coordinate was introduced in Section 3.17.) The vertical effective stress under hydrostatic conditions is therefore

$$\sigma' = (\varrho - \varrho_f) g (\zeta^* - \zeta) \tag{4.68}$$

where ζ^* is the ζ-coordinate for the top of the layer. We then get that the normal consolidation line (4.63) as a function of the ζ-coordinate is

$$e = e_0 - C_v \ln \left(\frac{\zeta^* - \zeta}{\zeta_0} \right) \tag{4.69}$$

where $\zeta_0 = \sigma_0' / ((\varrho - \varrho_f) g)$ is the reference ζ-thickness that corresponds to σ_0'. We would like to have the void ratio as a function of z rather than ζ, and a step in that direction is to have the z-coordinate as a function of ζ too. The z-coordinate is given by equation (3.153):

$$z = \int_\zeta^{\zeta^*} \frac{d\zeta}{1 - \phi(\zeta)} = \int_\zeta^{\zeta^*} \left(1 + e(\zeta) \right) d\zeta. \tag{4.70}$$

This integration is straightforward to carry out and the z-coordinate as a function of the ζ-coordinate is

$$z = (\zeta^* - \zeta) \left(1 + e_0 + C_v - C_v \ln \left(\frac{\zeta^* - \zeta}{\zeta_0} \right) \right). \tag{4.71}$$

Notice that both the void ratio (4.69) and the z-coordinate (4.71) are only functions of the ζ-depth from the surface $\zeta^* - \zeta$. The void ratio can now be plotted as a function of the z-coordinate by combining equations (4.69) and (4.71), since both equations are parameterized by the ζ-coordinate. Figure 4.3 shows the porosity of a clay layer as a function of depth (z-coordinate) when $C_v = 0.2$, $e_0 = 1.5$, $\sigma_0' = 1$ kPa and $(\varrho - \varrho_f) g = 10^4$ Pa/m. With these data we get that $\zeta_0 = 1$ m.

Exercise 4.6 Derive equation (4.71). Hint: $\int \ln x = -x + x \ln x$.

Exercise 4.7 Equation (4.69) for the normal consolidation line and equation (4.71) for the z-coordinate can be extended with a surface load σ_L. The equation for the void ratio is then

$$e = e_0 - C_v \ln \left(\frac{\zeta^* - \zeta}{\zeta_0} + \frac{\sigma_L}{\sigma_0'} \right) \tag{4.72}$$

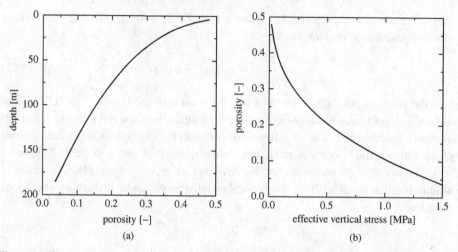

Figure 4.3. (a) *The porosity as a function of depth for hydrostatic sediments that follows the normal consolidation line. (b) The same data set, but the porosity is plotted as a function of effective vertical stress.*

or alternatively

$$e = e_0 - C_v \ln (u + u_0) \tag{4.73}$$

where $u = u(\zeta) = (\zeta^* - \zeta)/\zeta_0$ and $u_0 = \sigma_L/\sigma_0'$.

(a) Show that the z-coordinate as a function of the ζ-coordinate is

$$z = \zeta_0 u \left(1 + e_0 + C_v - C_v \ln(u + u_0)\right) + \zeta_0 u_0 C_v \ln(1 + u/u_0). \tag{4.74}$$

(b) Show that the difference in the z-coordinate caused by the load compared with the unloaded state at the same ζ-position is

$$\Delta z = \zeta_0 C_v u \ln(1 + u_0/u) + \zeta_0 C_v u_0 \ln(1 + u/u_0). \tag{4.75}$$

Recall that the top of the layer is $z = 0$ both before and after the load. The maximum difference in the z-coordinate is therefore found at the base of the layer where $\zeta = 0$ (or $u = \zeta^*/\zeta$). We see by inspection that $\Delta z \to 0$ when $\sigma_L \to 0$ (or equivalently $u_0 \to 0$).

(c) The relative importance of the numbers u_0 and u can be expressed by another number $N = u_0/u$, which is

$$N = \frac{\sigma_L}{(\varrho - \varrho_f)g(\zeta^* - \zeta)}. \tag{4.76}$$

The number N measures the pressure from the surface load relative to the pressure from the sedimentary layer itself (minus buoyancy). Show that Δz can be approximated as follows:

$$\Delta z \approx \zeta_0 C_v u \left(1 + \ln N\right), \quad \text{for } N \gg 1 \tag{4.77}$$
$$\Delta z \approx \zeta_0 C_v u \, N \left(1 - \ln N\right), \quad \text{for } N \ll 1 \tag{4.78}$$

depending on whether N is much larger or much less than 1.

Equation (4.75) gives the subsidence caused by a surface load like a heavy building on, for instance, a layer of clay. The subsidence will not be instantaneous, as we will see later, but gradual, and equation (4.75) gives the maximum subsidence that can be achieved after sufficiently long time.

4.7 Further reading

As already mentioned, the presentation of the compressibilities given here is based on Zimmerman *et al.* (1986). The concept of effective stress goes back to Terzaghi and his studies of compaction of soil (Terzaghi, 1925). Biot (1941) generalized the 1D effective stress concept of Terzaghi to a 3D poroelastic model, which is introduced in Section 13.3. Kümpel (1991) presents a short review of poroelasticity, and the alternative sets of poro-elastic parameters. Wang (2000) gives a rigorous treatment of poroelasticity, the poroelastic parameters, exact solutions of the poroelastic equations, applications and an introduction to numerical solutions with the finite-element method.

5

Burial histories

A sedimentary basin consists of strata of different lithologies deposited in different time intervals. The main data in the burial history are, therefore, the thickness and the lithology of each layer and the time of the horizons separating the layers. A *horizon* is taken to be a surface in the basin of a particular time, and the precise term is therefore *chronohorizon*. There is an extensive nomenclature for stratigraphic classification, but here we will not need more terms than horizon and formation. A formation is here simply the layer between two consecutive horizons.

A burial history usually has breaks or gaps in the stratigraphical record, either because of lack of deposition or because of erosion. Such a gap in a sequence of sedimentary rocks is a *hiatus*, and an erosion process can partly or completely remove several layers. It is often difficult to reconstruct what has been eroded; especially, regional erosion processes make it difficult find places where thickness information is preserved.

The histories of water depth, heat flow and surface temperature complement the burial history. All these elements go into the modeling of a sedimentary basin, and they apply at the boundaries of the basin. The development of the basin is modeled by adding layer on top of layer through the geohistory, eventually with periods of no deposition or periods with erosion. The deposition history gives the geometry and the material properties of the basin by processes on the basin surface. The water depth history, surface temperature history and the heat flow history become boundary conditions for the equations of fluid flow and heat flow. The geometry of basins is often complicated by additional processes, like faulting that breaks the geometrical continuity of the strata, or thrust faulting that may even mess up the chronological order of the strata. These topics are not covered here.

The porosity of the sediments is not constant through time and space. Even similar or the same lithology often shows a large scatter in the porosity, but lithologies have in common that the porosity is normally decreasing with time and depth. There are two main processes that reduce the porosity: *mechanical compaction* and *diagenesis*. The uppermost and unlithified sediments compact mechanically under the increasing weight of sediments being deposited. When the sediments become buried to the depths of sufficiently high temperatures they are lithified by diagenetic processes that fill in the pore space with cement. The loss of pore space leads to reduction of the layer thicknesses and compaction. We assume that diagenetic reactions are local and do not involve "long distance" transport of minerals by the pore fluid flow. The mass in a layer (or computational element) is therefore conserved.

A challenge with burial history modeling is the reconstruction of the thicknesses of each layer through the geohistory. This chapter looks at this particular problem.

5.1 Porosity as a function of net sediment thickness

We will show that a convenient way to deal with burial histories is to work with a Lagrangian vertical coordinate that is constant for each layer boundary through the geohistory. The natural choice for such a Lagrangian coordinate is the height up to a "grain" in the sedimentary column measured as net (fully compacted) rock. This height, denoted ζ, is a constant of each grain in the basin during deposition and compaction. See Section 3.17, where this Lagrangian coordinate is explained further.

In order to make some simple exact burial histories we assume that porosity decreases exponentially with the net (porosity-free) sediment depth as

$$\phi = (\phi_0 - \phi_{min}) \exp\left(-\frac{(\zeta^* - \zeta)}{\zeta_0}\right) + \phi_{min} \tag{5.1}$$

when expressed in the ζ-coordinate. The parameter ϕ_0 is the surface porosity, ϕ_{min} is the minimum porosity found at great depth, ζ^* is the current height (or total thickness) of the basin measured as porosity-free rock, and ζ_0 is a depth that characterizes the decreasing porosity. The porosity $\phi_0 - \phi_{min}$ is seen to be reduced to half at the ζ-depth $\zeta^* - \zeta = \ln 2\, \zeta_0 \approx 0.69\, \zeta_0$. The porosity function (5.1) is a modification of Athy's porosity function (Athy, 1930), who fitted porosity observations to the real sediment depth with an exponential function. (See Section 2.1.)

The real depth of the sediments from the basin surface is obtained from the ζ-coordinate and the porosity function (5.1) using equation (3.153), which is recaptured here:

$$z = \int_{\zeta}^{\zeta^*} \frac{d\zeta}{1 - \phi(\zeta)}. \tag{5.2}$$

When the porosity function (5.1) is inserted into the integral (5.2), and the integration is carried out, we get

$$z = \frac{1}{1 - \phi_{min}} \left((\zeta^* - \zeta) + \zeta_0 \ln\left(\frac{1 - \phi}{1 - \phi_0}\right)\right) \tag{5.3}$$

as shown in Exercise 5.1. The z-coordinate is as expected only a function of the ζ-depth from the surface, $\zeta^* - \zeta$, because the porosity function, ϕ, is only a function of the ζ-depth. We have now both the z-coordinate (5.3) and the porosity (5.1) as explicit functions of the ζ-coordinate. It is then possible to plot the porosity as a function of depth as shown in Figure 5.1. The figure also shows some other porosity–depth trends.

We have seen that each "grain" of the sediments is uniquely identified with a ζ-coordinate. Equation (5.3) can therefore be used to trace the real depth (relative to the

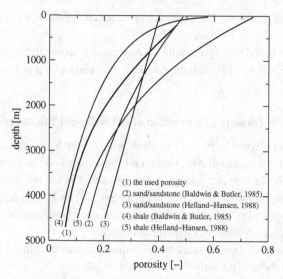

Figure 5.1. *The porosity function (5.1) is plotted as a function of depth for $\phi_0 = 0.5$, $\phi_{min} = 0.03$, $\zeta_0 = 1350$ m, and $\zeta^* = 3700$ m (see curve 1). Four other porosity–depth curves are also plotted.*

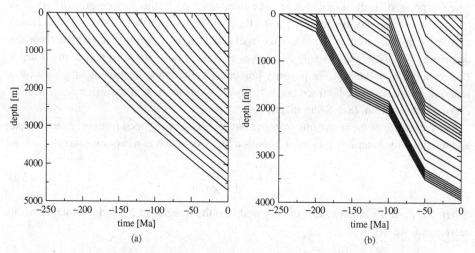

Figure 5.2. *(a) The horizons in a burial history where sediments are deposited at constant (net) rate. (b) A burial history with deposition at different constant rates in different time intervals.*

basin surface) of any sample through the burial history, given that we know $\zeta^* = \zeta^*(t)$, (the total amount of net sediment). The only limitation is that equation (5.3) does not apply if the burial history has erosion processes. The reason is that porosity reduction is an irreversible process. Burial histories with erosion are therefore not easily reconstructed with an analytical z-coordinate for the horizons.

Figure 5.2 shows the burial history of horizons for two simple deposition histories. Both plots are made using solution (5.3) for the z-coordinate through time for given constant

ζ-coordinates. The parameters in the porosity function (5.1) are $\phi_0 = 0.5$, $\phi_{min} = 0.03$ and $\zeta_0 = 1350$ m. The horizons in plot 5.2a are separated in space by 200 m net sediments, and show a burial history where deposition of net (porosity-free) sediment is at a constant rate of 3700 m over 250 Ma. The total amount of net rock as a function of time is therefore $\zeta^* = \omega t$, where $\omega = 14.8$ m Ma^{-1}. Such simple burial histories, with deposition of net sediment at a constant rate, are ideal to study the basic behavior of processes related to deposition, like for instance overpressure build-up. The plot 5.2b has horizons that are separated by 10 Ma in time, and shows a burial history where the deposition rates are piecewise constant in intervals of 50 Ma.

Burial histories shown in Figure 5.2 could be generalized by allowing for a separate porosity function for each layer. The solution (5.3) then has to be modified to an expression for a layer thickness. The real depths down to the horizons are obtained starting from the basin surface by computing the real layer thicknesses, layer by layer. Such an algorithm is straightforward to implement in a computer program.

Exercise 5.1 Show that the integral (5.2) becomes the function (5.3) when the porosity function (5.1) is inserted.
Solution: A change of integration variable from ζ to $u = \phi$ gives $du = (u - \phi_{min})(d\zeta/\zeta_0)$, and the integral (5.2) becomes

$$z = \int_{\phi}^{\phi_0} \frac{du}{(1 - u)(u - \phi_{min})}. \tag{5.4}$$

The function inside the integral is rewritten as

$$\frac{1}{(1 - u)(u - \phi_{min})} = \frac{a}{1 - u} + \frac{b}{u - \phi_{min}} \tag{5.5}$$

where $a = b = 1/(1 - \phi_{min})$. It is now straightforward to do the integration and the result is equation (5.3).

Exercise 5.2 Show that equation (5.3) for the z-coordinate can be rewritten as the following expression for the layer thickness:

$$dz = \frac{1}{1 - \phi_{min}} \left(d\zeta + \zeta_0 \ln\left(\frac{1 - \phi_1}{1 - \phi_2}\right) \right) \tag{5.6}$$

where $dz = z_2 - z_1$, $d\zeta = \zeta_2 - \zeta_1$ and where the subscripts 1 and 2 denote the horizon boundaries of the layer.

5.2 Pre-calibration of burial history calculations

It was shown in the previous section how simple burial histories could be made from knowledge of the net (porosity-free) thickness of rock in each layer, by assuming that the porosity is a function of net (porosity-free) depth. But we did not say anything about how one could obtain the net layer thicknesses from real thicknesses and the porosity function.

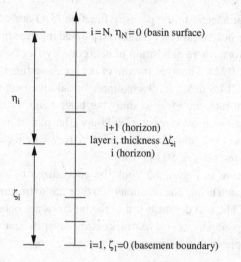

Figure 5.3. *The ζ-coordinate (ζ_i) of each horizon, the ζ-depth (η_i) of each horizon and the net layer thickness $\Delta\zeta_i$.*

We will now look at how net layer thicknesses can be initialized in order to reproduce the present-day formation thicknesses.

The net thicknesses can be obtained from the observed layer thicknesses starting from the basin surface by processing layer by layer. The real depth from the basin surface down to horizon i is z_i. The height from the basement up to horizon i, measured as net sediment, is ζ_i. The horizons are numbered starting from the basement ($i = 1$) up to the basin surface ($i = N$). The net thickness of each layer is denoted $\Delta\zeta_i$, and the real (observed) thickness of each layer is $\Delta z_i = z_{i+1} - z_i$, where the lowest layer is $i = 1$. The top layer is $i = N - 1$, because there is always one horizon more than there are layers (see Figure 5.3). The ζ-coordinates of the horizons are related as follows to the net layer thicknesses:

$$\zeta_{i+1} = \zeta_i + \Delta\zeta_i, \tag{5.7}$$

and using that $\zeta_1 = 0$ we get

$$\zeta_i = \sum_{k=1}^{i-1} \Delta\zeta_k. \tag{5.8}$$

The ζ-depth (the net depth) from the basin surface down to a horizon is denoted η_i, and we have

$$\eta_i = \eta_{i+1} + \Delta\zeta_i, \tag{5.9}$$

and using that $\eta_N = 0$ we get

$$\eta_i = \sum_{k=i}^{N-1} \Delta\zeta_k. \tag{5.10}$$

The total ζ-thickness of the column is

$$\zeta^* = \sum_{k=1}^{N-1} \Delta \zeta_k = \zeta_i + \eta_i \qquad (5.11)$$

for every i from 1 to N (see Figure 5.3). The net depths are found in an iterative manner, starting from the basin surface, where $\eta_N = 0$. Let us say that η_{i+1} is found and that we want to compute the next net thickness $\Delta \zeta_i$ in order to obtain $\eta_i = \eta_{i+1} + \Delta \zeta_i$. This computation can be done by approximating the average porosity of the layer with the porosity at the center of the layer,

$$\bar{\phi}_i = \phi(\eta_{i+1} + \Delta \zeta_i/2). \qquad (5.12)$$

We then have

$$\Delta \zeta_i = \left(1 - \bar{\phi}\right) \Delta z_i = \left(1 - \phi_i(\eta_{i+1} + \Delta \zeta_i/2)\right) \Delta z_i \qquad (5.13)$$

which is an equation for the net layer thickness $\Delta \zeta_i$. This equation is solved numerically by simple iteration, for nearly any porosity function. Each evaluation of the right-hand side yields a $\Delta \zeta_i$ that can be inserted back into the right-hand side to obtain an improved $\Delta \zeta_i$. This scheme can be initialized with the real layer thickness, and only a few iterations are needed to obtain an accurate numerical solution. Once all net thicknesses are found from the surface to the basement, we also have the ζ-coordinates of each horizon. A burial history simulation can then be done where the final thicknesses will match today's real thicknesses. This procedure works with any porosity function as long as it is a function of the net sediment depth, and the algorithm also remains unchanged if each layer has its own porosity function. We return in Section 5.4 to calibration of burial histories with erosion processes.

5.3 Porosity as a function of z

Porosity as a function of z, rather than the ζ-depth, simplifies the process of obtaining the net amount of porosity in each layer from today's real thicknesses. We simply get

$$\Delta \zeta_i = \left(1 - \phi(\bar{z} - z_N)\right) \Delta z_i, \qquad (5.14)$$

where $\bar{z} = (z_i + z_{i+1})/2$ is the present-day depth to the center of the layer, and where z_N is the present-day position of the basin surface. (The porosity function takes as argument the real depth relative to the basin surface.) We want the net thickness of each layer, because they are constant through the burial history. The computation of the real layer thicknesses from the ζ-thicknesses, at any time in the geohistory, is done with an iterative algorithm quite similar to the one in the previous section. We start at the basin surface z_N, and then proceed downwards layer by layer. When the depth z_{i+1} is found, the task is to find the next layer thickness Δz_i, and thereby the real depth to the horizon below, $z_i = z_{i+1} + \Delta z_i$. The thickness $\Delta \zeta_i$ and the porosity gives the real layer thickness

$$\Delta z_i = \frac{\Delta \zeta_i}{1 - \phi(z_i - z_N + \Delta z_i / 2)} \tag{5.15}$$

which is a function of the wanted layer thickness Δz_i. Equation (5.15) is straightforward to solve numerically using iteration, where an approximation for Δz_i can be inserted in the right-hand side to obtain an improved approximation. The iteration scheme can be started with $\Delta \zeta_i$ as the first guess, and it will converge towards an accurate solution after just a few iterations.

5.4 Erosion

The preceding sections show how the present-day layer thicknesses give the net amount of rock in each layer. It is still assumed that porosity is a function of depth from the basin surface, where depth is either real depth or net (porosity-free) depth. The net formation thicknesses are wanted because they give the net height (the ζ-coordinates) of the layer boundaries from the basement. The net thickness of the layers and the ζ-coordinate of the layer boundaries are constant through the geohistory, and it is the net amount of rock in each layer that allows the basin to be constructed at each time step through the geohistory. A burial history computation therefore goes through the wanted time steps, and the following tasks are done at each step:

1. deposit sediments and update the surface ζ-coordinate;
2. make the current porosity;
3. make the current real depths;
4. solve equations for the current temperature and pressure.

This approach to burial history modeling works fine as long as no erosion is involved. The problem with erosion is that we cannot obtain the net amount of rock in the layers directly from the present-day layer thicknesses. The porosity of a layer, when it is a function of depth, decreases with increasing burial. Porosity reduction is an irreversible process, and the porosity remains constant at the value of maximum burial, when erosion brings layers up to a shallower depth. The porosity of a layer keeps this minimum porosity until it is reburied to a depth that is deeper than the previous maximum.

It is therefore not possible to compute the porosity without doing a burial history, but at the same time it is not possible to compute the burial history without knowledge of the porosity. A solution to this deadlock is to simulate the full burial history iteratively until a sufficiently close match against the present-day layer thicknesses is achieved.

An erosion process appears in the present-day stratigraphy as a hiatus. This gap in the stratigraphy is modeled by first depositing the wanted amount of sediment and then eroding exactly the same amount of sediment. There may be several deposition and erosion processes that fill the time interval of a hiatus, as long as the net result is zero.

Table 5.1. *The burial history for the example shown*
in Figure 5.4. The formation "X" denotes erosion.

Time [Ma]	Thickness [m]	Formation	Comment
−250	500	A	(present-day thickness)
−225	500	B	(hiatus 1)
−200	500	C	(hiatus 1)
−175	500	D	(hiatus 1)
−150	500	E	(hiatus 1)
−125	−2000	X	(hiatus 1)
−100	1000	F	(present-day thickness)
−75	1000	G	(hiatus 2)
−50	−1000	X	(hiatus 2)
−25	1000	H	(present-day thickness)
0	−	−	

Table 5.1 shows an example of a burial history that includes erosion. Each line in the table represents a layer (formation) where the first column is the time when the deposition of the layer begins, the second column is the present-day thickness of the layer, and the third column is the formation name. A deposition process lasts until the start of the next deposition process, given by the next line in the table. The last line gives the end of the burial history, and therefore has no thickness or formation name. Lines with negative layer thicknesses are erosion processes. There are two erosions in Table 5.1. The first erosion process removes 2000 m, which are the four preceding layers of 500 m each. The second one removes one layer of 1000 m. Notice that the thicknesses of the eroded layers are not present-day thicknesses, because these thicknesses do not exist today. We will see that these real thicknesses are just a means to assign net thicknesses to eroded layers.

The net thicknesses for each layer need to be initialized before the simulation of the burial history can start. A simple way to do the initialization is to assume that the present-day porosity is 25% in all layers. The advantage of using the same initial porosity for all layers is that the initial guess for the net layer thicknesses adds up to zero for all layers involved in an erosion process. (Recall that the real thicknesses involved in a hiatus must add up to zero.)

The net amount of rock in each layer is updated at the end of each forward simulation of the burial history. A full forward simulation gives the present-day porosity (ϕ_i) of each layer (i) and the net thickness for the layer becomes $\Delta \zeta_i = (1 - \phi_i)\Delta z_i$, where Δz_i is the present-day layer thickness. The layers eroded during the time span of a hiatus do not have present-day thicknesses. The net thicknesses for these layers are not updated, and they therefore keep their initial net thicknesses. We must therefore keep in mind that the eroded

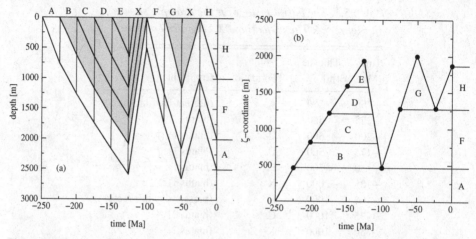

Figure 5.4. *The burial history in Table 5.1 is plotted in (a) as the real depths of the horizons, and in (b) by the ζ-coordinates of the horizons.*

layers always have net thicknesses that are 3/4 of their real thicknesses. (The number 3/4 is the factor $(1 - \phi)$ when $\phi = 25\%$.)

The horizons of the burial history can be represented by nodes in a numerical grid as shown in Figure 5.3. A new node will be added to the grid for each new horizon, and the grid will be expanding or contracting depending on the surface processes in the geohistory. An erosion process may remove several layers completely and thereby leave behind several nodes that follow the same horizon. Figure 5.4 shows an example where the nodes that once were the top of the formations B, C, D, and E all end up on top of formation A. Table 5.1 lists all layers in the burial history in Figure 5.4. There are only three layers that remain at the present time (A, F and H). All layers use the porosity function (5.1), with the parameters $\phi_0 = 0.5$, $\phi_{min} = 0.03$ and $\zeta_0 = 1350$ m. The porosity is kept constant when the depth is less than the maximum burial depth. For instance, formation A has nearly the same porosity from 125 Ma until the present time, because it is not buried any deeper later in the geohistory. Four iterative forward simulations were needed to match the present-day thickness of formations A, F and H. (The difference between the computed thickness and the observed thickness for these formations is less than 1 m.)

5.5 Numerical compaction computation

A simple means to model sediment compaction is to assume that the porosity is a function of net sediment depth (or real depth). A more realistic approach is to numerically compute the mechanical and the diagenetic porosity. The general porosity computation becomes a function of several parameters, for instance effective stress, maximum effective stress, temperature, the time step and parameters related to diagenesis. The representation of the layer thicknesses in terms of net sediment thickness is still a useful approach. It gives a Lagrangian grid, except for the surface layer, where sediments are deposited. We can compute the real layer thicknesses for any general porosity computation from the knowledge of the net amount of rock in each formation, by doing the following tasks at each time step:

1. compute the porosity using the z-coordinates, pressure (p), temperature (T), etc., from the previous time step: $\phi^{n+1} = f(\phi^n, z^n, T^n, p^n, t^{n+1} - t^n, \ldots)$;
2. compute current real layer thicknesses from the net-thicknesses and the current porosities: $\Delta z^{n+1} = \Delta\zeta/(1 - \phi^{n+1})$;
3. make current z-coordinates by adding the layer thicknesses: Δz^{n+1};
4. solve equations for temperature and pressure using the current z-coordinates.

The superscripts n and $n + 1$ denote the previous and the current time step, respectively. It is straightforward to iterate over the task list above, but experience shows that iterations are unnecessary with moderate time steps. The reason is that the porosity does not change very much from one time step to the next. The scheme above allows real thicknesses and the z-coordinate to be computed once we know the net amount of rock in each formation. The observed layer thicknesses are reproduced using repeated forward simulations of the burial history as shown in the preceding section. The net amount of rock in each layer is then updated at the end of each forward simulation using the computed present-day porosities and the present-day layer thicknesses. Notice that there are no restrictions on how we compute the porosities. It is difficult to prove rigorously that such a scheme converges, but experience tells us that less than ten forward simulations are needed to obtain a "good" match against present-day layer depths. Figure 5.5 shows an example where the compaction is initially mechanical, and then becomes purely chemical when more than 2% cement has filled the pore space. Six iterations was needed for this case and the difference between observed and computed present day thicknesses is less than 1 m.

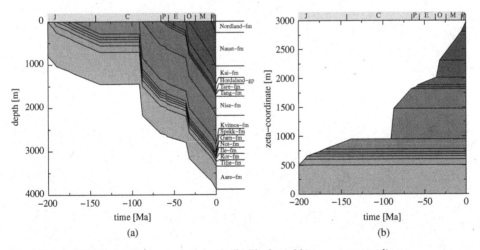

Figure 5.5. *(a) The burial history as real depth. (b) The burial history as ζ-coordinate.*

5.6 Further reading

A formulation of a temperature and a pressure equation for compacting sedimentary basins, similar to the one above, is proposed by Søreng (1989) and Wangen (1993). The burial history of a sedimentary basin can also be reconstructed in 1D with the so-called "back-stripping" method. This topic is discussed in Section 7.14 in connection with basin load and subsidence. There are also techniques that reconstruct the complex geometry of faulted basins in 2D and 3D. It should also be mentioned that there are iterative schemes that minimize the misfit between modeled and observed stratigraphic data by optimizing litho-spheric stretching factors and paleo-water depth (Poplavskii *et al.*, 2001, Bellingham and White, 2002, Rüpke *et al.*, 2008).

6

Heat flow

The Earth is losing heat through its surface, partly from cooling of the Earth and partly from heat generation by decay of radioactive isotopes. The heat loss from the Earth's surface is typically $0.05 \, \mathrm{W \, m^{-2}}$. This is much less than the influx from the Sun, which is typically $500 \, \mathrm{W \, m^{-2}}$ on a sunny day. However, almost all the energy received from the Sun is returned back into space as infrared radiation. The energy from the Sun powers the processes in the biosphere and in particular the water cycle (evaporation and rainfall), and it is therefore the energy source for erosion. On the other hand, the energy from the interior of the Earth drives large-scale geological phenomena like mantle convection, plate tectonics, volcanos, earthquakes and mountain building, see Figure 6.1.

In models of the heat flow and temperature in the subsurface it is important to distinguish between the crust and the mantle and also the lithosphere and the asthenosphere. The *crust* is the upper part of the Earth with a thickness in the range from 10 km to 70 km. It is made of more silica-rich and less-dense rocks than the mantle below. The crust has typical densities in the range from $2700 \, \mathrm{kg \, m^{-3}}$ to $2900 \, \mathrm{kg \, m^{-3}}$, and the *mantle* has a typical density $3300 \, \mathrm{kg \, m^{-3}}$. The crust is therefore both chemically and mechanically different from the mantle.

The *lithosphere* is the outermost part of the Earth which is considered rigid, and where heat transfer is by conduction. This part extends down to a mantle depth of 100 km to 250 km in continental areas. The mantle below the lithosphere is the *asthenosphere*, where the temperature is dominated by convective heat transfer. The transition from lithosphere to asthenosphere is a thermal boundary layer which is not sharp, although we assume that in the models.

Sedimentary basins are thin covers on top of the crust, and the temperature in a basin is to a large extent controlled by processes underneath the basin – in the crust and the mantle. Models for the heat flow through a sedimentary basin therefore involve both the crust and the lithospheric mantle.

This chapter presents the temperature equation and some solutions that are relevant for sedimentary basins. For most applications a simple temperature equation is sufficient, which is derived from a simple energy balance. We will therefore start with the derivation of a basic 1D temperature equation, and look at some applications, before a more complete temperature equation is derived.

Heat flow

Heat flow

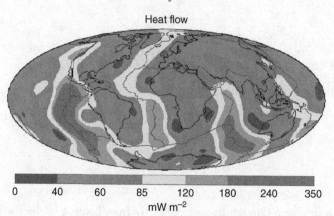

Figure 6.1. *The map of the world heat flux shows that the spreading ridges have the highest heat flow, and that oceanic plates that are formed at these ridges cool as they move away from the ridges. The continents have heat flow values in the interval from 40 Wm^{-2} to 80 Wm^{-2}, which is partly from heat generation in the crust and partly from cooling of the Earth. The figure is taken from International Heat Flow Commission (2008).*

6.1 The temperature equation

A temperature equation is an expression for conservation of energy. We consider energy conservation in a box of length Δx placed along the x-axis as shown in Figure 3.18. We then have that the increase in the internal energy ΔE_1 in the box is equal to the net inflow of energy ΔE_2. A change in the internal energy in the box is

$$\Delta E_1 = \Delta \left(\sum_i \phi_i \varrho_i e_i \right) A \Delta x \tag{6.1}$$

where A is the section area, ϕ_i is the volume fraction of each component (or phase) i, ϱ_i is the density of component i and e_i is the internal energy of component i per unit mass. The sum is over each component. A two-component system is typically a porous rock where one volume fraction $\phi_1 = \phi$ is the porosity and the other volume fraction $\phi_2 = 1 - \phi$ is the solid matrix. The summation is written explicitly with the summation sign instead of using the summation convention, because the summation is over three indices and not a pair of two identical indices. Net heat influx in the box is

$$\Delta E_2 = (F(x) - F(x + \Delta x)) A \Delta t + S A \Delta x \Delta t \tag{6.2}$$

where $F(x)$ is the flux of energy per unit area at position x and S is the heat generation per unit volume and unit time at the center of the box. The energy flux is

$$F = q + \sum_i \phi_i \varrho_i v_i e_i \tag{6.3}$$

where the first term q is the heat flux and the second term is heat transport by convection, and v_i is the velocity of component i. Fourier's law gives the heat flux

$$q = -\lambda \frac{dT}{dx} \tag{6.4}$$

as the product of the temperature gradient and the heat conductivity λ (see Section 2.12). Notice that the bulk heat conductivity is an average heat conductivity for all components in the system. The heat conductivity is often isotropic, but it can also be non-isotropic just like the permeability. The minus sign in Fourier's law is needed because heat flows from "high" to "low" temperatures. Energy conservation $\Delta E_1 = \Delta E_2$ requires that expressions (6.1) and (6.2) are equal, and we get

$$\frac{\partial}{\partial t} \left(\sum_i \phi_i \varrho_i e_i \right) + \frac{\partial}{\partial x} \left(\sum_i \phi_i \varrho_i v_i e_i \right) - \frac{\partial}{\partial x} \left(\lambda \frac{\partial T}{\partial x} \right) = S \tag{6.5}$$

when both sides are divided by $\Delta t \, \Delta x$. The t- and x-differentiations in equation (6.5) are carried out inside the parentheses and terms cancel because of mass conservation of each component

$$\frac{\partial}{\partial t} (\phi_i \varrho_i) + \frac{\partial}{\partial x} (\phi_i \varrho_i v_i) = 0 \tag{6.6}$$

(see section 3.18). Equation (6.5) then simplifies to

$$\sum_i \phi_i \varrho_i \frac{\partial e_i}{\partial t} + \sum_i \phi_i \varrho_i v_i \frac{\partial e_i}{\partial x} - \frac{\partial}{\partial x} \left(\lambda \frac{\partial T}{\partial x} \right) = S. \tag{6.7}$$

The specific internal energy e_i is a function of temperature and volume. Assuming that the volume is constant gives that

$$\frac{\partial e_i}{\partial t} = c_i \frac{\partial T}{\partial t} \quad \text{and} \quad \frac{\partial e_i}{\partial x} = c_i \frac{\partial T}{\partial x} \tag{6.8}$$

where

$$c_i = \left(\frac{\partial e_i}{\partial T} \right)_V \tag{6.9}$$

is the specific heat capacity at constant volume for component i. The rate of change of internal energy (6.8) gives that expression (6.7) for energy conversion becomes

$$\left(\sum_i \phi_i \varrho_i c_i \right) \frac{\partial T}{\partial t} + \left(\sum_i \phi_i \varrho_i c_i v_i \right) \frac{\partial T}{\partial x} - \frac{\partial}{\partial x} \left(\lambda \frac{\partial T}{\partial x} \right) = S. \tag{6.10}$$

Although the temperature equation is derived in 1D it is straightforward to generalize it to 3D. All we have to do is to replace $\partial/\partial x$ with the ∇-operator, and the temperature equation becomes

$$\left(\sum_i \phi_i \varrho_i c_i\right) \frac{\partial T}{\partial t} + \left(\sum_i \phi_i \varrho_i c_i \mathbf{v}_i\right) \cdot \nabla T - \nabla \cdot (\lambda \nabla T) = S \qquad (6.11)$$

where the velocity of component i is the vector $\mathbf{v}_i$.

We will now look at temperature equation (6.11) for a porous medium, which is a medium of two components: solid matrix and pore fluid. The coefficient before the term $\partial T/\partial t$ is the weighted mean of $\varrho_i c_i$ for each component of the medium, using the volume fractions ϕ_i as normalized weights. The weighted mean of $\varrho_f c_f$ and $\varrho_s c_s$ is written as

$$\varrho_b c_b = \phi \varrho_f c_f + (1 - \phi)\varrho_s c_s \qquad (6.12)$$

where ϱ_b and c_b are the bulk density and the bulk specific heat capacity, respectively. Be aware that the notation $\varrho_b c_b$ hides the fact that we do not know each of these factors separately, only their product.

The same averaging is carried out for the coefficient of the first-order term ∇T, assuming that $\mathbf{v}_i = \mathbf{v}_s$ for all components except the fluid ($i = f$). It is done by replacing the fluid velocity $\mathbf{v}_f$ by the Darcy flux $\mathbf{v}_D$ and the solid velocity $\mathbf{v}_s$. Recall from equation (2.31) that the Darcy velocity is a volume flux measured relative to the velocity of the porous medium,

$$\mathbf{v}_D = \phi \cdot (\mathbf{v}_f - \mathbf{v}_s) \quad \text{or} \quad \phi \mathbf{v}_f = \mathbf{v}_D + \phi \mathbf{v}_s. \qquad (6.13)$$

The coefficient before the ∇T-term is therefore

$$\phi \varrho_f c_f \mathbf{v}_f + (1 - \phi)\varrho_s c_s \mathbf{v}_s = \varrho_f c_f \mathbf{v}_D + \varrho_b c_b \mathbf{v}_s \qquad (6.14)$$

and the temperature equation (6.11) becomes

$$\varrho_b c_b \frac{DT}{dt} + \varrho_f c_f \mathbf{v}_D \cdot \nabla T - \nabla \cdot (\lambda \nabla T) = S \qquad (6.15)$$

using the coefficients (6.12) and (6.14). Notice that time differentiation is written as a material derivative with respect to the velocity of the (solid) sediment matrix.

The full temperature equation (6.15) is rarely needed, and most problems can be treated with a simplified version. The most common simplifications apply to problems in 1D with no source term for heat generation ($S = 0$) and where there is no heat convection ($\mathbf{v}_D = 0$). Furthermore, we rarely have to deal with problems where the solid material is moving. The material derivative therefore becomes the normal partial derivative, and the temperature equation simplifies to

$$\frac{\partial T}{\partial t} - \left(\frac{\lambda}{\varrho_b c_b}\right) \frac{\partial^2 T}{\partial z^2} = 0 \qquad (6.16)$$

in the vertical direction. The only parameter left in the temperature equation is the thermal diffusivity $\kappa = \lambda/(\varrho_b c_b)$. Furthermore, we can make a dimensionless version of the equation if there is a characteristic length l_0 and a characteristic temperature T_0 in the problem. Introducing the dimensionless quantities

$$\hat{z} = \frac{z}{l_0}, \quad \hat{T} = \frac{T}{T_0} \quad \text{and} \quad \hat{t} = \frac{t}{t_0} \qquad (6.17)$$

where the characteristic time is $t_0 = l_0^2/\kappa$, gives the dimensionless temperature equation

$$\frac{\partial \hat{T}}{\partial \hat{t}} - \frac{\partial^2 \hat{T}}{\partial \hat{z}^2} = 0. \tag{6.18}$$

We notice that the dimensionless temperature equation has no parameters, and it is therefore just one dimensionless solution of the problem. The dimensionless solution can be converted to a solution for any specific case using the scaling relationships (6.17).

6.2 Stationary 1D temperature solutions

A simple version of the temperature equation is a time-independent equation with zero velocity terms and no heat generation. The temperature equation is then (an elliptic equation)

$$\nabla \cdot (\lambda \nabla T) = 0. \tag{6.19}$$

A further simplification is to allow only heat flow in the vertical direction, which is a good assumption when a sedimentary basin shows little lateral variation. The temperature equation is then reduced to a simple 1D equation

$$\frac{d}{dz}\left(\lambda(z)\frac{dT}{dz}\right) = 0 \tag{6.20}$$

where the heat conductivity is a function of the depth. This z dependence is typically due to changing lithologies with depth, or from a changing porosity inside the same lithology. The porosity has normally a decreasing trend with depth. The average heat conductivity, therefore, increases with depth inside the same lithology, because the heat conductivity of the rock matrix is higher than the heat conductivity of water. Temperature equation (6.20) is solved by two integrations. The first integration recovers Fourier's law

$$\lambda(z)\frac{dT}{dz} = q_0 \tag{6.21}$$

with a constant (z-independent) heat flux q_0. Notice that the minus sign in Fourier's law is left out, because we want positive heat flow to be upwards, (although the z-axis points downwards). The next integration yields the temperature

$$T(z) = T_0 + q_0 \int_0^z \frac{dz}{\lambda(z)} \tag{6.22}$$

where T_0 is the temperature at the surface, $z = 0$. The temperature in the case of constant heat conductivity λ_0 is

$$T(z) = T_0 + \frac{q_0}{\lambda}z \tag{6.23}$$

where the temperature increases linearly with depth. The temperature as a function of the depth is called a *geotherm*.

The temperature increase over a depth interval Δz becomes $\Delta T = (q_0/\lambda_0)\Delta z$. The temperature at the base of a basin with several layers each with a constant heat conductivity λ_i and thickness Δz_i is therefore

$$T(z) = T_0 + q_0 \sum_i \frac{\Delta z_i}{\lambda_i} \tag{6.24}$$

where $z = \sum_i \Delta z_i$ is the total thickness of all layers. The temperature (6.24) is the discrete version of the temperature solution (6.22). The harmonic average (2.89) normal to a bedding plane is identified in the temperature solution (6.22). The same averaging applies to the heat conductivity as to the permeability, and the average heat conductivity in the depth interval from 0 to z is

$$\frac{1}{\lambda_{av}(z)} = \frac{1}{z} \int_0^z \frac{dv}{\lambda(v)} \tag{6.25}$$

and the temperature solution (6.22) can therefore be written

$$T(z) = T_0 + q_0 \frac{z}{\lambda_{av}(z)}. \tag{6.26}$$

The average heat conductivity is a useful property because it gives the (constant) heat flux through a vertical column. In the case that the temperature is T_m at the depth z_m the heat flux q_0 becomes

$$q_0 = \lambda_{av}(z_m) \frac{T_m}{z_m}. \tag{6.27}$$

The solution (6.26) then gives the temperature at any position in the depth interval 0 to z_m. This is a simpler way to find the temperature in a layered rock mass than the alternative, which is to glue together piecewise linear solutions.

Exercise 6.1 The temperature equation (6.20) becomes $d^2T/dz^2 = 0$, when the heat conductivity is independent of the z-coordinate. This equation has the general solution $T(z) = Az + B$, where the two coefficients A and B are given by two boundary conditions.

(a) Let the boundary conditions be the temperature T_0 at the surface $z = z_0$ and the temperature T_m at the depth z_m'. Find A and B and show that the temperature solution can be written

$$T(z) = \frac{(T_m - T_0)}{(z_m - z_0)}(z - z_0) + T_0 \tag{6.28}$$

which gives the constant heat flow

$$q = \lambda \frac{(T_m - T_0)}{(z_m - z_0)} \tag{6.29}$$

in the entire depth interval between z_0 and z_m.

(b) Let the first boundary conditions be the same as in (a), but use a constant heat flow q_m at the depth z_m as the second boundary condition. Find A and B and show that the solution can be written

$$T(z) = \frac{q_m}{\lambda}(z - z_0) + T_0 \tag{6.30}$$

where the temperature at the depth z_m becomes $T_m = T(z_m)$.

Exercise 6.2 Assume that the heat conductivity decreases linearly with porosity as

$$\lambda(\phi) = \lambda_s \cdot (1 - c\,\phi) \tag{6.31}$$

where λ_s is the heat conductivity of the pure solid (at $\phi = 0$). The coefficient c controls how fast the heat conductivity decreases with the porosity. Assume furthermore that the porosity decreases with depth as

$$\phi = \phi_0 \exp\left(-\frac{z}{z_0}\right) \tag{6.32}$$

where z_0 is a characteristic depth for porosity loss. There is little porosity reduction for depths $z \ll z_0$ and there is a substantial reduction for $z \gg z_0$. When the porosity–depth relationship (6.32) is inserted into the linear expression (6.31) for the bulk heat conductivity we get an expression for heat conductivity as a function of depth, $\lambda(z) = \lambda(\phi(z))$. Find the 1D stationary temperature solution with this z-dependent heat conductivity, when the surface temperature is zero.

Solution: The temperature is given by expression (6.22),

$$T(z) = \frac{q_0}{\lambda_s} \int_0^z \frac{dz}{1 - c\,\phi_0 \exp(-z/z_0)}. \tag{6.33}$$

A change of variable from z to $u = \phi_0 \exp(-z/z_0)$ leads to

$$T = \frac{q_0 z_0}{\lambda_s} \int_\phi^{\phi_0} \frac{du}{(1 - c\,u)\,u} = \frac{q_0 z_0}{\lambda_s} \left[\ln u - \ln(1 - cu)\right]_\phi^{\phi_0} \tag{6.34}$$

and replacing u by the porosity gives

$$T(z) = \frac{q_0}{\lambda_s} z + \frac{q_0}{\lambda_s} z_0 \ln\left(\frac{1 - c\,\phi_0 \exp(-z/z_0)}{1 - c\,\phi_0}\right). \tag{6.35}$$

The first part of the solution is the temperature in the case of constant heat conductivity of solid rock (for $\phi = 0$). The second term is therefore the difference between the wanted temperature solution and the linear solution for a constant heat conductivity. Equation (6.35) shows that the difference is increasing with depth, and that the maximum difference at infinite depth is

$$\Delta T_{max} = -\frac{q_0 z_0}{\lambda_s} \ln(1 - c\,\phi_0). \tag{6.36}$$

The solution (6.35) is plotted in Figure 6.2 for $\phi_0 = 0.5$, $z_0 = 1000$ m, $c = 1.0$, $q_0 = 0.05$ W m^{-2} and $\lambda_s = 2.5$ W m^{-1} K^{-1}. One sees that the temperature difference is close to $\Delta T_{max} = 14.0\,°$C for $z > 1500$ m.

Table 6.1. *The most important heat-producing radioactive elements in rocks. The data are taken from Van Schmus (1995).*

Index	Element	Heat generation H_i $\mu W\,kg^{-1}$	Half-life $t_{1/2}$ Ga	Decay constant λ year^{-1}
1	^{235}U	568.7	4.47	$9.9 \cdot 10^{-10}$
2	^{238}U	94.7	0.70	$1.5 \cdot 10^{-10}$
3	^{232}Th	26.4	14.0	$0.5 \cdot 10^{-10}$
4	^{40}K	29.2	1.25	$5.6 \cdot 10^{-10}$

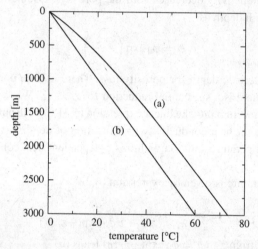

Figure 6.2. *(a) The temperature solution (6.35) and (b) the linear part of this temperature solution. See Exercise 6.2 for the numbers used.*

6.3 Heat generation

Radioactive elements emit α-particles (helium nuclei) and/or β-particles (electrons) when they decay. These emitted particles are brought to rest in the rock within a short range and their kinetic energy is converted to an increase in temperature. Radioactive heat generation is important because as much as half the surface heat flow may be attributed to decay of radioactive elements in the crust and sediments. The most important heat generating elements are the uranium isotopes ^{235}U and ^{238}U, the thorium isotope ^{232}Th and the potassium isotope ^{40}K, and Table 6.1 gives the heat production for each of these elements. The amount of a heat-generating isotope is usually given as a mass fraction (kg isotope per kg rock). The heat production per kilogram of rock can therefore be written

$$H_{\text{total}} = \sum_i C_i\, H_i \qquad (6.37)$$

where C_i is the mass fraction of each isotope i. (The index i and the corresponding heat production H_i are defined by Table 6.1.) Uranium is at present composed of 99.27% ^{238}U and 0.72% ^{235}U by weight. Thorium is 100% ^{232}Th, and potassium is composed of 0.0128% ^{40}K. It is customary to give the mass fractions of isotopes in terms of the fractions of U, Th and K. The total heat production then becomes

$$H_{total} = 0.72\% C_U H_1 + 99.27\% C_U H_2 + C_{Th} H_3 + 0.0128\% C_K H_4 \qquad (6.38)$$

where C_U, C_{Th} and C_K are the mass fractions of uranium, thorium and potassium, respectively. The total heat production can then be expressed in terms of C_U, C_{Th} and C_K as

$$H_{total} = H_U C_U + H_{Th} C_{Th} + H_K C_K \qquad (6.39)$$

where $H_U = (0.72\% H_1 + 99.27\% H_2) \cdot 10^{-6} = 9.8 \cdot 10^{-5} \ \mu W\,kg^{-1}$, $H_{Th} = H_3 \cdot 10^{-6} = 2.6 \cdot 10^{-5} \ \mu W\,kg^{-1}$ and $H_K = 0.0128\% H_4 \cdot 10^{-2} = 3.4 \cdot 10^{-5} \ \mu W\,kg^{-1}$. Notice that the factors 10^{-6} and 10^{-2} have been added because the mass fractions of U and Th are given in ppm (parts per million) and K in %. The expression of heat production per unit mass of rock can be multiplied with a typical crust density $\varrho = 2700 \ kg\,m^{-3}$, and it becomes an expression for heat production per unit volume:

$$S_{total} = S_U C_U + S_{Th} C_{Th} + S_K C_K \qquad (6.40)$$

where $S_U = \varrho H_U = 0.267 \ \mu W\,m^{-3}$, $S_{Th} = \varrho H_{Th} = 0.070 \ \mu W\,m^{-3}$ and $S_K = \varrho H_K = 0.091 \ \mu W\,m^{-3}$. Table 6.2 lists the heat production for some rock samples. We see that granite is the most heat-producing rock, and that mantle rocks (undepleted mantle and peridotite) have almost no heat production. Anther observation is that there is a considerable scatter in the heat production among rocks, even for rocks with similar types of minerals. We also observe the ratios $C_{Th}/C_U \approx 4$ and $C_K/C_U \approx 1.2$, which are often used to estimate the content of thorium and potassium once the content of uranium is known. These ratios can also be used to invert a measurement of heat production to an approximation for the contents of the elements U, Th and K. The concentrations are then

$$C_U \approx 1.5 S_{total}, \quad C_{Th} \approx 6 S_{total} \quad \text{and} \quad C_K \approx 1.8 S_{total} \qquad (6.41)$$

where the mass fractions C_U and C_{Th} are in ppm and the mass fraction C_K is in percent, when S_{total} has the units $\mu W\,m^{-3}$. (See Exercise 6.4.)

The heat generation from radioactive decay is not constant through time, because the decay to stable elements leaves less and less radioactive atoms. The number of decays per unit time is proportional to the number of atoms N of the radioactive isotope

$$\frac{dN}{dt} = -\lambda N \qquad (6.42)$$

where the disintegration constant λ tells how fast the decay is. We will see that a large λ leads to a short half-life. The decay constant can also be interpreted as a decay probability

Table 6.2. *The abundance of the elements U, Th and K in some rocks. The first data set is taken from Van Schmus (1995), the second is from Perry et al. (2006), the third is from Rudnick and Fountain (1995) and the fourth is from Turcotte and Schubert (1982).*

Name	U (ppm)	Th (ppm)	K (%)	S_{total} (μW m^{-3})	C_{Th}/C_U	C_K/C_U
Granite (1)	3.4	50	4.45	4.8	14.7	1.3
Granite (2)	3.0	24	3.67	2.8	8.0	1.2
Granite (low Ca)	4.7	20	4.2	3.0	4.3	0.9
Marine mud	2.8	12.2	2.96	1.9	4.4	1.1
Shale	3.1	9.5	2.2	1.7	3.0	0.7
Sandstone	1.7	5.5	1.07	0.9	3.2	0.6
Peridotite	0.005	0.01	0.001	0.002	2.0	0.2
Ultramafic intrusives	0.24	0.85	0.69	0.19	3.5	2.9
Gneiss, metasediments	1.86	6.89	1.93	1.15	3.7	1.0
Gabbro	0.23	0.9	0.27	0.15	3.9	1.2
Av. continental crust	1.6	5.8	2.0	1.0	3.6	1.3
Av. oceanic crust	0.9	2.7	0.4	0.47	3.0	0.4
Undepleted mantle	0.02	0.1	0.02	0.01	5.0	1.0

per unit time for an unstable nucleus. Equation (6.42) is straightforward to integrate, and gives the number of atoms as a function of time

$$N = N_0 \, e^{-\lambda t} \tag{6.43}$$

where N_0 is the number of atoms at time $t = 0$. It is often interesting to know how long it takes for the number of atoms to be reduced by half. We then have to solve the equation

$$e^{-\lambda t_{1/2}} = \frac{1}{2} = e^{-\ln 2} \tag{6.44}$$

which has the solution $t_{1/2}$ for the half-life

$$t_{1/2} = \frac{\ln 2}{\lambda} = \frac{0.693}{\lambda}. \tag{6.45}$$

The heat production for each radioactive isotope decays with its own time constant and the total heat production per unit mass of rock is

$$H(t) = \sum_i C_i \, H_i \, e^{-\lambda_i t} \tag{6.46}$$

where the index i is according to Table 6.1. Figure 6.3 shows the heat production for the average continental crust (see Table 6.2) for the past 4 Ga. The total heat production was 14% higher in the crust 500 Ma years ago, and 2 times higher 2.5 Ga ago. The decay in

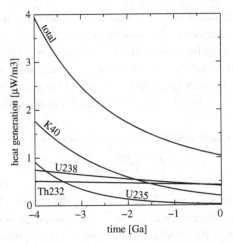

Figure 6.3. *Heat generation in the average continental crust in the past, where the present-day iso-tope concentrations are given in Table 6.2. Notice that the present-day heat production is dominated by* ^{238}U *and* ^{232}Th, *but the isotopes* ^{235}U *and* ^{40}K *were important in the Earth's early history.*

heat production is therefore ignored in models that do not go further back in time than a few hundred million years. Figure 6.3 shows also that the present-day heat production is dominated by ^{238}U and ^{232}Th, but that the ^{235}U and ^{40}K were important in the Earth's early history.

Exercise 6.3 What is the ratio C_{Th}/C_U when Th and U contribute equally to heat generation? And what is the ratio C_K/C_U when K and U have the same contribution to heat generation?

Exercise 6.4 Assume that $C_{Th}/C_U \approx 4$ and $C_K/C_U \approx 1.2$ and use equation (6.40) to derive the estimates (6.41) for C_U, C_{Th} and C_K as a function of heat production.

Exercise 6.5 Show that the number of radioactive atoms as a function of time (6.43) follows by integration of equation (6.42).
Solution: From equation (6.43) we get

$$\int_{N_0}^{N} \frac{dN}{N} = -\int_{0}^{t} \lambda dt = -\lambda t \tag{6.47}$$

which yields

$$\ln\left(\frac{N}{N_0}\right) = -\lambda t \quad \text{or} \quad N = N_0 \, e^{-\lambda t}. \tag{6.48}$$

6.4 Stationary 1D temperature solutions with heat generation

Solutions of the stationary (time-independent) temperature equation can be used to estimate geotherms when a geological environment does not change much over a long time span. One particular application is the estimation of geotherms in stable continental areas. There are no direct measurements of continental geotherms, and the estimates have to be based on surface (or near surface) observations. In particular observations of surface heat flow, heat production in surface rocks and geophysical measurements of the thickness of the crust are important. The heat production is negligible in mantle rocks, but important in crustal rocks (see Table 6.2). It is therefore important to know where the crust ends and the mantle begins.

The stationary temperature equation in the vertical direction is

$$\frac{d}{dz}\left(\lambda(z)\frac{dT}{dz}\right) = -S(z) \tag{6.49}$$

when there are zero velocity terms. Notice that both the heat conductivity $\lambda(z)$ and the heat production term $S(z)$ depend on the depth z. It is straightforward to solve the second-order temperature equation (6.49) with two integrations. Integration two times brings two integration constants into the solution, which are found by requiring the solution to fulfill two boundary conditions. One boundary condition is a constant temperature at the surface. The other boundary condition will either be the mantle heat flow at the base of the crust or the temperature at the base of the lithosphere. The latter boundary condition is preferred when there is some knowledge about the thickness of the lithosphere.

An integration of the temperature equation (6.49) from the depth z to the depth of the crust z_m gives the heat flux at any position in z in the crust:

$$q(z) = \lambda(z)\frac{dT}{dz} = \int_z^{z_m} S(u)du + q_m \tag{6.50}$$

where q_m is the mantle heat flow into the base of the crust (at $z = z_m$). The heat production is $S(z) \approx 0$ in the lithospheric mantle ($z > z_m$) and its contribution to the heat flow is negligible. The heat flow increases from the depth z_m towards the surface because of the integral over the heat source. The surface heat flow

$$q_s = \int_0^{z_m} S(u)\,du + q_m \tag{6.51}$$

is the mantle heat flow added to the contribution from heat production from the entire thickness of the crust, and it is the maximal heat flow. We will later see that the heat production in the crust is often the most important part of the surface heat flow.

The average heat production along a vertical through the crust is

$$S_{av} = \frac{1}{z_m}\int_0^{z_m} S(u)\,du \tag{6.52}$$

and we have from equation (6.51) that this average is simply

$$S_{av} = \frac{1}{z_m}(q_s - q_m). \tag{6.53}$$

An estimate of the mantle heat flux (q_m) combined with observation of surface heat flux (q_s) and the average thickness of the crust z_m gives the average heat production in the crust.

Equation (6.51) shows that the surface heat flow remains the same regardless of the vertical distribution of heat production $S(z)$ as long as average heat production is constant. For example the following three distributions of heat production:

$$S_1(z) = 2S_0 (1 - z/z_m) \qquad (6.54)$$

$$S_3(z) = S_0 \qquad (6.55)$$

$$S_2(z) = 2S_0 z/z_m \qquad (6.56)$$

have the same average heat production S_0 and they therefore give the same surface heat flow. But, we will soon see that the temperature at the base of the crust can be quite different for different distributions of heat production although the surface heat flow is the same.

An integration of the heat flow from the surface ($z = 0$) to the depth z yields the geotherm

$$T(z) = T_0 + q_m \int_0^z \frac{du}{\lambda(u)} + \int_0^z \frac{1}{\lambda(v)} \left(\int_v^{z_m} S(u)\, du \right) dv \qquad (6.57)$$

in the crust ($z < z_m$). There is negligible heat production in the mantle ($z > z_m$) and this part of the geotherm can therefore be written as

$$T(z) = T_m + q_m \int_{z_m}^z \frac{du}{\lambda(u)} \quad \text{for} \quad z > z_m \qquad (6.58)$$

where $T_m = T(z_m)$ is the temperature (6.57) at the base of the crust ($z = z_m$). The solution (6.57) uses the temperature T_0 at the surface $z = 0$ as one boundary condition and the mantle heat flux q_m at the base of the crust $z = z_m$ as the second boundary condition. The integrals in solution (6.57) are straightforward to calculate for constant heat conductivities λ_c and λ_m for the crust and the mantle, respectively, and a constant heat production S_0 in the crust. It gives the geotherm

$$T(z) = T_0 + \frac{q_m}{\lambda_c} z + \frac{S_0}{2\lambda_c} (2z_m z - z^2) \qquad \text{(crust)} \qquad (6.59)$$

$$T(z) = T_m + \frac{q_m}{\lambda_m} (z - z_m) \qquad \text{(mantle)} \qquad (6.60)$$

where T_m is the temperature

$$T_m = T_0 + \frac{q_m}{\lambda_c} z_m + \frac{S_0}{2\lambda_c} z_m^2 \qquad (6.61)$$

at the base of the crust ($z = z_m$). It often convenient to express the geotherm using the surface heat flow q_s instead of the heat production S_0 as shown by Exercise 6.6. The geotherm (6.60) is linear through the lithospheric mantle because of constant heat flow and the assumption of a constant heat conductivity. Figure 6.4 shows an example of the geotherm through a 25 km thick crust that has a constant heat generation $S_0 = 1\ \mu W\, m^{-3}$. The figure also shows the geotherm in the case of zero crustal heat generation, and the temperature difference at the base of the crust for the two geotherms is $\Delta T_m = S_0 z_m^2 / 2\lambda_c = 125\,°C$, in accordance with equation (6.61).

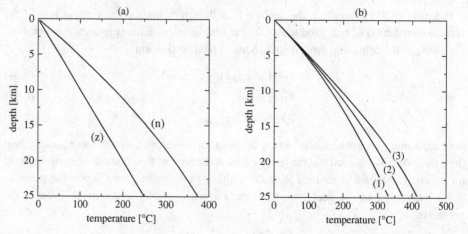

Figure 6.4. *(a) The geotherm (6.59) for* $q_m = 0.025\ Wm^{-2}$, $\lambda = 2.5\ Wm^{-1}K^{-1}$, $z_m = 25\ km$, $S_0 = 1 \cdot 10^{-6}\ Wm^{-3}$. *The linear geotherm in the case of zero heat production is also shown. (b) The geotherms in the case of the three different distributions (6.54) to (6.56) of the heat production in the crust. The average crustal heat production is the same for the three distributions.*

The temperature T_m at the base of the crust depends on the distribution of the heat production with depth. For example the linearly changing heat productions (6.54) to (6.56) give a temperature difference $\Delta T_m = \frac{1}{6}S_0 z_m^2 \approx 83\,°C$ between S_1 and S_3, as shown in Figure 6.4b. (See Exercise 6.10 for the details.)

The temperature $T_a \approx 1300\,°C$ specifies the transition zone between the lithosphere and the asthenosphere. The depth to the base of the lithosphere is the solution of $T(z_a) = T_a$, using the mantle geotherm (6.60), which gives

$$z_a = z_m + \frac{\lambda_m}{q_m}(T_a - T_0) - \frac{\lambda_m}{\lambda_c}\left(1 + \frac{S_0 z_m}{2q_m}\right)z_m. \qquad (6.62)$$

It is also possible to rewrite the depth to the asthenosphere using the surface heat flux instead of the heat production S_0 as shown in Exercise 6.6.

We will now look at the other situation where the temperature (T_a) at the base of the lithosphere $(z = z_a)$ is the boundary condition instead of the heat flow q_m. In the case of constant heat conductivities for the mantle and the crust and a constant crustal heat production we have

$$T_m = T_0 + \frac{q_m}{\lambda_c}z_m + \frac{S_0}{2\lambda_c}z_m^2 \qquad (6.63)$$

$$T_a = T_m + \frac{q_m}{\lambda_m}(z_a - z_m) \qquad (6.64)$$

which are two equations for the two unknowns T_m and q_m. We find that

$$T_m = \frac{\lambda_c T_0(z_a - z_m) + \lambda_m T_a z_m + \frac{1}{2}S_0 z_m^2(z_a - z_m)}{\lambda_c(z_a - z_m) + \lambda_m z_m} \qquad (6.65)$$

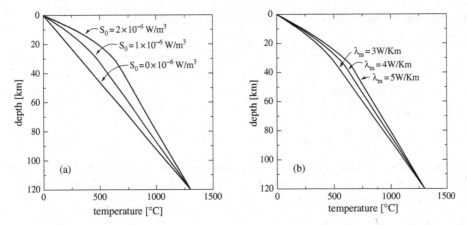

Figure 6.5. *The geotherm (6.59)–(6.60) is plotted for T_m and q_m given by equations (6.65) and (6.66), respectively, which is the case of a constant temperature T_a at the depth z_a. (a) The geotherm for different values of heat generation (S_0); (b) the geotherm for different mantle heat conductivities (λ_m). Other parameters are $S_0 = 1 \cdot 10^{-6}\ Wm^{-3}$, $\lambda_c = 3\ Wm^{-1}\ K^{-1}$, $\lambda_m = 3.5\ Wm^{-1}\ K^{-1}$, crustal thickness $z_m = 35\ km$, lithospheric thickness $z_a = 120\ km$ and the temperature $T_a = 1300\,^\circ C$ at the base of the lithosphere.*

and

$$q_m = \lambda_m \left(\frac{T_a - T_m}{z_a - z_m} \right) = \frac{\lambda_c(T_a - T_0) - \frac{1}{2}S_0 z_m^2}{(\lambda_c/\lambda_m)(z_a - z_m) + z_m}. \tag{6.66}$$

Figure 6.5 shows examples of the temperature solution (6.59)–(6.60) when the boundary conditions are fixed temperatures at the surface and at the base of the lithosphere. Not only the temperature at the base of the crust varies, but also the mantle heat flux varies, when a fixed temperature at the base of the lithosphere is used as a boundary condition. Note 6.1 shows an alternative and more direct solution of the stationary temperature equation (6.49) in the case of constant heat conductivities and heat production.

Note 6.1 The stationary temperature equation (6.49) can also be solved directly for the case of constant heat conductivities for the crust and mantle, and a constant crustal heat production. The temperature is first written as two parts, one for the crust and one for the mantle:

$$T(z) = \begin{cases} T_1(z) = -\frac{1}{2}\left(S_0/\lambda_c\right)z^2 + a_1 z + b_1, & z \text{ in crust} \\ T_2(z) = a_2 z + b_2, & z \text{ in mantle} \end{cases} \tag{6.67}$$

where there are four coefficients in the solution (a_1, b_1, a_2 and b_2). We have only two boundary conditions, so we need two more before we have four equations for the four unknown coefficients. The two remaining boundary conditions are for the interface between the crust and the mantle. The temperature at the interface is continuous and the heat flow is also continuous, which is written

$$T_1(z_m) = T_2(z_m) \quad \text{and} \quad \lambda_c \frac{dT_1}{dz}(z_m) = \lambda_m \frac{dT_2}{dz}(z_m). \tag{6.68}$$

The solution for the coefficients (from the boundary conditions) is

$$a_1 = \frac{(\lambda_m/2\lambda_c) S_0 z_m^2 - S_0 z_m (z_m - z_a) + \lambda_m (T_a - T_0)}{\lambda_m z_m - \lambda_c (z_m - z_a)} \tag{6.69}$$

$$a_2 = \frac{\lambda_c}{\lambda_m} a_1 - \frac{1}{\lambda_m} S_0 z_m \tag{6.70}$$

$$b_1 = T_0 \tag{6.71}$$

$$b_2 = T_a - a_2 z_a. \tag{6.72}$$

Coefficient a_1 is evaluated first, and it is then used to obtain coefficient a_2, before b_1 and b_2. It is left as an exercise to check that $T_m = a_2 z_m + b_2$ is the same as T_m (6.65) and that $q_m = \lambda_m a_2$ is the same as q_m (6.66).

Exercise 6.6 Replace the heat production S_0 by the average heat production $S_{av} = (q_s - q_m)/z_m$ and show that:
(a) the temperature at the base of the crust is

$$T_m = T_0 + \frac{z_m}{2\lambda_c} (q_s + q_m); \tag{6.73}$$

(b) the depth of the base of the lithosphere is

$$z_a = \frac{\lambda_m}{q_m} (T_a - T_0) + z_m - \frac{1}{2} z_m \frac{\lambda_m}{\lambda_c} \left(1 + \frac{q_s}{q_m}\right). \tag{6.74}$$

Exercise 6.7 Check that the solutions (6.59) and (6.60) give the same temperature and the same heat flow at the point $z = z_m$ where they are glued together.

Exercise 6.8 Assume that the surface temperature T_0 and the surface heat flow q_s are known. Show that the solution of the temperature equation (6.49) is

$$T(z) = T_0 + q_s \int_0^z \frac{du}{\lambda(u)} - \int_0^z \frac{1}{\lambda(v)} \left(\int_0^v S(u)\, du\right) dv. \tag{6.75}$$

Exercise 6.9 **(a)** Use equation (6.59) to show that the thickness of the crust has to be

$$z_m = \frac{q_m}{S} \tag{6.76}$$

for the mantle heat flow and a constant crustal heat generation to contribute equally to the surface heat flow.
(b) Show that the contribution from the linear term in the temperature solution (6.59) contributes the same as the heat source term at the base of the crust (at $z = z_m$), when the crust has the thickness

$$z_m = \frac{2q_m}{S}. \tag{6.77}$$

(c) What is z_m in (a) and (b) when $q_m = 0.02\ \mathrm{W\,m^{-2}}$ and $S = 10^{-6}\ \mathrm{W\,m^{-3}}$?

Exercise 6.10 Let the heat production in the crust be linearly dependent on depth;

$$S(z) = az + b. \qquad (6.78)$$

(a) Show that the geotherm in the crust ($z < z_m$) becomes

$$T(z) = T_0 + \frac{q_m}{\lambda_c}z + \frac{1}{\lambda_c}\left(\frac{1}{2}az_m^2 z + bz_m z - \frac{1}{6}az^3 - \frac{1}{2}bz^2\right). \qquad (6.79)$$

(b) Show that the geotherm in the mantle ($z > z_m$) is

$$T(z) = T_m + \frac{q_m}{\lambda_m}(z - z_m) \qquad (6.80)$$

where

$$T_m = T_0 + \frac{q_m}{\lambda_m}z_c + \frac{1}{\lambda_c}\left(\frac{1}{3}az_m^3 + \frac{1}{2}bz_m^2\right). \qquad (6.81)$$

(c) Show that T_m in the case of the three distributions (6.54) to (6.56) of heat production becomes

$$T_{m,1}(z) = T_0 + \frac{q_m}{\lambda_c}z_m + \frac{1}{3}\frac{S_0}{\lambda_c}z_m^2 \qquad (6.82)$$

$$T_{m,2}(z) = T_0 + \frac{q_m}{\lambda_c}z_m + \frac{1}{2}\frac{S_0}{\lambda_c}z_m^2 \qquad (6.83)$$

$$T_{m,3}(z) = T_0 + \frac{q_m}{\lambda_c}z_m + \frac{2}{3}\frac{S_0}{\lambda_c}z_m^2 \qquad (6.84)$$

which differ by $\frac{1}{6}S_0 z_m^2/\lambda_c$.

Exercise 6.11 Measurements of heat flow in continental areas like mountains sometimes suggest a linear relationship between the surface heat flow and the heat generation in the surface rocks. Figure 6.6 shows an example. An exponentially decreasing heat production with depth

$$S(z) = S_0\,e^{-z/z_0} \qquad (6.85)$$

can explain these observations. This depth-dependent crustal heat production is still used, although recent compilations of data show that it is not valid in general.
(a) Show that the heat generation (6.85) gives the surface heat flow

$$q_s \approx z_0 S_0 + q_m \qquad (6.86)$$

for $|z_m| \gg |z_0|$, where q_m is the mantle heat flow and z_m the thickness of the crust.
(b) Show that the geotherm is

$$T(z) = T_0 + \frac{q_m}{\lambda}z + \left(\frac{z_0^2 S_0}{\lambda}\right)\left(1 - e^{-z/z_0}\right) \qquad (6.87)$$

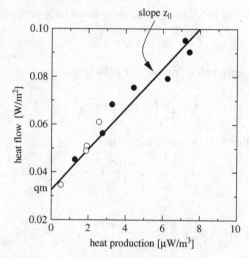

Figure 6.6. *Surface heat flow is plotted as a function of heat production in surface rocks with data from New England (filled circles) and from the Central Stable Region (open circles). The data sets are taken from Roy et al. (1968).*

where the first term is the surface temperature, the second term is a linear increase in the temperature caused by mantle heat flow, and the third term is the contribution from heat generation.

6.5 Heat flow and geotherms in stable continental areas

Surface heat flow and heat production in surface rocks have been mapped over large parts of the Earth and they are an important constraint for the modeling of the temperature–depth profiles. Jaupart *et al.* (2007) estimated the average continental heat flux to be ~ 65 mW m^{-2}. The heat flow average is biased towards a higher value because of geothermal activity, like certain areas in the Western USA. The average surface heat flow and the average surface heat production for the North American Craton are plotted in Figure 6.7 for five different provinces. The figure shows that the province average follows a linear relationship:

$$q_{av} = q_0 + H S_{av} \qquad (6.88)$$

where q_{av} and S_{av} are province-wide-averaged heat flow and heat production data. Jaupart *et al.* (2007) suggest that q_0 is a common heat flux at some intermediate depth in all provinces. The data in Figure 6.7 have $q_{av} = 33$ mW m^{-2} and $H = 9.1$ km. The linear relationship between the average surface heat flow and the average surface heat production resembles the one that has been suggested for local (non-averaged) values (see Exercise 6.11). This resemblance is just a coincidence, because there does not seem to be a relationship between local values of surface heat flow and surface heat production.

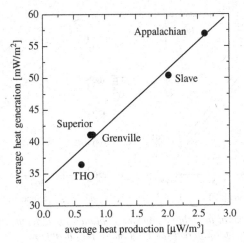

Figure 6.7. *Average heat flow and average heat production is plotted for five provinces in North America. The data are taken from Jaupart et al. (2007).*

We have from equation (6.51) that the surface heat flow is the heat flow from crustal heat production added to the mantle heat flow. Very low values of surface heat flow in stable continental areas provide limiting values for the mantle heat flux. Such areas have most likely little crustal heat production and the surface heat flow is therefore close to the mantle heat flow. The surface heat flow is as low as 22 mW m^{-2} in several areas of the Canadian shield (Jaupart *et al.*, 2007). This value therefore serves as an upper bound for the mantle heat flow. It appears that no crustal rock has a heat production that is less than $S_{min} = 0.1 \ \mu$W m^{-3}. An even better estimate for the mantle heat flow is therefore obtained by subtracting the minimum crustal heat production. Assuming the crustal thickness $z_m = 40$ km gives that $q_m = q_s - S_{min}z_m = 18$ mW m^{-2}. The mantle heat flow in stable continental areas is likely even lower. Swanberg *et al.* (1974) estimated a mantle heat flux as low as 11 mW m^{-2} in Norway, and Guillou *et al.* (1994) has more recently estimated that the mantle heat flux is in the range of 7–15 mW m^{-2}.

An estimate for the mantle heat flux is important because it allows the average crustal heat production to be estimated as $S_{av} = (q_s - q_m)/z_m$. The three observations q_s, q_m and z_m are therefore sufficient to make simple estimates of a geotherm using equations (6.59) and (6.60) – assuming that the heat conductivities of the crust and the mantle do not vary much. (We will look at the sensitivity of these parameters in a moment.)

As an example we will estimate a geotherm for the Kola–Karelian region where the surface heat flux is 45 mW m^{-2} and the crustal thickness is $z_m = 50$ km. These data are taken from Balling (1995), who discusses the surface heat flow data, heat production data and thermal modeling results for the Baltic shield and the Northern Tornquist zone. A mantle heat flux 15 mW m^{-2} is now assumed. These data give an average surface heat production $S_{av} = (q_s - q_m)/z_m = 0.6 \ \mu$W m^{-3}, a temperature $T_m = 600\,^\circ$C at the base of the crust and a depth to the base of the lithosphere that is $z_a = 185$ km. (The

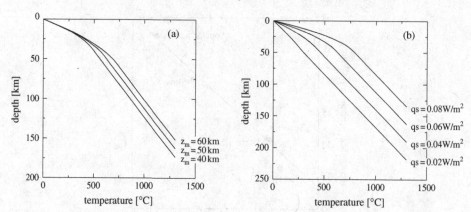

Figure 6.8. *(a) The geotherm in stable continents is plotted for different thicknesses of the crust. (b) The geotherm is plotted for different values of the surface heat flow.*

transition between the lithosphere and the asthenosphere is in a temperature interval around 1300 °C.) The heat conductivities of the crust and the mantle are $\lambda_c = 2.5 \ \mathrm{W \, m^{-1} \, K^{-1}}$ and $\lambda_m = 3.0 \ \mathrm{W \, m^{-1} \, K^{-1}}$. Figure 6.8 shows a series of geotherms with these crustal and mantle heat conductivities, but here the crustal thickness and the mantle heat flow are varied.

An advantage of such simple analytical estimates is that it is straightforward to analyze the sensitivity of the model with respect to the parameters. For instance an uncertainty Δz_m in the crustal thickness gives an uncertainty

$$\Delta z_a = \left(1 - \frac{1}{2} \frac{\lambda_m}{\lambda_c} \left(1 + \frac{q_s}{q_m}\right)\right) \Delta z_m \qquad (6.89)$$

in the lithospheric thickness.

6.6 Stationary geotherms in the lithospheric mantle

Heat conductivity is not constant over temperature intervals spanning several hundred °C. In fact most rocks have a decreasing heat conductivity with increasing temperature. This temperature dependence is often expressed as

$$\lambda_A(T) = \frac{\lambda_0}{1 + c_0 T} \qquad (6.90)$$

where λ_0 is the heat conductivity and the temperature is in °C (Seipold, 1995, 1998). (See also Table 2.3, which lists the heat conductivities of some crystalline rocks.) We see that $T = 1/c_0$ is the temperature increase from 0 °C that makes the heat conductivity decrease by half. The coefficient c_0 is typically of order $10^{-3} \ {}^\circ\mathrm{C}^{-1}$, which gives 1000 °C as an estimate of this temperature increase. Temperature dependent heat conductivity is of especial interest for the lithospheric mantle that spans a temperature range of roughly 1000 °C. We will now look at four alternative mantle heat conductivities and compare the corresponding geotherms. The first is function (6.90) and the second is the heat conductivity based on work by Hofmeister (1999) and given by McKenzie and Jackson (2005) as

$$\lambda_H(T) = \frac{b}{1+cT} + \sum_{n=1}^{3} d_n(T+273)^n \qquad (6.91)$$

where λ_H is in units $W\,m^{-1}\,K^{-1}$, T is in °C. The parameters are $b = 5.3$, $c = 0.0015$, $d_0 = 1.753 \cdot 10^{-2}$, $d_1 = -1.0365 \cdot 10^{-4}$, $d_2 = 2.2451 \cdot 10^{-7}$, $d_3 = -3.4071 \cdot 10^{-11}$. The third heat conductivity function is

$$\lambda_X(T) = \lambda_0 \left(\frac{298}{T+273}\right)^n \qquad (6.92)$$

which is proposed by Xu *et al.* (2004), where $\lambda_0 = 4.08\ W\,m^{-1}\,K^{-1}$ and $n = 0.406$. The fourth alternative is a constant heat conductivity $\lambda_B = 3\ W\,m^{-1}\,K^{-1}$. Figure 6.9a shows these heat conductivities for temperatures in the range from 200 °C to 1300 °C. The function λ_A is between the heat conductivities λ_H and λ_X, and λ_B is close to the average of λ_A and λ_H over the shown temperature interval.

The geotherm is once more given as a solution of the stationary temperature equation (6.20) in the vertical direction. It returns Fourier's law when it is integrated once:

$$\lambda(T)\frac{dT}{dz} = q_m \qquad (6.93)$$

where λ is the mantle heat conductivity and q_m is the mantle heat flow. It is now assumed that the mantle heat flow q_m is constant. The next step is an integration of Fourier's law

$$\int_{T_m}^{T} \lambda(T)\, dT = q_m \int_{z_m}^{z} dz \qquad (6.94)$$

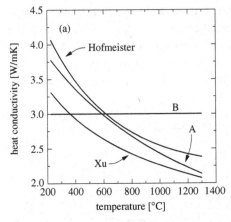
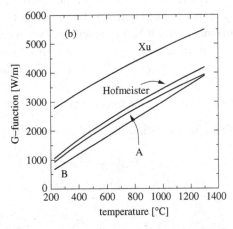

Figure 6.9. (a) Heat conductivity functions that depend on temperature. See the text for details. (b) The integral $G(T) = \int_{T_0}^{T} \lambda(T)\, dT$ is plotted for the temperature dependent heat conductivities in (a).

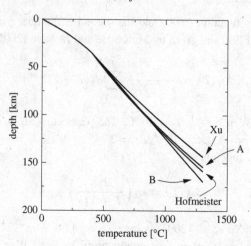

Figure 6.10. *The temperature profile is plotted through the lithosphere for the temperature dependent mantle heat conductivities in Figure 6.9. The mantle heat flow is $q_m = 0.02\ Wm^{-2}$. The crust has a thickness 35 km, heat conductivity $\lambda_c = 2.5\ Wm^{-1}\,K^{-1}$ and a heat production $S = 10^{-6}\ Wm^{-3}$. Profile B has a constant mantle heat conductivity $\lambda_m = 3\ Wm^{-1}\,K^{-1}$.*

which gives depth as a function of temperature:

$$z = z_m + \frac{1}{q_m}\left(G(T) - G(T_m)\right) \quad \text{where} \quad G(T) = \int_{T_m}^{T} \lambda(T)\,dT. \tag{6.95}$$

The difference in G-value for two temperatures gives the vertical distance between the temperatures. A constant heat conductivity λ_B gives a linear G-function. Figure 6.9b shows the G-functions for the heat conductivities in Figure 6.9a, and all G-functions have a quite similar steepness.

It is straightforward to plot the depth as a function of temperature once we have the G-function. (See Exercise 6.12 for details.) Figure 6.10 shows the geotherms in the case of the four different heat conductivities when the mantle heat flow is $q_m = 0.02\ \mathrm{W\,m^{-2}}$. The crust has a thickness $z_m = 35$ km with a constant heat generation $S_0 = 1 \cdot 10^{-6}\ \mathrm{W\,m^{-3}}$ and a heat conductivity $\lambda_c = 2.5\ \mathrm{W\,m^{-1}\,K^{-1}}$. The difference between the geotherms increases with increasing depth, and the heat conductivities λ_B and λ_H give similar results. The difference in depth for the temperature $1300\,^\circ$C is 30 km.

Exercise 6.12 Show that $G(T)$ for the heat conductivities λ_B, λ_A, λ_H and λ_X is

$$G_B(T) = \lambda_B\, T \tag{6.96}$$

$$G_A(T) = \frac{\lambda_0}{c_0}\ln(1 + c_0\, T) \tag{6.97}$$

$$G_H(T) = \frac{b}{c}\ln(1 + cT) + \sum_{n=1}^{3}\frac{d_n}{(n+1)}(T + 273)^{n+1} \tag{6.98}$$

$$G_X(T) = \lambda_0 \, 298^n \, \frac{(T+273)^{1-n}}{(1-n)}. \tag{6.99}$$

Exercise 6.13 We will now compute the stationary geotherm when the heat conductivity is both temperature- and pressure-dependent. The pressure dependence is normally given by the factor $(1 + \alpha_0 p)$, and the temperature- and pressure-dependent heat conductivity then becomes

$$\lambda(p, T) = \lambda(T)\,(1 + \alpha_0 p) \tag{6.100}$$

where $\lambda(T)$ may be any heat conductivity as a function of temperature. Assume that the pressure is $p = \varrho g z$ and show that z as a function of temperature then becomes

$$z = \frac{1}{\alpha_0 \varrho g}\left\{(1 + \alpha_0 \varrho g z_m)\exp\left(\frac{\alpha_0 \varrho g}{q_m}\,(G(T) - G(T_m))\right) - 1\right\} \tag{6.101}$$

where $G(T)$ is the function (6.95).

Solution: The starting point is Fourier's law (6.93), which now becomes

$$\lambda(T)\,dT = q_m \frac{dz}{(1 + \alpha_0 \varrho g z)}. \tag{6.102}$$

The integration (6.94) then gives

$$G(T) - G(T_m) = \frac{q_m}{\alpha_0 \varrho g}\ln\left(\frac{1 + \alpha_0 \varrho g z}{1 + \alpha_0 \varrho g z_m}\right) \tag{6.103}$$

and we are almost done.

6.7 Sediment maturity and vitrinite reflectance

Vitrinite is the most important thermal indicator in sedimentary basins and it is routinely measured in a large number of exploration wells by the oil companies. Vitrinite reflectance (VR) values can be grouped into intervals with respect to hydrocarbon generation like oil maturation and gas maturation, and then provide directly a measure of the maturity of sediment samples.

Vitrinite is one of the primary components of coals and it is common in sedimentary rocks that are rich in organic matter. It becomes thermally altered in response to the influence of temperature over time, and the degree of alteration is measured as reflectance (Tissot and Welte, 1978). The simplest approach to quantify the VR uses the TTI-concept introduced by Lopatin (1971), which is the time–temperature index defined for a temperature history $T(t)$ as

$$\mathrm{TTI}(t) = \int_0^t 2^{aT(t')+b}\,dt'. \tag{6.104}$$

TTI as an integral is a generalization of the observation made by Lopatin (1971) that reaction rates double by every step in temperature of $10\,°C$. The parameter a is therefore

$a = 0.1$, while the parameter $b = -10.5$ is just the scaling factor 2^b. VR (%Ro) is correlated to TTI by the functional relationship

$$\%Ro = \exp(p\ln(\text{TTI}) + r) \tag{6.105}$$

where the parameters p and r are calibrated against data sets. Several pairs of p and r parameters have been reported in the literature, see for instance Waples (1980) and Issler (1984) and Morrow and Issler (1993).

The TTI based computation of VR has been superseded by procedures based on Arrhenius kinetics (Burnham and Sweeney, 1989, Sweeney and Burnham, 1990). The vitrinite is divided into a series of components where each component has its own reaction rate. The decay of component i is given by a first-order reaction

$$\frac{dx_i}{dt} = -k_i x_i \tag{6.106}$$

where x_i is the amount of the component. An Arrhenius law gives the reaction rate

$$k_i = A_i \exp\left(-\frac{E_i}{RT}\right). \tag{6.107}$$

where A_i is the Arrhenius factor and E_i is the activation energy. The temperature depends on time and the reaction rate is therefore also time dependent. Integration of the first-order equation (6.106) gives that the component i is

$$x_i = x_{i,0} \exp\left(-\int_0^t k_i(t)\,dt\right) \tag{6.108}$$

where $x_{i,0}$ is the initial amount of component i in terms of normalized fractions. These initial fractions therefore add up to zero, $\sum_i x_{i,0} = 1$. The fractions $x_{i,0}$ are also referred to as stoichiometric factors, or weight fractions. The amount of each fraction that has reacted is $x_{i,0} - x_i$ and the total amount that has reacted is the transformation ratio

$$\text{Tr} = \sum_i (x_{i,0} - x_i) = 1 - \sum_i x_i \tag{6.109}$$

which is a number that increases from 0 towards 1 as the reactions go to completion. The transformation ratio is mapped to VR by the expression

$$\%Ro = \exp(c + d\,\text{Tr}). \tag{6.110}$$

The parameters c and d must be calibrated against data sets together with the parameters for the Arrhenius kinetics and the initial fractions. The most common parameter set for VR computations is the one suggested by Sweeney and Burnham (1990), see Table 6.3, which was one of the earliest models for VR based on Arrhenius kinetics. These parameters have become a "standard" in the computation of VR since they give a good match against a large number of data sets, both from sedimentary basins as well as laboratory experiments. There is also a need for a standard parameter set because it simplifies comparisons of maturity studies. The VR model using the data set from Sweeney and Burnham (1990) is

Table 6.3. *The kinetics for vitrinite reflection proposed by Sweeney and Burnham (1990).*

A_i	E_i	$x_{i,0}$
$[s^{-1}]$	$[kJ\,mole^{-1}]$	$[-]$
1e+13	142.351	0.03
1e+13	150.725	0.03
1e+13	159.098	0.04
1e+13	167.472	0.04
1e+13	175.846	0.05
1e+13	184.219	0.05
1e+13	192.593	0.06
1e+13	200.966	0.04
1e+13	209.340	0.04
1e+13	217.714	0.07
1e+13	226.087	0.06
1e+13	234.461	0.06
1e+13	242.834	0.06
1e+13	251.208	0.05
1e+13	259.582	0.05
1e+13	267.955	0.04
1e+13	276.329	0.03
1e+13	284.702	0.02
1e+13	293.076	0.02
1e+13	301.450	0.01

called *easy-ro*. Table 6.3 shows that easy-ro has 20 steps with the same Arrhenius prefactor $A = 10^{13}\,s^{-1}$, and that the activation energy comes in the step size $\Delta E = 8.374\,kJ\,mole^{-1}$ (2 kcal mole^{-1}) from $E = 142.4\,kJ\,mole^{-1}$ (34 kcal mole^{-1}) until $E = 301.5\,kJ\,mole^{-1}$ (72 kcal mole^{-1}). One particular feature of Table 6.3 is that the weighting factors add up to 0.85 and not 1. The reason was a lack of observations corresponding to high activation energies when the model was made. Therefore, there is an extra step $i = 21$ with initial fraction $x_i = 0.25$, which can be considered inert ($x_i(t) = x_{i,0} = 0.25$). When this extra step is subtracted from the transformation ratio (6.109) we get that $Tr = 0.85 - \sum_i x_i$ for easy-ro. The easy-ro model has the parameters $c = -1.6$ and $d = 3.7$, and the minimum and maximum values for VR are therefore $\%Ro = 0.2$ and $\%Ro = 4.7$, respectively. An alternative to the extra step is to normalize the steps in the easy-ro model by writing

$$d\,Tr = d\,(0.85 - \sum_i x_i) = 0.85\,d\,\left(1 - \sum_i (\frac{x_i}{0.85})\right) = d'\,(1 - \sum_i x_i') \qquad (6.111)$$

where the new d-parameter is $d' = 0.85\,d = 3.145$ and the normalized easy-ro fractions are $x_i' = x_i/0.85$.

Heat flow

Vitrinite is useful as a thermal indicator because it has similar kinetics to oil and gas generation. The values of VR are therefore divided into groups with respect to the different stages of hydrocarbon generation. One such grouping is

VR interval	Stage
0.5–0.7	early mature oil
0.7–1.0	mid mature oil
1.0–1.3	late mature oil
1.3–2.6	main gas generation

which shows that oil generation takes place in the interval 0.5–1.3. Different source rocks can behave quite differently with respect to oil and gas generation, and such a scheme is just a rule of thumb. Nevertheless, measurements of VR immediately give valuable information about the maturity of a sediment layer.

The simplest burial histories have deposition of sediments at a constant rate. If the heat conductivity is also constant then we derive that the temperature of a stratum increases linearly with time. Fourier's law gives

$$T(t) = T_{\text{surf}} + \frac{qz}{\lambda} = T_{\text{surf}} + \frac{q\omega t}{\lambda} \tag{6.112}$$

where T_{surf} is the surface temperature, q is the heat flow, λ is the heat conductivity and the ω is the burial rate, and the heating rate becomes $Q = q\omega/\lambda$. It is instructive to test the easy-ro model for heating at a constant rate. Figure 6.11a shows VR from the easy-ro model in the case of heating from 0 °C to 250 °C over time spans of 1, 10, 100 and

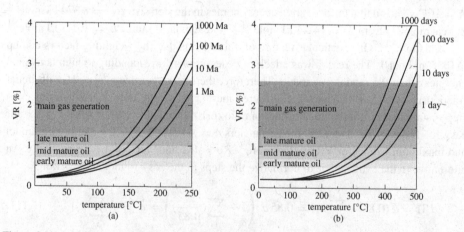

Figure 6.11. *(a) VR (easy-ro) for linear heating from 0 °C to 250 °C over four different time spans in million years. (b) VR (easy-ro) for linear heating from 0 °C to 500 °C over four different time spans in days.*

1000 million years. These rates are typical for deposition and burial in sedimentary basins and they give oil generation in the temperature interval from 100 °C to 150 °C. The easy-ro model can also be used on much larger heating rates – rates that are practical for laboratory experiments. Figure 6.11b shows the VR when the heating is from 0 °C to 500 °C over time spans of 1, 10, 100 and 1000 days. It is possible to generate a series of vitrinite measurements in some tens of days in the laboratory if the temperature goes up to 500 °C.

The decay of each fraction x_i as a function of temperature can be approximated as follows when the heating is at a constant rate

$$x_i(T) = x_{i,0} \exp\left[- N_i \left(F_i(T) - F_i(T_0) \right) \right] \tag{6.113}$$

where $F_i(T) = (T/T_i)^2 \exp(-T_i/T)$, $T_i = E_i/R$ and N is the (dimensionless) number

$$N_i = \frac{A_i T_i}{Q} \tag{6.114}$$

where Q is the heating rate. (See Note 6.2 for details.) The temperature at the beginning of the heating is T_0, which must not be confused with temperature T_i corresponding to activation energy E_i. Temperature and time are linearly related during heating at constant time-rate, and it is more convenient to use temperature instead of time, because time may vary by a large number of decades. The use of approximation (6.113) makes it simple to compute the VR for heating at a constant rate.

Vitrinite is important in oil exploration because it tells us whether a formation or stratum is in the oil window or not. It can answer similar questions such as where a stratum is in the oil window or when it was in the oil window. In immature areas where exploration wells have not been drilled and vitrinite has not been measured it is still possible to make assessments of the basin maturity by modeling the burial history – assuming that the most important formation boundaries have beeen mapped by seismic surveys. The temperature history can be computed numerically and with the associated VR using reasonable litho-logical properties and a reasonable heat flow history. Figure 6.12a shows an example of a burial history with the corresponding temperature history, and Figure 6.12b shows the present day VR computed with the easy-ro model. The modeled VR has a good match against VR observations in this case.

There have been some attempts to extract the heat flow history from VR observa-tions assuming that the burial history with lithological properties is sufficiently accurately known. These efforts have shown that it is very difficult (or nearly impossible) to obtain the heat flow history – except for around the temperature maximum.

Note 6.2 The decay of fraction i is given by the first-order equation (6.108). This equation can be rewritten as

$$\frac{dx}{x} = -\frac{A_i}{Q} \exp(-T_i/T) \, dT = -N \exp(-1/u) \, du \tag{6.115}$$

with $N = A_i T_i/Q$, using that $T(t) = T_0 + Qt$ and that $u = T/T_i$. The integration of $\exp(-1/u)$ is then approximated by $u^2 \exp(-1/u)$ as shown in Note 11.1.

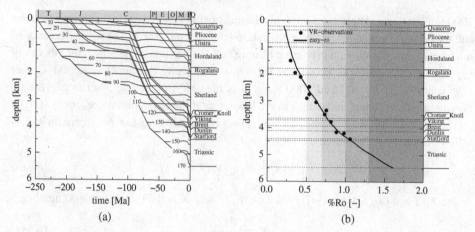

Figure 6.12. *(a) A burial history and its paleo-temperature. (b) VR observations and computed VR (easy-ro) for the temperature history in (a).*

The following two sections marked by ∗ can be considered as extended mathematical exercises and are not necessary reading.

6.8 Stationary heat flow in a sphere ∗

This section is a first attempt to show that conduction in a sphere is not a good model for the thermal state of the entire planet Earth. It turns out that heat transfer by conduction dominates only in the lithosphere, and that convection in the mantle is more efficient than conduction. Nevertheless, heat flow in a sphere is an interesting exercise that shows what the temperature at the center of the planet would have been if there had been heat conduction all the way to the center.

The equation for conservation of heat is now written for a thin shell rather than a box, see Figure 6.13. We then have that the radial transfer of energy into the shell at radius r added to the energy generated in the shell is equal to the radial transfer of energy out of the shell at radius $r + dr$. This can be written as

$$4\pi (r + dr)^2 q(r + dr) - 4\pi r^2 q(r) = 4\pi r^2 \, dr \, S(r) \qquad (6.116)$$

where $q(r)$ and $S(r)$ are the radial heat flow and heat production per unit volume, respectively, at radius r. To the first order in dr equation (6.116) becomes

$$4\pi (r + dr)^2 \left(q + \frac{dq}{dr} dr \right) - 4\pi r^2 q(r) = 4\pi r^2 \, dr \, S(r) \qquad (6.117)$$

which can be further simplified by expanding the term in parentheses and collecting only terms to first order in dr. We are then left with

$$\frac{dq(r)}{dr} + \frac{2q(r)}{r} = S(r). \qquad (6.118)$$

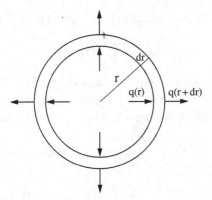

Figure 6.13. *Heat flow through a spherical shell.*

Energy conservation (6.118) in a spherical geometry can also be written as

$$\frac{1}{r^2}\frac{d}{dr}\left(r^2 q(r)\right) = S(r) \tag{6.119}$$

which is verified by carrying out the differentiation. The radial heat flow q is related to the radial temperature gradient by Fourier's law

$$q = -\lambda(r)\frac{dT}{dr}, \tag{6.120}$$

where the heat conductivity is assumed to be a function of r. Inserting Fourier's law into expression (6.119) for energy conservation gives the temperature equation for radial heat flow

$$\frac{1}{r^2}\frac{d}{dr}\left(r^2\lambda(r)\frac{dT}{dr}\right) = -S(r). \tag{6.121}$$

The temperature equation (6.121) is straightforward to integrate when both the heat conductivity λ and the heat production per unit volume S are constants. Two integrations then yield

$$T(r) = -\frac{1}{6}\left(\frac{S}{\lambda}\right)r^2 - \frac{c_1}{r} + c_2 \tag{6.122}$$

where c_1 and c_2 are integration constants that will be found from boundary conditions. One boundary condition is that $dT/dr = 0$ at the center of the sphere ($r = 0$). We see that the only way to have a finite temperature gradient at the center is to let $c_1 = 0$. The other boundary condition is to let $T = T_0$ be the temperature at the surface, $r = r_0$. The temperature solution is then

$$T(r) = \frac{1}{6}\left(\frac{S}{\lambda}\right)\left(r_0^2 - r^2\right) + T_0 \tag{6.123}$$

and the maximum temperature at the center of the sphere ($r = 0$) becomes

$$T_{\max} = \frac{1}{6}\left(\frac{S}{\lambda}\right) r_0^2 + T_0. \tag{6.124}$$

It would be interesting to see what this simple model predicts for the temperature at the center of the Earth. Before we can do that we need to know the heat production per unit volume, S. One way to obtain S is to replace it by an expression involving the surface heat flow, which is

$$q_0 = -\lambda \left.\frac{dT}{dr}\right|_{r=r_0} = \frac{S\,r_0}{3}. \tag{6.125}$$

The heat generation is therefore $S = 3q_0/r_0$, when expressed using the surface heat flow, and the maximum temperature at the center ($r = 0$) becomes

$$T_{\max} = \frac{q_0 r_0}{3\lambda} + T_0. \tag{6.126}$$

Using the values $r_0 = 6378$ km, $q_0 = 0.05$ W m^{-2}, $\lambda = 2.5$ W m^{-1} K^{-1} and $T_0 = 0\,°$C we get that $S = 2.4 \cdot 10^{-8}$ W m^{-3} and $T_{\max} = 63780\,°$C. This temperature is probably at least one order of magnitude too large. Heat conduction alone cannot therefore explain the thermal state of the Earth.

6.9 Transient cooling of a sphere *

The purpose of this section is to show that the cooling of the Earth by conduction through its history is not a good model.

We have so far only looked at stationary (time-independent) solutions of the temperature equation. The simplest time-dependent version of the full temperature equation (6.15) is

$$\varrho_b c_b \frac{\partial T}{\partial t} - \nabla\left(\lambda \nabla T\right) = 0 \tag{6.127}$$

where the convective term and heat source term are both zero. (The material derivative is replaced by an ordinary derivative because the rock matrix does not move relative to the coordinate system.)

We will now solve temperature equation (6.127) for a cooling sphere. Although it is too simplified as a model for the cooling of the Earth, it is nevertheless an interesting exercise in heat conduction. Realistic models for the cooling of the Earth have to account for heat lost by mantle convection. It follows from equation (6.121) that temperature equation (6.127) becomes

$$\varrho_b c_b \frac{\partial T}{\partial t} - \frac{1}{r^2}\frac{\partial}{\partial r}\left(r^2 \lambda \frac{\partial T}{\partial r}\right) = 0, \tag{6.128}$$

using spherical symmetry, $T = T(r)$. The boundary conditions for a cooling sphere with radius $r = r_0$ will be a zero temperature at the surface of the sphere, $T(r=r_0, t) = 0$, and a zero temperature gradient at the center of the sphere, $\partial T/\partial r(r=0, t) = 0$. (Symmetry requires the heat flow at the center of the sphere to be zero.) An initial condition is also needed before the temperature equation can be solved, and it is a constant temperature $T(r, t{=}0) = T_0$ throughout the sphere at $t = 0$.

We introduce the dimensionless radius $\hat{r} = r/r_0$ and the dimensionless temperature $\hat{T} = T/T_0$, where both dimensionless variables are numbers between 0 and 1. A step towards a dimensionless temperature equation (6.128) is

$$\frac{r_0^2 \varrho_b c_b}{\lambda} \frac{\partial \hat{T}}{\partial t} - \frac{1}{\hat{r}^2} \frac{\partial}{\partial \hat{r}} \left(\hat{r}^2 \lambda \frac{\partial \hat{T}}{\partial \hat{r}} \right) = 0, \tag{6.129}$$

where the coefficient

$$t_0 = \frac{r_0^2 \varrho_b c_b}{\lambda} \tag{6.130}$$

has units of time. We will see that the time t_0 characterizes the temperature transient, and that it is the natural choice for a quantity to scale time. Dimensionless time is therefore $\hat{t} = t/t_0$ and the dimensionless temperature equation becomes

$$\frac{\partial \hat{T}}{\partial \hat{t}} - \frac{1}{\hat{r}^2} \frac{\partial}{\partial \hat{r}} \left(\hat{r}^2 \lambda \frac{\partial \hat{T}}{\partial \hat{r}} \right) = 0, \tag{6.131}$$

which has no (explicit) parameters. The boundary conditions in dimensionless form are

$$\hat{T}(\hat{r}{=}1, \hat{t}) = 0 \quad \text{and} \quad \frac{\partial \hat{T}}{\partial \hat{r}}(\hat{r}{=}0, \hat{t}) = 0 \tag{6.132}$$

and the dimensionless initial condition is

$$\hat{T}(\hat{r}, \hat{t}{=}0) = 1. \tag{6.133}$$

The solution of the temperature equation (6.137) is the Fourier series

$$\hat{T}(\hat{r}, \hat{t}) = \hat{r} \sum_{n_1}^{\infty} a_n \sin(n\pi\hat{r}) \exp\left(-(n\pi)^2 \hat{t}\right) \tag{6.134}$$

as shown in Note 6.3, where the Fourier coefficients are

$$a_n = \frac{2(-1)^{n+1}}{n\pi}. \tag{6.135}$$

Figure 6.14 shows the solution (6.134) for the time steps 0.001, 0.02, 0.04, 0.08, 0.16 and ∞, and how the unit sphere cools down from the initial temperature 1 to the stationary temperature 0.

We see that the terms in the Fourier series decay to zero with a half-life $\hat{t}_{1/2} = \ln 2/(n\pi)^2$. The first term ($n = 1$) in the series has the longest half-life, and the half-lives then decrease with decreasing n. It is also seen that the absolute value of the Fourier coefficients decrease as $|a_n| \sim 1/n$ with increasing n. The first term, with the largest absolute value and the longest half-life, can be used to estimate the half-life of the cooling transient

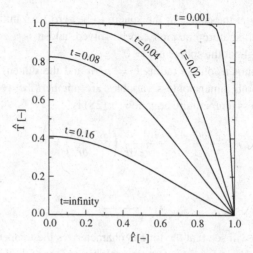

Figure 6.14. *The cooling of a unit sphere with an initial temperature equal to 1.*

as $\hat{t}_{1/2} \approx \ln 2/(\pi)^2 \approx 0.07$. Figure 6.14 shows that this is a good estimate. The real half-life becomes $t_{1/2} = t_0 \hat{t}_{1/2}$, which explains why t_0 is a characteristic time. The duration of the transients for conductive heat flow problems are typically $t' \sim t_0$, and the temperature will have reached a stationary state when $t \gg t_0$. The characteristic time t_0 is simple to calculate and it tells us right away how long we can expect transients to live. It follows that the temperature for times much larger than t_0 is close to the stationary solution. Stationary solutions are normally much simpler to obtain than the transient solution, and an inspection of t_0 therefore tells us if a transient temperature solution is needed.

The temperature solution with units is obtained directly from the definition of the dimensionless variables as

$$T(r, t) = T_0 \hat{T} (r/r_0, t/t_0) . \tag{6.136}$$

The usefulness of the dimensionless formulation is that it is a general formulation (regardless of the actual dimensions). We have noticed that the dimensionless temperature equation for conductive cooling is parameterless, which means that its solution is universal regardless of quantities like the radius of the sphere r_0 or the initial temperature T_0. The time constant becomes $t_0 = 1287$ Ga when using the radius of the Earth $r_0 = 6371$ km and the thermal diffusivity $\varrho_b c_b / \lambda = 10^{-6}$ m^2 s^{-1}. This is a very long time compared to the age 4.54 Ga of the Earth, and heat conduction alone cannot explain its cooling. We need mantle convection to do that.

Note 6.3 Temperature equation (6.131) can also be written as

$$\frac{\partial \hat{T}}{\partial \hat{t}} - \frac{2}{\hat{r}} \frac{\partial \hat{T}}{\partial \hat{r}} - \frac{\partial^2 \hat{T}}{\partial \hat{r}^2} = 0 \tag{6.137}$$

when the differentiations are carried out. The solution of temperature equation (6.137) becomes simplified by replacing $\hat{T}$ by $\hat{T} = u/\hat{r}$, and the equation for u is

$$\frac{\partial u}{\partial \hat{t}} - \frac{\partial^2 u}{\partial \hat{r}^2} = 0 \qquad (6.138)$$

which is the temperature equation for 1D conductive heat transport. The boundary condition for equation (6.138) at $\hat{r} = 1$ is $u(\hat{r}=0, \hat{t}) = 0$, but the other boundary condition $\partial \hat{T} / \partial \hat{r} = 0$ at $\hat{r} = 0$ is difficult to express in a simple way using u. The trick is now to avoid the boundary condition at $\hat{r} = 0$ by instead solving the equation for u through the entire sphere from $\hat{r} = -1$ to $\hat{r} = 1$. The boundary conditions are then $u = 0$ at both ends, ($\hat{r} = -1$ and $\hat{r} = 1$), and the initial condition $\hat{T}(\hat{r}, \hat{t}=0) = 1$ becomes $u(\hat{r}, \hat{t}=0) = \hat{r}$. Equation (6.138) is solved by the method called *separation of variables*, where u is written as the product of a function U of $\hat{t}$ and a function V of $\hat{r}$ as $u(\hat{r}, \hat{t}) = U(\hat{t}) \cdot V(\hat{r})$. When $U \cdot V$ is inserted into equation (6.138) we get

$$U'V - UV'' = 0 \qquad (6.139)$$

or

$$\frac{U'}{U} = \frac{V''}{V} = -k^2. \qquad (6.140)$$

Since U'/U is a function only of $\hat{t}$ and V''/V is a function of only $\hat{r}$ these two fractions must be equal to a constant. This constant is written $-k^2$, where it is anticipated that it has to be a negative number. The equation for U becomes $U' = -k^2 U$, which has the solution (see Exercise 6.5)

$$U(\hat{t}) = U_0 \exp(-k^2 \hat{t}). \qquad (6.141)$$

The equation for V is $V'' + k^2 V = 0$, which has as solution any linear combination of $\sin(k\hat{r})$ and $\cos(k\hat{r})$. The equation for V and the boundary conditions for u are fulfilled for all $V_n = \sin(n\pi\hat{r})$. There is not just one number k, but $k_n = n\pi$ for each integer n, and there is therefore a function U for each k_n as well, $U_n = U_0 \exp(-k_n^2 \hat{t})$. Temperature equation (6.138) for u is linear, and any linear combination of $U_n(\hat{t}) V_n(\hat{r})$ is therefore a solution too, and u can therefore be written as a Fourier series

$$u(\hat{r}, \hat{t}) = \sum_{n=1}^{\infty} a_n \sin(k_n \hat{r}) \exp(-k_n^2 \hat{t}). \qquad (6.142)$$

The initial condition $u(\hat{r}, t=0) = \hat{r}$ is needed for the remaining task of finding the Fourier coefficients a_n. It is straightforward to show that the functions $V_n(\hat{r}) = \sin(n\pi\hat{r})$ form an orthogonal basis with respect to the inner product $(V_n, V_m) = \int_{-1}^{1} V_n V_m \, d\hat{r}$. We therefore have that $(V_n, V_m) = \frac{1}{2}\delta_{nm}$. The Fourier coefficients are obtained from the initial condition $u(\hat{r}, \hat{t}=0) = \hat{r}$, which is

$$\sum_{n=1}^{\infty} a_n \sin(k_n \hat{r}) = \hat{r}. \qquad (6.143)$$

Multiplication of both sides by $\sin(n\pi\hat{r})$ followed by integration over $\hat{r}$ from -1 to 1 gives

$$a_n = \int_{-1}^{1} \hat{r} \sin(n\pi\hat{r}) \, d\hat{r} = \frac{2(-1)^{n+1}}{n\pi}. \qquad (6.144)$$

6.10 Heat flow and salt domes

Salt structures are found in a large number of sedimentary basins, for instance the Gulf of Mexico, the Persian Gulf and the North Sea (Hudec and Jackson, 2007). Large bodies of salt are important reasons for non-vertical heat flow for at least two reasons – salt has a heat conductivity that may be as much as six times as high as the heat conductivity of saturated clay and shales, and the salt bodies often have complex geometries. Table 2.4 gives the heat conductivity of salt as ~ 6 W m^{-1} K^{-1} at surface conditions, while shale may have a heat conductivity as low as ~ 1 W m^{-1} K^{-1} when the porosity is $\sim 40\%$. Although the heat conductivity of salt decreases with increasing temperature from 6 W m^{-1} K^{-1} at 5 °C to 4.1 W m^{-1} K^{-1} at 100 °C according to the data in Table 2.4, it is still considerably more than for shaly rocks. Salt begins as evaporites at the surface, which become buried as a sheet-like structure. It behaves as a fluid and it can be viewed as a fluid under lithostatic pressure. Complex salt structures may form as the salt is pressed up as domes and diapirs by the weight of the overlaying rocks (Hudec and Jackson, 2007, Gemmer *et al.*, 2004). Figure 6.15a shows an example of a simple salt diapir that has its roots in a sheet-like structure. We see that the diapir pushes the isotherms away. It becomes hotter above the diapir and cooler below the diapir, than at similar depths away from the diapir. It is possible to make a few general comments with regard to heat flow and salt structures. Firstly, we notice that the salt does not bend the isotherms where the salt is sheet-like. The isotherms become more separated because the salt has a higher heat conductivity compared to the surrounding shale, but heat flow is vertical. A salt diapir seems to affect its thermal environment a distance laterally that is roughly the same as its height. The heat flow conditions at the base of the specific case of Figure 6.15a are also affected by both the sandstone and the salt layer. The diapir alters the thermal conditions vertically above and underneath a distance that is also roughly the height of the diapir.

The hotter conditions above the diapir may lead to an increase in the surface heat flow, if the diapir is not too deeply buried. Figure 6.15b shows the surface heat flux for the diapir in Figure 6.15a. The surface heat flux illustrates the point that the diapir alters the isotherms laterally a distance that is similar to its height. There is a clear peak in the surface heat flux above the diapir, and the heat flux increases from 42 mW m^{-2} close to the left boundary to the maximum 63 mW m^{-2} above the center of the diapir. It is possible to make a simple assessment of this increase in the heat flow in terms of the average heat conductivities. This estimate is based on the observation that the heat flow is nearly vertical at the center of the diapir. The geotherm through the center of the diapir is compared with the geotherm close to the left boundary in Figure 6.16. The two geotherms are at x-coordinates $x_1 = 0.5$ km and $x_2 = 4.9$ km, respectively, in Figure 6.15. We notice from the figure 6.15a that the isotherm for 74 °C would have passed through the middle of the diapir nearly unbent. The depth of this isotherm is where the geotherms in Figure 6.16 cross. This depth and the corresponding temperature are $z_c = 1.7$ km and $T_c = 74$ °C, respectively. The heat flow along the two geotherms can be written as $q_i = \bar{\lambda}_i T_c / z_c$, where $\bar{\lambda}_i$ is the average heat

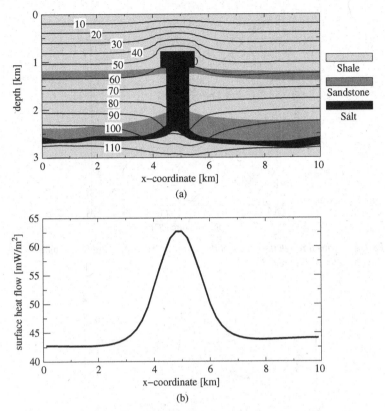

Figure 6.15. *(a) The temperature field around a salt diapir. (b) The surface heat flow from the salt diapir in (a).*

conductivity at position x_i from the depth z_c to the surface. The ratio of the heat fluxes becomes

$$\frac{q_2}{q_1} = \frac{\bar{\lambda}_2}{\bar{\lambda}_1}. \tag{6.145}$$

The maximum increase in the surface heat flow is therefore given by the ratio of the two average heat conductivities. We let the heat conductivities for shale, sandstone and salt be $1.1\ \mathrm{W\,m^{-1}\,K^{-1}}$, $2.5\ \mathrm{W\,m^{-1}\,K^{-1}}$ and $4.5\ \mathrm{W\,m^{-1}\,K^{-1}}$, respectively. These conductivities combined with the heights h_a, h_b and h_c from Figure 6.16 give the average heat conductivities $\bar{\lambda}_1 = 1.17\ \mathrm{W\,m^{-1}\,K^{-1}}$ and $\bar{\lambda}_2 = 1.85\ \mathrm{W\,m^{-1}\,K^{-1}}$, and the ratio $\bar{\lambda}_2/\bar{\lambda}_1 = 1.58$. This result is in agreement with the ratio of the corresponding heat fluxes from Figure 6.15b, which is $63/42 = 1.5$. The match is not exact because the heat flow is not strictly vertical at the center of the diapir. The heat flow is strictly vertical only along axes that are symmetric with respect to heat flow properties.

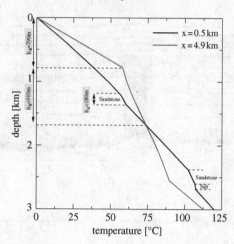

Figure 6.16. *Geotherms through the center of the diapir and at the left side of the model. The positions of the geotherms refer to Figure 6.15.*

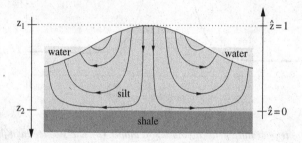

Figure 6.17. *Meteoric fluid flow in an island of permeable sediments. Notice that the dimensionless $\hat{z}$-axis points upwards.*

Hudec and Jackson (2007) presents the field salt tectonics in terms of observations and mechanisms. How salt structures alter the thermal field has been studied by several authors, for instance Lerche (1990b), Mello *et al.* (1995) and Petersen and Lerche (1995).

Exercise 6.14 Verify the average heat conductivities λ_1 and λ_2 using the data from Figure 6.16 and the heat conductivities for shale, sandstone and salt given above.

6.11 Forced convective heat transfer

Areas with sufficiently "strong" fluid flow have heat convected by the fluid in addition to transfer by conduction through the rock. This kind of heat transfer is called *forced heat convection*. The fluid flow is important for the temperature, but the temperature has a negligible feedback on the fluid flow. We will now look at forced heat convection by vertical fluid flow, for instance vertically into the ground from a high water table (see Figure 6.17), and we are looking for a criterion that can decide when the fluid flow is sufficiently strong to influence the temperature. The starting point is temperature equation (6.15), which combines conductive and convective heat flow. In the vertical direction the temperature equation (6.15) simplifies to

$$\varrho_b c_b \frac{\partial T}{\partial t} + \varrho_f c_f v_D \frac{\partial T}{\partial z} - \frac{\partial}{\partial z}\left(\lambda \frac{\partial T}{\partial z}\right) = 0, \tag{6.146}$$

where ϱ_b and ϱ_f are the average bulk density and the fluid density, respectively, v_D is the Darcy flux and λ is the heat conductivity. The second term (with the Darcy flux v_D) represents convective heat transport and the third term (with the heat conductivity λ) represents conductive heat flow. The rock is assumed to be at rest, ($v_s = 0$), and the material derivative in equation (6.15) is replaced by a (normal) partial derivative in equation (6.146). The relative importance of convective heat flow relative to conductive heat flow is best analyzed by making a dimensionless version of the equation. The temperature equation (6.146) is now rewritten as

$$\frac{l_0^2 c_b \varrho_b}{\lambda} \frac{\partial \hat{T}}{\partial t} + \frac{l_0 \varrho_f c_f v_D}{\lambda} \frac{\partial \hat{T}}{\partial \hat{z}} - \frac{\partial^2 \hat{T}}{\partial \hat{z}^2} = 0 \tag{6.147}$$

using the dimensionless temperature $\hat{T} = (T - T_1)/(T_2 - T_1)$ and dimensionless vertical position $\hat{z} = (z_2 - z)/(z_2 - z_1)$, where the length of the system is $l_0 = z_2 - z_1$. The temperature is scaled using the difference between the highest temperature T_2 at the deepest point along the aquifer and the surface temperature T_1. The vertical coordinate is scaled with the distance from the surface down to the low-permeable base (see Figure 6.17). Notice that a hat above the symbol denotes a dimensionless quantity, and that the $\hat{z}$-axis is pointing upwards. A positive Darcy flux is therefore upwards along the $\hat{z}$-axis. We see that both the dimensionless $\hat{T}$ and $\hat{z}$ variables are numbers between 0 and 1 (where $\hat{z} = 0$ at the base and $\hat{z} = 1$ at the top of the system). The coefficient of the time derivative is identified as the characteristic time for the process

$$t_0 = \frac{l_0^2 c_b \varrho_b}{\lambda}. \tag{6.148}$$

We will see in the next section that the temperature solution is close to a stationary solution for $t \gg t_0$, and t_0 therefore characterizes the time span of the thermal transients. The other coefficient (before the convective) term is the dimensionless Peclet number

$$\text{Pe} = \frac{l_0 \varrho_f c_f v_D}{\lambda}. \tag{6.149}$$

The dimensionless time is $\hat{t} = t/t_0$ and the dimensionless version of temperature equation (6.147) becomes

$$\frac{\partial \hat{T}}{\partial \hat{t}} + \text{Pe} \frac{\partial \hat{T}}{\partial \hat{z}} - \frac{\partial^2 \hat{T}}{\partial \hat{z}^2} = 0 \tag{6.150}$$

where the Pe-number is the only parameter left. The Pe-number measures convective heat transport relative to conductive heat transport, which we see by writing the Pe-number as follows:

$$\text{Pe} = \frac{\text{Convective heat transport}}{\text{Conductive heat transport}} = \frac{\varrho_f c_f v_D (T_2 - T_1)}{\lambda \, (T_2 - T_1)/l_0}. \tag{6.151}$$

The numerator is the heat flux from a Darcy flux v_D with a temperature difference $T_2 - T_1$. (Recall that v_D is a volume flux – it is the volume of fluid per time through a unit area.) The denominator is the heat flux resulting from the conductivity λ multiplied by the thermal gradient $(T_2 - T_1)/l_0$. The stationary (time independent) version of equation (6.150) is studied first:

$$\text{Pe} \frac{d\hat{T}}{d\hat{z}} - \frac{d^2 \hat{T}}{d\hat{z}^2} = 0 \tag{6.152}$$

where the boundary conditions in dimensionless form are $\hat{T}(\hat{z}{=}0) = 1$ and $\hat{T}(\hat{z}{=}1) = 0$. Equation (6.152) is straightforward to integrate two times and the stationary temperature solution is

$$\hat{T}(\hat{z}) = c_1 e^{\text{Pe}\,\hat{z}} + c_2. \tag{6.153}$$

This solution is also easily verified by insertion into the stationary equation (6.152). The two boundary conditions specify the two integration constants, and the stationary temperature solution becomes

$$\hat{T}(\hat{z}) = \frac{e^{\text{Pe}\,\hat{z}} - e^{\text{Pe}}}{1 - e^{\text{Pe}}}. \tag{6.154}$$

The (dimensionless) stationary temperature solution (6.154) is plotted in Figure 6.18 for $\text{Pe} = -10, -1, -0.1, 0, 0.1, 1$ and 10. The temperature regime $|\text{Pe}| \ll 1$ gives a geotherm that is nearly a straight line and it is therefore conduction-dominated. The other regime, $|\text{Pe}| \gg 1$, has a geotherm that departs from the conduction-dominated straight line because it is convection-dominated. Notice that the sign of the Pe-number reflects the direction of the Darcy flux. The Pe-number is positive for flow upwards and negative for flow downwards. The temperature at the base is brought towards the top for $\text{Pe} = 10$, and the temperature at the top is brought towards the base for $\text{Pe} = -10$. The regime $|\text{Pe}| \approx 1$ is where convection and conduction are roughly equally important. The scaling relations give the temperature solution with units

$$T(z) = T_1 + (T_2 - T_1) \, \hat{T} \left(\frac{z_2 - z}{z_2 - z_1} \right). \tag{6.155}$$

Deviations from a near linear trend in subsurface temperature data are often explained by groundwater flow. The Pe-number is useful because it tells us to what degree heat convection is important. It also gives an estimate of the vertical Darcy flux once it is estimated, for example by matching the curve (6.154) against temperature observations. Furthermore, the Darcy flux combined with pressure measurements allows for permeability estimations.

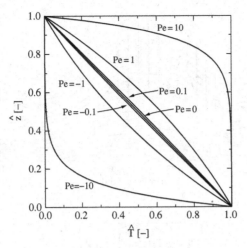

Figure 6.18. *The stationary temperature solution (6.154) for* Pe-*numbers* -10, -1, -0.1, 0, 0.1, 1 *and* 10.

Figure 6.19 shows an example of basin scale meteoric water flow on an island. (We will return to meteoric fluid flow in more detail in Chapter 14.) The fluid flow is driven by a water table that follows the ground surface, which is the part of the surface that is above sea level. The fluid flow is strongest vertically into the ground at the top of the island. Figure 6.19 shows two versions of the case, where the silt in panel (b) has a permeability that is a factor of 10 larger than in panel (a). The corresponding fluid flow also differs by an order of magnitude. The isotherms appear undisturbed by fluid flow in case (a), but they are noticeably depressed in case (b). The maximum vertical fluid flow at the top of the island is $7 \cdot 10^3$ m Ma^{-1} and $7 \cdot 10^4$ m Ma^{-1} in panels (a) and (b), respectively. The size of the system is 1000 m from the top of the island down to the low permeable base, and the thermal diffusivity is in both cases $\kappa = 3.6 \cdot 10^{-7}$ m^2 s^{-1}. The Pe-numbers are Pe $= 0.6$ and 6 for cases (a) and (b), respectively, which is consistent with the interpretation from the simple dimensionless model (6.154). The interpretation works fine although the vertical fluid flow in this example is not constant – it decreases almost linearly from the top towards the base. We could have used the average Darcy flux, which is half the maximum, and we would still have the same results in terms of the Pe-number. This example therefore shows that it is reasonable to expect the Pe-number to serve as an indicator for when fluid flow is important, even if it is based on the average Darcy flux. We notice that the temperature at the depth of the shale layer is different in cases (a) and (b). A constant heat flux is a better boundary condition than a constant temperature at the base. It is possible to make an analytical model that accounts for these two points, the linearly decreasing Darcy flux with depth and the constant heat flux at the base, as shown in Note 6.4. Such a 1D model can be calibrated against the geotherm through the top of the island as shown in Figure 6.20. Domenico and Palciauskas (1973) analyzed forced heat convection from regional groundwater flow

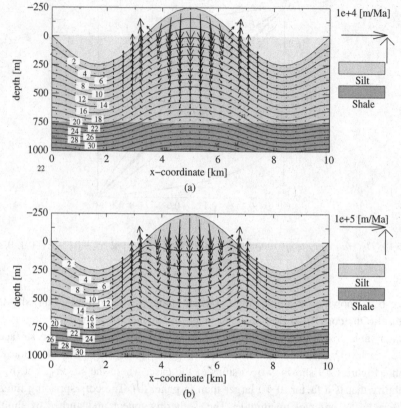

Figure 6.19. *Meteoric fluid flow and heat convection. (a) The fluid flow has a negligible impact on the temperature in this case. (b) The permeability of "Silt" is 10 times larger than in (a) and heat convection becomes noticeable.*

with an analytical model. Their model has a similar geometry as in the example shown in Figure 6.19.

The groundwater flow may be estimated indirectly from temperature measurements. Stallman (1963) discusses the problem of calculating the hydraulic head that drives the groundwater flow from temperature observations. Bredehoeft and Papadopolous (1965) appear to have been the first to apply the Pe-number in groundwater flow analysis based on temperature studies. The Pe-number has later been a reference point for a number of studies of groundwater flow (Ge, 1998, Manning and Igebritsen, 1999, Reiter, 2001, Ferguson *et al.*, 2003, Anderson, 2005, Lubis and Sakura, 2008, Verdoya *et al.*, 2008). The simple model above, which is restricted to vertical flow, becomes insufficient when horizontal flow is also important. For instance Reiter (2001) has worked with generalizations of simple 1D models in order to account for both vertical and horizontal flow.

Note 6.4 The temperature equation in the case of linearly decreasing fluid flow with depth becomes

$$v_0 \hat{z} \frac{d\hat{T}}{d\hat{z}} - \frac{d^2\hat{T}}{d\hat{z}^2} = 0 \qquad (6.156)$$

where the velocity term is $v_0\hat{z}$. The fluid flux is not constant but it decreases from v_0 at the top ($\hat{z} = 1$) towards zero at the base ($\hat{z} = 0$). Setting $f = d\hat{T}/d\hat{z}$ in the temperature equation (6.156) gives after integration that

$$\frac{d\hat{T}}{d\hat{z}} = c_1 \exp\left(\frac{1}{2}v_0\hat{z}^2\right) \qquad (6.157)$$

and a second time of integration gives

$$\hat{T}(\hat{z}) = c_1 \int_0^{\hat{z}} \exp\left(\frac{1}{2}v_0\mu^2\right) d\mu + c_2. \qquad (6.158)$$

The boundary conditions are $d\hat{T}/d\hat{z} = \hat{q}$ at $\hat{z} = 0$ and $\hat{T} = 0$ at $\hat{z} = 1$. The first boundary condition gives $c_1 = \hat{q}$ and the second boundary condition gives

$$c_2 = -\hat{q} \int_0^1 \exp\left(\frac{1}{2}v_0\mu^2\right) d\mu \qquad (6.159)$$

and the solution for temperature becomes

$$\hat{T}(\hat{z}) = -\hat{q} \int_{\hat{z}}^1 \exp\left(\frac{1}{2}v_0\mu^2\right) d\mu. \qquad (6.160)$$

It is possible to rewrite solution (6.160) for the case where $v_0 < 0$ using the error-function defined as

$$\text{erf}(u) = \frac{2}{\sqrt{\pi}} \int_0^u \exp(-\mu^2) d\mu. \qquad (6.161)$$

The velocity v_0 is written as $v_0 = -v_1$ (where v_1 is positive) and the solution becomes

$$\hat{T}(\hat{z}) = \hat{q}\sqrt{\frac{\pi}{2v_1}} \left(\text{erf}\left(\sqrt{\frac{v_1}{2}}\hat{z}\right) - \text{erf}\left(\sqrt{\frac{v_1}{2}}\right) \right). \qquad (6.162)$$

The scaled heat flux is $\hat{q} = -1$ in Figure 6.20 and the optimal values for v_1 are then $v_1 = 1.07$ and $v_1 = 4.5$. The parameter v_1 is the maximum of the linearly decreasing fluid flow term, and its optimal values are similar to the Pe-numbers obtained from the maximum vertical Darcy flow. Other numbers used in Figure 6.20 are $T_1 = 0\,°C$, $T_2 = 33\,°C$ and $z = 1000\hat{z} - 750$ in units of meters. The temperature solution with units is therefore

$$T(z) = T_1 + (T_2 - T_1)\,\hat{T}\left(\frac{z + 750}{1000}\right) \qquad (6.163)$$

which is the analytical temperature plotted in Figure 6.20. The optimal value for v_1 can be used to estimate the Darcy flux, and we get that $v_1 = 1.1$ gives the Darcy flux $v_D = 1 \cdot 10^4\ \text{m}\,\text{Ma}^{-1}$ and that $v_1 = 4.5$ gives $v_D = 4 \cdot 10^4\ \text{m}\,\text{Ma}^{-1}$, assuming that $l_0 = 1000\ \text{m}$ and $\kappa = 3 \cdot 10^{-7}\ \text{m}^2\,\text{s}^{-1}$.

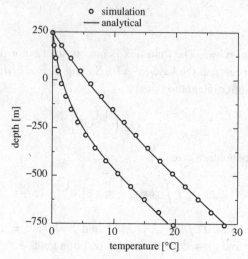

Figure 6.20. *The geotherms for cases (a) and (b) in Figure 6.19 are shown for the center of the island. See Note 6.4 for the 1D temperature solution.*

Exercise 6.15

(a) How large must the Darcy flux v_D be for heat convection to become noticeable (Pe $= 1$), when the other parameters are $\lambda = 2 \text{ W m}^{-1} \text{ K}^{-1}$, $l_0 = 300 \text{ m}$, $\varrho_f = 1000 \text{ kg m}^{-3}$, $c_f = 1000$ and $T_2 - T_1 = 50 \text{ °C}$?

(b) Let the fluid pressure in an aquifer be 0.5 MPa above the hydrostatic fluid pressure. What is the permeability of the rock above the aquifer?

Exercise 6.16

(a) Integrate the stationary temperature equation (6.152) two times and obtain the solution (6.153).

(b) Use the boundary conditions $\hat{T}(\hat{z}{=}0) = 1$ and $\hat{T}(\hat{z}{=}1) = 0$ to show that

$$c_1 = \frac{1}{1 - e^{\text{Pe}}} \quad \text{and} \quad c_2 = -\frac{e^{\text{Pe}}}{1 - e^{\text{Pe}}}. \tag{6.164}$$

(c) Show that the stationary solution of temperature equation (6.152) for Pe $= 0$ is $\hat{T}(\hat{z}) = 1 - \hat{z}$, using the boundary conditions from (b).

(d) The stationary temperature solution (6.154) is not defined for Pe $= 0$, but show that the solution $\hat{T}(\hat{z}) \to 1 - \hat{z}$ when Pe $\to 0$. Use that $e^x \approx 1 + x$ for small x.

Exercise 6.17

This exercise builds on note 6.4, which treats the case of a vertical fluid flux that is increasing linearly with depth. We will now look at a constant temperature $\hat{T}(\hat{z} = 0) = 1$ at the base of the model instead of a constant heat flux. Show that the temperature solution (6.158) then becomes

$$\hat{T}(\hat{z}) = 1 - \frac{\mathrm{erf}(\sqrt{v_1/2}\,\hat{z})}{\mathrm{erf}(\sqrt{v_1/2})} \tag{6.165}$$

for downwards flow.

Exercise 6.18 Show that the temperature equation (6.150) is linear. (Hint: show that if T_1 and T_2 are solutions of the temperature equation then so also is any linear combination $c_1 T_1 + c_2 T_2$.)

6.12 Transient convective heat flow

Solving the time-dependent temperature equation (6.150) gives how fast the temperature will reach a stationary state. We need an initial condition before we can do that. It is simply $\hat{T}(\hat{z}, \hat{t}{=}0) = 1 - \hat{z}$ in this case, which is the stationary temperature solution in the absence of heat convection. Notice that the initial condition has to respect the boundary conditions, $\hat{T}(\hat{z}{=}0, \hat{t}) = 1$ and $\hat{T}(\hat{z}{=}1, \hat{t}) = 0$. The temperature solution is now written as a sum of two parts

$$\hat{T}(\hat{z}, \hat{t}) = \hat{T}_{\mathrm{stat}}(\hat{z}) + \hat{T}_{\mathrm{trans}}(\hat{z}, \hat{t}) \tag{6.166}$$

where the first part, $\hat{T}_{\mathrm{stat}}(\hat{z})$, is the stationary temperature solution (6.154), and the second part $\hat{T}_{\mathrm{trans}}(\hat{z}, \hat{t})$ is the transient. The temperature solution approaches the stationary part (time-independent part) for $\hat{t} \to \infty$, when the transient decays to zero. The transient part at $\hat{t} = 0$ has to be

$$\hat{T}_{\mathrm{trans}}(\hat{z}, \hat{t} = 0) = 1 - \hat{z} - \hat{T}_{\mathrm{stat}}(\hat{z}), \tag{6.167}$$

in order for the temperature solution to fulfill the initial condition. The transient solution becomes the Fourier series (as shown in Note 6.5)

$$\hat{T}_{\mathrm{trans}}(\hat{z}, \hat{t}) = \exp\left(\frac{1}{2}\mathrm{Pe}\hat{z}\right) \sum_{n=1}^{\infty} a_n \exp\left(-k_n \hat{t}\right) \sin(n\pi\hat{z}), \tag{6.168}$$

where each term decays to zero with the exponential factor $\exp(-k_n \hat{t})$, and where

$$k_n = n^2 \pi^2 + \frac{1}{4}\mathrm{Pe}^2. \tag{6.169}$$

The initial condition (6.167) gives the Fourier coefficients

$$a_n = \frac{2n\pi\,\mathrm{Pe}}{k_n^2}\left((-1)^n e^{\mathrm{Pe}/2} - 1\right). \tag{6.170}$$

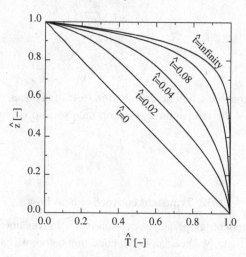

Figure 6.21. *The transient temperature solution (6.154) for* Pe $= 10$ *as a function of time.*

Each term in the Fourier series (6.168) decays to zero with a half-life $\hat{t}_{1/2} = \ln(2)/k_n$, and the longest half-life is for the term $n = 1$. The longest half-life serves as an estimate for the half-life of the (complete) temperature transient, and it is approximated as

$$\hat{t}_{1/2} = \frac{\ln 2}{\pi^2 + \text{Pe}^2/4} \approx \begin{cases} \frac{\ln 2}{\pi^2}, & \text{Pe} \ll 1 \\ \frac{4\ln 2}{\text{Pe}^2}, & \text{Pe} \gg 1. \end{cases} \tag{6.171}$$

The half-life (6.171) is almost independent of the Pe-number for the conduction-dominated regime, Pe $\ll 1$. It decreases with increasing Pe-number, and in the convection-dominated regime, Pe $\gg 1$, we see that transients decrease as $\sim 1/\text{Pe}^2$.

The full temperature solution (6.166) is plotted in Figure 6.21 for the time steps $\hat{t} = 0.0$, 0.02, 0.04, 0.08 and ∞, when Pe $= 10$. The difference between the initial temperature and the stationary temperature is almost zero for the conduction-dominated regime Pe $\ll 1$. Figure 6.21 shows that the transient temperature solution is roughly halfway between the initial condition and the stationary solution for $\hat{t} = 0.04$, which is in accordance with $\hat{t}_{1/2} \approx 4\ln 2/\text{Pe}^2 \approx 0.03$ in this case (with Pe $= 10$).

Note 6.5 The transient part of the temperature solution for equation (6.150) is obtained by separation of variables. The transient temperature is then written as a product of two functions

$$\hat{T}_{\text{trans}}(\hat{z}, \hat{t}) = U(\hat{t})V(\hat{z}), \tag{6.172}$$

where U is only a function of $\hat{t}$ and V is only a function of $\hat{z}$. When the product UV is inserted into the temperature equation (6.150) we get

$$U'V + \text{Pe}\,UV' - UV'' = 0, \tag{6.173}$$

where the prime denotes differentiation. This expression is rewritten as

$$\frac{U'}{U} = -\text{Pe}\,\frac{V'}{V} + \frac{V''}{V}, \tag{6.174}$$

where the left-hand side of the equality is dependent only on $\hat{t}$ and the right-hand side is dependent only on $\hat{z}$. The equality must therefore be equal to a constant, which is written $-k$. The minus sign before k anticipates that it is a positive number. The equation for the factor U becomes

$$\frac{U'}{U} = -k, \tag{6.175}$$

and the solution for U is (see Exercise 6.5)

$$U(\hat{t}) = U_0 e^{-k\hat{t}}, \tag{6.176}$$

where U_0 is an arbitrary constant. We cannot allow $U(\hat{t})$ to grow exponentially with time, which explains why the constant k has to be a positive number. The equation for V is

$$V'' - \text{Pe}\,V' + kV = 0, \tag{6.177}$$

and it is factorized in the following way:

$$\left(\frac{d}{d\hat{z}} - r_1\right)\left(\frac{d}{d\hat{z}} - r_2\right)V = 0 \tag{6.178}$$

where $r_{1,2}$ are the roots

$$r_{1,2} = \frac{1}{2}\left(\text{Pe} \pm \sqrt{\text{Pe}^2 - 4k}\right) \tag{6.179}$$

of the quadratic equation $r^2 - \text{Pe}\,r + k = 0$. We see that functions of the form $\exp(r_i\hat{z})$, $i = 1, 2$, are solutions of equation (6.178). The function V can therefore be written as any linear combination of these two solutions

$$V(\hat{z}) = c_1 e^{r_1\hat{z}} + c_2 e^{r_2\hat{z}}. \tag{6.180}$$

The boundary conditions for the transient solution require that $V(\hat{z}{=}0) = 0$ and $V(\hat{z}{=}1) = 0$, which is the same as $c_1 + c_2 = 0$ and $c_1 \exp(-D) + c_2 \exp(D) = 0$, respectively, where $D = \sqrt{(\text{Pe}/2)^2 - k}$. The boundary conditions therefore require that $c_2 = -c_1$ and that $\exp(2D) = 1$, where the latter equation has the non-trivial solution $D = i\pi n$, with $i = \sqrt{-1}$, and where n is any integer. From $D^2 = -(\pi n)^2$ it follows that $k_n = (\text{Pe}/2)^2 + (\pi n)^2$ and that $r_{1,2} = \frac{1}{2}\text{Pe} \pm i\pi n$. There is not just one k, but one for each integer n. Since D is a complex number it follows from the formula for the exponentiation of a complex number, $\exp(i\phi) = \cos\phi + i\sin\phi$, where ϕ is real that

$$V_n(\hat{z}) = c_3 \exp(\text{Pe}\,\hat{z}/2)\sin(\pi n\hat{z}), \tag{6.181}$$

where c_3 is an arbitrary constant and where the subscript n tells us that there is one V for each integer n. (Only the complex part $i\sin\phi$ fulfills the boundary conditions. Only this part is therefore kept, but the factor i is dropped.) There is also a U_n associated with each n because there is one k_n for each n. Any linear combination of the products $U_n(\hat{t})V_n(\hat{z})$ solves the transient equation with the given boundary conditions, because the temperature equation is linear. The transient temperature can therefore be written as the Fourier series

$$\hat{T}_{\text{trans}}(\hat{z}, \hat{t}) = \exp(\text{Pe}\,\hat{z}/2) \sum_{n=1}^{\infty} a_n \exp(-k_n \hat{t}) \sin(\pi n \hat{z}) \qquad (6.182)$$

where the Fourier coefficients a_n have to be obtained from the initial condition. The Fourier series (6.168) for $\hat{t} = 0$ must therefore be equal to the initial condition (6.167), which implies that

$$\exp(\text{Pe}\,\hat{z}/2) \sum_{n=1}^{\infty} a_n \sin(\pi n \hat{z}) = 1 - \hat{z} - \frac{\exp(\text{Pe}\,\hat{z}) - \exp(\text{Pe})}{1 - \exp(\text{Pe})} \qquad (6.183)$$

or when rewritten

$$\sum_{n=1}^{\infty} a_n \sin(\pi n \hat{z}) = (1 - \hat{z}) \exp(-\text{Pe}\,\hat{z}/2) + \frac{\sinh(\text{Pe}(1 - \hat{z})/2)}{\sinh(\text{Pe}/2)}. \qquad (6.184)$$

The sine functions are orthogonal with respect to the inner product defined by integration of $\hat{z}$ from 0 to 1:

$$\int_0^1 \sin(\pi n \hat{z}) \sin(\pi m \hat{z}) \, d\hat{z} = \begin{cases} 0, & n \neq m \\ \frac{1}{2} & n = m. \end{cases} \qquad (6.185)$$

The Fourier coefficients a_n are therefore obtained by multiplication of equation (6.184) with $\sin(\pi n \hat{z})$ followed by an integration of $\hat{z}$ from 0 to 1. The following definite integrals simplify the remaining task of finding the coefficients a_n:

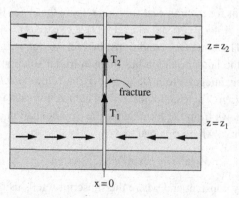

Figure 6.22. *A fracture connecting two aquifers.*

$$\int_0^1 \frac{\sinh(\text{Pe}(1 - \hat{z})/2)}{\sinh(\text{Pe}/2)} \sin(\pi n \hat{z})\, d\hat{z} = -\frac{n\pi}{(n\pi)^2 + (\text{Pe}/2)^2} \tag{6.186}$$

$$\int_0^1 (1 - \hat{z}) \exp(-\text{Pe}\,\hat{z}/2) \sin(\pi n \hat{z})\, d\hat{z}$$

$$= \frac{n\pi \left((n\pi)^2 - \text{Pe} + (\text{Pe}/2)^2 + (-1)^n \text{Pe}\exp(\text{Pe}/2)\right)}{\left((n\pi)^2 + (\text{Pe}/2)^2\right)^2}. \tag{6.187}$$

These integrals can be obtained after some work, or more easily with the help of tables of integrals.

6.13 Heat flow in fractures

Figure 6.22 shows a fracture connecting two aquifers. The fracture drains fluid from the lowest aquifer, which is discharged into the highest aquifer. This situation is quite similar to the one in Figure 6.17, except that the fluid now flows upwards. Another difference is that the vertical fluid flow, which took place over a wide area in Figure 6.17, is now assumed to be confined to a narrow fracture. We saw in Section 6.11 that fluid flow could increase the temperature beyond the conductive temperature solution. The parameter that controls the amount of convective heat transfer was shown in Section 6.12 to be the Peclet number. The regime defined by Pe $\ll$ 1 is conduction-dominated, while the other regime Pe $\gg$ 1 is convection-dominated.

Heat transfer in and around the rectangular fracture in Figure 6.22 differs from the 1D vertical heat flow in Figure 6.17 by being a 2D problem. The conductive heat flow in the rock surrounding the fracture has a horizontal component when the fracture is heated above

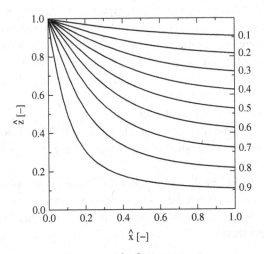

Figure 6.23. *The stationary temperature around a fracture.*

the conduction-dominated level. This turns out to imply that the two regimes of conductive and convective heat flow are controlled by a different Peclet number than the one defined in Section 6.11. We can see this by looking at the energy balance for a small element dz along the fracture. The energy transported into the element minus the energy transported out of the element during a time step dt is

$$\Delta E_1 = (c_f \varrho_f T v_0 w\, dt)_z - (c_f \varrho_f T v_0 w\, dt)_{z+dz} \approx -c_f \varrho_f v_0 w \frac{\partial T}{\partial z} dz\, dt \qquad (6.188)$$

where c_f is the heat capacity of the fluid, ϱ_f is the fluid density, v_0 is the average fluid velocity in the fracture and w is the width of the fracture. The energy ΔE_1 has to be equal to the energy ΔE_2 lost through the vertical sides,

$$\Delta E_2 = 2\lambda \frac{\partial T}{\partial x} dz\, dt \qquad (6.189)$$

where λ is the heat conductivity and the factor 2 accounts for the two sides of the fracture. Energy conservation then requires that $\Delta E_1 = \Delta E_2$, which implies that

$$\mathrm{Pe}_f \frac{\partial T}{\partial z} = 2\frac{\partial T}{\partial x} \qquad (6.190)$$

where

$$\mathrm{Pe}_f = \frac{c_f \varrho_f v_0 w}{\lambda} \qquad (6.191)$$

is the Peclet number for heat transport in the fracture. The Pe_f-number defines two regimes, one for conduction- and another for convection-dominated heat transport, in the same way as the Pe-number from Section 6.11. We can see this by looking at an approximate temperature solution for the two cases. The conduction-dominated regime has $\hat{T} \approx 1 - \hat{z}$, and therefore $\partial \hat{T}/\partial \hat{x} \approx 0$ and $\partial \hat{T}/\partial \hat{z} \approx 1$, and from equation (6.190) we get that Pe $\ll 1$.

In the convection-dominated regime we can approximate the temperature along most of the fracture by the temperature at the inlet, $\hat{T} \approx 1$, and we therefore have that $\partial \hat{T}/\partial \hat{z} \approx 0$ for most of the fracture. We will see below that $\partial \hat{T}/\partial \hat{x} \approx 1$ for the stationary temperature field around a fracture when the temperature is $\hat{T} = 1$ along the fracture. The relation (6.190) therefore requires that $\mathrm{Pe}_f \gg 1$ for the convection-dominated case.

The difference between the Pe-number for 1D heat flow and the Pe_f-number for 2D heat flow around fractures is the characteristic length.

The length of the system is the characteristic length for 1D heat flow, but for 2D heat flow around the fracture it is the width of the fracture that enters the Peclet number as a characteristic length.

The temperature solution in the regime $\mathrm{Pe}_f \gg 1$ can be approximated by the constant temperature of the inlet along the fracture. The surroundings of the fracture become heated to this temperature. Solving the stationary temperature equation in 2D in the case of constant heat conductivity

$$\nabla^2 T = 0 \qquad (6.192)$$

gives the maximum impact on the surroundings of the fracture. It is sufficient to solve the Laplace equation along the positive x-axis ($x \geq 0$) and in the interval $z_1 < z < z_2$ because of symmetry (see Figure 6.22). The boundary condition along the fracture is $T(x{=}0, z) = T_2$, and the boundary conditions along the horizontal sides are $T(x, z = z_1) = T_1$ and $T(x, z = z_2) = T_2$ for $x \geq 0$.

A dimensionless version of the Laplace equation is made by scaling x and z with the characteristic length of the system, $l_0 = z_2 - z_1$, and temperature is scaled with the temperature difference $T_2 - T_1$. The dimensionless variables are then

$$\hat{x} = \frac{x}{l_0}, \quad \hat{z} = \frac{z_2 - z}{l_0} \quad \text{and} \quad \hat{T} = \frac{T - T_1}{T_2 - T_1}. \tag{6.193}$$

The boundary conditions in dimensionless form are $\hat{T}(\hat{x}{=}0, \hat{z}) = 1$, $\hat{T}(\hat{x}, \hat{z} = 0) = 1$ and $\hat{T}(\hat{x}, \hat{z} = 1) = 0$. Note 6.6 shows that the solution of the dimensionless Laplace equation with these boundary conditions is

$$\hat{T}(\hat{x}, \hat{z}) = 1 - \hat{z} + \sum_{n=1}^{\infty} a_n \exp(-n\pi\hat{x}) \sin(n\pi\hat{z}), \tag{6.194}$$

where the Fourier coefficients a_n are

$$a_n = -\frac{2(-1)^n}{n\pi}. \tag{6.195}$$

The first part of the solution (6.194), $1 - \hat{z}$, represents conductive heat flow, and the Fourier series is therefore the thermal impact of the "hot" vertical fracture. The solution (6.194) is plotted in Figure 6.23, where we see how isotherms bend up along the fracture. We also see that the isotherms are almost horizontal a distance $x = l_0$ (or $\hat{x} = 1$) away from the fracture. The observation that the temperature solution is $\hat{T}(\hat{x}, \hat{z}) \approx 1 - \hat{z}$ for $\hat{x} \gg 1$ follows from the exponential decay of each Fourier component with $\hat{x}$.

The characteristic time (6.148) associated with the length l_0

$$t_0 = \frac{l_0^2 c_b \varrho_b}{\lambda} \tag{6.196}$$

estimates the time span needed for the fracture to heat up its surroundings. The system will be close to the stationary state when $t \gg t_0$, which is when the stationary solution (6.194) applies. As an example we see that t_0 becomes 3000 years for $l_0 = 100\,\text{M}$ and the thermal diffusivity $\lambda/\varrho_b c_b = 10^{-6} \text{M}^2 \text{s}^{-1}$.

Note 6.6 The solution (6.194) for the Laplace equation with the boundary conditions $\hat{T}(\hat{x}{=}0, \hat{z}) = 1$, $\hat{T}(\hat{x}, \hat{z} = 0) = 1$ and $\hat{T}(\hat{x}, \hat{z} = 1) = 0$ is written as the sum of two parts $\hat{T}(\hat{x}, \hat{z}) = 1 - \hat{z} + \hat{T}_b$ (where both parts are solutions of the Laplace equation). The first part, $1 - \hat{z}$, is the temperature solution for heat conduction between the lower horizontal boundary at $\hat{T} = 1$ and the upper horizontal boundary at $\hat{T} = 0$. The second part, $\hat{T}_b(\hat{x}, \hat{z})$, is the temperature perturbation by the fracture, and we see that the boundary conditions for $\hat{T}_b(\hat{x}, \hat{z})$ are $\hat{T}_b(\hat{x}{=}0, \hat{z}) = \hat{z}$, $\hat{T}_b(\hat{x}, \hat{z} = 0) = 0$ and $\hat{T}_b(\hat{x}, \hat{z} = 1) = 0$.

The Laplace equation is solved for $\hat{T}_b$ by separation of variables, where $\hat{T}_b$ is written as $\hat{T}_b(\hat{x}, \hat{z}) = U(\hat{x}) \, V(\hat{z})$. When the product $U \, V$ is inserted into the Laplace equation we get

$$\frac{U''}{U} = -\frac{V''}{V} = k^2 \qquad (6.197)$$

where k^2 is a constant. We have that U''/U is a function only of $\hat{x}$ and that V''/V is a function only of $\hat{z}$, and because these two expressions are equal then they must be equal to a constant. The equation for U is $U'' - k^2 U = 0$, which has as solution any linear combination of $U(\hat{x}) = \exp(-k\hat{x})$ and $U(\hat{x}) = \exp(k\hat{x})$. Since we cannot allow the solution to grow exponentially with increasing $\hat{x}$ the solution has to be $U_0 \exp(-k\hat{x})$ where U_0 is a constant. The equation for V is $V'' + k^2 V = 0$ which has solutions any linear combination of $\sin(k\hat{z})$ and $\cos(k\hat{z})$. The boundary conditions $\hat{T}_b = 0$ along the upper and lower horizontal boundaries are fulfilled for every $\sin(k\hat{z})$ where $k = n\pi$. There is therefore one $k_n = n\pi$ for each (positive) integer n, and there is also a U_n and a V_n for each (positive) n. Each product $U_n(\hat{x}) V_n(\hat{z})$ is a solution of the Laplace equation for $\hat{T}_b$, and the solution is the Fourier series

$$\hat{T}_b(\hat{x}, \hat{z}) = \sum_{n=1}^{\infty} a_n \exp(-k_n \hat{x}) \, \sin(k_n \hat{z}). \qquad (6.198)$$

The boundary condition $\hat{T}_b(\hat{x}{=}0, \hat{z}) = \hat{z}$, which becomes

$$\sum_{n=1}^{\infty} a_n \sin(k_n \hat{z}) = \hat{z} \qquad (6.199)$$

leads to the Fourier coefficients a_n. Using that V_n are orthogonal with respect to the inner product $(V_n, V_m) = \int_0^1 V_n V_m \, d\hat{z} = \frac{1}{2}\delta_{mn}$, we have

$$\frac{1}{2} a_n = \frac{1}{2} \int_0^1 \hat{z} \sin(n\pi\hat{z}) \, d\hat{z} = -\frac{(-1)^n}{n\pi}, \qquad (6.200)$$

which gives the Fourier coefficients.

Exercise 6.19 Use the data given in Exercise 6.15.
(a) Estimate the time needed for the fracture to heat up its surroundings.
(b) Assume that a fracture in impermeable rock drains the same amount of fluid as in Exercise 6.15. Show that the width of the area with vertical fluid flow in Exercise 6.15 has to be $W = w^3/12k$, when the permeability of the rock in Exercise 6.15 is k, and the width of the fracture is w. Hint: give that the permeability of the fracture is $w^2/12$.
(c) What is the width W when $w = 0.1$ mm?

6.14 Instantaneous heating or cooling of semi-infinite half-space

The change in temperature of surface rocks caused by lava flows (flood basalt) or glacier flows can be calculated by taking the temperature change at the surface to be instantaneous and the crust to be infinitely deep. It also turns out that the temperature solution for the instantaneous heating or cooling of the semi-infinite half-space is a useful starting point for other similar problems, like cooling sills or dikes, and the thermal structure of the lithosphere.

The semi-infinite half-space has a surface at $z = 0$ and it covers the entire z-axis below the surface. The temperature of the half-space is initially at T_0 (for $t < 0$), but at $t = 0$ the surface temperature is suddenly changed to T_s. The general temperature equation (6.15) simplifies in this case to

$$\frac{\partial T}{\partial t} - \kappa \frac{\partial^2 T}{\partial z^2} = 0 \tag{6.201}$$

where $\kappa = \lambda/(c_b \varrho_b)$ is the thermal diffusivity. The boundary conditions are the surface temperature $T(z{=}0, t) = T_s$ and the temperature $T = T_0$ at infinite depth (for $z \to \infty$). Note 6.7 shows that the solution of the temperature equation becomes

$$T(z, t) = T_0 + (T_s - T_0) \, \text{erfc} \left(\frac{z}{2\sqrt{\kappa t}} \right), \tag{6.202}$$

where $\text{erfc}(\eta)$ is a special function called the *complementary error function*. It is defined as

$$\text{erfc}(\eta) = 1 - \text{erf}(\eta) \tag{6.203}$$

in terms of the *error-function* erf, which is given by the integral

$$\text{erf}(\eta) = \frac{2}{\sqrt{\pi}} \int_0^\eta e^{-x^2} dx. \tag{6.204}$$

The error function and the complementary error function are shown in Figure 6.24, see also Exercise 6.22. The temperature solution could also have been written

$$T(z, t) = T_s + (T_0 - T_s) \, \text{erf} \left(\frac{z}{2\sqrt{\kappa t}} \right) \tag{6.205}$$

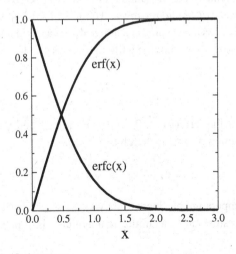

Figure 6.24. *The erf and erfc functions.*

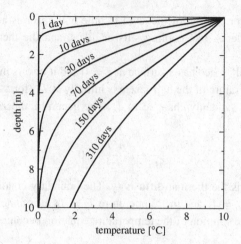

Figure 6.25. *The heating of the ground after the surface temperature has suddenly increased by* 10 °C. *The thermal diffusivity of the ground is* $\kappa = 10^{-6}\,m^2\,s^{-1}$.

which is seen by inserting the definition erfc $= 1 -$ erf into equation (6.202). The temperature solution is plotted in Figure 6.25 for heating 10 m into the ground after the surface temperature has increased by 10 °C.

The initial temperature of the subsurface is the constant T_0, which is a sufficiently good approximation for shallow depths. A typical thermal gradient could be 35 °C/km, which implies that the temperature increase over 10 m is only 0.35 °C. However, we could have added the stationary temperature solution $T(z) = Az + T_0$ to the transient solution (6.202) where A is the thermal gradient, because the temperature equation is linear. Any linear combination of temperature solutions that together fulfill the boundary conditions is also a solution.

The depth into the ground where the temperature has changed by 10% of the difference $T_s - T_0$ is called the *thermal boundary layer*. The use of 10% is arbitrary – it is just a suitable number that serves as a measure for a noticeable change in the ground temperature. The temperature for heating of the infinite half-space gives

$$\frac{T(z,t) - T_0}{T_s - T_0} = \mathrm{erfc}(\eta) = 0.1, \tag{6.206}$$

where Figure 6.24 gives that erfc$(\eta) = 0.1$ for $\eta = 1.16$. The equation for the thermal boundary layer as a function of time is therefore

$$z_T(t) = 2.32\sqrt{\kappa t}. \tag{6.207}$$

The depth where the temperature has increased by 10% of $T_s - T_0$ (the difference between the new surface temperature and the initial temperature) is proportional to the square root of t.

Fourier's law gives that the surface heat flow is

$$q = -\lambda \frac{\partial T}{\partial z}\bigg|_{z=0}$$

$$= -\lambda(T_s - T_0) \frac{\partial}{\partial \eta}(\mathrm{erfc}(\eta))_{\eta=0} \frac{\partial \eta}{\partial z}$$

$$= \lambda(T_s - T_0)\left(\frac{2}{\sqrt{\pi}}e^{-\eta^2}\right)_{\eta=0}\frac{1}{2\sqrt{\kappa t}}$$

$$= \frac{\lambda(T_s - T_0)}{\sqrt{\pi \kappa t}}. \tag{6.208}$$

The differentiation of the temperature solution (6.202) with respect to z simplifies by use of the chain rule. The surface heat flow becomes infinite at $t = 0$ and it then decreases as $1/\sqrt{t}$. The heat flow can be used to compute the energy E that has passed into the ground through an area A as a function of time. We have

$$E(t) = A \int_0^t q(t')dt' = A\lambda(T_s - T_0)\sqrt{\frac{t}{\pi \kappa}} \tag{6.209}$$

which shows that the energy transferred to the ground increases as $\sqrt{t}$. (Recall that heat flow has units of energy per unit area and unit time.)

The temperature solution (6.202) gives the temperature of the subsurface at time t in response to a step-increase in the surface temperature at $t = 0$. We can also consider the step rise in temperature as an event that happened a time t ago. We know that the temperature equation (6.201) is linear and we can therefore add step functions for changes in the surface temperature. If for instance the temperature increased by a step T_2 at time t_2 and then decreased with the step $-T_2$ at a later time t_1 the temperature into the ground is

$$T(z) = T_2\left[\mathrm{erfc}\left(\frac{z}{2\sqrt{\kappa t_2}}\right) - \mathrm{erfc}\left(\frac{z}{2\sqrt{\kappa t_1}}\right)\right]. \tag{6.210}$$

This is the temperature transient of the subsurface from a rectangular pulse in the surface temperature from t_2 to t_1 with the size T_2. We can continue in this way, by adding pulses, and then generate the thermal response of the subsurface to a piecewise constant surface temperature. The temperature becomes

$$T(z) = \sum_{n=1}^{N} T_n\left[\mathrm{erfc}\left(\frac{z}{2\sqrt{\kappa t_n}}\right) - \mathrm{erfc}\left(\frac{z}{2\sqrt{\kappa t_{n-1}}}\right)\right] \tag{6.211}$$

when the temperature is T_n in the interval from t_n to t_{n-1}. The surface temperature is first changed a time t_N ago and it is then changed at the steps t_n until the present time $t_0 = 0$. Expression (6.211) has been used in the reconstruction of the past surface temperature from borehole temperature measurements. Observations of the temperature at a large number of vertical positions along a borehole makes it possible to obtain the optimal piecewise surface temperature as demonstrated by Beltrami *et al.* (1997) and Beltrami (2001).

Note 6.7 It is normally a good idea to make the temperature equation dimensionless by scaling time and distance by characteristic quantities before it is solved. But, there is not obvious length scale for an infinitely deep ground. We have already seen that the characteristic time t_0 is related to the characteristic length l_0 by $t_0 = l_0^2/\kappa$ (or alternatively $l_0 = \sqrt{\kappa t_0}$). It now turns out that the temperature for the semi-infinite half-space can be expressed as a function of the dimensionless ratio

$$\eta = \frac{z}{2\sqrt{\kappa t}} \tag{6.212}$$

where it is anticipated that a factor $1/2$ simplifies the matter. The dimensionless temperature as a function of η is

$$\frac{T(z,t) - T_0}{T_s - T_0} = \Theta(\eta) \quad \text{or} \quad T(z,t) = T_0 + (T_s - T_0)\Theta(\eta) \tag{6.213}$$

where the function Θ measures the heating of the half-space from the initial temperature T_0 to the temperature T_s using the unit interval (the interval from 0 to 1). The boundary conditions for heating of the semi-infinite half-space can therefore be written

$$\Theta(\eta = 0) = 1 \quad \text{and} \quad \Theta(\eta \to \infty) = 0 \tag{6.214}$$

because $z = 0$ implies $\eta = 0$ and $z = \infty$ implies $\eta = \infty$. Differentiation of $T(z,t)$ (6.213) leads to

$$\frac{\partial T}{\partial t} = -\Theta' \cdot \left(\frac{z}{2\sqrt{\kappa t}}\right) \cdot \left(\frac{1}{2t}\right) \tag{6.215}$$

$$\frac{\partial T}{\partial z} = \Theta' \cdot \left(\frac{1}{2\sqrt{\kappa t}}\right) \tag{6.216}$$

$$\frac{\partial^2 T}{\partial z^2} = \Theta'' \cdot \left(\frac{1}{4\kappa t}\right). \tag{6.217}$$

Insertion of these parts into the temperature equation (6.201) yields

$$\Theta''(\eta) = -2\eta\,\Theta'(\eta). \tag{6.218}$$

Letting $f = \Theta'$ gives

$$\frac{df}{f} = -2\eta\,d\eta \tag{6.219}$$

which is integrated to

$$\Theta(\eta)' = c_1 \exp(-\eta^2) \tag{6.220}$$

where c_1 is an integration constant. One more integration then yields the temperature solution

$$\Theta(\eta) = c_1 \int_0^\eta e^{-x^2}\,dx + c_2 \tag{6.221}$$

where c_2 is the second integration constant. The first boundary condition, $\Theta(\eta=0) = 1$, implies that $c_2 = 1$. The second boundary condition, $\Theta(\eta \to \infty) = 0$, then implies that

$$c_1 \int_0^\infty e^{-x^2} dx = -1. \tag{6.222}$$

Tables of definite integrals give that $\int_0^\infty e^{-x^2} dx = \sqrt{\pi}/2$ and therefore $c_1 = -2/\sqrt{\pi}$, which shows that $\Theta(\eta) = 1 - \text{erf}(\eta) = \text{erfc}(\eta)$ is the solution (6.202). The variable η is called a *similarity* variable, and it reduces the partial differential equation into an ordinary differential equation, which in this case is straightforward to integrate.

Cooling of the semi-infinite half-space leads to the same solution in a slightly different manner. We have that $T_s < T_0$ and the dimensionless temperature Θ is therefore

$$\frac{T(z,t) - T_s}{T_0 - T_s} = \Theta(\eta) \quad \text{or} \quad T(z,t) = T_0 + (T_s - T_0)\Theta(\eta), \tag{6.223}$$

because we want the dimensionless temperature Θ to be a number from 0 to 1. The boundary conditions for cooling of the semi-infinite half-space are

$$\Theta(\eta = 0) = 0 \quad \text{and} \quad \Theta(\eta \to \infty) = 1. \tag{6.224}$$

The expression for $\Theta(\eta)$ becomes the same as in equation (6.221), where the first of the two boundary conditions in (6.224) leads to $c_2 = 0$. The second of these boundary conditions leads to $c_1 = 2/\sqrt{\pi}$, and the function Θ is therefore $\Theta(\eta) = \text{erf}(\eta)$, which yields the solution (6.205). This solution is as we have already seen equal to the solution (6.202).

Exercise 6.20 The surface temperature is suddenly increased by $10\,^\circ$C. How much time is needed for the temperature to increase by $1\,^\circ$C at 0.1 m, 1 m and 10 m?

Exercise 6.21 The surface temperature is increased from the initial temperature T_0 to $T_s = T_0 + \Delta T$. Find the depth where the temperature has increased by $\Delta T/2$ as a function of time. Hint: $\text{erfc}(0.4769) = 0.5$.

Exercise 6.22 Use definition (6.203) of the complementary error function and definition (6.204) of the error function and show that

$$\text{erfc}(\eta) = \frac{2}{\sqrt{\pi}} \int_\eta^\infty e^{-x^2} dx. \tag{6.225}$$

6.15 Cooling sills and dikes

A sill is a tabular igneous intrusion that is oriented parallel to the planar structure of the surrounding rock, and a dike is a near-vertical planar igneous intrusion that cuts through the bedding of the country rock, see Figure 6.26. These intrusions are assumed to be emplaced instantaneously with a high temperature ($\sim 1000\,^\circ$C). The thermal impact on the country rock from such intrusions is found by solving the temperature equation (6.201) with

Heat flow

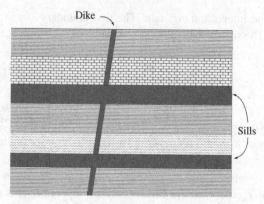

Dike

Sills

Figure 6.26. *Sills are tabular igneous intrusions that are fed by near-vertical dikes.*

boundary conditions and an initial condition that represents the hot intrusion. The difference between the temperature equation for sills and dikes is that the temperature of a sill depends on the z-coordinate and the temperature of a dike depends on the x-coordinate. We will assume that the country rock has initially the temperature $T = 0\,°C$ and that the sill or the dike has a width $2a$ and an initial temperature T_0. The thermal diffusivity (κ) of the intrusion and the country rock do not differ much and can therefore be assumed to be the same.

Sills and dikes have half the thickness a as a characteristic length, and the sill/dike temperature at emplacement T_0 as a characteristic temperature. We can scale the length by a and the temperature by T_0 to make a dimensionless version of the time-dependent temperature equation (6.226). For sills we get the following dimensionless temperature equation:

$$\frac{\partial \hat{T}}{\partial \hat{t}} - \frac{\partial^2 \hat{T}}{\partial \hat{z}^2} = 0 \tag{6.226}$$

where $\hat{T} = T/T_0$ and $\hat{z} = z/a$, and where t is scaled with the characteristic time $t_0 = a^2/\kappa$. We will let $\hat{z} = 0$ be the center of the sill, which means that the sill extends from $\hat{z} = -1$ to $\hat{z} = 1$. Boundary conditions for the temperature equation are

$$\hat{T}(\hat{z}{=}{-}\infty, \hat{t}) = 0 \quad \text{and} \quad \hat{T}(\hat{z}{=}\infty, \hat{t}) = 0 \tag{6.227}$$

which say that there is no thermal impact on the country rock "far" away from the sill. The initial condition is the temperature $\hat{T} = 1$ inside the sill and $\hat{T} = 0$ outside the sill. The solution of the temperature equation (6.226) with these boundary conditions and the initial condition is

$$\hat{T}(\hat{z}, \hat{t}) = \frac{1}{2}\left(\operatorname{erf}\left(\frac{\hat{z}+1}{2\sqrt{\hat{t}}}\right) - \operatorname{erf}\left(\frac{\hat{z}-1}{2\sqrt{\hat{t}}}\right)\right), \tag{6.228}$$

as explained in Note 6.8. The solution is plotted in Figure 6.27, which shows how the sill cools. For $\hat{t} = 1$ the temperature has decreased to $\hat{T} \approx 0.5$ at the center of the sill, and for

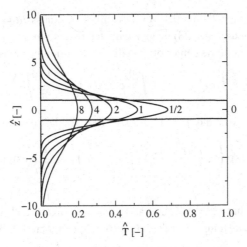

Figure 6.27. *The temperature of a cooling sill at* $\hat{t} = 0, 1/2, 1, 2, 4$ *and* 8.

$\hat{t} = 4$ the temperature has decreased further to $\hat{T} \approx 0.25$. The decrease in temperature at the center of the sill is then as follows for a longer time span:

$\hat{t}$	$\hat{T}$
0	1
1	$5.2 \cdot 10^{-1}$
10	$1.8 \cdot 10^{-1}$
100	$5.6 \cdot 10^{-2}$
1000	$1.8 \cdot 10^{-2}$
10^4	$5.6 \cdot 10^{-3}$
10^5	$1.8 \cdot 10^{-3}$
10^6	$5.6 \cdot 10^{-4}$

We see that we have to wait more than $1000 \times t_0$ before the dimensionless temperature falls to less than 2% at the center of the sill. A sill emplaced with the temperature $T_0 = 1000\,°C$ has then the temperature $18\,°C$. The table below shows how long time t_0 is for sills of different thicknesses:

Sill thickness	a	t_0
1 m	1/2 m	2.9 days
10 m	5 m	289 days
100 m	50 m	79 years
1000 m	500 m	7927 years

The time span $1000 \times t_0$ is then ~ 0.1 Ma for a sill of thickness $2a = 100$ m.

The temperature solution (6.228) is approximated by a simpler expression for distances far away from the sill or at the center of the sill. The solution becomes

$$\hat{T}(\hat{z}, \hat{t}) = \frac{1}{\sqrt{\pi}} \left(\int_0^{\eta_1} e^{-x^2} dx - \int_0^{\eta_2} e^{-x^2} dx \right) = \frac{1}{\sqrt{\pi}} \int_{\eta_1}^{\eta_2} e^{-x^2} dx \qquad (6.229)$$

where $\eta_1 = (\hat{z} - 1)/(2\sqrt{\hat{t}})$ and $\eta_2 = (\hat{z} + 1)/(2\sqrt{\hat{t}})$. The last integral is approximated as

$$\hat{T}(\hat{z}, \hat{t}) \approx \frac{1}{\sqrt{\pi}} \exp\left(-\left(\frac{\eta_1 + \eta_2}{2} \right)^2 \right) (\eta_1 - \eta_2) = \frac{1}{\sqrt{\pi \hat{t}}} \exp\left(-\frac{\hat{z}^2}{4\hat{t}} \right) \qquad (6.230)$$

for $\hat{z} \gg 1$.

At $\hat{z} = 0$ we have that $\eta_1 = 1/(2\sqrt{t})$ and $\eta_2 = -1/(2\sqrt{t})$, and for $\hat{t} \gg 1$ we get that $\eta_1 \ll 1$ and $\eta_2 \ll 1$. The temperature solution (6.229) can therefore be approximated as

$$\hat{T}(\hat{z}, \hat{t}) \approx \frac{1}{\sqrt{\pi}} \int_{\eta_1}^{\eta_2} dx = \frac{1}{\sqrt{\pi \hat{t}}} \qquad (6.231)$$

for $\hat{t} \gg 1$, because the integrand becomes $e^{-x^2} \approx 1$ for $x \approx 0$.

The sill will be felt as an increase in the temperature followed by a decrease in the temperature at a position $\hat{z}$ outside the sill. Solving the equation $\partial \hat{T}/\partial \hat{t} = 0$ for t gives the time when the temperature impact is at its maximum at a position $\hat{z}$. A straightforward differentiation of temperature equation (6.229) yields

$$\frac{\partial \hat{T}}{\partial \hat{t}} = -\frac{1}{4\pi} \left[e^{-\eta_1^2}(\hat{z} + 1) - e^{-\eta_2^2}(\hat{z} - 1) \right] \hat{t}^{-3/2} = 0 \qquad (6.232)$$

which implies that

$$e^{-\eta_1^2}(\hat{z} + 1) = e^{-\eta_2^2}(\hat{z} - 1). \qquad (6.233)$$

Further simplification leads to

$$\exp\left(\frac{\hat{z}}{\hat{t}} \right) = \frac{\hat{z} + 1}{\hat{z} - 1} \qquad (6.234)$$

or

$$\hat{t}_{\text{max}} = \frac{\hat{z}}{\ln\left(\frac{\hat{z}+1}{\hat{z}-1} \right)}. \qquad (6.235)$$

Equation (6.235) for $\hat{t}_{\text{max}}$ is approximated by the simple expression

$$\hat{t}_{\text{max}} \approx \frac{\hat{z}^2}{2} \qquad (6.236)$$

for $\hat{z} \gg 1$ (see Exercise 6.26). Inserting $\hat{t}_{\text{max}}$ into the temperature solution (6.228) gives the temperature maximum $\hat{T}_{\text{max}}$. Using the approximation (6.236) $\hat{t}_{\text{max}}$ gives

$$\hat{T}_{\text{max}} \approx \sqrt{\frac{2}{\pi e} \frac{1}{\hat{z}}}. \qquad (6.237)$$

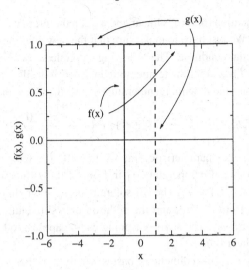

Figure 6.28. *The functions* $f(x)$ *and* $g(x)$ *are plotted. Notice how the initial condition for a sill is fulfilled for the linear combination* $\frac{1}{2}(f(x) + g(x))$.

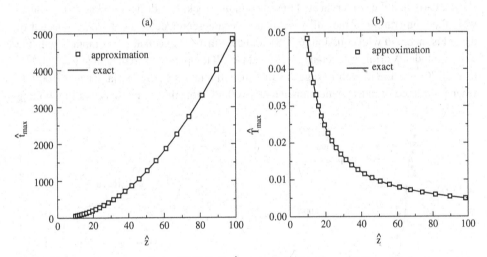

Figure 6.29. *(a) The exact solution (6.235) for* $\hat{t}_{max}$ *is plotted as the solid line, and the approximation (6.236) to* $\hat{t}_{max}$ *is plotted with the square markers. (b) The solid line is the exact temperature solution (6.228) plotted at time* $\hat{t}_{max}$ *given by (6.235). The square markers are the approximation (6.237) to* $\hat{T}_{max}$.

Exercise 6.27 shows how this approximation is derived. The exact time of temperature maximum $\hat{t}_{max}$ is plotted in Figure 6.29a, and the exact maximum temperature $\hat{T}_{max}$ is plotted in Figure 6.29b. We see that approximations (6.236) and (6.237) of $\hat{t}_{max}$ and $\hat{T}_{max}$, respectively, are accurate.

The temperature equation (6.226) is linear and we can therefore add together solutions to make a new solution. We just have to make sure that the new solution fulfills the boundary conditions and the initial condition. The following example shows how a sill solution can be added to a solution $T_n(z, t)$ for the temperature in absence of sills. If a sill of thickness a_1 intrudes the depth z_1 at time t_1 with a temperature T_1 we then have that the temperature is

$$T(z, t) = T_n(z, t) + \left[T_1 - T_n(z_1, t_1)\right] \hat{T} \left(\frac{(z - z_1)}{a}, \frac{(t - t_1)}{t_0}\right) \qquad (6.238)$$

for $t \geq t_1$, where t_0 is the characteristic time for the sill. The temperature $T_n(z_1, t_1)$ has to be subtracted from the sill temperature T_1 in front of the $\hat{T}$-function in order to fulfill the initial condition $T(z_1, t_1) = T_1$. The sill-solution has zero (or almost zero) temperature a long distance away from the sill, and the addition of a sill solution does not alter any boundary condition far away. The same reasoning can be applied to handle several sills at the same time or at a later time.

Sills sometimes show up in sedimentary basins where their thermal impact is recorded by vitrinite reflectance in the neighboring sediments. Figure 6.30a shows an example of a sill with the thickness 92 m that intruded into the sediments 54 Ma before the present time. The sill was later buried by 1.2 km of sediments. The vitrinite observations in Figure 6.30a show clearly that the sill has heated the surrounding rock. It is difficult to know in advance what the temperature of the sill was when it became emplaced. The thermal transient of the sill is modeled with equation (6.238), and the vitrinite reflectance is computed with the easy-ro model. Vitrinite reflectance is plotted for sill temperatures in the range from 500 °C to 1250 °C, and the modeling suggests that the sill had a temperature larger than 1000 °C when it intruded. It must be mentioned that a 1000 °C hot sill may cause fracturing, boiling

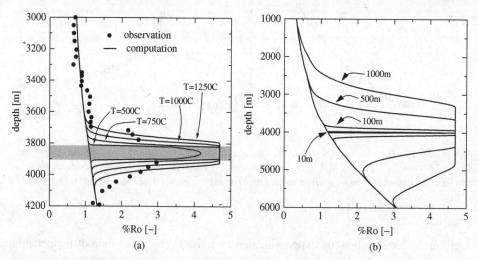

Figure 6.30. *(a) Vitrinite observations along a well that goes through a 92 m thick sill. The sill is also modeled with different initial temperatures. (b) The vitrinite above and below a sill is altered in a distance that is roughly the sill thickness. The initial sill temperature is 1000 °C.*

and thermal convection in its surrounding water saturated sediments when it is pressed into the rock. This could be an explanation for the mismatch between the modeled and the observed vitrinite reflectance. A model for sill cooling by conduction is therefore best suited for dry rocks.

Figure 6.30a shows that the distance away from the sill with increased vitrinite reflectance is roughly the thickness of the sill. This observation holds over several orders of magnitude as seen from Figure 6.30b.

Note 6.8 A simple way to derive the temperature solution (6.228) is to begin with the error function solution $\mathrm{erf}(\hat{z}/(2\sqrt{\hat{t}}))$. We have already seen in Section 6.14 that it is a solution of the temperature equation (6.226). A translation of this error function solution at a distance ± 1 is also a solution. We will now define the following two functions:

$$f(\hat{z}) = \mathrm{erf}\left(\frac{\hat{z}+1}{2\sqrt{\hat{t}}}\right) \quad \text{and} \quad g(\hat{z}) = -\mathrm{erf}\left(\frac{\hat{z}-1}{2\sqrt{\hat{t}}}\right) \tag{6.239}$$

for $t = 0^+$. The argument of the error functions are either $-\infty$ or ∞, and the functions $f(\hat{z})$ and $g(\hat{z})$ become a step function from -1 to 1 and from 1 to -1, respectively. The two step functions are both 1 in the $\hat{z}$-interval from -1 to 1. The linear combination $(1/2)(f(\hat{z}) + g(\hat{z}))$ will therefore fulfill the initial condition. See Figure 6.28 where $f(\hat{z})$ and $g(\hat{z})$ are plotted. The temperature solution (6.228) is therefore the wanted solution. The method of adding together linear combinations of error-function-solutions, where the sum at $t = 0^+$ fulfills both the initial condition and the boundary conditions, is a simple and powerful way to solve the temperature equation.

Exercise 6.23
(a) How much time is needed for the temperature at the center of a 2 m, 20 m and 200 m thick sill to be reduced by half? Use $\kappa = 10^{-6}$ m^2 s^{-1}.
(b) How much time is needed for the temperature at the center of a 2 m, 20 m and 200 m thick sill to be reduced to 5%?

Exercise 6.24 How far away from the sill will the maximum temperature increase by 10%, 5% and 2.5% of the initial sill temperature?

Exercise 6.25
(a) At what time will the temperature maximum take place at a distance 1000 m away from a 2 m, 20 m and 200 m thick sill? Use $\kappa = 10^{-6}$ m^2 s^{-1}.
(b) What is the temperature at the maximum in (a)?

Exercise 6.26 Derive approximation (6.236).
Solution: We have

$$\frac{\hat{z}+1}{\hat{z}-1} = \frac{1+\frac{1}{\hat{z}}}{1-\frac{1}{\hat{z}}} \approx \left(1+\frac{1}{\hat{z}}\right)\left(1+\frac{1}{\hat{z}}\right) \approx 1+\frac{2}{\hat{z}} \tag{6.240}$$

to first order in $1/\hat{z}$ when $\hat{z} \gg 1$. The log function can also be approximated to first order, $\ln(1 + x) \approx x$, for $x \approx 0$. We then get that

$$\ln\left(\frac{\hat{z} + 1}{\hat{z} - 1}\right) \approx \frac{2}{\hat{z}} \tag{6.241}$$

when $\hat{z} \gg 1$.

Exercise 6.27 Derive approximation (6.237).
Solution: The temperature at a given time and position is given by the integral (6.229), where the integration limits are $\eta_1 = (\hat{z} + 1)/(2\sqrt{\hat{t}})$ and $\eta_2 = (\hat{z} - 1)/(2\sqrt{\hat{t}})$. Using that $\hat{t}_{max} \approx \hat{z}^2/2$ for $\hat{z} \gg 1$ gives

$$\eta_1 = \frac{1}{\sqrt{2}}\left(1 + \frac{1}{z}\right) \quad \text{and} \quad \eta_2 = \frac{1}{\sqrt{2}}\left(1 - \frac{1}{z}\right). \tag{6.242}$$

The integral (6.229) can then be approximated as

$$\hat{T}_{max} = \frac{1}{\sqrt{\pi}} e^{-1/2} \frac{2}{\sqrt{2}\hat{z}} \tag{6.243}$$

which is the approximation (6.237).

Exercise 6.28 Show that the solution (6.229) for the sill temperature can be approximated by

$$\hat{T}(\hat{z}, \hat{t}) = \frac{1}{\sqrt{\pi\hat{t}}}\left(1 - \frac{3\hat{z} + 1}{12\hat{t}}\right) \tag{6.244}$$

when $|\hat{z}| \ll \hat{t}$. Hint: use that $e^{-x^2} \approx 1 - x^2$ for $x \approx 0$.

Exercise 6.29 Show that the following temperature solution gives the cooling of a sill emplaced at a finite depth h with zero temperature at the surface,

$$\hat{T}(\hat{z}, \hat{t}) = F(\hat{z}, \hat{t}) - F(2h - \hat{z}, \hat{t}) \tag{6.245}$$

where

$$F(\hat{z}, \hat{t}) = \frac{1}{2}\left(\text{erf}\left(\frac{(\hat{z} + 1)}{2\sqrt{\hat{t}}}\right) - \text{erf}\left(\frac{(\hat{z} - 1)}{2\sqrt{\hat{t}}}\right)\right). \tag{6.246}$$

Notice that the center of the sill is at $\hat{z} = 0$ and the surface is at $\hat{z} = h$. The solution is shown in Figure 6.31.
Hint: show that the solution has the symmetry $\hat{T}(h + l, \hat{t}) = -\hat{T}(h - l, \hat{t})$ around $z = h$. The symmetry implies that $\hat{T} = 0$ at $\hat{z} = h$ for all $\hat{t}$, which assures the boundary condition at the surface.

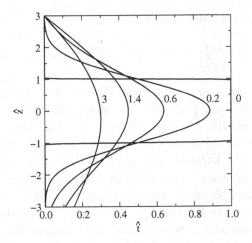

Figure 6.31. *The cooling of a sill at a finite depth $h = 3$ for $\hat{t} = 0, 0.2, 0.6, 1.4$ and 3.0.*

6.16 Solidification and latent heat of fusion

When liquid magma is cooled to its melting temperature it starts to solidify. A change of phase from liquid to solid takes place at the melting temperature and heat is released. The heat released by the phase change is called the latent heat of fusion. We will now study the solidification of thick magmatic layers from above like flood basalts and lava lakes. There is also solidification of the magma from below, but as long as there is melted rock in between the two solidified regions they can be treated separately. Solidification from below is treated in the next section. Solidification from above is the same as the freezing of water on a lake. Such problems involving the growth of a solidified region are named Stefan problems after the German physicist who first calculated the thickness of freezing ice on water through time. The problem with solidification, like the freezing of a lake, is that there is no characteristic length scale of the problem. The frozen (solid) part grows with time starting from zero thickness, and it is not characterized by any fixed length scale. This suggests that the temperature solutions for the infinite half-space might apply. The heat released by the solidification is transported by conduction through the solid part. The temperature equation (6.201) for heat conduction has to be solved in the solid upper part for a growing solid area. Boundary conditions are the surface temperature (T_s) and the temperature at the base of the solid part, the melting temperature, (T_m). A challenge with transient heat flow during solidification is that the solution has to account for the growth of the solidified part. It turns out that the error function solution applies to this problem:

$$T(z, t) = T_s + (T_m - T_s)\frac{\text{erf}(\eta)}{\text{erf}(\eta_m)} \quad \text{with} \quad \eta = \frac{z}{2\sqrt{\kappa t}}, \qquad (6.247)$$

where the solidification boundary $z_m(t)$ is given by

$$\frac{z_m(t)}{2\sqrt{\kappa t}} = \eta_m \quad \text{or} \quad z_m(t) = 2\eta_m\sqrt{\kappa t}. \tag{6.248}$$

We already know that the error function solution solves the temperature equation, and it is straightforward to verify that the boundary conditions are fulfilled. However, it remains to be shown that it is possible to find a (constant) number η_m, which controls the growth of the solidified layer. We will see that energy conservation at the base of the solid layer gives the number η_m. The latent heat of fusion (L) is the energy released per unit mass being solidified, and the energy released per unit volume during solidification becomes ϱL. The energy released per unit area when the depth interval dz_m is frozen is $\varrho L\, dz_m$. This energy has to be transported away by conduction in the time interval dt of freezing. Energy conservation requires that the energy released by solidification is exactly the energy transported away by heat conduction

$$\varrho L\frac{\partial z_m}{\partial t} = \lambda\left.\frac{\partial T}{\partial z}\right|_{\eta_m} \tag{6.249}$$

where λ is the heat conductivity of the solid part. The left-hand side of boundary condition (6.249) is obtained from equation (6.248) for the liquid/solid boundary

$$\varrho L\frac{\partial z_m}{\partial t} = \varrho L\eta_m\sqrt{\frac{\kappa}{t}} \tag{6.250}$$

and the right-hand side is obtained from the temperature solution (6.247)

$$\lambda\frac{\partial T}{\partial z} = \lambda(T_m - T_s)\frac{e^{-\eta_m^2}}{\text{erf}(\eta_m)}\frac{1}{\sqrt{\pi\kappa t}}. \tag{6.251}$$

When the parts (6.250) and (6.251) are inserted into the boundary condition (6.249) we get

$$\frac{e^{-\eta_m^2}}{\eta_m\text{erf}(\eta_m)} = \frac{\sqrt{\pi}L}{c(T_m - T_0)}, \tag{6.252}$$

which is the equation for η_m. Notice that the thermal diffusivity is $\kappa = \lambda/(\varrho c)$, where c is the specific heat capacity. It is not possible to solve equation (6.252) exactly, but it is simple to estimate η_m by using the plot of the function $f(x) = e^{-x^2}/(x\,\text{erf}(x))$ in Figure 6.32.

It is interesting to know what the number η_m might be. Typical parameters for magma are $L = 400\ \text{kJ kg}^{-1}$, $T_m - T_s = 1000\,°\text{C}$, and $c = 1.0\ \text{kJ kg}^{-1}\text{K}^{-1}$, which makes the function $f(\eta_m) = 0.7$. Figure 6.32 gives that $\eta_m \approx 0.85$. Figure 6.33 shows an example of the temperature solution (6.247) through the solidified magma using these parameter values and that $T_s = 0\,°\text{C}$. In the case of freezing water typical parameters are $T_s = -10\,°\text{C}$, $L = 320\ \text{kJ kg}^{-1}$ and $c = 4\ \text{kJ kg}^{-1}\text{K}^{-1}$, which makes $f(\eta_m) = 14$. From Figure 6.32 we then get that $\eta_m = 0.25$.

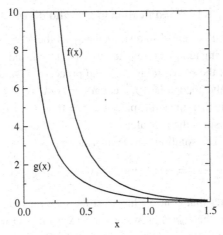

Figure 6.32. $f(x) = e^{-x^2}/(x\,\mathrm{erf}(x))$ and $g(x) = e^{-x^2}/(x(1+\mathrm{erf}(x)))$.

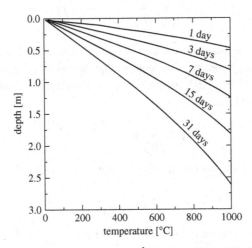

Figure 6.33. *The temperature in solidified magma at $\hat{t} = 1, 3, 7, 15$ and 31 days.*

Exercise 6.30 How thick is the layer of solidified magma after 1 day, 10 days and 100 days? Assume that magma has $L = 400$ kJ kg^{-1}, $c = 1$ kJ kg^{-1} K^{-1} and a temperature difference $T_m - T_0 = 1000$ K, where T_m is the melting temperature and T_0 is initial temperature of the surrounding rock.

Exercise 6.31 How much time is needed for the ice on a lake to become 1 m thick when the surface temperature is $-10\,^{\circ}$C? Use that $L = 320$ kJ kg^{-1} and $c = 4$ kJ kg^{-1} K^{-1} for water.

6.17 Solidification of sills and dikes

The solidification of a sill is treated in a similar way as the freezing of a lake. The sill is assumed to be at the melting temperature when it is emplaced, and solidification then starts at the boundaries against the surrounding rock and proceeds towards the center of the sill. We have once more to solve the temperature equation for the solid part that is now growing into the sill. The top surface of the sill is placed at $z = 0$ with the z-axis pointing upwards. The solidified part of the sill is then the area $z_m < z < 0$, where z_m is the interface between liquid and solid magma. The solution for the interface z_m is

$$\frac{z_m}{2\sqrt{\kappa t}} = -\eta_m \quad \text{or} \quad z_m(t) = -2\eta_m \sqrt{\kappa t} \tag{6.253}$$

where η_m is the parameter that controls how fast the magma solidifies. Notice the minus sign on η_m which tells us that $z_m(t)$ grows downwards from $z = 0$. (The parameter η_m is a positive number.) Boundary conditions for the temperature equation are the temperature T_0 far away from the sill ($z = \infty$) and $T = T_m$, the melting temperature at $z = z_m$. The temperature solution is then once again an erf-function solution

$$T(z, t) = T_0 + (T_m - T_0)\frac{\text{erfc}(z/2\sqrt{\kappa t})}{\text{erfc}(-\eta_m)}, \tag{6.254}$$

where the variable η_m is obtained from energy conservation at the interface between liquid and solid magma. Energy conservation requires that the rate of energy released by solidification is exactly the rate of energy transported away by conduction, which is written in the same way as in the previous section:

$$\varrho L \frac{\partial z_m}{\partial t} = \lambda \frac{\partial T}{\partial z}\bigg|_{\eta_m}. \tag{6.255}$$

The velocity of the solid/liquid interface (6.253) becomes

$$\frac{\partial z_m}{\partial t} = -\eta_m \sqrt{\frac{\kappa}{t}} \tag{6.256}$$

and the gradient of the temperature solution (6.254) is

$$\frac{\partial T}{\partial z} = -(T_m - T_0)\frac{e^{-\eta_m^2}}{(1 + \text{erf}(\eta_m))}\frac{1}{\sqrt{\pi \kappa t}}. \tag{6.257}$$

This assumes that the complementary error function is $\text{erfc}(-x) = 1 - \text{erf}(-x)$, and that $\text{erf}(-x) = -\text{erf}(x)$. When expressions for $\partial z_m/\partial t$ and $\partial T/\partial z$ are inserted into the boundary condition (6.255) we get an equation for the parameter η_m (where both z_m and t drop out)

$$\frac{e^{-\eta_m^2}}{\eta_m(1 + \text{erf}(\eta_m))} = \frac{\sqrt{\pi} L}{(T_m - T_0)c}. \tag{6.258}$$

This equation is most easily solved graphically using a plot of the function $g(x) = e^{-x^2}/(x(1 + \text{erf}(x)))$, see Figure 6.32. The time it takes for the sill to solidify is obtained directly from equation (6.253). If the sill has the thickness $2a$ then we have that the sill is solidified when $z_m(t_s) = -a$, which gives

$$t_s = \frac{a^2}{4\kappa\eta_m^2}. \tag{6.259}$$

A sill has the characteristic length scale, a, and the characteristic time scale $t_0 = a^2/\kappa$. The dimensionless time when solidification is complete is therefore

$$\hat{t}_s = \frac{1}{4\eta_m^2}. \tag{6.260}$$

Both the temperature solution and the position of the liquid/solid interface can be written in a dimensionless form. The dimensionless temperature solution is

$$\hat{T} = \frac{T - T_0}{T_m - T_0} = \frac{\text{erfc}(\hat{z}/2\sqrt{\hat{t}})}{\text{erfc}(-\eta_m)} \tag{6.261}$$

where the dimensionless version of z_m is

$$\hat{z}_m(\hat{t}) = -2\eta_m\sqrt{\hat{t}}. \tag{6.262}$$

Figure 6.34a shows a plot of solution (6.261) at different time steps during solidification. One should notice that the solutions (6.254) and (6.261) for solidification apply only as long as there is molten magma. We must also remember that the solution represents a magma that solidifies at a temperature T_m, and not over a temperature interval. Figure 6.34b shows the solidification of a 100 m thick sill and also the subsequent cooling. The sill has

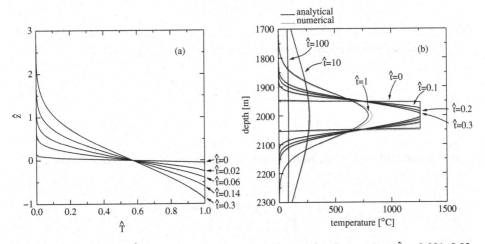

Figure 6.34. *(a) The temperature in a sill during solidification of magma at times* $\hat{t} = 0.001, 0.02,$ *0.06, 0.14 and 0.3, when the parameter* $\eta_m = 0.8.$ *(The center of the sill is at* $\hat{z} = -1.$*) Equation (6.260) gives that the sill is solidified at* $\hat{t}_s = 0.39.$ *(b) The solidification of a 100 m thick sill and the subsequent cooling.*

a latent heat $L = 350\ \mathrm{kJ\,kg^{-1}}$, a heat capacity $c = 1.3\ \mathrm{kJ\,kg^{-1}\,K^{-1}}$ and a melting temperature $T_m = 1261\,°\mathrm{C}$. The background temperature is $T_0 = 0\,°\mathrm{C}$, and the η_m-parameter becomes 0.8. The thermal diffusivity of both the rock and the melt is $\kappa = 1 \cdot 10^{-6}\ \mathrm{m^2\,s^{-1}}$. The characteristic time for the process becomes $t_0 = 79.2$ years, and the sill has solidified after $t_s = 0.39\,t_0 = 31$ years. The cooling of the sill after solidification can be modeled with the temperature solution (6.228). But we have to make an adjustment of the initial sill temperature with respect to the energy liberated by solidification. The energy released when a volume V solidifies is $E_m = \varrho L\,V$, and the energy stored in the same volume due to the temperature difference $T_m - T_0$ is $E = \varrho\,c\,(T_m - T_0)V$. The total energy $E' = E_m + E$ can be represented by an increase in the temperature from T_m to $T'_m = T_m + L/c$. Using the temperature T'_m with the $\hat{T}$-function (6.228) the temperature solution for cooling of the sill becomes $T(z,t) = T'_m\,\hat{T}(z/a, t/t_0)$. This is not the exact temperature solution for the cooling of a sill after it has solidified, but it is a good approximation. Figure 6.34b shows a comparison of this solution and a numerical solution for solidification and cooling. It turns out that the approximation is accurate, except for a short time interval right after solidification, where there is a slight mismatch (see $\hat{t} = 1$). The numerical solution is based on an "effective heat capacity" method (see Note 6.9), where the sill solidifies over a $25\,°\mathrm{C}$ temperature interval from $1236\,°\mathrm{C}$ to $1261\,°\mathrm{C}$.

Note 6.9 *Effective heat capacity.* Solidification of a volume V liberates the energy $E = \varrho L V$, where L is the latent heat of fusion. (The term specific latent heat can also be used for L because it is energy per mass.) If the solidification takes place over a temperature interval from T_1 to T_2 the energy liberated, when the temperature decreases by ΔT, is

$$\Delta E = \frac{\varrho L V}{T_2 - T_1}\,\Delta T. \tag{6.263}$$

The temperatures T_1 and T_2 are the solidus and liquidus, respectively. The rate at which energy is released can be added as a source term in the energy equation

$$\varrho c\,\frac{\partial T}{\partial t} - \lambda\,\frac{\partial^2 T}{\partial z^2} = -\frac{\varrho L}{T_2 - T_1}\,\frac{\partial T}{\partial t} \tag{6.264}$$

where the minus sign is needed before the source term in order to make it positive. ($\partial T/\partial t$ is negative when the temperature is decreasing.) The heat source term can be brought over to the left-hand side, where the heat capacity is replaced by effective heat capacity

$$c_{\mathrm{eff}}(T) = \begin{cases} c, & T \le T_1 \\ c + \frac{L}{T_2 - T_1}, & T_1 < T < T_2 \\ c, & T_2 \le T. \end{cases} \tag{6.265}$$

Water goes from liquid to solid at $0\,°\mathrm{C}$, but most rocks solidify over a temperature interval that may be several hundred degrees. The effective heat capacity method is not suited for water, unless the melting temperature can be approximated by a temperature interval.

6.18 Periodic heating of the surface

The surface temperature of the Earth changes regularly on several different time scales. On a short time scale it changes with night and day, and on the time scale of a year it changes with the seasons. Finally, on a time scale of several hundred years to millions of years the surface temperature is also controlled by climatic changes. The impact of periodic changes on the surface temperature is found by solving the temperature equation

$$\frac{\partial T}{\partial t} - \kappa \frac{\partial^2 T}{\partial z^2} = 0 \tag{6.266}$$

with a periodically changing surface temperature

$$T(z = 0, t) = \Delta T \cos(\omega t) \tag{6.267}$$

as a boundary condition on the surface $z = 0$. The amplitude of the temperature change is ΔT and the circular frequency is ω. The second boundary condition is an unchanged temperature at great depth

$$T(z = -\infty, t) = 0. \tag{6.268}$$

We see that the time it takes to complete one cycle, or one period, is

$$t_p = \frac{2\pi}{\omega}. \tag{6.269}$$

The inverse of circular frequency, $t_0 = 1/\omega$, is a natural choice for a characteristic time, and a characteristic length then follows from the characteristic time combined with the thermal diffusivity, $l_0 = \sqrt{\kappa t_0} = \sqrt{\kappa/\omega}$. We have already seen several examples where a characteristic length is used in combination with the thermal diffusivity to obtain a characteristic time. The characteristic length is "large" for a "small" frequency, and "small" for a "large" frequency, which reflects the fact that slow oscillations in the surface temperature penetrate deeper than fast oscillations. We will soon see how this follows from the temperature solution. The temperature equation (6.266) can then be written in dimensionless form as

$$\frac{\partial \hat{T}}{\partial \hat{t}} - \frac{\partial^2 \hat{T}}{\partial \hat{z}^2} = 0 \tag{6.270}$$

where $\hat{T} = T/\Delta T, \hat{t} = t/t_0$ and $\hat{z} = z/l_0$. The dimensionless boundary conditions become

$$\hat{T}(\hat{z} = 0, \hat{t}) = \cos(\hat{t}) \quad \text{and} \quad \hat{T}(\hat{z} = \infty, \hat{t}) = 0. \tag{6.271}$$

The scaled temperature equation (6.270) with boundary conditions (6.271) has no (explicit) parameters, and the solution is

$$\hat{T}(\hat{z}, \hat{t}) = \exp\left(-\frac{\hat{z}}{\sqrt{2}}\right) \cos\left(\hat{t} - \frac{1}{\sqrt{2}}\hat{z}\right), \tag{6.272}$$

as shown in Note 6.10. A first thing to notice about the temperature solution is that the temperature decays exponentially with depth, because it is proportional to the factor

$\exp(-\hat{z}/\sqrt{2})$. It follows that the amplitude is reduced by the factor $1/e \approx 0.36$ at the depth $\hat{z} = \sqrt{2} \approx 1.41$ and by the factor 0.01 at the depth $\hat{z} = -\sqrt{2}\ln(0.01) \approx 6.5$. The depth $\hat{z} = \sqrt{2}$ is called the skin depth. Another point is that the subsurface lags behind the surface. The depth z reaches its temperature maximum at time $\Delta\hat{t} = \hat{z}/\sqrt{2}$ after the surface, and the temperature at the depth $\hat{z} = \sqrt{2}\pi \approx 4.44$ is in opposite phase to the surface temperature. The temperature is at its maximum at this depth when the temperature at the surface is at its minimum.

The temperature solution is plotted in Figures 6.35a, and b, which show how the temperature variations decay with depth. The figures also show how the phase changes with the depth.

Some examples of the characteristic time and length for different periods are shown in the following table when the thermal diffusivity is $\kappa = 10^{-6} \text{ m}^2 \text{ s}^{-1}$:

Period	l_0 [m]	Skin depth [m]
24 hours	0.1	0.2
1 month	0.6	0.9
1 year	2.2	3.2
10 years	7.1	10.0
100 years	22	32
1000 years	71	100
10^4 years	224	316
10^5 years	708	1002
10^6 years	2240	3168

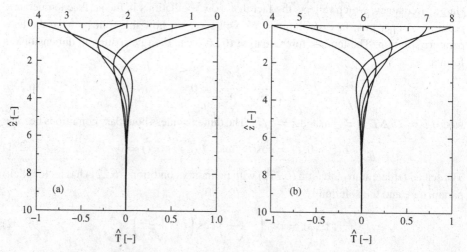

Figure 6.35. *The figures show the temperature into the subsurface for time steps $\hat{t}_n = n\pi/4$ through one period, where n is an integer from 0 to 8.*

The skin depth for daily variations of the surface temperature is 0.17 m and for yearly variations 3.2 m.

The temperature with units follows directly from the dimensionless solution. We just insert the definitions for $\hat{T}$, $\hat{z}$ and $\hat{t}$ in equation (6.272) and get

$$T(z,t) = \Delta T \exp\left(-\sqrt{\frac{\omega}{2\kappa}}z\right) \cos\left(\omega t - \sqrt{\frac{\omega}{2\kappa}}z\right). \tag{6.273}$$

The solution in this form shows that variations with high frequencies die out with depth much faster than variations with low frequencies. Depth therefore acts as a filter that removes high-frequency variations in the surface temperature.

We have already seen that we can add temperature solutions to obtain a new solution, because the temperature equation is linear. We can for instance add the stationary solution $az + b$, given by a constant thermal gradient a and constant average surface temperature b. Another possibility is to add together the solutions with different amplitude ΔT_i and frequency ω_i,

$$T(z,t) = \sum_i \Delta T_i \exp\left(-\sqrt{\frac{\omega_i}{2\kappa}}z\right) \cos\left(\omega_i t - \sqrt{\frac{\omega_i}{2\kappa}}z\right). \tag{6.274}$$

This solution solves the problem with periodic surface temperature variations of the form

$$f(t) = \sum_i \Delta T_i \cos(\omega_i t) \tag{6.275}$$

and such a general Fourier series can be used to approximate almost any kind of periodic surface temperature.

Figure 6.36 shows observations of the temperature at the depths 1.2 m, 4 m and 9 m over a time span of 2 years and 4 months (Isaksen and Sollid, 2002). The data set is from a site with permafrost on the island of Svaldbard. We notice that there are weekly variations in the temperature data at 1.2 m depth, and that these high frequency variations have died out at 4 m where the temperature signal is smooth. We have that $l_0 \le 0.6$ m for variations with a period up to 1 month, and these variations are reduced to nearly zero (by a factor $1/100$) at the depth $6\,l_0 = 3.9$ m. The model (6.273) is plotted for $\kappa = 1 \cdot 10^{-6}$ m^2 s^{-1}, $\Delta T = 9.2\,°$C, and a constant offset $-6.3\,°$C has been added. The model does not match the data at 4 m and 9 m, although weekly variations are filtered away, because there are variations in the observations with a period longer than a year.

Note 6.10 The temperature equation (6.270) with boundary conditions (6.271) is solved by the separation of variables, $\hat{T}(\hat{z}, \hat{t}) = U(\hat{t})V(\hat{z})$. When the product $U(\hat{t})V(\hat{z})$ is inserted into the temperature equation (6.270) we get

$$\frac{U'}{U} = \frac{V''}{V} = a + ib. \tag{6.276}$$

These two ratios have to be equal to a constant $(a+ib)$, because U'/U is only a function of $\hat{t}$ and V''/V is only a function of $\hat{z}$. The constant is the complex number $a+ib$ where a and

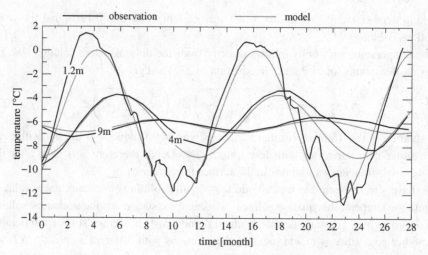

Figure 6.36. *The model for periodic heating of the surface is matched against temperature observation at the depths 1.2 m, 4 m and 9 m over a time span of three years (Isaksen and Sollid, 2002).*

b are real numbers and $i = \sqrt{-1}$. Integration of $U'/U = a + ib$, as shown in Exercise 6.5, yields

$$U(\hat{t}) = U_0 e^{(a+ib)\hat{t}} \tag{6.277}$$

where $a = 0$. We cannot have $a > 0$ because that leads to a temperature that grows exponentially with time, which is not the case. The alternative $a < 0$ is not possible either, because it leads to a temperature that decays exponentially to zero. We therefore have

$$U(\hat{t}) = c e^{ib\hat{t}} \tag{6.278}$$

where c is an integration constant. The equation for $V(\hat{z})$ becomes $V'' - ibV = 0$, which has as a solution linear combinations of $\exp(-\sqrt{ib}\hat{z})$ and $\exp(+\sqrt{ib}\hat{z})$. The square root of i is $\sqrt{i} = \pm(1 + i)/\sqrt{2}$. Recall from the complex plane that $e^{i\phi} = \cos\phi + i\sin\phi$, and we have that $i = e^{i\pi/2} = (e^{i\pi/4})^2 = (1 + i)^2/2$. The solution for $V(\hat{z})$ is then $V(\hat{z}) = \exp(-(1 + i)\hat{z}/\sqrt{2})$. The other possible solution $V(\hat{z}) = \exp(+(1 + i)\hat{z}/\sqrt{2})$ cannot be used because it implies that the temperature increases exponentially with depth, which is not possible. The solution of the temperature equation is now

$$\hat{T}(\hat{z}, \hat{t}) = U(\hat{t})V(\hat{z})$$

$$= c \exp\left(ib\hat{t} - (1 + i)\sqrt{\frac{b}{2}}\hat{z}\right)$$

$$= c \exp\left(-\sqrt{\frac{b}{2}}\hat{z}\right)\exp\left(i\left(b\hat{t} - \sqrt{\frac{b}{2}}\hat{z}\right)\right). \tag{6.279}$$

The last factor, the exponential of a complex number, is written as a real and an imaginary part using that $e^{i\phi} = \cos\phi + i\sin\phi$. We only need the real part, and the temperature solution is therefore

$$\hat{T}(\hat{z},\hat{t}) = c \exp\left(-\sqrt{\frac{b}{2}}\hat{z}\right) \cos\left(b\hat{t} - \sqrt{\frac{b}{2}}\hat{z}\right). \tag{6.280}$$

The integration constants b and c are found by comparing $\hat{T}(\hat{z}=0,\hat{t}) = c\cos(b\hat{t})$ with the boundary condition (6.267), and we then get that $b = 1$ and $c = \Delta T$.

6.19 Variable surface temperature

The previous section showed how the temperature varies with depth when there is a periodic variation in the surface temperature, and Section 6.14 showed the thermal response of the subsurface to a piecewise constant surface temperature history. We will now see how we can obtain the temperature at a depth at a specific time for any variation in the surface temperature. We recall that a linear combination of solutions of the temperature equation is also a solution. The temperature can therefore be written as a sum of two parts:

$$T(z,t) = T_s(z) + \Delta T(z,t) \tag{6.281}$$

where $T_s(z)$ is the stationary solution and $\Delta T(z,t)$ is the transient solution. Note 6.11 shows that the transient part can be written as

$$\Delta T(z,t) = \int_{z/2\sqrt{\kappa t}}^{t} T_{\text{surf}}\left(t - \frac{z^2}{4\kappa\mu}\right) \exp(-\mu^2)\,d\mu \tag{6.282}$$

when the surface temperature at time t is $T_{\text{surf}}(t)$. It is assumed that the surface temperature is zero until $t = 0$, before it starts to vary. The transient solution is therefore zero before $t = 0$. The thermal transient (6.282) can easily be integrated numerically for any given surface temperature $T_{\text{surf}}(t)$, because the factor $\exp(-\mu^2)$ goes rapidly towards zero. The integration can also be carried out exactly for simple forms of the function T_{surf}.

The simplest application of expression (6.282) is to assume that the surface temperature makes a step from zero to a constant T_{surf} at $t = 0$. We then obtain directly the solution for instantaneous heating or cooling of the semi-infinite half-space. Another application is a surface temperature that increases linearly with time as $T_{\text{surf}}(t) = ct$ from $t = 0$, where c is the heating rate. The transient part of the temperature solution then becomes

$$\Delta T(z,t) = ct\left\{\left(1 + \frac{z^2}{2\kappa t}\right) \text{erfc}\left(\frac{z}{2\sqrt{\kappa t}}\right) - \frac{z}{\sqrt{\pi\kappa t}}\exp\left(-\frac{z^2}{4\kappa t}\right)\right\} \tag{6.283}$$

where Note 6.12 shows how it follows from the general expression (6.282). The solution (6.272) for periodic heating of the surface can also be derived starting with equation (6.282) as shown in Section 2.6 of Carslaw and Jaeger (1959).

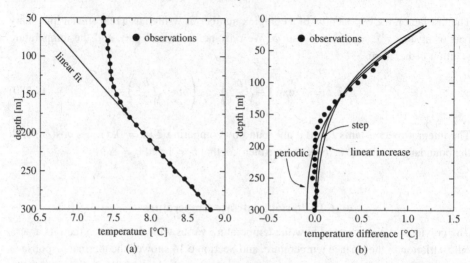

Figure 6.37. *(a) Observations of temperature against depth and a linear fit against the stationary part of the geotherm. (b) Optimization of three different models against the transient part of the data set.*

Figure 6.37a shows temperature observations that indicates a higher surface temperature in a period leading up to the present time. The data set is from a borehole site in Canada (Mareschal, 2005). The stationary state is obtained by a linear fit of the data below 170 m, and it gives the surface temperature 6.07 °C, and the thermal gradient $dT/dz = 9.15$ °C/km. Figure 6.37b shows the difference between the data and the stationary geotherm, which is therefore the transient part. The figure also shows the best match against the data for three simple models of variation in the surface temperature.

The first model is equation (6.202) for instantaneous heating of the surface. The surface temperature makes a step $\Delta T = 1.3$ °C at $t = 0$ and we are seeking the time t that gives the best match against the data. By trying out t in the interval from 0 years to 500 years with a step of 10 years we find that the optimal value is $t_1 = 150$ years.

The second model is the thermal transient (6.273) from periodic heating of the surface, where the surface temperature is $T_{surf}(t) = \Delta T \cos(\omega t)$ with $\Delta T = 1.3$ °C. The geotherm at $t = 2\pi t_0$, where $t_0 = 1/\omega$, is optimized with the result $t_0 = 257$ years. The full period becomes $2\pi t_0 = 1614$ years, which implies that there has been an increase in the surface temperature from 0 to ΔT over a time span of $\frac{1}{2}\pi t_0 = 403$ years.

The third model is the temperature solution (6.283) for a linear increase in the surface temperature. The temperature increases from 0 °C at time $t = 0$ until $\Delta T = 1.3$ °C at time t. The optimal time span is $t = 340$ years for this model.

These models suggest that the previous 400 years at the boresite has had a period with a hotter climate than normal. This result is in accordance with a more elaborate surface temperature estimate for the same borehole observations (Mareschal, 2005). The match against the data for these models in terms of the root-mean-square deviation are close to each other, as we can also see from Figure 6.37b. These models are simple to match against a data set when just one parameter is considered. An improved match requires a more complex surface temperature history with more parameters, and a non-trivial optimization method.

Measurements of the borehole temperatures are an important source of information about the climate in the past few centuries (Vasseur *et al.*, 1983, Beltrami and Mareschal, 1992, 1995, Huang *et al.*, 2000). A large number of boreholes have been drilled and measured (Mareschal, 2005) all over the planet. Numerical techniques have been developed and applied in the reconstruction of the past surface temperature (Vasseur *et al.*, 1983, Wang, 1992, Beltrami and Mareschal, 1995, Chouinard and Mareschal, 2007). An extensive discussion of methods and results of borehole climatology is given by Gonzalez-Rouco *et al.* (2008).

Note 6.11 *Duhamel's theorem.* The transient solution (6.282) is derived from *Duhamel's theorem*. It gives the transient solution of the temperature equation

$$\frac{\partial T}{\partial t} - \kappa \frac{\partial^2 T}{\partial z^2} = 0 \tag{6.284}$$

for the infinite half-space ($z \geq 0$), when the boundary condition at $z = 0$ is the surface temperature $T_{\text{surf}}(t)$ as a function of time. It assumes that the initial condition is $T = 0$ and that $T_{\text{surf}}(t) = 0$ for $t < 0$. The second boundary condition is $T = 0$ at infinite depth. The solution T is written in terms of the simpler solution U of the same problem, where U is the temperature in the case of instantaneous heating of the surface with a unit step-function, $T_{\text{surf}} = 1$ for $t \geq 0$. Duhamel's theorem then says that the temperature solution T is

$$T(z, t) = \int_0^t T_{\text{surf}}(\lambda) \frac{\partial U}{\partial t}(z, t - \lambda) \, d\lambda. \tag{6.285}$$

A direct way to verify the theorem is to insert the solution (6.285) into temperature equation (6.284). (See Exercise 6.32.) The remaining task is to verify that the solution (6.285) fulfills the boundary condition at the surface, where we have

$$T(z{=}0, t) = \int_0^t T_{\text{surf}}(\lambda) \frac{\partial U}{\partial t}(z{=}0, t - \lambda) \, d\lambda$$
$$= \int_0^t T_{\text{surf}}(\lambda) \, \delta(t - \lambda) \, d\lambda = T_{\text{surf}}(t). \tag{6.286}$$

The function U is the unit step function of time at $z = 0$, and we used that $\partial U / \partial t$ is equal to Dirac's delta function $\delta(t - \lambda)$ for $z = 0$. The derivation of the step function and Dirac's delta function are covered in detail in text books on mathematical methods in physics, see for instance Riley *et al.* (1998). From Section 6.14 we have that function U is

$$U(z, t) = \text{erfc}\left(\frac{z}{2\sqrt{\kappa t}}\right) = \frac{2}{\sqrt{\pi}} \int_{z/2\sqrt{\kappa t}}^{\infty} e^{-\mu^2} \, d\mu \tag{6.287}$$

which gives

$$\frac{\partial U}{\partial t}(z, t) = \frac{z}{2\sqrt{\pi \kappa t^3}} \exp\left(-\frac{z^2}{4\kappa t}\right).$$ (6.288)

When $\partial U / \partial t$ is inserted into solution (6.285) we get

$$T(z, t) = \frac{z}{2\sqrt{\pi \kappa}} \int_0^t \frac{T_{\text{surf}}(\lambda)}{(t - \lambda)^{3/2}} \exp\left(-\frac{z^2}{4\kappa(t - \lambda)}\right) d\lambda.$$ (6.289)

The remaining step towards the temperature solution (6.282) is a change in integration variable from λ to

$$\mu(\lambda) = \frac{z}{2\sqrt{\kappa(t - \lambda)}} \quad \text{or} \quad t - \lambda = \frac{z^2}{4\kappa \mu^2}$$ (6.290)

which gives that $d\lambda = (z^2/2\kappa \mu^3) d\mu$. The new integration limits are $\mu(\lambda{=}0) = z/2\sqrt{\kappa t}$ and $\mu(\lambda{=}t) = \infty$.

Note 6.12 *Linearly increasing surface temperature.* The temperature solution (6.282) is

$$T(z, t) = \frac{2c}{\sqrt{\pi}} \int_{z/2\sqrt{\kappa t}}^{\infty} (t - \frac{z^2}{4\kappa \mu^2}) e^{-\mu^2} d\mu$$ (6.291)

when $T_{\text{surf}}(t) = ct$. The next step is integration by parts

$$\int \frac{e^{-\mu^2}}{\mu^2} d\mu = -\frac{e^{-\mu^2}}{\mu} - 2 \int e^{-\mu^2} d\mu$$ (6.292)

and then use the definition of the complementary error function (6.203).

Exercise 6.32

(a) Let the temperature be written as $T(z, t) = \int_0^t f(z, t - \lambda) d\lambda$ and show that

$$\frac{\partial T}{\partial t} = f(z, t - \lambda)|_{\lambda=t} + \int_0^t \frac{\partial f}{\partial t}(z, t - \lambda) d\lambda.$$ (6.293)

(b) Insert temperature (6.285) from Duhamel's theorem into the temperature equation and show that

$$\frac{\partial T}{\partial t} - \kappa \frac{\partial^2 T}{\partial z^2} = T_{\text{surf}}(t) \left[\frac{\partial U}{\partial t}(z, t - \lambda)\right]_{\lambda=t} + \int_0^t T_{\text{surf}}(\lambda) \frac{\partial}{\partial t} \left(\frac{\partial U}{\partial t} - \kappa \frac{\partial^2 U}{\partial z^2}\right) d\lambda.$$ (6.294)

The factor $\partial U / \partial t$ is zero at $t = 0$ when $z > 0$, but it is undefined for $t = 0$ and $z = 0$. The integral is zero because U is a solution of the temperature equation, and expression (6.285) is therefore a solution of the temperature equation for $z > 0$.

6.20 Temperature transients from sediment deposition or erosion

The deposition of sediments may cause a suppression of the surface heat flow and the temperature in a sedimentary basin. In order to see this we will consider the following case – let a sedimentary layer be deposited instantaneously. The temperature in the layer is the same as on the surface, and the temperature underneath the layer is unchanged. The thermal state of the subsurface is shifted downwards by the thickness of the layer, since it has not had time to adapt to the new surface. This effect is often called "thermal blanketing." The blanketing effect is enhanced by the low heat conductivity of uncompacted sediments near the surface. (The heat conductivity of pores filled with brine is "low" compared to the sediment matrix.) The thermal gradient is zero in the layer right after deposition and it then starts to increase as the temperature moves towards a new stationary state. The deposition of sedimentary layers is not instantaneous, but may take millions of years. We will in this section see that the suppression of the thermal gradient is controlled by the deposition rate and also by the time span of the deposition period.

In order to handle the effect of deposition on the temperature we must first find a temperature equation that describes the process. We have that the heat flow by conduction is in a coordinate system that is attached to the sediments and therefore follows the burial during deposition. The z-coordinate in this system is now denoted z^* and the temperature equation along the z^*-axis is

$$\frac{\partial T}{\partial t} - \kappa \frac{\partial^2 T}{\partial z^{*2}} = 0. \tag{6.295}$$

We would like to work in a coordinate system with the z-axis attached to the surface, where the surface remains as $z = 0$ during deposition. The transformation between these two coordinate systems is

$$z = z^* + vt \tag{6.296}$$

where v is the deposition rate. We then have that $z^* = 0$ corresponds to $z = vt$ during burial, and that the two z-axes overlap at $t = 0$. Differentiation with respect to time along the z^*-axis gives

$$\left(\frac{\partial T}{\partial t}\right)_{z^*} = \left(\frac{\partial T}{\partial t}\right)_z + \frac{\partial T}{\partial z}\frac{\partial z}{\partial t} = \left(\frac{\partial T}{\partial t}\right)_z + v\frac{\partial T}{\partial z} \tag{6.297}$$

when it is transformed to the z-axis. The temperature equation along the z-axis is therefore

$$\frac{\partial T}{\partial t} + v\frac{\partial T}{\partial z} - \kappa \frac{\partial^2 T}{\partial z^2} = 0 \tag{6.298}$$

where a convection term represents the thermal effect of deposition. Compaction of the sediments is for simplicity ignored in the following, and burial is therefore at the same rate as deposition. We need two boundary conditions before we can solve (6.298), where one is a constant temperature at the surface. We will look at two alternatives for the other boundary condition, which will be either (1) a constant thermal gradient at infinite depth or (2) a constant temperature at the base of the lithosphere. These two alternative boundary conditions give two different solutions that overlap for reasonable assumptions about a sedimentary basin. The first case, of constant thermal gradient at infinite depth, gives the temperature

$$T(z, t) = T_0 + A(z + vt) - \frac{A}{2}(z + vt)\,\mathrm{erfc}\left(\frac{z + vt}{2\sqrt{\kappa t}}\right)$$
$$+ \frac{A}{2}(z - vt)\exp\left(-\frac{vz}{\kappa}\right)\mathrm{erfc}\left(\frac{z - vt}{2\sqrt{\kappa t}}\right) \tag{6.299}$$

where T_0 is the surface temperature and A is the thermal gradient. This solution is not straightforward to obtain and the reader is referred to Section 15.2 in Carslaw and Jaeger (1959). It was obtained by Benfield (1949) in a study of thermal transients due to erosion. Erosion gives the opposite effect of deposition – sediments are brought up towards the surface faster than they are able to cool and reach a thermal steady state. We notice that the temperature (6.299) is proportional to the gradient A when the surface temperature is zero. Figure 6.38a shows the temperature as a function of depth for different deposition rates, and the deposition rate must be larger than $100\,\mathrm{m\,Ma^{-1}}$ for the geotherm to depart from the stationary state. The surface heat flow is shown in Figure 6.38b, where we see that the deposition rate $100\,\mathrm{m\,Ma^{-1}}$ gives a 10% reduction when 5 km of sediments have been deposited. Slow burial histories, where for example 5 km is deposited over a time span of several hundred million years, do not give any noticeable deviation from the stationary state. But rapid deposition of sediments, for instance 1 km of sediments during a time span of 0.1 Ma, reduces the surface heat flow by a half.

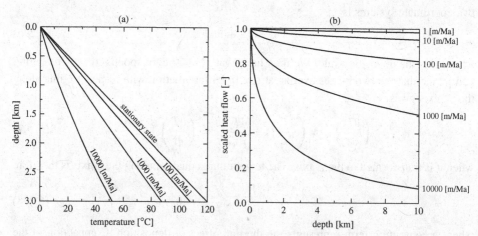

Figure 6.38. (a) The geotherms are shown for deposition rates $10^2\ mMa^{-1}$, $10^3\ mMa^{-1}$ and $10^4\ mMa^{-1}$ at the time when 3 km of sediments have been deposited. The corresponding times are therefore 30 Ma, 3 Ma and 0.3 Ma. (b) The surface heat flow is shown for a range of deposition rates as a function of the thickness of the basin sediments. (The basin thickness is vt, where v is the deposition rate and t is time.)

Benfield's temperature (6.299) solves the convection–conduction equation (6.298) assuming that the temperature gradient is constant at infinite depth. We recall from Section 6.12 that we have already obtained a Fourier series solution of equation (6.298), but then for a model with a finite depth. The boundary condition at the surface is the same for these two solutions, but the second boundary condition is a fixed temperature at a given depth for the Fourier series solution – for instance at the base of the lithosphere. Note 6.14 shows how it is possible to apply the Fourier series solution (6.166) to thermal blanketing. A condition for the validity of this solution is that the depth to the base of the lithosphere is much larger than the thickness of the deposited sediments. The Fourier series solution (6.166) and Benfield's solution (6.299) are then completely overlapping under this condition.

The reduction in temperature from a recent and rapid deposition process may have an impact on the vitrinite reflectance. This is exemplified with a numerical study shown in Figure 6.39a, where deposition of 1000 m of sediments took place over a period of 1.5 Ma in the Pliocene (from 3.5 Ma to 2 Ma). There was also a rapid deposition process in the Cretaceous, where 600 m is deposited over 9 Ma (from 85 Ma to 79 Ma), and the depression of isotherms is clearly seen. The thermal transient from the deposition of the formation Naust in the Pliocene is shown in Figure 6.39b. We see that the thermal depression increases linearly with depth as expected from the geotherms in Figure 6.38a. The deposition rate is 667 m/Ma and the depression in the temperature becomes 20 °C at the depth of 3 km, which is reasonable according to Figure 6.38a. Figure 6.40a shows that the difference in the computed vitrinite reflectance is noticeable between stationary and transient temperature solutions. The stationary solution is insensitive to deposition processes and it overestimates the temperature, and the corresponding vitrinite reflectance is therefore also overestimated. The necessity of transient temperature solutions in thermal studies of sedimentary basin were observed already in early work that included vitrinite

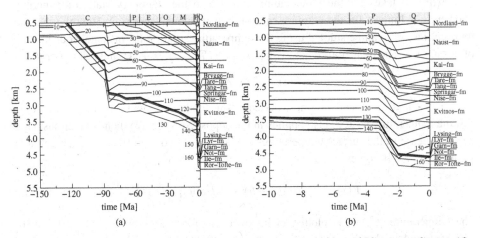

Figure 6.39. *(a) A burial history that has a deposition of nearly 1000 m of Pliocene sediments (the formation "Naust"). (b) The thermal transient is shown for the deposition of the formation "Naust."*

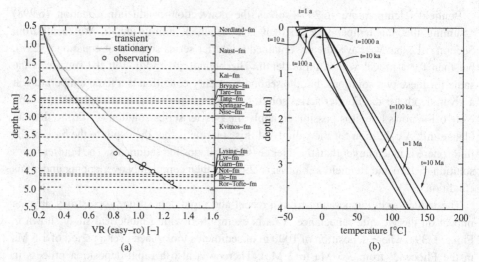

Figure 6.40. *(a) The computed maturity is shown for the burial history in Figure 6.39. The maturity is substantially larger in the case of a stationary temperature solution than for a solution that accounts for the thermal transient from deposition. (b) The time it takes for the subsurface to adapt to a new thermal stationary state in the case of a sudden deposition of a 1000 m thick layer.*

reflectance, see for instance Lerche (1990a). Figure 6.39b shows that the temperature does not return to the same temperature as before the deposition of Naust during the following 2 Ma until present time. A point concerning numerical simulation of such thermal transients is that the model should include a large part of the lithosphere below the basin (if not the entire lithosphere). A boundary condition far below the basin will not disturb the thermal effect from deposition or erosion.

The time it takes for the temperature to return to the stationary state after thermal blanketing is longest for the extreme case of instantaneous deposition. How the thermal transient dies out can be studied using the temperature solution for instantaneous heating of the surface. The temperature before deposition is $T(z) = T_0 + Az$ and right after it is

$$T(z) = T_0 + A(z - h) \tag{6.300}$$

for $z > 0$. The thermal state is simply buried by the depth h. We will for simplicity assume that the temperature (6.300) also applies in the deposited layer. We then have an instantaneous increase in the surface temperature from $T = -Ah$ to $T = 0$, and the temperature solution (6.202) becomes

$$T(z) = T_0 + A(z - h) + A\, h\, \mathrm{erfc}\left(\frac{z}{2\sqrt{\kappa t}}\right). \tag{6.301}$$

The temperature (6.301) is plotted in Figure 6.40b for deposition of $h = 1000$ m of sediments, when the thermal gradient is $A = 40\ ^\circ\mathrm{C/km}$ and the thermal diffusivity $\kappa = 10^{-6}\ \mathrm{m^2\,s^{-1}}$. We see that the temperature is close to the more reasonable initial condition, with the surface temperature $T = 0$ in the new layer, after $t = 10^4$ years. More than 1 Ma is needed for the temperature to approach the new thermal state. We can use the same reasoning as with the thermal boundary layer in Section 6.14 to study the transient. The depth where the temperature has regained 90% of the of depression $\Delta T = A h$ is $z = 0.176\sqrt{\kappa t}$, because erfc(0.088) = 0.9. For instance a time span $t = 16$ Ma is needed for 90% of the perturbation to have died out at the depth $z = 4$ km.

Deposition or erosion of sediments may bring sedimentary basins away from a thermal steady state for tens to hundreds of million years. The importance of thermal transients from deposition or erosion of sediments has been studied by several authors after the pioneering work of Benfield (1949) – among others Hutchison (1985), Lucazeau and Douaran (1985), Karner (1991) and Wangen (1995).

Note 6.13 *Dimensionless formulation.* It is possible to scale the temperature equation (6.298) and Benfield's solution (6.299) using the characteristic length $l_1 = \kappa/v$, the characteristic time $t_1 = \kappa/v^2$ and characteristic temperature $T_1 = A\kappa/v$. Both the equation and the solution then become parameterless. The dimensionless deposition rate is 1. It is then possible to compare time and depth with t_1 and l_1, respectively, in order to predict the amount of thermal blanketing.

Note 6.14 *Fourier series solution.* The Fourier series (6.166) solves the dimensionless convection–diffusion equation (6.150), when the temperatures at the boundaries are fixed. The Peclet number now represents the deposition rate. The dimensionless solution (6.166) can be written with units as

$$T(z, t) = T_0 + (T_{\text{bot}} - T_0)\, \hat{T}\left(\frac{l_0 - z}{l_0}, \frac{t}{t_0}\right) \tag{6.302}$$

where l_0 is the depth to the base of the lithosphere and T_{bot} is the temperature at the depth l_0. Time is scaled with characteristic time $t_0 = l_0^2/\kappa$ and the Peclet number becomes Pe = v/v_0, where v_0 is the characteristic velocity $v_0 = \kappa/l_0$. The solution (6.302) gives the same results as Benfield's solution as long as the amount of deposited sediments $s = vt$ is much less than the length scale l_0. The dimensionless formulation (6.150) has a stationary part controlled by the Pe-number, which tells us that Pe $\ll 1$ gives negligible thermal blanketing and that the opposite regime, Pe $\gg 1$, may give strong thermal blanketing. The regime in between, which gives moderate blanketing, is characterized by a deposition rate $v_0 \sim 250$ m Ma^{-1} when $\kappa = 10^{-6}$ m^2 s^{-1} and $l_0 = 125$ km. The time it takes to reach a steady state with the Fourier solution can be estimated by the half-life of the first term in the series. Each term in the Fourier series goes to zero as $\exp(-k_n\hat{t})$ where $k_n = n^2\pi^2 + \text{Pe}^2/4$ for $n \geq 1$. The first term, which dies out most slowly, has the half-life

$$\hat{t}_{1/2} = \frac{\ln 2}{\pi^2 + \frac{1}{4}\text{Pe}^2}. \tag{6.303}$$

The condition for strong thermal blanketing is that Pe should be larger than 1 and that $\hat{t}$ should go beyond $\hat{t}_{1/2}$. Furthermore, the time must not be so large that the amount of deposited sediments becomes unreasonable. To assure that the last condition is fulfilled we require that $vt_{1/2} \ll l_0$, which is the same as

$$f(\text{Pe}) = \frac{\text{Pe}\ln 2}{\pi^2 + \frac{1}{4}\text{Pe}^2} \ll 1. \tag{6.304}$$

The function $f(\text{Pe})$ has a maximum at $\text{Pe} = 2\pi$ where it is $\ln 2/\pi \approx 0.22$. Since the function f is rather large in the interval from $\text{Pe} = 0.5$ to $\text{Pe} = 100$ we cannot expect the reduction in the temperature to reach a steady state for moderate deposition rates. The Fourier series solution (6.166) of the convection–diffusion equation was used by Wangen (1995) in a study of thermal blanketing.

6.21 Conservation of energy once more

A more general temperature equation than equation (6.15) can be derived by starting with a more complete expression for energy conservation than what was used in Section 6.1. The following derivation is not only more complete, but also more formal. It makes use of Reynolds transport theorem, the continuity equation, Newton's second law and some thermodynamics. The final result is the same temperature equation except for two terms – one term that accounts for the work done on the system and a new term for the rate of change of internal energy. We could have just added the first term to the previous temperature equation, since it has a simple and direct interpretation. However, the formal derivation shows important parts of continuum mechanics, and also provides better insight into the foundations of the temperature equation. The first law of thermodynamics expresses energy conservation as

$$E = W + Q \tag{6.305}$$

where the energy E of the system is equal to the work W done on the system and the heat Q added to the system. The energy E is the internal energy E_I plus the kinetic energy E_K and the first law becomes

$$\frac{d}{dt}(E_K + E_I) = \frac{d}{dt}(W + Q) \tag{6.306}$$

when written in terms of rates. The kinetic energy is

$$E_K = \frac{1}{2} \int_V \varrho \, v^2 \, dV \tag{6.307}$$

where the velocity squared is $v^2 = \mathbf{v} \cdot \mathbf{v} = v_i v_i$, and the volume of the system is V. The internal energy inside the same volume V is

$$E_I = \int_V \varrho \, e \, dV \tag{6.308}$$

where e is the internal energy per unit mass. The time rate of change of work done on the system, dW/dt, which was zero in the previous derivation of the temperature equation, is now

$$\frac{dW}{dt} = \int_{\partial V} \mathbf{f} \cdot \mathbf{v} \, dA + \int_V \mathbf{b} \cdot \mathbf{v} \, dV \qquad (6.309)$$

where $\mathbf{f}$ is the force on the surface ∂V of the volume V, and where $\mathbf{b} = \varrho g \mathbf{n}_z$ is the gravitational body force per unit volume. The time-rate of heat input to the system is

$$\frac{dQ}{dt} = -\int_{\partial V} \mathbf{q} \cdot \mathbf{n} \, dA + \int_V S \, dV \qquad (6.310)$$

where the term dQ/dt is equal to the heat flux through the boundaries of the system added to the heat generation inside the system. Energy conservation has so far been expressed for only one component (or phase), but a generalization to a multicomponent system is straightforward and it is done towards the end of the section. The time differentiation on the left-hand side of equation (6.306) is brought through the integration signs using Reynolds transport theorem (see Exercise 3.27). Reynolds transport theorem is needed because the volume V of the system cannot be assumed constant, but changes as a function of time. The left-hand side of the first law of thermodynamics (6.306) becomes

$$\frac{d}{dt}(E_K + E_I) = \int_V \left\{ \frac{\partial}{\partial t}\left(\varrho\,(e + \frac{1}{2}v^2)\right) + \nabla \cdot \left(\mathbf{v}\varrho\,(e + \frac{1}{2}v^2)\right) \right\} dV. \qquad (6.311)$$

The differentiations are carried through the parentheses and the terms are regrouped using the material derivative as follows

$$\frac{d}{dt}(E_K + E_I) = \int_V \left\{ \frac{D}{dt}\left(\varrho\,(e + \frac{1}{2}v^2)\right) + \mathbf{v} \cdot \nabla\left(\varrho\,(e + \frac{1}{2}v^2)\right) \right\} dV$$

$$= \int_V \left\{ \varrho\frac{De}{dt} + \frac{1}{2}\varrho\frac{Dv^2}{dt} + \left(e + \frac{1}{2}v^2\right)\left(\frac{D\varrho}{dt} + \varrho\nabla \cdot \mathbf{v}\right) \right\} dV$$

$$= \int_V \left\{ \varrho\frac{De}{dt} + \frac{1}{2}\varrho\frac{Dv^2}{dt} \right\} dV \qquad (6.312)$$

where the continuity equation (3.161) is identified in the last factor in the second step. This factor is therefore zero, and we are left with equation (6.312).

The right-hand side of the first law of thermodynamics (6.306) can also be written as an integral over the volume V. The surface integral of the heat flow crossing the surface ∂V is converted to a volume integral using Gauss's theorem (3.171). The surface integral over the force $\mathbf{f}$ is converted to a volume integral by first expressing the force on the surface as $f_i = \sigma_{ij} n_j$, using the stress tensor σ_{ij} and the (outward) normal vector n_j to the surface,

$$\int_{\partial V} f_i v_i \, dV = \int_{\partial V} \sigma_{ij} n_j v_i \, dV = \int_{\partial V} \sigma_{ij} v_j n_i \, dV = \int_V \frac{\partial(\sigma_{ij} v_j)}{\partial x_i} \, dV. \qquad (6.313)$$

(Notice that the symmetry of the stress tensor was used between the second and the third equality.) We therefore have

$$\frac{d}{dt}(W+Q) = \int_V (\mathbf{b}\cdot\mathbf{v} + \nabla\cdot(\sigma\mathbf{v}) - \nabla\cdot\mathbf{q} + S)\,dV \qquad (6.314)$$

where $\sigma_{ij}v_j$ is written as the matrix–vector product $\sigma\mathbf{v}$. Combining the left-hand side (6.312) and the right-hand side (6.314) of the first law of thermodynamics (6.306) gives the temperature equation

$$\varrho\frac{De}{dt} + \nabla\cdot\mathbf{q} - S = -\frac{1}{2}\varrho\frac{Dv^2}{dt} + \nabla\cdot(\sigma\mathbf{v}) + \mathbf{b}\cdot\mathbf{v}$$

$$= -v_i\left(\varrho\frac{Dv_i}{dt} - \frac{\partial\sigma_{ij}}{\partial x_j} - b_i\right) + \sigma_{ij}\frac{\partial v_i}{\partial x_j}$$

$$= \sigma_{ij}\frac{\partial v_i}{\partial x_j}. \qquad (6.315)$$

Newton's second law is identified in the second step, and the contribution from work done on the system is the term $\sigma_{ij}\,\partial v_i/\partial x_j$.

We are seeking an equation for temperature. The next step is therefore to replace the internal energy by an expression in temperature, and to do that some thermodynamics is needed. We begin by splitting the stress σ_{ij} into deviatoric stress σ'_{ij} and pressure p:

$$\sigma_{ij} = \sigma'_{ij} + \sigma_m\delta_{ij} = \sigma'_{ij} - p\,\delta_{ij} \qquad (6.316)$$

where the pressure is defined as the negative of the mean stress $\sigma_m = \frac{1}{3}\sigma_{ii}$. (See Section 3.10 for deviatoric stress.) The right-hand side (6.315) becomes $\sigma'_{ij}\,\partial v_i/\partial x_j$ when $-p\,\partial v_i/\partial x_i$ is moved to the left-hand side. On the left-hand side we now look at the change in the internal energy and work during a time interval dt

$$\varrho\,de + p\,d\epsilon_{ii} = \varrho\left(de + p\,d\left(\frac{1}{\varrho}\right)\right) \qquad (6.317)$$

$$= \varrho\,(de + p\,dv) \qquad (6.318)$$

$$= \varrho\,T\,ds \qquad (6.319)$$

where $d\epsilon_{ii} = (\partial v_i/\partial x_i)\,dt$ is the strain increment during the time step, dv is the change in specific volume and ds is the change in specific entropy. We will look more closely at the steps above, and by bringing to mind equation (3.13) we see that the strain increment $d\epsilon_{ii}$ is the same as a relative volume change dV/V. Specific volume is defined as volume per unit mass, and it is therefore the same as inverse density, $v = 1/\varrho$, which gives

$$\frac{d\epsilon_{ii}}{\varrho} = \frac{1}{\varrho}\frac{dV}{V} = -\frac{d\varrho}{\varrho^2} = d\left(\frac{1}{\varrho}\right) = dv. \qquad (6.320)$$

The relation $dV/V = -d\varrho/\varrho$ follows from mass conservation in the volume V since $m = \varrho V$ is constant. The pressure p was needed in order to introduce the change in specific entropy, which is defined as $ds = de + p\,dv$ in thermodynamics. The reason for introducing the specific entropy is that thermodynamics gives

$$ds = \frac{c_p}{T}\,dT - \frac{\alpha}{\varrho}\,dp \tag{6.321}$$

where c_p is the heat capacity at constant pressure, T is the absolute temperature in kelvin and α is the thermal expansibility (see Notes 6.15, 6.16 and 6.17). The temperature equation for a single phase medium is therefore

$$\varrho c_p \frac{DT}{dt} - \alpha T \frac{Dp}{dt} = \nabla \cdot (\lambda \nabla T) + S + \sigma'_{ij}\frac{\partial v_i}{\partial x_j} \tag{6.322}$$

which includes a term for the rate of change of pressure. This term can normally be ignored as justified by Exercise 6.33.

The derivation above generalizes to a multicomponent system, like a porous rock saturated by a fluid, by expressing the internal energy (E), the kinetic energy (K) and the work done on the system (W) as sums over each component. The derivation above is then carried out for each component, where the continuity equation and Newton's second law are used separately for each component to simplify the expression. The only difference is that the density ϱ is now replaced by $\phi_k \varrho_k$, where ϕ_k and ϱ_k are the volume fraction and the density of each component, respectively. The term $Q = -\lambda \nabla T$ for heat conduction is not written as a sum over each component, because it is a bulk property. The heat generation S is also treated as a bulk property, but it could have been divided between each component as $S = \phi_k S_k$. The multicomponent version of the single component equation (6.322) is then

$$\sum_k \phi_k \varrho_k c_k \left(\frac{DT}{dt}\right)_k - \nabla \cdot (\lambda \nabla T) = S + \sum_k \left(\sigma'_{ij}\frac{\partial v_i}{\partial x_j}\right)_k. \tag{6.323}$$

The subscript k on the material derivative says that the velocity of component k is used. This is exactly the same temperature equation as equation (6.11), except for the last term on the right-hand side that involves the work done on the system per unit time for each component. The temperature equation for a fluid saturated porous rock is

$$\varrho_b c_b \frac{DT}{dt} + \varrho_f c_f \mathbf{v}_D \cdot \nabla T - \nabla \cdot (\lambda \nabla T) = S + \left(\sigma'_{ij}\frac{\partial v_i}{\partial x_j}\right)_s \tag{6.324}$$

by introducing the bulk properties from equation (6.12) and the Darcy flux from equation (6.14). (The subscript b denotes the bulk and the subscript f the fluid.) The material derivative of equation (6.324) is with respect to the solid velocity. The term $(\sigma'_{ij}\partial v_i/\partial x_j)_s$ represents the work per unit time in the solid phase, and the corresponding term in the fluid phase is ignored.

Note 6.15 Entropy can be taken as a function of temperature and pressure, $s = s(T, p)$, and the increment ds is

$$ds(T, p) = \left(\frac{\partial s}{\partial T}\right)_p dT + \left(\frac{\partial s}{\partial p}\right)_T dp. \tag{6.325}$$

The specific heat capacity at constant pressure is defined as

$$c_p = T \left(\frac{\partial s}{\partial T} \right)_p$$ (6.326)

which gives

$$\left(\frac{\partial s}{\partial T} \right)_p = \frac{c_p}{T}.$$ (6.327)

The thermal expansibility is defined as

$$\alpha = \frac{1}{v} \left(\frac{\partial v}{\partial T} \right)_p$$ (6.328)

and a Maxwell relation of thermodynamics gives

$$\left(\frac{\partial s}{\partial p} \right)_T = - \left(\frac{\partial v}{\partial T} \right)_p$$ (6.329)

and we have

$$\left(\frac{\partial s}{\partial p} \right)_T = -v\alpha = -\frac{\alpha}{\varrho}.$$ (6.330)

Recall that the specific volume is the inverse density, $v = 1/\varrho$. Inserting the partial derivatives (6.327) and (6.330) into the differential (6.325) gives expression (6.321). These relationships are a part of thermodynamics and they are thoroughly explained in any standard text book on the subject.

Note 6.16 The temperature equation introduced in section 6.1 takes specific heat capacity at constant *volume*, while the temperature equation above takes specific heat capacity at constant *pressure*. Using thermodynamics it is possible to show that the difference between these two heat capacities is

$$c_p - c_v = \frac{T\alpha^2}{\varrho\gamma}$$ (6.331)

where α is the thermal expansibility and γ is the compressibility. This difference (6.331) is normally of no practical importance. Typical parameters for a rock, $\alpha = 3 \cdot 10^{-5} \text{ K}^{-1}$, $\gamma = 3 \cdot 10^{-10} \text{ Pa}^{-1}$, $\varrho = 3 \cdot 10^3 \text{ kg m}^{-3}$ and $T = 1000 \text{ K}$ give that $c_p - c_v = 1 \cdot 10^{-3} \text{ kJ kg}^{-1} \text{ K}^{-1}$, which is much less than a typical value $c_p \approx 1 \text{ kJ kg}^{-1} \text{ K}^{-1}$.

Note 6.17 The temperature does not have to be the main variable in an equation for energy conservation. Alternatives are, for instance, the specific entropy s and the specific enthalpy $h = e + pv$. A different version of the left-hand side of the temperature equation (6.322) is

$$\varrho c_p \frac{DT}{dt} - \alpha T \frac{Dp}{dt} = \text{rhs} \tag{6.332}$$

$$\varrho T \frac{Ds}{dt} = \text{rhs} \tag{6.333}$$

$$\varrho \frac{Dh}{dt} - \frac{Dp}{dt} = \text{rhs} \tag{6.334}$$

where rhs denotes the right-hand side of equation (6.322).

Exercise 6.33 Assume that the rate of change of temperature and pressure is equally important in the temperature equation (6.322). How large is the pressure step dp that corresponds to a temperature step $dT = 1\,°C$? Let $c_p = 1\,\text{kJ kg}^{-1}\,\text{K}^{-1}$, $\alpha = 2.5 \cdot 10^{-5}\,\text{K}^{-1}$, $\varrho = 2500\,\text{kg m}^{-3}$ and $T = 1000\,\text{K}$.
Solution: The step in temperature and pressure are equally important when $\varrho c_p\, dT = \alpha T\, dp$, which gives

$$dp = \frac{\varrho c_p}{\alpha T} dT = 100\,\text{MPa}. \tag{6.335}$$

Exercise 6.34 A body is composed of parts of mass m_k and volume V_k. The bulk density is $\varrho = m/V$ where $m = \sum_k m_k$ and $V = \sum_k V_k$. Show that

$$\varrho = \frac{m}{V} = \sum \phi_k \varrho_k \tag{6.336}$$

where $\phi_k = V_k/V$ is the volume fraction of component k and $\varrho_k = m_k/V_k$ is the density of component k.

6.22 Mantle adiabat

There are convection cells in the mantle. Let us assume that the mantle upwells or downwells sufficiently fast for almost no heat to be lost from a small volume of mantle rock to its surroundings, when it follows the flow. Such a process, without any loss of heat, is called an adiabatic process. It is tempting to think that the temperature would remain constant in a volume that does not exchange heat with its surroundings. That would also have been the case if the mantle had been incompressible. To see that we look at the temperature equation (6.322), which gives

$$\varrho c_p \frac{DT}{dt} - \alpha T \frac{Dp}{dt} = 0 \tag{6.337}$$

when there is no conductive heat transfer. An adiabatic process in thermodynamics is a process with constant entropy, and we see that equation (6.337) gives that $ds = 0$ when compared with expression (6.321) for the change in entropy. The adiabat is therefore given as

$$\frac{dT}{dp} = \frac{T\alpha}{\varrho c_p} \quad \text{or} \quad \frac{\partial T}{\partial z} = \frac{T\alpha g}{c_p} \tag{6.338}$$

when using $p = \varrho g \, z$. The adiabatic temperature gradient becomes $dT/dz \approx 0.5\,°C/km$ using the parameters $T = 1300\,°C$ (or 1573 K), $\alpha = 3 \cdot 10^{-5}\,K^{-1}$ and $c_p = 1\,kJ\,kg^{-1}\,K^{-1}$. The gradient (6.338) can be rewritten as

$$\frac{dT}{T} = \frac{\alpha g}{c_p}dz \tag{6.339}$$

which is straightforward to integrate:

$$T = T_0 \exp\left(\frac{\alpha g}{c_p}(z - z_0)\right) = T_0 \exp\left(\frac{z - z_0}{l_0}\right) \tag{6.340}$$

where T_0 is the temperature at the reference position z_0. (Both T and T_0 are defined as absolute temperatures in kelvin.) The second equality expresses the adiabat in terms of the characteristic length scale $l_0 = c_p/(\alpha g)$. We can approximate the adiabat as a linear geotherm

$$T = T_0\left(1 + \frac{z - z_0}{l_0}\right) \tag{6.341}$$

as long as $|z - z_0| \ll l_0$. The numbers above give that the length scale is typically $l_0 \approx 3 \cdot 10^5$ km, which suggests that the geotherm is linear over several hundred km. Figure 6.41 shows a mantle adiabat in the upper asthenosphere and a geotherm through the lithosphere. The geotherm in the lithosphere is at 1300 °C at the depth 120 km, through radioactive heat

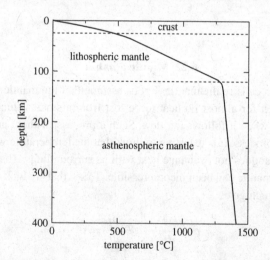

Figure 6.41. *The figure shows the geotherm through the lithosphere and into the upper part of the asthenosphere.*

production in the crust (10^{-6} W m^{-3}). The geotherm in Figure 6.41 is the combination of the temperature solution (6.59)–(6.60) and the linear mantle geotherm (6.341).

6.23 Further reading

Carslaw and Jaeger (1959) solve the temperature equation for a large number of problems with respect to boundary conditions, initial conditions, spatial dimensions and geometry. They treat among other topics stationary problems, time-dependent problems, convection and heat production. The mathematical foundations for the solutions are also covered. This is probably the most comprehensive collection of solutions of the temperature equation that has been published. Jaeger (1964, 1968) presents several models for the cooling and solidification of sills and dikes. Turcotte and Schubert (1982) has a chapter on heat flow that covers a wide range of thermal phenomena. Bickle and McKenzie (1987) discuss the transport by heat and matter in terms of the Peclet number. They derive dimensionless equations and show analytical solutions. A starting point for further reading about convective and conductive heat flow in fractures is Vasseur and Demongodin (1995).

7

Subsidence

This chapter deals with the mechanisms responsible for subsidence and basin formation. Before we look at these mechanisms we will recapture a little bit of the plate tectonic setting of sedimentary basins.

The surface of the Earth is made of two types of rigid plates – those beneath the oceans and those beneath the continents. The oceanic plates are produced at mid-ocean ridges and are consumed by subduction beneath other plates at ocean trenches. The continents are placed on stable plates that are much older than the oceanic plates. *Plate tectonics* is the theory of these plates and their movements. The rigid part of the Earth's interior is called the *lithosphere* and the lower boundary for the lithosphere is given by an isotherm ($\sim$1300 °C) in the mantle. The mantle below this isotherm is called the *asthenosphere*, and it is sufficiently hot to behave like a fluid on a geological time scale. The lithosphere is viewed as a body that floats on a fluid-like mantle asthenosphere, and it behaves like a rigid body that can transmit stress over large distances. The forces that push a plate away from the ridge and pull the plate down into the Earth at an ocean trench are carried through the plate. The lithosphere is divided into the crust and the lithospheric mantle by the boundary named the Moho after the Croatian seismologist Mohorovičić who discovered the discontinuity in 1909. The Moho is a material contrast between a crust that is less dense than the lithospheric mantle below. The crust of the continental lithosphere is older, less dense, more heterogeneous and thicker than the oceanic crust (see Figure 7.1). We will see that the reason why large parts of the continental plates are above sea level is because the continental crust is thicker than the oceanic crust.

In this chapter we will look at basin subsidence related to isostatic equilibrium, crustal stretching and thermal subsidence from decay of temperature transients caused by stretching.

7.1 Isostatic subsidence

The lithosphere is a rigid plate floating on the ductile asthenosphere like a block of wood floating on water, and when it is loaded it floats deeper. The lithosphere is, for instance, loaded when sediments are deposited in a water-filled basin. The pressure remains the

The Thickness of the Earth's Crust

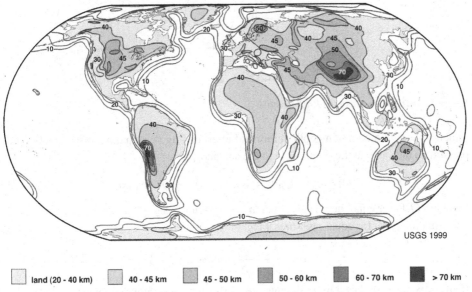

| land (20 - 40 km) | 40 - 45 km | 45 - 50 km | 50 - 60 km | 60 - 70 km | > 70 km |

Figure 7.1. *A plot of the Earth's crustal thickness made by USGS (1999). The continental crust is normally thick at the center of the continents and under high mountains, like the Andes and Tibet, and thin at the margins.*

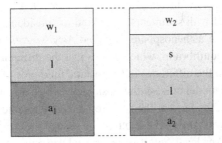

Figure 7.2. *The pressure at the same depth in the ductile asthenosphere remains the same after loading the rigid lithosphere with sediments. The sediments replace some of the water (a=asthenosphere, l=lithosphere, s=sediments and w=water).*

same at the same depth in the ductile asthenosphere after loading, because of its fluid-like behavior. See Figure 7.2. The same depth before and after deposition of sediments is

$$w_1 + l + a_1 = w_2 + s + l + a_2 \qquad (7.1)$$

where w_1 is the water depth before loading, w_2 is the water depth after loading, s is the sediment thickness and l is the thickness of the lithosphere. The thickness of a part of the

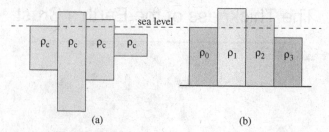

Figure 7.3. *(a) Airy isostasy (b) Pratt isostasy.*

asthenosphere to a common depth before and after loading is a_1 and a_2, respectively. The pressure is the same at the depth (7.1) in the asthenosphere

$$\varrho_w w_1 + \varrho_l l + \varrho_a a_1 = \varrho_w w_2 + \varrho_s s + \varrho_l l + \varrho_a a_2. \tag{7.2}$$

Equation (7.1) for equal depths gives that $a_2 = a_1 + w_1 - w_2 - s$, and when a_2 is inserted into equation (7.2) for equal pressures we get the subsidence

$$s = \frac{(\varrho_a - \varrho_w)}{(\varrho_a - \varrho_s)}(w_1 - w_2). \tag{7.3}$$

If all the water is replaced by sediments, $w_2 = 0$, we get a sediment thickness s that is the initial water depth w_1 multiplied by the factor $f = (\varrho_a - \varrho_w)/(\varrho_a - \varrho_s)$. The factor is greater than 1 because sediments have a higher density than water ($\varrho_s > \varrho_w$), and therefore makes the lithosphere float deeper. If we let $\varrho_w = 1000 \text{ kg m}^{-3}$, $\varrho_s = 2300 \text{ kg m}^{-3}$ and $\varrho_a = 3300 \text{ kg m}^{-3}$ we get the factor $f = 2.3$. A water-filled basin becomes more than twice as deep when filled with sediments. This kind of subsidence, which is caused by a lithosphere floating on the asthenosphere, is called *Airy isostatic subsidence*. Isostasy is the same as hydrostatic equilibrium, which means that the difference in pressure between any two points is equal to the difference in hydrostatic fluid pressure.

There is an alternative model of isostatic subsidence, which assumes lateral density variations in such a way that the base of the crust remains at the same depth. See Figure 7.3. This model is called *Pratt isostasy*. The depth of the base of the crust is b at the lateral position where the crust is at sea level, and ϱ_0 is the crustal density at this position. Hydrostatic equilibrium implies that the weight of all columns in Figure 7.3b have the same weight. The crustal density is denoted ϱ_p at the position where the crust has a height h above sea level. We have that $\varrho_p(b + h) = \varrho_0 b$, and the crustal density becomes

$$\varrho_p = \varrho_0 \left(\frac{b}{b+h}\right). \tag{7.4}$$

The lateral density of the crust is therefore a function of the crustal height above sea level. In a similar way, if the crust is at a depth h below sea level the weight of the columns are $\varrho_p(b - h) + \varrho_w h = \varrho_0 b$, which gives

$$\varrho_p = \frac{\varrho_0 b - \varrho_w h}{b - h}. \tag{7.5}$$

Pratt isostasy is less used than Airy isostasy, probably because of the assumption of a crustal base at the same depth regardless of elevation. There are few observations that support such an assumption. On the other hand, Airy isostasy can easily be extended with lateral variations in density.

Exercise 7.1 Basins compact even if there is no sediment infill, because of porosity reduction from chemical processes. Let a sedimentary basin have the initial thickness S_0, the initial average porosity ϕ_0, and the initial water depth w_0. The average basin density is $\varrho = \phi \varrho_w + (1 - \phi)\varrho_s$, where ϱ_w is the water density and ϱ_s is the sediment matrix density.
(a) Assume Airy isostasy and find an expression for the water depth as a function of the average basin porosity. What is the maximum water depth when the average basin porosity is zero?
(b) Find the total compaction of the basin as a function of the average basin porosity.
Solution: (a) The Airy isostasy gives that pressure in the ductile mantle

$$w_0\varrho_w + S_0\varrho_0 + a_0\varrho_m = w\varrho_w + S\varrho + a_1\varrho_m \tag{7.6}$$

is the same at the same depth:

$$w_0 + S_0 + a_0 = w + S + a_1. \tag{7.7}$$

The basin has the thickness S and the water depth is w when the porosity is ϕ. The mantle thickness before and after the change of porosity is a and a_1, respectively. Elimination of a and a_1 gives

$$w = w_0 + \frac{S\varrho - S_0\varrho_0 - (S - S_0)\varrho_m}{\varrho_m - \varrho_w}. \tag{7.8}$$

The weight of the basin becomes

$$S\varrho = \left(\phi\varrho_w + (1 - \phi)\varrho_s\right)\frac{\zeta_0}{1 - \phi} = (e\varrho_w + \varrho_s)\zeta_0 \tag{7.9}$$

where $\zeta_0 = (1 - \phi_0)S_0$ is the net (porosity-free) thickness of the basin, and where $e = \phi/(1 - \phi)$ is the void ratio. The same calculation gives that the initial weight of the basin is $S_0\varrho_0 = (e_0\varrho_w + \varrho_s)\zeta_0$. The basin thickness is $S = (1 + e)\zeta_0$ in terms of the void ratio. When these are inserted into equation (7.8) we get

$$w = w_0 + (e_0 - e)\zeta_0. \tag{7.10}$$

The water depth grows linearly with decreasing void ratio. The maximum increase in water depth is therefore $\Delta w = e_0\zeta_0$. Expression (7.10) for the water depth is reasonable because $e\zeta_0$ measures the amount of water in the sedimentary basin with void ratio e. The difference $(e_0 - e)\zeta_0$ is therefore the amount of water lost from the basin when the void ratio goes from e to e_0.
(b) The increase in water depth from compaction of the basin is

$$w = w_0 + S_0 - S = w_0 + (e_0 - e)\zeta_0 \tag{7.11}$$

using that $S_0 = (1 + e_0)\zeta_0$ and $S = (1 + e)\zeta_0$.

Exercise 7.2

(a) Equation (7.3) gives the water depth that corresponds to the sediment thickness S. Find an expression for the uncertainty Δw in the water depth when there is an uncertainty in the average (bulk) sediment density $\Delta \varrho_s$. Express the uncertainty as a relative uncertainty $\Delta w/w$.

(b) Calculate the relative uncertainty for $\Delta \varrho_s = 250 \text{ kg m}^{-3}$, when $\varrho_s = 2300 \text{ kg m}^{-3}$ and $\varrho_a = 3300 \text{ kg m}^{-3}$.

Solution: **(a)** The water depth becomes $w = S(\varrho_a - \varrho_s)/(\varrho_a - \varrho_w)$ when a sedimentary basin of thickness S and density ϱ_s is replaced by water. The uncertainty Δw caused by the uncertainty $\Delta \varrho_s$ is then $\Delta w = -S \Delta \varrho_s/(\varrho_a - \varrho_w)$, and the relative uncertainty is $\Delta w/w = -\Delta \varrho_s/(\varrho_a - \varrho_s)$, which is independent of the sediment thickness.

(b) The numbers above for the densities give that $\Delta w/w = -0.25$. In other words, if the average bulk sediment density $\Delta \varrho_s = 250 \text{ kg m}^{-3}$ is too high then it implies that the water depth is 25% too low.

Exercise 7.3 The bulk sediment density is $\varrho_b = \phi \varrho_w + (1 - \phi)\varrho_s$, where ϱ_w is the water density, ϱ_s is the sediment matrix density and ϕ the porosity. Assume that the Athy function, $\phi = \phi_0 \exp(-z/z_0)$, gives the basin porosity as a function of depth, where ϕ_0 is the porosity at the basin surface and z_0 is a depth that characterizes the compaction. The bulk density averaged over the basin is $\bar{\varrho}_b = (1/s) \int_0^s \varrho_b \, dz$, where the basin surface is at $z = 0$ and the s is the thickness of the basin.

(a) Show that an uncertainty Δz_0 in the parameter z_0 implies an uncertainty $\Delta \bar{\varrho}_b = -(\varrho_s - \varrho_w)\bar{\phi} \, \Delta z_0/z_0$ in the basin average of the sediment density, where $\bar{\phi}$ is the basin average of the porosity.

(b) Use the result of Exercise 7.2 to compute relative uncertainty in the water depth that corresponds to the sediments when $\Delta z_0/z_0 = 0.2$, $\bar{\phi} = 0.15$, $\varrho_s = 2300 \text{ kg m}^{-3}$.

7.2 Thickness of crustal roots

Isostasy can be used to obtain the depth of the crustal roots of the continents. We already know that the continents are made of thicker, older and also less dense crustal rocks than the oceanic crust. It is now assumed for simplicity that the continental and the oceanic crust have the same density ϱ_c. The oceanic crust has the thickness c and it is covered by water of depth w, see Figure 7.4. The principle of isostasy requires that the pressures at the same depth in the mantle beneath the oceanic crust and the continental crust are the same. We therefore have the following equality:

$$w\varrho_w + c\varrho_c + d\varrho_m = (d + c + w + h)\varrho_c \tag{7.12}$$

where the left-hand side is the pressure in the mantle beneath the oceanic crust and the right-hand side is the pressure beneath the continent. The height of the continent above the sea level is h and the depth of the crustal root below the base of the oceanic crust is d (see Figure 7.4). Rearranging equation (7.12) gives

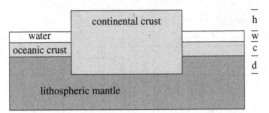

Figure 7.4. *The continent has a height h above sea level and a crustal root of thickness d below the oceanic crust. The oceanic crust has a thickness c with water of depth w above.*

$$h\varrho_c = d\,(\varrho_m - \varrho_c) - w\,(\varrho_c - \varrho_w),\qquad (7.13)$$

which says that weight of the continent above sea level is equal to the buoyancy from the mantle displaced by the root of the continental crust minus the buoyancy from the water being displaced by the continental crust. The thickness of the crustal root of the continent becomes

$$d = \frac{\varrho_c}{(\varrho_m - \varrho_c)}h + \frac{(\varrho_c - \varrho_w)}{(\varrho_m - \varrho_c)}w.\qquad (7.14)$$

As an example we can calculate the crustal root of a mountain range with an elevation 3 km above sea level when there is an ocean with depth 1 km over the oceanic crust. The densities of the crust and mantle are $\varrho_c = 2900$ kg m^{-3} and $\varrho_m = 3300$ kg m^{-3}, respectively. The thickness of the crustal root is then $d = 26.5$ km. The total thickness of the continental crust becomes $h_{\text{tot}} = h + w + c + d = 37$ km, when an average thickness $c = 6$ km is used for the oceanic crust. This estimate is close to the average thickness of the continental crust, which is 35 km.

Exercise 7.4 A floating iceberg is similar to a continent floating on the asthenosphere. Use the same reasoning as above to show that the submerged base of an iceberg is nine times the size of its top when the density of ice is 90% of the water density.

Exercise 7.5 A mountain erodes and becomes 1 km lower. How much has then been eroded from the mountain?
Solution: A mountain of height h has a crustal root of depth b. Airy isostasy gives that $(h + b)\varrho_c = b\varrho_m$, where ϱ_c and ϱ_m are the crustal and mantle densities, respectively. The crustal root becomes $b = h\varrho_c/(\varrho_m - \varrho_c)$. The height of the mountain is reduced with Δh, and the crustal root is therefore reduced with $\Delta b = \Delta h\,\varrho_c/(\varrho_m - \varrho_c)$. The total amount of crust that is eroded from the mountain is then

$$\Delta h + \Delta b = \left(\frac{\varrho_m}{\varrho_m - \varrho_c}\right)\Delta h.\qquad (7.15)$$

A crustal density $\varrho_c = 2800$ kg m^{-3} and a mantle density $\varrho_c = 3300$ kg m^{-3} give that 6.6 km must be eroded from the mountain to reduce its height by $\Delta h = 1$ km.

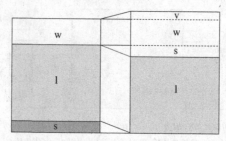

Figure 7.5. *An increase in the sea level creates isostatic subsidence.*

7.3 Subsidence from eustatic sea level changes

Eustacy is world-wide fluctuations in the sea level due to a changing capacity of the world basins or a changing volume of ocean water. Global sea level changes have consequences for basin subsidence when isostasy is assumed. Figure 7.5 shows a situation where the sea level has increased by an amount v. The basin is simply filled with more water until the sea level has risen with a distance v relative to the initial sea level. The assumption of isostasy says that pressures at the same depths in the asthenosphere are the same. We therefore have

$$\varrho_w w + \varrho_l l + \varrho_a s = \varrho_w (v + w + s) + \varrho_l l \tag{7.16}$$

where s is the subsidence caused by the sea level rise, and l is the thickness of the lithosphere (see Figure 7.5). The densities of water, lithosphere and asthenosphere are ϱ_w, ϱ_l and ϱ_a, respectively. Equation (7.16) for isostatic equilibrium gives a basin subsidence

$$s = \frac{\varrho_w}{\varrho_a - \varrho_w} v \tag{7.17}$$

from a sea level rise v, and the total increase in water depth after the sea level rise becomes

$$v + s = \frac{\varrho_a}{\varrho_a - \varrho_w} v. \tag{7.18}$$

If the asthenosphere and the water have the densities $\varrho_m = 3300 \text{ kg m}^{-3}$ and $\varrho_w = 1000 \text{ kg m}^{-3}$, respectively, the isostatic subsidence s is 0.43 times the sea level change and the change in total water depth $v + s$ is 1.43 times the sea level change. Attempts to map eustatic sea level fluctuations suggest that they are a maximum of 200 m (Haq *et al.*, 1987), which means that they are not very important for the subsidence of a basin.

7.4 Basin subsidence by crustal thinning

Many sedimentary basins, for instance those in the North Sea and the Mid-Norwegian Margin, were formed by stretching and thinning of the continental crust. These basins have gone through several periods of extension, and the most recent one was at the end of the Cretaceous period. The crust is less dense than the mantle and crustal thinning therefore

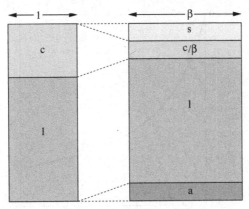

Figure 7.6. *A basin of depth s forms when the crust is stretched by a factor β. The initial thickness of the crust is c, the initial thickness of lithospheric mantle is l and a is the thickness of the asthenosphere that moves upwards and becomes the lower part of the stretched lithosphere.*

leads to a heavier lithosphere that floats deeper on the ductile mantle. The stretching and uniform thinning of the crust is shown in Figure 7.6. An initially unstretched lithosphere is shown to the left and the same lithosphere after uniform stretching and subsidence is shown to the right. From mass conservation we get that the initial thickness of the crust c becomes reduced to c/β when it is stretched uniformly by a factor β. Figure 7.6 also shows the upwelling asthenospheric mantle that has cooled and formed new lithospheric mantle. The subsidence of the stretched lithosphere is found by isostasy. We will assume that the lithosphere is stretched so slowly that the temperature through the lithosphere stays stationary even though there is a flow of mantle rocks upwards. Isostasy says that the pressure at the same depth in the asthenosphere remains the same:

$$\varrho_c c + \varrho_m l = \varrho_s s + \varrho_c (c/\beta) + \varrho_m l + \varrho_m a. \tag{7.19}$$

We have that l is the initial thickness of the lithospheric mantle, s is the thickness of the sedimentary basin, and a is the thickness of the asthenosphere that moves upwards due to stretching and becomes part of the new lithosphere. The pressure in equation (7.19) is at the top of the asthenosphere, which is at the depth

$$c + l = s + (c/\beta) + l + a. \tag{7.20}$$

We see that all terms with l drop out in both equations (7.19) and (7.20), and by inserting $a = c - (c/\beta) - s$ from equation (7.20) into equation (7.19) we get that the subsidence s as a function of the stretching factor β is

$$s = \frac{(\varrho_m - \varrho_c)}{(\varrho_m - \varrho_s)} \left(1 - \frac{1}{\beta}\right) c. \tag{7.21}$$

The basin subsidence s divided by the initial thickness of the crust is plotted as a function of β in Figure 7.7 for $\varrho_s = 2300 \text{ kg m}^{-3}$, $\varrho_c = 2900 \text{ kg m}^{-3}$ and $\varrho_m = 3300 \text{ kg m}^{-3}$. The

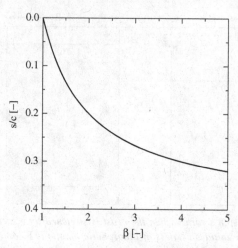

Figure 7.7. *The subsidence given by equation (7.21) divided by the initial thickness of the crust is plotted as a function of the β-factor. The following densities are used $\varrho_s = 2300\ kg\,m^{-3}$, $\varrho_c = 2900\ kg\,m^{-3}$ and $\varrho_m = 3300\ kg\,m^{-3}$.*

factor $(\varrho_m - \varrho_c)/(\varrho_m - \varrho_s)$ becomes 0.4 with these densities and the subsidence becomes 0.2 of the initial thickness of the crust when the crust is stretched by a factor $\beta = 2$. A basin of thickness $s = 6$ km can therefore be formed by stretching a continental crust of thickness $c = 30$ km with the β-factor 2.

7.5 The McKenzie model of basin subsidence

We will now take a closer look at the subsidence caused by stretching of the lithosphere. The result of stretching is that the crust gets thinned and that hot asthenospheric mantle moves upwards. The resulting model is the so-called McKenzie model, named after McKenzie (1978) who proposed this model. The McKenzie model explains basin formation by subsidence caused by stretching and thinning of the continental crust as shown in the preceding section. The crust is less dense than the mantle and crustal thinning therefore leads to a heavier lithosphere that floats deeper on the ductile mantle. The stretching of the lithosphere is illustrated in Figure 7.8, where a cross-section of the lithosphere with unit length is stretched by a factor β. The stretching factors are often named β-factors. The subsidence causing the thinning of the crust is counteracted by the uplift from a hotter lithosphere. The hot mantle that moves upwards has a lower density due to thermal expansion, which is the reason why the temperature of the lithosphere becomes important. Although the stretching of the lithosphere takes millions of years it is often sufficiently fast for instantaneous stretching to be a good approximation. (We will later estimate how fast the stretching must be for the assumption of instantaneous stretching to be valid.)

We see from Figure 7.8 that the stretched lithosphere moves upwards and that the lower part of the lithosphere is replaced by the asthenosphere. The temperature follows the

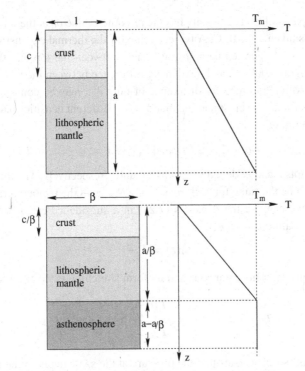

Figure 7.8. *An almost instantaneous and uniform stretching of the lithosphere leads to upwelling of the hot asthenosphere and a steeper thermal gradient in the lithosphere. The basin part is considered thin compared to the thickness of the lithosphere and its contribution to the temperature is therefore omitted.*

lithospheric material as it is instantaneously stretched and the geotherm becomes steeper after stretching. The initial geothermal gradient is $dT/dz = T_a/a$, where T_a is the temperature at the asthenosphere–lithosphere boundary and a is the initial thickness of the lithosphere. The temperature gradient immediately after stretching becomes a factor β-higher ($dT/dz = \beta T_a/a$). The asthenosphere that replaces the stretched lithosphere is assumed to have the temperature T_a. The elevated temperature of the lithosphere after stretching is a transient state, and the temperature will return to its initial state by conductive cooling. The asthenosphere–lithosphere boundary therefore moves back to its initial position as a consequence of conductive cooling.

The temperature dependence of a rock density can be approximated to first order as

$$\varrho = \varrho_0 \left(1 - \alpha(T - T_0) \right) \tag{7.22}$$

where

$$\alpha = \frac{1}{\varrho} \left(\frac{\partial \varrho}{\partial T} \right)_p \tag{7.23}$$

is the thermal expansibility. The density has the reference value ϱ_0 at the reference temperature T_0, which is taken to be $0\,°C$. A typical value for the thermal expansibility of mantle rocks is $3 \cdot 10^{-5}\,K^{-1}$. The temperature difference between the top and the base of the lithosphere is roughly $1000\,°C$, and the density difference between the base and the top is therefore $\sim 3\%$. Although a density difference of only 3% may be considered small it is nevertheless important. The crust and the mantle have different densities, and they are as a function of temperature:

$$\varrho_c(T) = \varrho_{c,0}(1 - \alpha T) \quad \text{and} \quad \varrho_m(T) = \varrho_{m,0}(1 - \alpha T) \tag{7.24}$$

where the subscripts c and m denotes crust and mantle, respectively. The thermal expansibility α is taken to be the same for both rock types. We need the temperature as a function of depth before we can use the densities (7.24) in a subsidence calculation. The initial lithospheric temperature is

$$T_0(z) = T_a \frac{z}{a} \tag{7.25}$$

and the temperature after instantaneous and uniform stretching is (see Figure 7.8)

$$T_I(z, t = 0^+) = \begin{cases} T_a \beta \dfrac{z}{a}, & 0 \le z \le \dfrac{a}{\beta} \\ T_a, & \dfrac{a}{\beta} < z \le a. \end{cases} \tag{7.26}$$

The assumption of isostasy states that the pressure at the same depth in the asthenosphere remains the same:

$$\int_0^c \varrho_c \, dz + \int_c^a \varrho_m \, dz =$$
$$\int_0^{c/\beta} \varrho_c dz + \int_{c/\beta}^{a/\beta} \varrho_m dz + \left(a - \frac{a}{\beta} - s_I\right)\varrho_m(T_a) + s_I \varrho_s. \tag{7.27}$$

The right-hand side is the pressure in the asthenosphere beneath the unstretched lithosphere, and the left-hand side is the pressure beneath the stretched lithosphere. The initial temperature (7.25) is therefore used in the densities on the left-hand side of equation (7.27), and the transient temperature (7.26) is used on the right-hand side. The contribution $\varrho_s s$ from a sedimentary basin of thickness s, where ϱ_s is the average density of the sediments, is added on the right-hand side. The sedimentary cover is considered so thin that any temperature dependence can be neglected. Isostatic equilibrium (7.27) gives the basin subsidence

$$s_I = a\left(1 - \frac{1}{\beta}\right) \frac{\left(\varrho_{m,0} - \varrho_{c,0}\right)\left(\frac{c}{a}\right)\left(1 - \frac{1}{2}\alpha T_a \frac{c}{a}\right) - \frac{1}{2}\varrho_{m,0}\alpha T_a}{\varrho_{m,0}\left(1 - \frac{1}{2}\alpha T_a\right) - \varrho_s} \tag{7.28}$$

after the integrations over the densities are carried out (see Note 7.1 for details). The subsidence (7.28) accounts for the hot lithosphere that results from the uniform and instantaneous stretching. Thermal expansion of the lithospheric mantle leads to reduced subsidence, which can be seen from equation (7.28). We notice that the last term $\varrho_{m,0}\alpha T_a/2$ can

make the subsidence negative, which means uplift. It follows from equation (7.28) that the
condition for subsidence ($s > 0$) is

$$\left(\varrho_{m,0} - \varrho_{c,0}\right)\left(\frac{c}{a}\right)\left(1 - \frac{1}{2}\alpha\, T_a \frac{c}{a}\right) > \frac{1}{2}\varrho_{m,0}\alpha\, T_a \tag{7.29}$$

which gives

$$\frac{c}{a} > \frac{1}{2}\frac{\varrho_{m,0}}{(\varrho_{m,0} - \varrho_{c,0})}\alpha\, T_a. \tag{7.30}$$

(The factor $1 - \frac{1}{2}\alpha\, T_a$ which is typically 0.97 has been approximated by 1.) Crustal thinning
implies subsidence, while thermal expansion is the cause for uplift, and condition (7.30)
states when the crust is sufficiently thick for subsidence from crustal thinning to be stronger
than thermal uplift. We get from inequality (7.30) that the condition for subsidence is
$c/a > 0.13$, when the densities for the crust and the mantle are $\varrho_c = 2800\ \mathrm{kg\,m^{-3}}$ and
$\varrho_m = 3300\ \mathrm{kg\,m^{-3}}$, respectively, thermal expansibility is $\alpha = 3\cdot 10^{-5}\ \mathrm{K^{-1}}$ and the temperature at the base of the lithosphere is $T_a = 1300\,°\mathrm{C}$. A typical thickness of the lithosphere is
120 km, which makes the minimum thickness of the continental crust 15 km for subsidence
to take place. If the continental crust is less thick than $\sim$15 km it becomes temporarily thermally uplifted. The average thickness of the continental crust is 35 km, which is sufficiently
thick for crustal thinning to imply subsidence.

Figure 7.9 shows the subsidence (7.28) for instantaneous and uniform stretching. The
subsidence is plotted as a function of the stretching factor β and the thickness of the crust;
the two parameters that control subsidence. A dashed line shows the subsidence given by
equation (7.21) for the case of no thermal expansion and uplift. The latter subsidence is
the maximum possible subsidence for a given β-factor, and it is therefore denoted s_{max}.

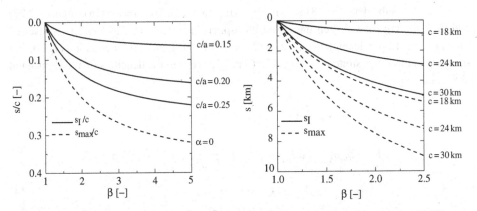

Figure 7.9. *The solid curves in both plots show the subsidence s_I given by equation (7.28) as a
function of the stretching factor β, and the thickness of the crust c. The dashed curve is the maximum
subsidence s_{max} given by equation (7.21), where thermal uplift is absent. The curve to the left shows
the subsidence relative to the crust and the curve to the right shows the subsidence in units km. It
is seen from the plot to the left, as well as equation (7.21), that s_{max}/c is only a function of the
stretching factor β.*

It is shown in the next section how conductive cooling leads to further (thermal) subsidence through time, and that the maximum amount of thermal subsidence is the difference between initial subsidence (with thermal expansion and thermal uplift) and the subsidence without thermal expansion. The maximum thermal subsidence in Figure 7.9 is therefore the difference between the solid line and the dashed line for the same crustal thickness. The initial subsidence (7.28) can be rewritten as

$$s_I = s_{max} - s_{T,max} \tag{7.31}$$

where

$$s_{max} = \left(1 - \frac{1}{\beta}\right)\frac{(\varrho_{m,0} - \varrho_{c,0})}{(\varrho_{m,0} - \varrho_s)}c \tag{7.32}$$

is the maximum subsidence and where

$$s_{T,max} = \frac{1}{2}\left(1 - \frac{1}{\beta}\right)\frac{\varrho_{m,0}}{(\varrho_{m,0} - \varrho_s)}\alpha T_a a \tag{7.33}$$

is the maximal thermal subsidence. Expression (7.31) was obtained by a slight simplification of expression (7.28) using the approximation $1 - \alpha T_a/2 \approx 1$. The maximum thermal subsidence is the same as the maximum uplift caused by the upwelling hot mantle. In a later section we will see how the initial thermal uplift dies out with the thermal transient. The condition (7.30) for initial subsidence to be positive is simply an expression for when the maximum subsidence (s_{max}) is larger than maximum thermal subsidence ($s_{T,max}$). The maximum thermal subsidence becomes $s_{T,max}/c = 0.07, 0.11$ and 0.13 for the three stretching factors $\beta = 1.5, 2$ and 2.5 used in Figure 7.9.

Note 7.1 The subsidence (7.28) is found by carrying out the integrations in equation (7.27) and then solving for s. When the initial lithospheric temperature $T_0(z)$ (7.25) is inserted into the temperature dependent densities (7.24) on the left-hand side of equation (7.28), and the temperature after stretching $T_I(z)$ (7.26) is inserted into the densities on the right-hand side, and we get

$$\int_0^c \varrho_{c,0}\left(1 - \alpha T_a \frac{z}{a}\right) dz + \int_c^a \varrho_{m,0}\left(1 - \alpha T_a \frac{z}{a}\right) dz$$
$$= \int_0^{c/\beta} \varrho_{c,0}\left(1 - \alpha T_a \beta \frac{z}{a}\right) dz + \int_{c/\beta}^{a/\beta} \varrho_{m,0}\left(1 - \alpha T_a \beta \frac{z}{a}\right) dz$$
$$+ \left(a - \frac{a}{\beta} - s\right)\varrho_{m,0}(1 - \alpha T_a) + s\varrho_s. \tag{7.34}$$

The integrations are straightforward to carry out, and when the terms with the subsidence are collected on the left-hand side, we arrive at the subsidence (7.28).

Exercise 7.6 Verify that the initial subsidence (7.28) (s_I) becomes expression (7.21) for maximum subsidence (s_{max}) with constant densities when the thermal expansibility is zero.

Exercise 7.7 The initial subsidence caused by crustal thinning is s_I, the maximum subsidence is s_{max} and the maximum thermal subsidence is $s_{T,max}$.
(a) Show that $s_I > s_{T,max}$ implies that $s_{max} > 2 s_{T,max}$.
(b) Show that $s_{max} > 2 s_{T,max}$ implies that

$$\frac{c}{a} > \frac{\rho_{m,0}}{\rho_{m,0} - \rho_{c,0}} \alpha T_a. \tag{7.35}$$

(c) Compare condition (7.30) for when the initial subsidence is larger than the thermal subsidence with condition (7.35) for when the initial subsidence is positive.

7.6 The thermal transient of the McKenzie model

Instantaneous and uniform stretching of the lithosphere leads to upwelling of the hot asthenosphere of temperature T_a and heating of the lithosphere by the mantle and crust that move upwards. The entire lithosphere gets hotter as shown in Figure 7.10. The temperature after stretching $T_I(z)$ then decays to the steady state temperature $T_0(z)$ by conductive cooling. The temperature equation is now solved for the transient cooling back to the steady state temperature. The temperature at any time after stretching is written

$$T(z, t) = T_0(z) + U(z, t), \tag{7.36}$$

where $U(z, t)$ is the thermal transient that decays to zero, and where the steady state temperature is $T_0(z) = T_a z/a$. The temperature equation for conductive cooling is

$$\frac{\partial T}{\partial t} - \kappa \frac{\partial^2 T}{\partial z^2} = 0 \tag{7.37}$$

as shown in Section 6.1. When the temperature (7.36) is inserted into the temperature equation (7.37) we get that U is also a solution of the temperature equation,

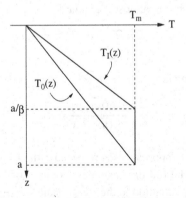

Figure 7.10. *The steady state temperature $T_0(z)$ and the temperature $T_I(z)$ immediately after instantaneous and uniform stretching.*

$$\frac{\partial U}{\partial t} - \kappa \frac{\partial^2 U}{\partial z^2} = 0. \tag{7.38}$$

The initial condition for $U(z, t)$ is the difference between the temperature immediately after stretching and the steady state temperature,

$$U(z, t = 0) = T_I(z) - T_0(z) = \begin{cases} T_a\left(\beta - 1\right)\frac{z}{a}, & 0 \le z < a/\beta \\ T_a\left(1 - \frac{z}{a}\right), & a/\beta \le z < a. \end{cases} \tag{7.39}$$

Boundary conditions for $U(z, t)$ are

$$U(z = 0, t) = 0 \quad \text{and} \quad U(z = a, t) = 0 \tag{7.40}$$

because $T_0(z) = T_I(z)$ for $z = 0$ and $z = a$. The solution of the temperature equation (7.38) is simplified when it is turned into dimensionless form. The dimensionless temperatures are $\hat{T} = T/T_a$ and $\hat{U} = U/T_a$, because T_a is a natural choice for a characteristic temperature. The thickness of the lithosphere is the characteristic length and the dimensionless z-coordinate becomes $\hat{z} = z/a$. The scaled version of the steady state temperature $T_0(z)$ is simply $\hat{T}(\hat{z}) = \hat{z}$. The characteristic length a and the thermal diffusivity gives the characteristic time

$$t_0 = \frac{a^2}{\kappa}. \tag{7.41}$$

The dimensionless temperature equation for $\hat{U}$ becomes the parameterless equation

$$\frac{\partial \hat{U}}{\partial \hat{t}} - \frac{\partial^2 \hat{U}}{\partial \hat{z}^2} = 0 \tag{7.42}$$

when time is scaled as $\hat{t} = t/t_0$. The initial condition becomes simplified to

$$\hat{U}(\hat{z}, t = 0) = \begin{cases} (\beta - 1)\hat{z}, & 0 \le \hat{z} < 1/\beta \\ (1 - \hat{z}), & 1/\beta \le \hat{z} < 1 \end{cases} \tag{7.43}$$

while the boundary conditions are almost the same

$$\hat{U}(\hat{z} = 0, \hat{t}) = 0 \quad \text{and} \quad \hat{U}(\hat{z} = 1, \hat{t}) = 0 \tag{7.44}$$

in dimensionless variables. The solution of the temperature equation (7.42) with the initial condition (7.43) and the boundary conditions (7.44) is (as shown in Note 7.2)

$$\hat{U}(\hat{z}; \hat{t}) = 2\beta \sum_{n=1}^{\infty} \frac{\sin(n\pi/\beta)}{(n\pi)^2} \sin(n\pi\hat{z}) \, e^{-(n\pi)^2 \hat{t}}. \tag{7.45}$$

The transient solution $\hat{U}(\hat{z}, \hat{t})$ approaches zero because the factors $e^{-(n\pi)^2 \hat{t}}$ approach zero as time $\hat{t}$ goes to infinity. The full dimensionless solution of the temperature equation is found by adding the transient solution to the steady state solution,

$$\hat{T}(\hat{z}, \hat{t}) = \hat{z} + \hat{U}(\hat{z}, \hat{t}). \tag{7.46}$$

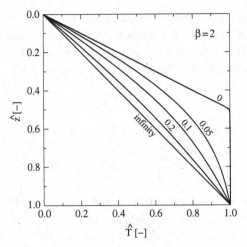

Figure 7.11. *The transient decay of the temperature towards the steady state temperature after instantaneous and uniform stretching with the factor $\beta = 2$ at the times $\hat{t} = 0$, 0.05, 0.1, 0.2 and ∞. The Fourier series (7.45) gives the transient part of the time-dependent temperature solution (7.46).*

The temperature with units is straightforward to obtain using the scaling relations that defined the dimensionless variables, and we have

$$T(z, t) = T_a \, \hat{T}\left(\frac{z}{a}, \frac{t}{t_0}\right). \tag{7.47}$$

The transient solution (7.46) is shown in Figure 7.11, where we notice that the temperature is roughly halfway between the initial temperature $\hat{T}_I(\hat{z})$ and the steady state temperature $\hat{T}_0(\hat{z})$ at $\hat{t} \approx 0.1$. This observation also follows from the transient part (7.45) of the temperature solution, where the factor $e^{-(n\pi)^2\hat{t}}$ has the slowest decay to zero for the first term ($n = 1$). This factor also controls the largest Fourier coefficient, and the term $n = 1$ can therefore be used to estimate the half-life $\hat{t}_{1/2}$ of the transient. It gives that

$$e^{-\pi^2 \hat{t}_{1/2}} = \frac{1}{2} = e^{-\ln 2} \tag{7.48}$$

which yields the half-life

$$\hat{t}_{1/2} = \frac{\ln 2}{\pi^2} = 0.07. \tag{7.49}$$

The estimated half-life is close to what Figure 7.11 gives. The half-life with units is found by multiplication with the characteristic time t_0 from equation (7.41). A lithosphere that has a thickness $a = 120$ km and a thermal diffusivity $\kappa = 1 \cdot 10^{-6} \mathrm{m^2 \, s^{-1}}$ has the characteristic time $t_0 = a^2/\kappa = 470$ Ma. The half-life is then $t_{1/2} = 33$ Ma. The transient has nearly died out at $\hat{t} = 0.2$ or $t = 94$ Ma. The thickness of the lithosphere is important for the characteristic time t_0, and the thermal transients last less time for a thinner lithosphere and longer for a thicker lithosphere.

Note 7.2 The solution of the temperature equation (7.42) for the transient $\hat{U}(\hat{z}, \hat{t})$ is found by separation of variables. The transient temperature is then written as $\hat{U}(\hat{z}, \hat{t}) = V(\hat{t}) W(\hat{z})$, where $V(\hat{t})$ is only a function of $\hat{t}$ and the function $W(\hat{z})$ is only a function of $\hat{z}$. When $V(\hat{t}) W(\hat{z})$ is inserted into the temperature equation (7.42) we get

$$V'W - VW'' = 0. \tag{7.50}$$

(A prime denotes differentiation.) Equation (7.50) can be rewritten as

$$\frac{V'}{V} = \frac{W''}{W} = -\lambda^2 \text{ (constant)}. \tag{7.51}$$

The equality $V'/V = W''/W$ implies that the ratios V'/V and W''/W are equal to the same constant, because V'/V is only a function of $\hat{t}$, and W''/W is only a function of $\hat{z}$. The constant is written as $-\lambda^2$, where it is anticipated that the constant is a negative number. Equation (7.51) is two equations, $V'/V = -\lambda^2$ for $V(\hat{t})$, and $W''/W = -\lambda^2$ for $W(\hat{z})$. The solution of equation $V' = -\lambda^2 V$ is $V(\hat{t}) = b \exp(-\lambda^2 \hat{t})$ as shown in Exercise 6.5, where coefficient b is the value of V at $\hat{t} = 0$. It is seen by inspection that $\cos(\lambda \hat{z})$ and $\sin(\lambda \hat{z})$ are solutions of the equation $W'' = -\lambda^2 W$. Only the sine function fulfills the boundary condition (7.44) at $\hat{z} = 0$, and can therefore be used. The other boundary condition requires $W(\hat{z} = 1) = 0$, which is only possible for $\lambda = n\pi$ when n is an integer larger than zero. It is not just one product $V(\hat{t})W(\hat{z})$ that solves the temperature equation, but every product:

$$V_n(\hat{t})W_n(\hat{z}) = b_n \sin(n\pi\hat{z}) e^{-(n\pi)\hat{t}}, \quad n \geq 1 \tag{7.52}$$

is a solution of the temperature equation. The transient temperature can therefore be written as a linear combination of all such products as the Fourier series

$$\hat{U}(\hat{z}, \hat{t}) = \sum_{n=1}^{\infty} b_n \sin(n\pi\hat{z}) e^{-(n\pi)\hat{t}}. \tag{7.53}$$

The coefficients b_n in the Fourier series are obtained from the initial condition (7.43), which says that

$$\sum_{n=1}^{\infty} b_n \sin(n\pi\hat{z}) = \begin{cases} (\beta - 1)z, & 0 \leq z < 1/\beta \\ (1 - z), & 1/\beta \leq z < 1. \end{cases} \tag{7.54}$$

The unknown coefficients b_n are found after both sides of equation (7.54) are multiplied with $\sin(m\pi)$ and then integrated over $\hat{z}$ from 0 to 1. We can then utilize that the functions $W_n(\hat{z}) = \sin(n\pi)$ are orthogonal with respect to an inner product defined by integration from 0 to 1,

$$(W_n, W_m) = \int_0^1 \sin(n\pi\hat{z}) \sin(m\pi\hat{z})d\hat{z} = \begin{cases} 1/2, & n = m \\ 0, & n \neq m. \end{cases} \tag{7.55}$$

The Fourier coefficients are therefore given by

$$\frac{1}{2} b_n = \int_0^{1/\beta} (\beta - 1)\hat{z} \sin(n\pi\hat{z})\, d\hat{z} + \int_{1/\beta}^1 (1 - \hat{z}) \sin(n\pi\hat{z})\, d\hat{z}, \tag{7.56}$$

where the integrations on the right-hand side finally lead to

$$b_n = 2\beta \frac{\sin(n\pi/\beta)}{(n\pi)^2}. \tag{7.57}$$

The three integrations needed to find the Fourier coefficients are given as Exercise 7.9.

Exercise 7.8 Derive the orthogonality relationship (7.55).

Exercise 7.9
(a) Derive the following integrals:

$$\int_0^1 \hat{z} \sin(n\pi\hat{z})\, d\hat{z} = \frac{(-1)^n}{n\pi} \tag{7.58}$$

$$\int_0^{1/\beta} \hat{z} \sin(n\pi\hat{z})\, d\hat{z} = -\frac{\cos(n\pi/\beta)}{n\pi\beta} + \frac{\sin(n\pi/\beta)}{(n\pi)^2} \tag{7.59}$$

$$\int_{1/\beta}^1 \sin(n\pi\hat{z})\, d\hat{z} = \frac{(-1)^n}{n\pi} + \frac{\cos(n\pi/\beta)}{n\pi}. \tag{7.60}$$

(b) Use the integrals to verify that equation (7.56) gives the Fourier coefficients b_n (7.57).

7.7 The surface heat flow of the McKenzie model

Estimates for the surface heat flow are interesting because the heat flow can be measured at the surface, as opposed to the heat flow at large depths. The temperature solution (7.46) gives the surface heat flow

$$q_s = \lambda \left. \frac{\partial T}{\partial z} \right|_{z=0} = \lambda \frac{T_a}{a} \left. \frac{\partial \hat{T}}{\partial \hat{z}} \right|_{\hat{z}=0} \tag{7.61}$$

when it is expressed by the dimensionless temperature solution, and where λ is the heat conductivity. The heat flow through the lithosphere before stretching is $q_0 = \lambda T_a/a$, where T_a/a is the thermal gradient of the steady state temperature. The surface heat flow relative to the steady state heat flow is therefore

$$\hat{q} = \frac{q_s}{q_0} = \left. \frac{\partial \hat{T}}{\partial \hat{z}} \right|_{\hat{z}=0} = 1 + 2\beta \sum_{n=1}^{\infty} \frac{\sin(n\pi/\beta)}{n\pi} e^{-(n\pi)^2 \hat{t}}. \tag{7.62}$$

The first term 1 corresponds to the steady-state heat flow and the sum over n is the transient part of the surface heat flow, which decays to zero with time. Figure 7.12 shows the surface heat flow for $\beta = 2$. It is at its maximum immediately after instantaneous stretching, when

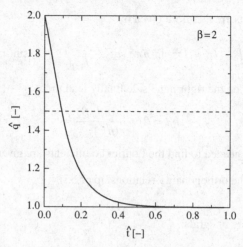

Figure 7.12. *The scaled surface heat flow after instantaneous stretching with a factor β = 2.*

the surface heat flow is a factor β higher than the steady state heat flow. Only 20% is left of the transient after $\hat{t} \approx 0.2$, and it has almost died out at $\hat{t} \sim 0.5$. The half-life of the series (7.62) was estimated in the previous section to be $\hat{t} = 0.07$, which is in good agreement with Figure 7.12. The characteristic time for the lithosphere is roughly 470 Ma, and the time $\hat{t} \sim 0.2$ is therefore ~94 Ma. The increased surface heat flow from rapid stretching of the lithosphere with a factor $\beta = 2$ will therefore last up to 100 Ma after a stretching event.

The surface heat flow (6.50) in the case of radioactive heat generation in the crust is

$$q_s = S c + q_m \tag{7.63}$$

where S is the heat production per volume of rock, c is the thickness of the crust and q_m is the heat flow from the mantle into the base of the crust. It is now possible to make simple estimates of what the surface heat flow is right after and a long time after a stretching event, when relation (7.63) gives the surface heat flow before stretching. The surface heat flow right after instantaneous stretching becomes

$$q_s \approx S \frac{c}{\beta} + \beta q_m \tag{7.64}$$

because the heat producing crust is thinned by the β-factor and the mantle heat flow increases by factor β. A long time after, when the thermal transient has died out, the surface heat flow becomes

$$q_s \approx S \frac{c}{\beta} + q_m. \tag{7.65}$$

The stationary surface heat flow is therefore reduced by a rift phase because a stretched and thinned crust produces less heat than the unstretched crust. The loss in heat production from the crust is often counteracted by heat production in the sediments that fill the basin.

It is also possible for the transient surface heat flow to be reduced right after the stretching event too, if the crust is sufficiently heat producing, as shown in Exercise 7.10.

Exercise 7.10
(a) Show that the surface heat flow (7.64) after instantaneous stretching increases compared to the initial heat flow (7.63) only if

$$\beta > \frac{Sc}{q_m}. \tag{7.66}$$

We see that we will always have an increased heat flow after stretching when $q_m > Sc$ because β is never less than 1.
(b) How large must β be for the heat flow to become larger right after stretching when $S = 10^{-6} \, \text{W m}^{-3}$, $c = 30 \, \text{km}$ and $q_m = 15 \, \text{mW m}^{-2}$?

Note 7.3 (Mathematical curiosity) The surface heat flow immediately after stretching relative to the steady state heat flow is $\hat{q} = \beta$. (It is simply the thermal gradient of the temperature immediately after stretching $T_I(z)$ over the gradient of steady state temperature $T_0(z)$.) From equation (7.62) we also have

$$\hat{q} = 1 + 2\beta \sum_{n=1}^{\infty} \frac{\sin(n\pi/\beta)}{n\pi} \tag{7.67}$$

for time $t = 0$. It then follows that the sum of the series has to be

$$\sum_{n=1}^{\infty} \frac{\sin(n\pi/\beta)}{n\pi} = \frac{\beta - 1}{2\beta}. \tag{7.68}$$

Expression (7.68) can be used to find out how many terms are needed in the Fourier series (7.62) to obtain a certain accuracy.

7.8 The thermal subsidence of the McKenzie model

A lithosphere that is hot after instantaneous stretching returns slowly (over more than 100 Ma) to its steady state temperature by conductive cooling. The lithosphere contracts (gets more dense) because of the cooling, which leads to thermal subsidence. The amount of thermal subsidence is found assuming isostasy, and using solution (7.46) for the transient temperature after rifting. The principle of isostasy states that the pressure at the same depth in the asthenosphere remains the same through time. The pressure at the depth $a + s_T$ is

$$\int_0^a \varrho_m(T(z,t)) \, dz + \varrho_s s_T(t) = \int_0^a \varrho_m(T_I(z)) \, dz + \varrho_m(T_a)s_T(t) \tag{7.69}$$

where $s_T(t)$ is thermal subsidence as a function of time. The mantle density ϱ_m is a function of temperature as given by equation (7.24), and the temperature is a function of depth. The sediment density ϱ_s is constant, and the sedimentary basin is taken to be sufficiently thin compared to the thickness of the lithosphere so that the basin can be ignored in the

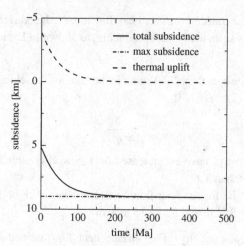

Figure 7.13. *The subsidence of the McKenzie model is plotted as a function of time for stretching with a factor* $\beta = 2$. *The subsidence is plotted as the initial subsidence* s_I *given by equation (7.28) added to the thermal subsidence* $s_T(t)$ *given by equation (7.70).*

temperature solution. We notice that the crust is absent in equation (7.69) for thermal subsidence. The assumption that the crust has the same density as the mantle is a simplification that applies as long as the crust is much thinner than the lithosphere. The thermal subsidence (7.69) becomes

$$s_T(t) = a\,\frac{4\beta\alpha T_a\varrho_{m,0}}{(\varrho_{m,0} - \varrho_s)} \sum_{m=0}^{\infty} \frac{\sin((2m+1)\pi/\beta)}{((2m+1)\pi)^3}\left(1 - e^{-((2m+1)\pi)^2 t/t_0}\right) \qquad (7.70)$$

once the integrations have been carried out, as shown in Note 7.4. Figure 7.13 shows the thermal subsidence (7.70) for $\beta = 2$, when the crust is $c = 36$ km thick, the densities of the sediments, crust and the mantle are $\varrho_s = 2300$ kg m^{-3}, $\varrho_c = 2800$ kg m^{-3} and $\varrho_m = 3300$ kg m^{-3}, respectively, the thermal expansibility is $\alpha = 3 \cdot 10^{-5}$ K^{-1} and the temperature at the base of the lithosphere is $T_a = 1300\,°$C. The initial (and instantaneous) subsidence s_I given by equation (7.28) is taken as the starting point for the thermal subsidence. The subsidence $s = s_I + s_T(t)$ is therefore the total subsidence of the McKenzie model, where the thermal subsidence approaches the maximum thermal subsidence, $s_T(t) \to s_{T,\max}$, with increasing time. (See Exercise 7.11.)

Note 7.4 Equation (7.69) can be rewritten as

$$(\varrho_{m,0} - \varrho_s)s_T = \varrho_{m,0}\,\alpha T_a a \int_0^1 \left(\hat{U}(\hat{z}, \hat{t} = 0) - \hat{U}(\hat{z}, \hat{t})\right) d\hat{z} \qquad (7.71)$$

using the dimensionless expression (7.45) for the thermal transient. When the series for $\hat{U}$ is inserted and the integration is carried out we get

$$(\varrho_{m,0} - \varrho_s)s_T = 2\varrho_{m,0}\,\alpha T_a a \sum_{n=1}^{} \frac{\sin(n\pi/\beta)}{(n\pi)^3}\left(1 - e^{-(n\pi)^2\hat{t}}\right)(1 - \cos(n\pi)). \qquad (7.72)$$

The last factor $1 - \cos(n\pi)$ is 2 for $n = 1, 3, 5$ and all other odd indices n, and it is 0 for $n = 2, 4$ and all even indices n. Only the odd indices are accounted for by replacing n by $2m + 1$ for $m = 0, 1, 2, \ldots$. We then get expression (7.70) for thermal subsidence.

Exercise 7.11 *Maximum thermal subsidence.* Derive expression (7.33) for maximum thermal subsidence using the steady state temperature (7.25) $T_0(z)$ and the initial temperature (7.26) $T_I(z)$ immediately after stretching.
Solution: From equation (7.69) we get

$$(\varrho_{m,0} - \varrho_s) s_{T,\max} = \varrho_{m,0}\,\alpha\,T_a a \int_0^1 \left(\hat{T}_I(\hat{z}) - \hat{T}_0(\hat{z}) \right) d\hat{z} \qquad (7.73)$$

where the integral is

$$\int_0^{1/\beta} (\beta - 1)\hat{z}\, d\hat{z} + \int_{1/\beta}^1 (1 - \hat{z})\, d\hat{z} = \frac{1}{2}\left(1 - \frac{1}{\beta} \right) \qquad (7.74)$$

which is equation (7.33).

Exercise 7.12 (Mathematical curiosity) Show that

$$\sum_{m=0}^{\infty} \frac{\sin((2m+1)\pi/\beta)}{((2m+1)\pi)^3} = \frac{1}{8\beta}\left(1 - \frac{1}{\beta} \right). \qquad (7.75)$$

Hint: let the maximum thermal subsidence expressed by equation (7.70) be equal to the maximum thermal subsidence given by equation (7.33).

7.9 Lithospheric stretching of finite duration

We have seen how the β-factor measures the amount of stretching of the lithosphere. The length of a rectangular-shaped section increases by a factor β and the (vertical) depth thins by a factor $1/\beta$ (see Figure 7.6). The McKenzie model in its simplest version approximates the stretching by an instantaneous event, although the extension process may take several tens of million years. We will now look at the McKenzie model for finite duration stretching, where a realistic strain rate gives the extension. Strain rate is a local property defined at any point by

$$G(t) = \frac{1}{l}\frac{dl}{dt} \qquad (7.76)$$

where l is a small (infinitesimal) distance at the point. The rate (7.76) becomes constant for any vertical length l when the strain rate is independent of the vertical position (see Exercise 7.15). We therefore have that $G = -(1/z)(dz/dt)$, where a minus sign has been added because dz/dt is negative during extension. The strain rate may be time dependent, and it is normally different from zero only during rifting phases. This equation is straightforward to integrate (see Exercise 6.5), and we get

$$z(t) = z_0 \exp\left(-\int_0^t G(t')\, dt'\right). \tag{7.77}$$

The distance $z(t)$ is the depth of a point in the lithosphere during stretching that had the initial depth z_0 before stretching started at $t = 0$. The vertical velocity at the depth z during stretching becomes

$$v_z = \frac{dz}{dt} = -Gz \tag{7.78}$$

when the vertical velocity at the surface $(z = 0)$ is set to zero. Mass conservation, in case of a constant density, requires that $\nabla \cdot \mathbf{v} = 0$, where $\mathbf{v} = (v_x, v_z)$. We therefore have

$$\frac{\partial v_x}{\partial x} = -\frac{\partial v_z}{\partial z} = G \tag{7.79}$$

and the lateral velocity becomes after integration

$$v_x = \frac{dx}{dt} = G\,(x - x_{\text{ref}}) \tag{7.80}$$

where x_{ref} is an arbitrary reference position where the lateral velocity is zero. Integration of the velocity (7.80) in the same way as shown in Exercise (6.5) gives the lateral position

$$x(t) = x_{\text{ref}} + (x_0 - x_{\text{ref}}) \exp\left(\int_0^t G(t')\, dt'\right) \tag{7.81}$$

where x_0 is the initial position at $t = 0$. The path

$$\mathbf{r}(t) = \Big(x(t), z(t)\Big) \tag{7.82}$$

tells where a particle (point in the lithosphere) moves during stretching. The path is also a streamline, when the strain rate G is constant and the velocity field is independent of time. It is then a solution of the equations for streamlines

$$\frac{d\mathbf{r}}{dt} = \mathbf{v}(\mathbf{r}) \tag{7.83}$$

and the position vector $\mathbf{r}(t)$ becomes

$$(x(t), z(t)) = (x_0 e^{Gt},\ z_0 e^{-Gt}) \tag{7.84}$$

where the initial position at $t = 0$ is (x_0, z_0). The corresponding velocity is

$$(v_x, v_z) = (Gx, -Gz) \tag{7.85}$$

when the velocity is measured relative to the origin. Figures 7.14a and 7.14b show the velocity field and the streamlines, respectively, as dimensionless quantities, where the dimensionless coordinates are $\hat{x} = x/l_x$ and $\hat{z} = z/l_z$, where $2l_x$ is the lateral extent of the rectangle and l_z is the depth. Time is scaled with $t_0 = 1/G$, which is the characteristic time for the stretching process.

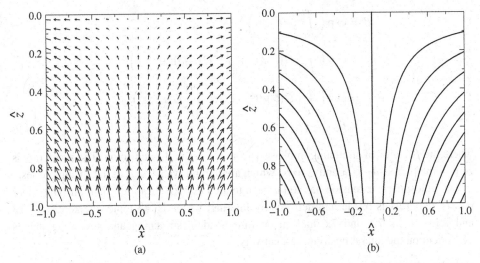

Figure 7.14. *Lithospheric flow during uniform stretching. (a) The flow field (7.85). (b) The streamlines (7.84).*

The β-factor measures how much a rectangular shaped part of the lithosphere becomes stretched,

$$\beta(t) = \frac{x(t)}{x_0} = \exp\left(\int_0^t G(t')dt'\right) \tag{7.86}$$

when the strain rate is independent of position (equal to G everywhere). The stretching factor increases exponentially with time in the case of a constant strain rate. Relation (7.86) can be inverted to give the strain rate as a function of the stretching factor

$$G = \frac{1}{\beta}\frac{d\beta}{dt}. \tag{7.87}$$

This relationship also follows directly from the definition of strain rate (see Exercise 7.16). If we only know the total stretching factor β_s, and can assume that the stretching took place with a constant strain rate during a time interval t_s, then it follows from equation (7.86) that the strain rate is

$$G = \frac{\ln\beta_s}{t_s}. \tag{7.88}$$

Stretching of the lithosphere is normally restricted to certain rift phases, where the strain rate is zero in the periods between. It then follows from equation (7.86) that the total β-factor is the product of the β-factors of each rift phase, because equation (7.86) can be written as

$$\beta = \exp\left(\int_0^t G(t')dt'\right)$$

$$= \exp\left(\int_{t_1}^{t_2} G(t')dt' + \int_{t_3}^{t_4} G(t')dt' + \cdots\right)$$

$$= \exp\left(\int_{t_1}^{t_2} G(t')dt'\right) \cdot \exp\left(\int_{t_3}^{t_4} G(t')dt'\right) \cdots$$

$$= \beta_1 \cdot \beta_2 \cdot \beta_3 \cdots \tag{7.89}$$

where the β-factor of stretching in phase 1 from t_1 to t_2 is β_1, and so on. The strain rate is zero in the periods t_2 to t_3 and t_4 to t_5, which are the intervals between the rifting phases.

The lithospheric extension has so far been treated as pure shear deformation. The upper part of the crust is brittle and deforms by faulting, which is the topic of Exercises 7.13 and 7.14. Notes 7.5 and 7.6 look at depth-dependent stretching and stretching that is dependent on the lateral position, respectively.

Note 7.5 *Depth-dependent stretching*: In order to better calibrate models it has been proposed that the crust and the mantle are stretched with different factors (Royden and Keen, 1980). The crust is then stretched with a factor β and the mantle with a different factor δ. Different β- and δ-factors give depth-dependent stretching. The stretching of the mantle and the crust by different factors allows the thermal transient of the mantle to be calibrated independently of the thinning of the crust. One might for instance create substantial uplift (and erosion) by stretching the mantle by a large δ-factor, while the thinning of the crust is by a moderate β-factor. One problem with such a simple approach to depth-dependent stretching is mass conservation (see Figure 7.15). It also implies that the Moho (crust mantle boundary) becomes a large detachment zone. The streamlines for lithospheric flow become discontinuous across the Moho. Depth-dependent stretching is possible, as demonstrated by numerical simulation of ductile flow, but it then involves quite complex flow patterns (Huismans and Beaumont, 2008).

Note 7.6 We have so far assumed that stretching takes place with a strain rate that is independent of x- and z-positions across an entire profile, only dependent on time. This is not the case for real extensional basins where different parts of the basin and the lithosphere underneath have undergone different amount of stretching. It is often sufficient to assume that the stretching is only laterally dependent (only dependent on the x-position). An important consequence of an x-dependent strain rate is that vertical lines remain vertical during stretching, because v_x is the same for all positions along a vertical line. The simplest

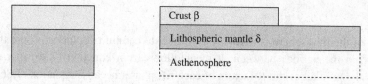

Figure 7.15. *The crust is stretched with a factor β and the lithospheric mantle by a factor δ.*

way to do modeling with a laterally varying strain rate is to approximate the strain rate as a piecewise constant. The lithosphere is then made discrete using columns, where each one has its own (time-dependent) strain rate.

An x-dependent strain rate implies that the stretching factor β also becomes x-dependent. In a Lagrangian coordinate frame (that follows the lateral movements) both the strain rate and the β-factor are only dependent on t. We have that $G = (1/\beta)(D\beta/dt)$ in a Lagrangian coordinate frame. When the time derivative is brought from the Lagrangian frame to the Euler coordinate x we get

$$G = \frac{1}{\beta}\frac{D\beta}{dt} = \frac{1}{\beta}\left(\frac{\partial\beta}{\partial t} + \frac{\partial\beta}{\partial x}\frac{\partial x}{\partial t}\right) \tag{7.90}$$

where $\partial x/\partial t = v_x$. From mass conservation (7.79) we also have that $G = \partial v_x/\partial x$, which gives the following equation for the β-factor:

$$\frac{\partial\beta}{\partial t} + \frac{\partial\beta}{\partial x}v_x - \beta\frac{\partial v_x}{\partial x} = 0. \tag{7.91}$$

More complicated flow patterns result if the strain rate is allowed to vary with both the x- and the z-coordinates.

Exercise 7.13 The upper and brittle part of the lithosphere may deform by normal faulting and rotation of fault blocks during extension. Figure 7.16 shows how the "domino-type" of fault block rotation accommodates the stretching. We notice that large displacement along a few faults is the same as little displacement along many faults.
(a) Show that the β-factor is

$$\beta = \frac{1}{\sin\phi} \tag{7.92}$$

when the blocks make an angle ϕ with the horizontal.
(b) What is the angle ϕ when β is 2? (Answer: 30°)
(c) Assume that the fault blocks rotate to an angle ϕ_1 during a first rift phase and then further to an angle $\phi_2 < \phi_1$ during a second rift phase. (The angle ϕ decreases from $\pi/2$

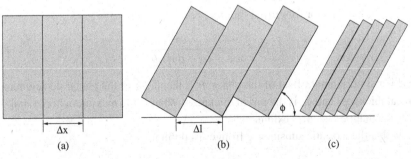

(a)   (b)   (c)

Figure 7.16. *(a) Normal faulting. (b) Domino style rotation of the fault blocks. (c) A large number of smaller faults makes the displacement at each fault less.*

220   *Subsidence*

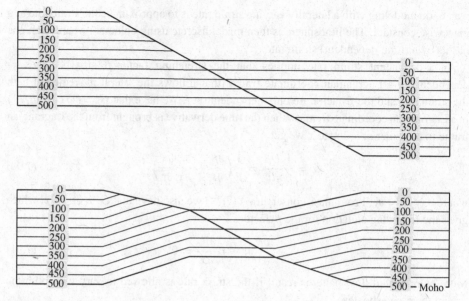

Figure 7.17. *(a) Wernicke (1985) simple shear model for extension and crustal thinning by a low-angle detachment fault. (b) The crustal configuration in the case of isostatic equilibrium. The temperature contours are shown for the case of instantaneous extension.*

towards zero with increasing extension.) Show that the β-factors for the two rift phases become

$$\beta_1 = \frac{1}{\sin \phi_1} \quad \text{and} \quad \beta_2 = \frac{\sin \phi_1}{\sin \phi_2}. \tag{7.93}$$

Exercise 7.14 Extension takes place along a low-angle detachment fault in Wernicke's simple shear model, as shown in Figure 7.17, where the fault penetrates into the lithospheric mantle (Wernicke, 1985). The detachment fault is an asymmetric tectonic structure. Both the pure and simple shear models predict crustal thinning of the rifted continental margins.

(a) Show that the β-factor for crustal thinning becomes

$$\beta = \frac{c_0}{c} = \frac{c_0}{c_0 - d \tan \theta} \tag{7.94}$$

where d is the horizontal offset of the plates, θ is the angle of the planar detachment, c is the crustal thickness and c_0 is the initial thickness. What is the maximum horizontal offset d before the mantle becomes exhumed?

(b) Show that the isostatic subsidence from crustal thinning is

$$s = \frac{(\varrho_m - \varrho_c)}{(\varrho_m - \varrho_s)} d \tan \theta. \tag{7.95}$$

Solution: (b) Use the subsidence (7.21). Subsidence of the simple shear model was studied in great detail by Weissel and Karner (1989).

Exercise 7.15 Show that the strain rate (7.76) applies for vertical line segments of any length when it is constant.
Solution: A long line l can be divided into small (infinitesimal) segments l_i as $l = \sum_i l_i$ where G is the same for each line segment. We therefore have

$$G \sum_i l_i = \frac{d}{dt} \sum_i l_i \quad \text{or} \quad Gl = \frac{dl}{dt}. \tag{7.96}$$

Exercise 7.16 Show that expression (7.87) for the strain rate as a function of the stretching factor follows directly from the definition of strain rate.
Solution: The strain rate is defined as $G = (1/l)(\Delta l/\Delta t)$, where $l = \beta l_0$ is a small line segment that has been stretched a factor β from the initial length l_0. The line segment is stretched the step $\Delta l = \Delta \beta \, l_0$ during the time interval Δt, which then gives equation (7.87).

Exercise 7.17 Let us consider deposition of sediments during a rift phase. Show that the net thickness of the synrift formation becomes

$$\zeta_1 = \frac{1}{\ln\beta}\left(1 - \frac{1}{\beta}\right)\zeta_0 \tag{7.97}$$

where β is the stretching factor and ζ_0 is the thickness of the unstretched formation as (net) porosity-free sediments. Assume that the deposition rate (as net sediments) and the strain rate are constant.
Solution: The duration of extension is first divided into time steps Δt, and a thickness $\Delta \zeta = \omega \Delta t$ is deposited during each time step, where the deposition rate is ω. The β-factor is $\beta = e^{Gt}$ as a function of time, where G is the strain rate, and $t = 0$ is the beginning of the rift phase. The thickness $\Delta \zeta$ is therefore reduced to $\Delta \zeta \exp(-Gt)$ after a time span t. The sum of the thicknesses through the entire rifting phase becomes

$$\zeta_1 = \int_0^{t_0} e^{-Gt}\, d\zeta = \omega \int_0^{t_0} e^{-Gt} dt = \frac{\omega}{G}\left(1 - e^{-Gt_0}\right) \tag{7.98}$$

where the time $t = t_0$ is the end of the formation. The strain rate is $G = \ln\beta/t_0$ and the unstretched formation thickness is $\zeta_0 = \omega t_0$. Replacing the strain rate by the β-factor at the end of rifting then gives expression (7.97).

7.10 Finite duration stretching and temperature

The velocity field controls the temperature of the mantle when heat conduction is negligible compared to heat convection by mantle flow. The temperature equation for convective heat flow is then

$$\frac{\partial T}{\partial t} + \mathbf{v} \cdot \nabla T = 0 \tag{7.99}$$

where the mantle velocity is given by equation (7.85) as $\mathbf{v} = (Gx, -Gz)$, The streamlines of mantle flow can be used to solve the convective temperature equation, because the temperature follows the movement of the mantle. The temperature is initially $T(z) = T_a z / l_m$ at $t = 0$, where T_a is the temperature at the depth l_m at the base of the lithosphere. In order to find the temperature at position (x, z) at the time t, we have to follow a streamline backwards in time until $t = 0$, and see what the temperature was at the initial position. From the streamline solution (7.84) it follows that a particle at position (x, z) at time t had the initial vertical position $z_0 = z e^{Gt}$ (at time $t = 0$). The initial temperature at position z_0 (at time $t = 0$) is $T = (T_a / l_m) z_0$, and the temperature (x, z) at time t is therefore

$$T(x, z, t) = \begin{cases} \dfrac{T_a z e^{Gt}}{l_m}, & z e^{Gt} \leq l_m \\ T_a, & z e^{Gt} > l_m. \end{cases} \qquad (7.100)$$

It is straightforward to verify the solution by inserting it into the temperature equation (7.99). (See Exercise 7.18.) The factor $\beta = z \exp(Gt)$ is also identified, and the temperature (7.100) is therefore equal to the temperature (7.26) for instantaneous stretching.

The assumption that convective heat flow dominates over heat conduction during stretching is important, because the flow of mantle and crust can then be approximated by instantaneous stretching. How rapid stretching then has to be is estimated using the half-life for decay of thermal transients by heat conduction. The half-life is approximated by

$$t_s = \frac{\ln 2\, l_0^2}{\pi^2 \kappa} \qquad (7.101)$$

where l_0 is the characteristic length scale, as shown by equation (7.49). The characteristic length scale is now taken to be $l_0 = l_m / \beta$, which is the depth to the asthenosphere after instantaneous stretching by a factor β. The duration of stretching has to be shorter than this half-life for convective heat flow to dominate. We therefore have that the duration of stretching must be shorter than

$$t_s = \frac{\ln 2\, l_m^2}{\pi^2 \kappa \beta^2} = \frac{t_{1/2}}{\beta^2} \qquad (7.102)$$

where $t_{1/2}$ is the half-life for the length scale of the entire lithosphere. A lithospheric thickness $l_m = 120$ km and a thermal diffusivity $\kappa = 1 \cdot 10^{-6}$ m^2 s^{-1} gives that $t_{1/2} = 32$ Ma. The limitation of the duration of stretching can therefore be estimated as

$$t_s < \frac{32 \text{ Ma}}{\beta^2}. \qquad (7.103)$$

Jarvis and McKenzie (1980), who first made such an estimate, suggested $t_s < 60$ Ma$/\beta^2$.

Heat conduction cannot be ignored if stretching is over a time span of several tens of million years or more. The convection–conduction equation for the temperature must then be solved. This equation is

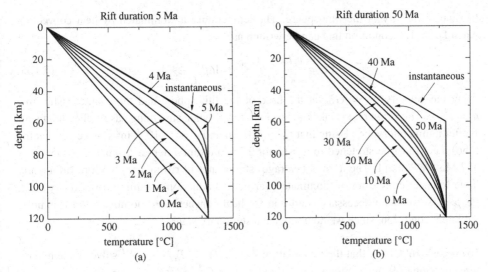

Figure 7.18. *The temperature of the lithosphere is shown during finite duration rifting, where the final β-factor is* 2. *Left: rifting during 5 Ma. Right: rifting during 50 Ma.*

$$\frac{\partial T}{\partial t} + Gz\frac{\partial T}{\partial z} - \kappa\frac{\partial^2 T}{\partial z^2} = 0 \qquad (7.104)$$

in the vertical direction, where the vertical velocity is $v_z = Gz$. The temperature equation (7.104) does not only apply at $x = 0$, where $v_x = 0$. It applies for $x \neq 0$ too, as long as the initial isotherms are horizontal and the vertical velocity is independent of the lateral position. It is not straightforward to give an exact solution to this equation, because of the z-dependent velocity term, see Jarvis and McKenzie (1980). But a numerical (finite-difference) solution of equation (7.104) is a simple means to explore the temperature during finite duration rifting. Figures 7.18a and 7.18b show numerical solutions where the boundary conditions are 0 °C at the surface and 1300 °C at the base of the lithosphere. Both figures show stretching with a final β-factor 2. Figure 7.18a shows stretching for the time interval 5 Ma, which is sufficiently short for the upper bound (7.103) to apply, and instantaneous stretching is a good approximation. Figure 7.18b shows stretching over a time interval 50 Ma that is longer than the upper bound (7.103), and heat conduction becomes important. Instantaneous stretching cannot be assumed in this case.

Note 7.7 The Peclet number can be used to decide when flow is sufficiently fast for heat convection to dominate over heat conduction. The Pe-number was introduced in Section 6.11, where it was applied to processes that had reached a stationary state. The stretching of the lithosphere will not last long enough to reach a steady state. Nevertheless, the Pe-number can be used to check whether convection would have dominated over heat conduction if stretching had lasted for a sufficiently long time for the temperature to be nearly stationary. The Pe-number for vertical flow is $\mathrm{Pe} = v_z l_m/\kappa$, where κ is the thermal diffusivity of the lithosphere. The temperature is dominated by heat convection when $Pe \gg 1$, a condition that can be written as

$$Pe = \frac{v_z l_m}{\kappa} = \frac{G l_m^2}{\kappa} = \ln\beta \, \frac{t_0}{t_s} \gg 1 \qquad (7.105)$$

using the velocity $v_z = G l_m$ at the base of the lithosphere and the characteristic time $t_0 = l_m^2 / \kappa$. The strain rate is taken to be $G = \ln\beta / t_s$ when stretching has reached the factor β after a time interval t_s. Using that $l_m = 120$ km and $\kappa = 1 \cdot 10^{-6} \text{ m}^2 \text{ s}^{-1}$ we get that the lithosphere must be stretched to a β-factor 2.71 in a time interval t_s much less than $t_0 = 457$ Ma for $Pe \gg 1$. The normal (average) strain rates of rifting are therefore sufficiently large for heat convection to dominate over heat conduction assuming a stationary state. $Pe \gg 1$ serves as a necessary condition for heat convection to dominate, but it is not a sufficient condition, since we do not reach a stationary state.

Exercise 7.18 Check that the temperature $T(x, z, t) = (T_a / l_m) \, z \, e^{Gt}$ solves the temperature equation (7.99) when the velocity field is $\mathbf{v} = (Gx, -Gz)$.

7.11 Lithospheric extension, phase changes and subsidence/uplift

We will in this section look at subsidence or uplift caused by density changes in the crust and the mantle. For example the formation of eclogite in the crust leads to a larger density, which again implies subsidence. The assumption of isostatic equilibrium can be used to estimate the subsidence caused by a change in the density. We get the subsidence

$$\Delta s = \frac{\Delta \varrho_c}{(\varrho_m - \varrho_w)} c \qquad (7.106)$$

when the average density of the crust is changed by $\Delta \varrho_c$, the crust has the thickness c, the mantle has the density ϱ_m and the depression is filled with water with a density ϱ_w. A small change $\Delta \varrho_c = 23 \text{ kg m}^{-3}$ gives just $\Delta s = 100$ m of subsidence for a 10 km thick crust, when the mantle and water densities are $\varrho_m = 3300 \text{ kg m}^{-3}$ and $\varrho_w = 1000 \text{ kg m}^{-3}$, respectively. A more substantial change of the average crustal density, $\Delta \varrho_c = 230 \text{ kg m}^{-3}$, leads to 1 km of subsidence. It is the difference in the average density that controls the subsidence, and it is therefore important to know both how much it has changed and in which depth interval the change took place.

Similarly, a change in the average density of the lithospheric mantle leads to either subsidence or uplift. (An increase in the average density gives subsidence and a decrease gives uplift.) The assumption of isostasy implies that a change $\Delta \varrho_m$ in the average mantle density gives the subsidence

$$\Delta s = \frac{\Delta \varrho_m}{(\varrho_m - \varrho_w)} a \qquad (7.107)$$

where the thickness of the lithospheric mantle is a. We looked in Section 7.8 at the thermal transient created by lithospheric extension and the associated thermal expansion of the mantle. The change in the average mantle density after instantaneous stretching with a factor β is

$$\Delta \varrho_m = -\frac{1}{2}\varrho_{m,0}\,\alpha\,T_a\left(1 - \frac{1}{\beta}\right) \tag{7.108}$$

and it becomes $\Delta\varrho_m = -32\,\mathrm{kg\,m^{-3}}$ for $\beta = 2$, using the parameters $\varrho_{m,0} = 3300\,\mathrm{kg\,m^{-3}}$, $\alpha = 3 \cdot 10^{-5}\,\mathrm{kg\,m^{-3}}$ and $T_a = 1300\,°\mathrm{C}$. The associated thermal uplift becomes $\Delta s = 1670$ m. (See equation (7.33).) A change in the average mantle density gives much more uplift or subsidence than in the crust because the lithospheric mantle is an order of magnitude thicker than the crust.

It turns out that phase changes might take place in the lithospheric mantle when hot mantle moves upwards during extension. There is a change in density when a material enters a new phase. Phase changes in the mantle are dependent on the composition, and we will in the following look at just one possible phase transition – the spinel–peridotite to plagioclase–peridotite transformation. Kaus *et al.* (2005) shows that the spinel–plagioclase transformation is the most important one during lithospheric extension because it gives the largest density difference.

The example with thermal uplift shows that a moderate alteration of the average mantle density leads to a sizable vertical movement. The possible additional subsidence/uplift from a phase change in the mantle during extension is now studied with a simple model based on a study of Simon and Podladchikov (2008). Podladchikov *et al.* (1994) and Yamasaki and Nakada (1997) have presented similar models. Figure 7.19 shows a simplified phase diagram where the spinel–plagioclase phase transition is represented by a straight line. Kaus *et al.* (2005) and Simon and Podladchikov (2008) discuss phase diagrams with three phases and the associated density changes for different types of mantle

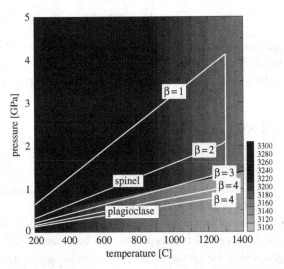

Figure 7.19. *A simplified picture of mantle density and the spinel–plagioclase phase transition.*

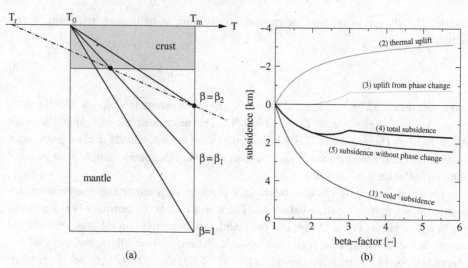

Figure 7.20. *(a) The mantle enters the plagioclase–peridotite phase when the geotherm (for* $\beta = \beta_1$*) crosses the phase boundary (dashed line) at the base of the crust. The depth of the plagioclase–peridotite phase decreases with extension until the temperature reaches the astheno-sphere temperature* T_a *(for* $\beta = \beta_2$*). (b) The subsidence during extension is shown when the uplift from a spinel-peridotite to plagioclase-peridotite phase transition is included.*

compositions. A series of geotherms for instantaneous stretching with different β-factors are plotted on the phase diagram (Figure 7.19). We see that the geotherm for $\beta = 3$ has crossed the line from spinel-peridotite to plagioclase-peridotite. The density decreases across the phase transition, and it therefore leads to additional uplift. The temperature for the phase transition is linearly approximated as a function of depth

$$T = T_r + A_r z \tag{7.109}$$

where A_r is the steepness of the phase boundary (see Figure 7.20a). The phase bound-ary can also be written in terms of pressure (p) as a function of temperature using that $p = \varrho_m g z$. The densities on each side of the phase boundary are linearly approximated in temperature as

$$\rho_M(T) = \begin{cases} \varrho_{m,0} \cdot (1 - \alpha_m T) & \text{for spinel–peridotite} \\ \varrho_{x,0} \cdot (1 - \alpha_x T) & \text{for plagioclase–peridotite.} \end{cases} \tag{7.110}$$

The pressure could also have been included in the first-order approximation of the density as shown by Kaus *et al.* (2005). Notice that ϱ_M now denotes the mantle density at a given temperature, while ϱ_m is the average mantle density.

The mantle along the initial geotherm ($\beta_1 = 1$) is on the spinel–peridotite side of the phase boundary because the crust is too deep. The crust becomes thinned with progressing extension and the gradient of the geotherm increases until the geotherm crosses the phase boundary at the base of the crust (see Figure 7.20a). This is written as

$$T_r + A_r z = T_0 + A_0 \beta z \quad \text{and} \quad z = \frac{c_0}{\beta} \tag{7.111}$$

which has as solution the minimum β-factor needed for the top of the mantle to enter the plagioclase phase

$$\beta_1 = \frac{A_0 c_0}{T_0 - T_r + A_0 c_0}. \tag{7.112}$$

The mantle is then plagioclase–peridotite from the top of the crust down to the depth where the geotherm crosses the phase boundary, which is the interval

$$z_{\text{top}} = \frac{c_0}{\beta} \quad \text{to} \quad z_{\text{bot}} = \frac{T_0 - T_r}{A_r - A_0 \beta}. \tag{7.113}$$

The depth where the geotherm crosses the phase boundary increases with increasing β-factor until it reaches the asthenosphere temperature T_a (see Figure 7.20a) for the β-factor

$$\beta_2 = \left(\frac{T_a - T_r}{T_a - T_0}\right)\left(\frac{A_r}{A_0}\right). \tag{7.114}$$

The subsidence during extension has two components as we have already seen – subsidence from thinning of the crust and subsidence/uplift from a change $\Delta \varrho_m$ in the average mantle density:

$$s = \left(\frac{\varrho_{m,\text{ref}} - \varrho_c}{\varrho_{m,\text{ref}} - \varrho_w}\right)\left(1 - \frac{1}{\beta}\right) c_0 + \left(\frac{\Delta \varrho_m}{\varrho_{m,\text{ref}} - \varrho_w}\right) a. \tag{7.115}$$

The last part is what gives thermal uplift by the density difference (7.108). The average mantle density is initially $\varrho_{m,\text{ref}}$ and it becomes $\varrho_m = \varrho_{m,\text{ref}} + \Delta \varrho_m$ with increasing stretching. We can now include the phase change in the average mantle density ϱ_m and we get

$$\Delta \varrho_m = -\frac{1}{2}\varrho_{m,0}\, \alpha_m \left(1 - \frac{1}{\beta}\right) T_a \tag{7.116}$$

$$+ \frac{1}{a}(\varrho_{x,0} - \varrho_{m,0})(z_{\text{top}} - z_{\text{bot}}) \tag{7.117}$$

$$- \frac{1}{2a^2}(\varrho_{x,0}\alpha_x - \varrho_{m,0}\alpha_m)(z_{\text{top}}^2 - z_{\text{bot}}^2)\beta T_a \tag{7.118}$$

where equation (7.113) gives how z_{bot} and z_{top} depend on the β-factor. The first term (7.116) is the (familiar) change in mantle density from thermal expansion, when we have only one phase, and it is this term that causes thermal uplift. The second term (7.117) is the change in average mantle density for plagioclase–peridotite in the depth interval from z_{top} to z_{bot}. The third term (7.118) is a correction to the first term that accounts for thermal expansion in the plagioclase-peridotite interval, and it is normally insignificant compared to the other two terms.

Curve (4) in Figure 7.20b shows the subsidence during extension as a function of the β-factor. The density parameters are $\varrho_{m,0} = 3300$ kg m^{-3}, $\varrho_{x,0} = 3250$ kg m^{-3}, $\alpha_m = \alpha_x = 3 \cdot 10^{-11}$ K^{-1}, which are also used in the phase diagram in Figure 7.19.

Simon and Podladchikov (2008) estimate the decrease in density to be in the range from 10 kg m^{-3} to 90 kg m^{-3} depending on the composition. The crust has an initial thickness $c_0 = 35$ km and density $\varrho_c = 2800 \text{ kg m}^{-3}$, and the phase boundary is given by $T_r = -300 \,°\text{C}$ and $A_r = 0.04 \,°\text{C/m}$. The lithosphere has the thickness $a = 120$ km, the surface temperature $T_0 = 0 \,°\text{C}$ and the temperature $T_a = 1300 \,°\text{C}$ at the base of the lithosphere.

The mantle geotherm starts to cross the phase boundary at $\beta_1 \approx 2$ and we see that we then get uplift. The uplift increases until $\beta_2 = 3$ where the geotherm crosses the phase boundary at the asthenosphere temperature. The uplift from phase change is then increasing less. Curve (5) shows what the subsidence would have been without change of phase. It is the sum of subsidence from crustal thinning and thermal uplift, curves (1) and (2), respectively. The uplift by only phase change is curve (3). The total subsidence is therefore the sum of the curves (1), (2) and (3).

This kind of uplift from a change of phase is controlled by a thermal transient and it will therefore gradually disappear after extension, when the geotherm returns to a steady state. The final state after a sufficiently long time is the permanent subsidence from crustal thinning alone, given by curve (1).

There have been observed anomalous subsidence patterns that could be explained by such phase changes (Kaus *et al.*, 2005). It should also be mentioned that there are vertical movements that are difficult to explain, like for instance the uplift of East Greenland. Density changes associated with change of phase have been suggested among several other possibilities – see Japsen and Chalmers (2000) for a review of the uplift around the North Atlantic. This uplift is probably not related to lithospheric extension and it might be of a permanent nature.

Note 7.8 Expression (7.115) for subsidence follows from isostasy by considering the weight of a column before and after extension

$$\varrho_c c_0 + \varrho_{m,\text{ref}} a_0 = \varrho_w s + \varrho_c c + \varrho_m a \qquad (7.119)$$

when they have the same height

$$c_0 + a_0 = s + c + a. \qquad (7.120)$$

The subsidence, the thickness of the crust, the thickness of the lithospheric mantle are s, c and a, respectively. The subscript 0 denotes before extension. We can use equation (7.120) to eliminate either a_0 or a. If we eliminate a_0 we get the subsidence (7.115). Alternatively, if we eliminate a instead of a_0 we get the same subsidence, except that it is written as the sum of two slightly different terms. The average density ϱ_m appears now in the expression for subsidence instead of the reference average density $\varrho_{m,\text{ref}}$. The average ϱ_m is time dependent since it includes the thermal transient from extension. Expression (7.115) is therefore preferred although a does not stay constant during extension.

Exercise 7.19 Consider a vertical column of the lithosphere that has a crust of thickness c and density ϱ_c and a lithospheric mantle of thickness a and density ϱ_m. Assume isostasy and show that equations (7.106) and (7.107) give the subsidence/uplift caused by a change in the average crust or mantle density, respectively.

Exercise 7.20 How large a reduction in the average mantle density is needed to create 2 km of uplift when the initial average mantle density is $\varrho_m = 3300 \text{ kg m}^{-3}$, the thickness of the lithospheric mantle is $a = 150$ km and the uplift takes place in the air (above sea level)?

Exercise 7.21
(a) Verify expressions (7.112) to (7.114) for β_1, z_{bot}, z_{top} and β_2.
(b) Verify the density difference (7.116) to (7.118).

Exercise 7.22
(a) Show that a change ΔA_r in the gradient of the phase boundary makes a change

$$\Delta s = -\frac{(\varrho_{x,0} - \varrho_{x,0})}{(\varrho_{m,ref} - \varrho_m)} \frac{(T_0 - T_r)}{(A_r - A_0\beta)^2} \Delta A_r \tag{7.121}$$

in the subsidence. Hint: use the chain rule of differentiation

$$\Delta s = \frac{ds}{d\varrho_m} \frac{d\varrho_m}{dz_2} \frac{dz_2}{dA_r} \Delta A_r. \tag{7.122}$$

(b) What is Δs when $\Delta A_r = 0.01$? Use the numbers in the text above for the other parameters.

Exercise 7.23 Radioactive heat production makes the lithosphere hotter than it otherwise would have been. Lithospheric extension thins the heat producing crust, and the lithospheric mantle becomes less hot when it returns to a steady state after a sufficiently long time. The reduction in the steady-state mantle temperature implies thermal contraction and subsidence. This exercise looks at the amount of subsidence one could expect.

In case the radioactive heat production should disappear the corresponding increase in the average mantle density from thermal contraction becomes

$$\Delta\rho_m = \frac{1}{z_a}\varrho_m\alpha \int_0^{z_a} \Delta T(z)\,dz \tag{7.123}$$

where $\Delta T(z)$ is the difference in the geotherms for non-zero and zero radioactive heat production. The subsidence from the increase in density is $s = \Delta\rho_m z_a/(\varrho_m - \varrho_w)$.
(a) Use the geotherm (6.59)–(6.60) and show that the integral over the temperature difference is

$$\int_0^a \Delta T(z)\,dz = \frac{S_0 z_m^2}{\lambda_c}\left(-\frac{1}{6}z_m + \frac{1}{4}z_a + \frac{1}{4}z_1\right) \tag{7.124}$$

where

$$z_1 = z_m + \frac{\lambda_m}{\lambda_c}\left(T_a - T_m\right) \tag{7.125}$$

is the depth to the asthenosphere for the geotherm with heat production. The geotherm without heat production has the larger depth z_a to the asthenosphere. The asthenosphere is assumed to be isothermal.

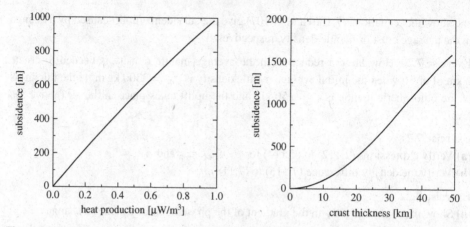

Figure 7.21. *(a) The subsidence corresponding to radioactive heat production. (b) The subsidence as a function of the thickness of the heat-producing crust. See Exercise 7.23 for the details.*

(b) Insert the expression (6.61) for T_m into z_1 and then z_1 into the integral (7.124), and show that it can be written

$$\int \Delta T \, dz = \frac{S_0 z_m^2}{\lambda_c} \left(\left(\frac{1}{12} - \frac{1}{4} \left(1 + \frac{S_0 z_m}{2 q_m} \right) \frac{\lambda_m}{\lambda_c} \right) z_m + \frac{1}{4} \frac{\lambda_m}{q_m} (T_a - T_0) + \frac{1}{4} z_a \right). \tag{7.126}$$

Figure 7.21a shows the subsidence as a function of heat production S_0 and Figure 7.21b shows the subsidence as a function of the thickness of the crust z_m. The other parameters are $T_0 = 0$ °C, $T_a = 1300$ °C, $z_a = 120$ km, $z_m = 35$ km, $S_0 = 1$ μW m^{-3}, $q_m = 20$ mW m^{-2}, $\lambda_c = 2.5$ W m^{-1} K^{-1}, $\lambda_m = 3$ W m^{-1} K^{-1}, $\varrho_c = 3300$ kg m^{-3}, $\varrho_m = 2800$ kg m^{-3} and $\alpha = 3 \cdot 10^{-5}$ K^{-1}.

(c) The thickness of the crust is reduced during lithospheric extension. How much extra subsidence could one expect when a $z_m = 35$ km thick crust is stretched and thinned with $\beta = 2$?

Comments: The geotherm used in this exercise has the mantle heat flow q_m as a boundary condition at the base of the lithosphere. The alternative geotherm (6.65)–(6.66), where the asthenosphere temperature T_a is the boundary condition at the base of the lithosphere, gives less temperature difference between the geotherms for non-zero and zero heat generation. Another point is that the loss in heat production from a thinned crust may be counterbalanced by infill of a basin with heat producing sediments.

7.12 Lithospheric extension and decompression melting

We have seen that extension and crustal thinning lead to upwelling of hot mantle rocks. If mantle rocks are brought up to sufficiently shallow depths the temperature may cross the *solidus*, which is the temperature where the rocks start to melt. Figure 7.22 shows a situation where rapid lithospheric extension has caused mantle rocks to cross the solidus. Mantle

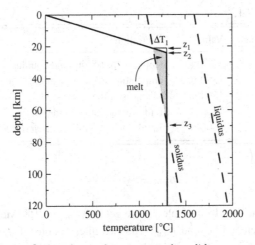

Figure 7.22. *Melt is generated when the geotherm crosses the solidus.*

rocks do not melt completely, only partially, after the solidus is crossed. The temperature must increase by several hundred degrees more before the rock melts completely. The temperature (and pressure) where the last fraction of the rock melts is the *liquidus*. We notice that melt is not generated because the rock becomes heated, but because it's brought upwards. The rock is slightly cooled when it moves upwards, and it is the decreasing pressure from decreasing depth that causes the rock to cross the solidus. Melting is for this reason called *pressure release melting* or *decompression melting*. The solidus and the liquidus are now approximated by linear functions of depth as shown in Figure 7.22, and they are written as

$$T_s(z) = T_{s,0} + A z \quad \text{and} \quad T_l(z) = T_{l,0} + A z. \tag{7.127}$$

The temperature of the liquidus and the solidus at surface conditions are $T_{s,0}$ and $T_{l,0}$, respectively, and the steepness A is a constant parameter. Knowledge of the solidus allows us to find the minimum amount of stretching required for melting to start, which is when the geotherm touches the solidus. We assume that the McKenzie model based on rapid and uniform stretching provides a good approximation for the geotherm. The temperature is assumed unchanged for a volume of mantle that follows the flow of the asthenosphere when it moves upwards during extension. Notes 7.11 and 7.12 look at this assumption by estimating the temperature change along (adiabatic) geotherms when compressibility and melting are accounted for. Melting starts when the temperature T_a at the base of the lithosphere is brought up to the depth where it crosses the solidus

$$T_{s,0} + A \frac{a}{\beta_{\min}} = T_a \tag{7.128}$$

and where a is the thickness of the lithosphere at steady state conditions. The minimum amount of stretching becomes

Table 7.1.

Parameter	Value	Description
A	3 °C/km	Steepness for solidus and liquidus
T_a	1300 °C	Temperature at the base of the lithosphere
a	120 km	Thickness of the lithosphere
$T_{s,0}$	1100 °C	Solidus temperature at the surface
$T_{l,0}$	1600 °C	Liquidus temperature at the surface

$$\beta_{min} = \frac{a\,A}{T_a - T_{s,0}} \qquad (7.129)$$

and if we are using the data from Table 7.1 we get that $\beta_{min} = 1.8$. Melt is produced when the lithosphere is stretched beyond β_{min}. The two intersections of the geotherm with the solidus give the depth interval where melt production takes place, see Figure 7.22. The uppermost intersection of the geotherm with the solidus is given by

$$T_{s,0} + A\,z_1' = \beta T_a \frac{z_1}{a}, \qquad (7.130)$$

or

$$z_1 = \frac{T_{s,0}}{\beta T_a/a - A}. \qquad (7.131)$$

The lowermost intersection between the geotherm and the solidus is simply

$$T_{s,0} + A\,z_3 = T_a \qquad (7.132)$$

or

$$z_3 = \frac{T_a - T_{s,0}}{A}. \qquad (7.133)$$

We notice that the uppermost intersection z_1 is a function of the stretching factor β, while the lowermost intersection z_3 is not. This is also seen from the plot in Figure 7.23, where the melt zone is shown for different stretching factors. The fraction of melt produced can be approximated by linear melting between the solidus and the liquidus, and the melt fraction is then

$$f = \frac{T - T_s}{T_l - T_s}. \qquad (7.134)$$

The densities of melt and solid are taken to be the same, and the amount of melt in terms of mass fraction and volume fraction are therefore equal (see Exercise 7.27). The melt fraction for different stretching factors is shown in Figure 7.23, where it is seen that melt generation beyond 10% requires a β-factor larger than 2. Integration of the melt fraction over the depth interval with melt generation gives the total amount of melt produced

$$M = \int_{z_3}^{z_1} f(z)\,dz \qquad (7.135)$$

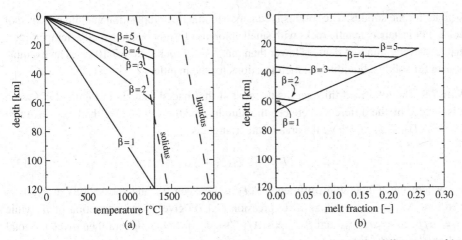

Figure 7.23. *Melt generation for different stretching factors. (a) the geotherms for different stretching factors are plotted together with the solidus. (b) the melt fraction as a function of depth for different stretching factors. (Data from Table 7.1 are used.)*

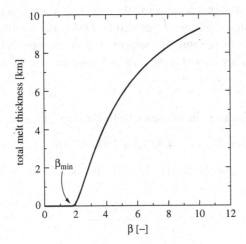

Figure 7.24. *The total melt thickness is plotted as a function of the stretching factor β. (Data from Table 7.1 are used.)*

which becomes

$$M(\beta) = \frac{T_a a^2 A \beta}{2(T_{l,0} - T_{s,0})(\beta T_a - aA)} \left(\frac{1}{\beta_{min}} - \frac{1}{\beta} \right)^2 \tag{7.136}$$

when it is expressed as a function of the stretching factor. (See Note 7.9 for the details of the derivation.) Figure 7.24 shows the total thickness of melt production as a function of the stretching factor β, and to generate more than 3000 m with melt requires stretching factors larger than 3. The amount of melt shown in Figure 7.24 should be handled with care, not only because the model is very simple, but also because there are great uncertainties in the data for the solidus. The solidus is sensitive to mantle composition and the presence of fluids. For instance, mantle rocks with small amounts of water begin to melt at temperatures that are several hundred degrees lower than that for dry rocks. Note 7.10 gives an example of data for solidus, liquidus and melt fractions for the mantle.

Note 7.9 The total melt thickness is the area of the shaded triangle in Figure 7.22, when it is divided by the difference between the liquidus and the solidus. We therefore have that $M = I/(T_{l,0} - T_{s,0})$, where the area of the triangle is

$$I = \frac{1}{2}\Delta T_1(z_3 - z_2) \tag{7.137}$$

and where the temperature difference ΔT_1 is between T_a and the solidus at z_1. It is therefore $\Delta T_1 = T_a - \beta T_a z_1/a$. Expression (7.131) gives z_1 as a function of β, while $z_2 = a/\beta$, $z_3 = a/\beta_{\min}$ and $\beta_{\min} = aA/(T_a - T_{s,0})$. Some algebra then gives the total melt thickness (7.136).

Note 7.10 *Solidus and liquidus of the mantle*: We have assumed that the liquidus and solidus are straight lines. This simplifying assumption is often sufficient considering the large number of uncertainties that appear in the modeling of magma generation. But there are alternatives. An often used one is provided by McKenzie and Bickle (1988), who fitted empirical functions of pressure and temperature to experimental observations of the melting of peridotite. They arrived at pressure as a function of solidus T_s (in °C)

$$p = (T_s - 1100)/136 + 4.968 \times 10^{-4} \exp\left(1.2 \times 10^{-2}(T_s - 1100)\right) \tag{7.138}$$

where p is the mantle pressure in GPa, and the liquidus (in °C) as a function of pressure

$$T_l = 1736.2 + 4.343\,p + 180\tan^{-1}(p/2.2169). \tag{7.139}$$

The melt fraction between the solidus and liquidus is a function of the dimensionless temperature

$$\hat{T} = \frac{T - \frac{1}{2}(T_s + T_l)}{T_l - T_s} \tag{7.140}$$

as

$$X = 0.5 + \hat{T} + (\hat{T}^2 - 0.25)(0.4256 + 2.988\hat{T}). \tag{7.141}$$

The dimensionless temperature $\hat{T} = -1/2$ is the solidus and $\hat{T} = 1/2$ is the liquidus. Both solidus and liquidus are only functions of pressure, but the solidus T_s appears as the solution of equation (7.138) that gives the pressure. For a given pressure we therefore have to solve equation (7.138) for T_s. The solution has to be found numerically, for instance by Newton's method, which works fine with convergence after less than five iterations. When both T_s and T_l are found from a given pressure then the next step is to use the temperature to obtain the dimensionless temperature and finally the melt fraction X. Figure 7.25a shows the solidus, the liquidus and the geotherms for melt fractions in steps of 10%. The depth in Figure 7.25a is linearly related to pressure as $p = \varrho_m g z$, where $\varrho_m = 3300\ \mathrm{kg\,m^{-3}}$. The

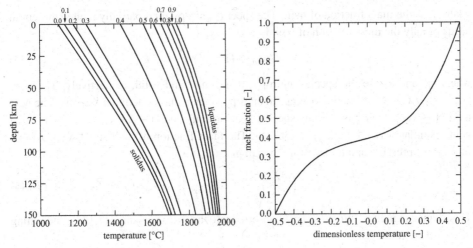

Figure 7.25. *(a) The solidus and liquids as functions of depth. (b) Melt fraction as a function of dimensionless temperature.*

solidus appears as quite straight down to a depth of ~100 km. But Figure 7.25b shows that the melt fraction does not change linearly between the solidus and the liquidus. The melt fraction increases steeply from 0 to 0.3 with increasing temperature (at a given pressure) before it enters a plateau from 0.3 to 0.6, where it increases slowly with increasing temperature. The remaining solid then melts over a relatively short temperature interval with increasing temperature. McKenzie and Bickle (1988) discuss the various melt fractions in terms of mantle composition.

Note 7.11 *Mantle adiabats and melting:* The temperature of the mantle is found in Section 6.22 for a piece of rock that is brought up to a shallower depth during lithospheric extension. It is assumed that the rock does not exchange heat with its surroundings, and the temperature decrease is then solely due to thermal expansion. The temperature reduction becomes even stronger if the rock begins to melt during its rise during extension. The melting process requires heat which leads to a lowered temperature. A process that does not exchange heat with its surroundings is an adiabatic process – a process where the entropy is constant. The thermodynamic expression for an adiabatic process is $ds = 0$, when s is the specific entropy of the bulk rock. We have from equation (6.321) that the change of entropy in a single phase system is

$$ds = \frac{c_p}{T}\,dT - \frac{\alpha}{\varrho}\,dp. \tag{7.142}$$

For a two-phase system (solid and melt) this expression becomes

$$ds = (s_m - s_s)\,dX + \frac{1}{T}\left((1-X)c_p^s + Xc_p^m\right)dT - \left((1-X)\frac{\alpha_s}{\varrho_s} + X\frac{\alpha_m}{\varrho_m}\right)dp \tag{7.143}$$

where X is the mass fraction of melt. The specific entropy for a partially melted system in terms of only the mass fraction of melt is

$$s = Xs_m + (1 - X)s_s \tag{7.144}$$

where s_m and s_s are the specific entropy of the melt and solid, respectively. The term $(s_m - s_s)\,dX$ is therefore the increase in entropy when a mass fraction dX melts. The next term $(1 - X)c_p^s + Xc_p^m$ is the bulk specific heat capacity and $(1 - X)(\alpha_s/\varrho_s) + X(\alpha_m/\varrho_m)$ is in a similar way the bulk value of α/ϱ. The mass fraction of melt $X = X(p, T)$ is a function of pressure and temperature, which gives that

$$dX = \left(\frac{\partial X}{\partial p}\right)_T dp + \left(\frac{\partial X}{\partial T}\right)_p dT. \tag{7.145}$$

This change in melt fraction dX can be inserted into the entropy change (7.143), and setting $ds = 0$ then gives

$$\frac{dT}{dp} = \frac{(s_m - s_s)\left(\frac{\partial X}{\partial p}\right)_T - \left((1 - X)\frac{\alpha_s}{\varrho_s} + X\frac{\alpha_m}{\varrho_m}\right)}{(s_m - s_s)\left(\frac{\partial X}{\partial T}\right)_p - \frac{1}{T}\left((1 - X)c_p^s + Xc_p^m\right)}. \tag{7.146}$$

This version of the adiabat becomes the same as the previous one (6.338) in the case of zero melt fraction. It has the form $dT/dp = F(p, T)$ since the melt fraction is a function of p and T. This equation has to be integrated numerically in the case of a general melt fraction, for instance with a simple Runge–Kutta scheme. McKenzie (1984) discusses melting and magma generation and shows several examples of melt fractions and the adiabat (7.146). Note 7.12 studies the adiabat (7.146) in the special case of linear solidus, liquidus and melt fraction.

Note 7.12 *Mantle adiabats and linear melt fraction*: It is instructive to look at the adiabat (7.146) in the case of linear solidus and liquidus

$$T_s(p) = T_{s,0} + Bp \quad \text{and} \quad T_l(p) = T_{l,0} + Bp \tag{7.147}$$

and also a linear melt fraction

$$X(p, T) = \frac{T - T_s(p)}{T_l(p) - T_s(p)} = \frac{T - T_{s,0} - Bp}{T_{l,0} - T_{s,0}}. \tag{7.148}$$

The parameter B is $B = A/(\varrho g)$, where A is the steepness of the solidus and liquidus as in equation (7.127). The linear melt fraction (7.148) gives

$$\left(\frac{\partial X}{\partial T}\right)_p = \frac{1}{T_{l,0} - T_{s,0}} \quad \text{and} \quad \left(\frac{\partial X}{\partial p}\right)_T = -\frac{B}{T_{l,0} - T_{s,0}}. \tag{7.149}$$

We assume that the density, heat capacity and the thermal expansibility is the same for both melt and solid. The adiabat (7.146) then becomes

$$\frac{dp}{dT} = \frac{a_1}{T} + a_2 \tag{7.150}$$

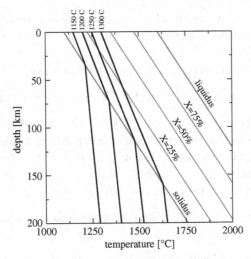

Figure 7.26. *Adiabats are shown for the case of linear solidus, liquidus and a linear melt fraction. Pressure and depth are related by* $p = \varrho g z$.

where

$$a_1 = \frac{c_p\,(T_{l,0} - T_{s,0})}{(\alpha/\varrho)(T_{l,0} - T_{s,0}) + \Delta s\, B} \quad \text{and} \quad a_2 = \frac{\Delta s}{(\alpha/\varrho)(T_{l,0} - T_{s,0}) + \Delta s\, B} \tag{7.151}$$

which can be integrated to

$$p(T) = p_0 + a_1 \ln\left(\frac{T}{T_0}\right) + a_2(T - T_0) \tag{7.152}$$

$$\approx p_0 + \left(\frac{a_1}{T_0} + a_2\right)(T - T_0) \tag{7.153}$$

where (T_0, p_0) is a reference point. The approximation (7.153) is valid as long as $|T - T_0| \ll T_0$, which is normally the case (see Exercise 7.25). Figure 7.26 shows some examples of adiabats that cross the solidus. The solidus and the liquidus are the same as in Table 7.1 and the difference in specific entropy is $\Delta s = s_m - s_s = 356\ \mathrm{J\,kg^{-1}\,K^{-1}}$ after McKenzie (1984). We notice that the adiabats become less steep when they cross the solidus, by roughly a factor $1/2$, which affect estimates of melt generated. Equation (7.152) with $\Delta s > 0$ is the part of the adiabat that is above the solidus. The adiabat below the solidus has $\Delta s = 0$ and it therefore has $a_1 = c_p \varrho / \alpha$ and $a_2 = 0$. An easy way to plot an adiabat is to select a reference point (T_0, p_0) on the solidus, and then use the linear expression (7.153) to plot the two parts of the adiabat – the one with $\Delta s > 0$ above the solidus and the other with $\Delta s = 0$ below the solidus.

Exercise 7.24 The aim of this exercise is a numerical implementation of the empirical melt fraction $X = X(p, T)$ from Note 7.10.

(a) Find an expression for the step in a Newton solver of equation (7.138).

(b) Use (a) to make a numerical implementation of function $X = X(p, T)$ in your favorite programming language.

Solution: Equation (7.138) can be rewritten as $F(U) = 0$ where

$$F(U) = p - a_1 U - a_2 \exp(a_3 U) \quad \text{and} \quad F'(U) = -a_1 - a_2 a_3 \exp(a_3 U) \quad (7.154)$$

where $U = T_s - 1100$, $a_1 = 1/136$, $a_2 = 4.968 \times 10^{-4}$ and $a_3 = 1.2 \times 10^{-2}$. If U is a reasonable guess for a solution then we search a step ΔU such that $F(U + \Delta U) = 0$. The latter expression gives to first order that $F(U) + F'(U)\Delta U = 0$ and the Newton step becomes $\Delta U = -F(U)/F'(U)$. A guess for U gives a step ΔU and then an improved $U \leftarrow U + \Delta U$, which is used to make a new step and so on.

Exercise 7.25 Show that adiabat (7.152) can be approximated by equation (7.153). Hint: use that $\ln(1 + x) \approx x$ for $|x| \ll 1$.

Exercise 7.26 Assume that we have observations of underplating that suggest a total melt of thickness $M = 2$ km has been generated under lithospheric extension. How large a β-factor is needed to produce this melt thickness when we apply the parameters in Table 7.1 (Hint: Figure 7.24.)

Exercise 7.27 The mass fraction of melt is denoted by X.

(a) Show that the volume fraction of melt is related to the mass fraction by

$$\phi = \frac{X \varrho_s}{(1 - X)\varrho_m + X \varrho_s}. \quad (7.155)$$

(b) Show the inverse – that the mass fraction of melt is related to the volume fraction as

$$X = \frac{\phi \varrho_m}{(1 - \phi)\varrho_s + \phi \varrho_m}. \quad (7.156)$$

Solution: (a) The volumes of melt and solid are $V_m = XM/\varrho_m$ and $V_s = (1 - X)M/\varrho_s$, respectively, where M is the total mass. The volume fraction $\phi = V_m/(V_m + V_s)$ then becomes (7.155).

(b) The relation (7.155) is straightforward to invert to the relation (7.156). Alternatively, the mass of melt and solid are $M_m = \phi V \varrho_m$ and $M_s = (1 - \phi)V \varrho_s$, respectively, when V is the total volume. The mass fraction $X = M_m/(M_m + M_s)$ then becomes (7.156).

7.13 Thermal subsidence of the oceanic lithosphere

Oceanic lithospheric plates are created at mid-ocean ridges, where the upwelling hot asthenosphere cools and forms new plates. The asthenosphere moves the plates apart from the ridge at the same rate as they are produced. The cooling of a plate becomes a 1D (vertical) problem for a column of rock that follows the movement of the plate. See Figure 7.27 where the same column of rock is shown at two different times. The temperature in a column at the ridge, where the asthenosphere reaches the surface, can be assumed to have

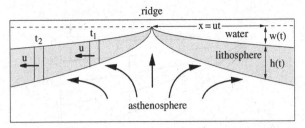

Figure 7.27. *The oceanic lithosphere is created at the mid-ocean ridge, and the plates subside as they move away from the ridge and cool.*

the same mantle temperature from the surface to a "large" depth. As the column moves away from the ridge it starts to cool from the surface. This is precisely the problem of a heating/cooling of a semi-infinite half-space discussed in Section 6.14. The temperature in the column is therefore given by equation (6.205),

$$T(z) = T_s + (T_a - T_s) \operatorname{erf}\left(\frac{z}{2\sqrt{\kappa t}}\right), \tag{7.157}$$

where T_s is the surface temperature, T_a is the temperature of the upwelling hot mantle, and where κ is the thermal diffusivity. The base of the lithospheric plate is now approximated by the isotherm $T = 0.9 \times T_a$, which allows us to find the thickness of the plate as a function of time. We have that $\operatorname{erf}(1.2) \approx 0.9$, and the surface temperature $T_s = 0$ gives the thickness $h(t)$ of the lithosphere as

$$\frac{h(t)}{2\sqrt{\kappa t}} = 1.2 \tag{7.158}$$

or

$$h(t) = 2.4\sqrt{\kappa t}. \tag{7.159}$$

The thickness of the oceanic lithosphere is controlled by conductive cooling by means of only one parameter, the thermal diffusivity κ. Using that $\kappa = 1 \cdot 10^{-6} \text{ m}^2\text{ s}^{-1}$ gives the plate thicknesses $h = 13.4$ km for $t = 1$ Ma, $h = 42$ km for $t = 10$ Ma and $h = 134$ km for $t = 100$ Ma.

The oceanic lithosphere subsides as it cools and gets denser. It gets denser because it contracts thermally, which follows from the mantle density function $\varrho_m(T) = \varrho_{m,0}(1 - \alpha T)$. The thermal subsidence is found by assuming isostatic equilibrium, which says that the pressure is the same at the same depths in the ductile mantle. The pressure in the mantle below the oceanic lithosphere is therefore equal to the pressure at the same depth below the ridge

$$w\varrho_w + \int_0^h \varrho_m(T(z', t))\, dz' = (w + h)\varrho_m(T_a) \tag{7.160}$$

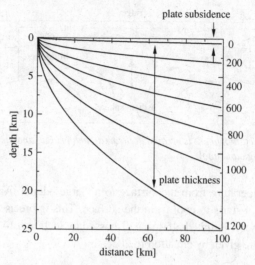

Figure 7.28. *The thickness of the lithospheric plate is roughly 25 km at a distance 100 km away from the ridge, and the subsidence of the plate is roughly 1 km at the same distance.*

where the water density ϱ_w is constant. Isostatic equilibrium (7.160) gives the subsidence w as a function of time as

$$w(t) = \frac{2\varrho_{m,0}\alpha T_a}{(\varrho_{m,0} - \varrho_w)} \left(\frac{\kappa t}{\pi}\right)^{1/2} \tag{7.161}$$

where Note 7.13 provides the details of the derivation. The rate of sea floor spreading can be approximated by a constant (average) spreading velocity u. The age of the oceanic lithosphere is then $t = x/u$, where x is the distance from the ridge. The subsidence at a distance x from the ridge is therefore

$$w(x) = \frac{2\varrho_{m,0}\alpha T_a}{(\varrho_{m,0} - \varrho_w)} \left(\frac{\kappa x}{\pi u}\right)^{1/2}. \tag{7.162}$$

In the same way, the thickness of the plate can also be expressed in terms of the distance x instead of the age t. Figure 7.28 shows the subsidence and the thickness of an oceanic plate that moves away from the ridge with a velocity $u = 2$ cm/year. The other parameters are $\kappa = 1 \cdot 10^{-6}$ m^2 s^{-1}, $\varrho_{m,0} = 3300$ kg m^{-3}, $\varrho_w = 1000$ kg m^{-3}, $T_a = 1300\,°$C, $\alpha = 3 \cdot 10^{-5}$ K^{-1}. The subsidence (7.162) can be written compactly as $w(x) = w_0\sqrt{x/x_0}$ where $w_0 = 2\varrho_{m,0}\alpha T_a/(\varrho_{m,0} - \varrho_w)$ and $x_0 = \pi u/\kappa$. Figure 7.29 shows the fitting of this function to a data set across a ridge. The numbers above give that $w_0 = 0.11$ m and the optimal value for x_0 is $x_0 = 1.55 \cdot 10^{-3}$ m. The water depth at the ridge is 1511 m and it is slightly more than the peak value from the observations. The depth at the ridge could have been added as a second optimization parameter. The optimal value for x_0 gives that the average spreading rate is $u = \kappa x_0/\pi$, or $u = 0.015$ m/year when $\kappa = 1 \cdot 10^{-6}$ m^2 s^{-1}. The spreading velocity gives $\sim$800 km of oceanic plate at each side of the ridge after 50 Ma.

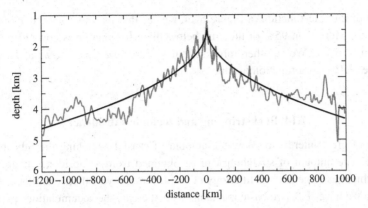

Figure 7.29. *The subsidence $w(x)$ is fitted to observations of the bathymetry across a spreading ridge.*

This is in agreement with the dating of the plate where 50 Ma is ∼850 km away from the ridge. We notice that the topography of the seafloor is not smooth, and there is also a certain degree of asymmetry across mid-ocean ridges. It has been observed that ocean floor older than 70 Ma does not fit very well the $\sqrt{t}$-behavior of the subsidence (7.161).

Note 7.13 The expression (7.160) for isostatic subsidence of the oceanic lithosphere can be rewritten as

$$w(t)\left(\varrho_m(T_a) - \varrho_w\right) = \int_0^h \left(\varrho_m(T(z',t)) - \varrho_m(T_a)\right) dz' \qquad (7.163)$$

and using that $\varrho_m(T) = \varrho_{m,0}\left(1 - \alpha T\right)$ gives

$$w(t)\left(\varrho_m(T_a) - \varrho_w\right) = -\varrho_{m,0}\,\alpha \int_0^h \left(T(z',t) - T_a\right) dz'. \qquad (7.164)$$

The temperature (7.157) in an oceanic plate is

$$\frac{T(z,t) - T_s}{T_a - T_s} = \mathrm{erf}\left(\frac{z}{2\sqrt{\kappa t}}\right), \qquad (7.165)$$

where T_s is the surface temperature. We therefore have

$$\frac{T(z,t) - T_a}{T_a - T_s} = \frac{T(z,t) - T_s + T_s - T_a}{T_a - T_s} = \mathrm{erf}(\eta) - 1 = -\mathrm{erfc}(\eta) \qquad (7.166)$$

where $\eta = z/2\sqrt{\kappa t}$, and where $\mathrm{erfc}(\eta) = 1 - \mathrm{erf}(\eta)$ is the complementary error-function. A change of integration variable on the right-hand side of equation (7.164) from z' to η gives

$$w(t)\left(\varrho_m(T_a) - \varrho_w\right) = 2\varrho_{m,0}\,\alpha\,T_a\sqrt{\kappa t} \int_0^{\eta_h} \mathrm{erfc}(\eta)\,d\eta, \qquad (7.167)$$

when the surface temperature is set to zero. The integral over η is from 0 to $\eta_h = 1.2$, because the base of the lithosphere is defined by the isotherm $T = 0.9 \times T_a$. By looking at the plot of the erfc-function in Figure 6.24, we see that $\eta_h = 1.2$ can be approximated by $\eta_h = \infty$. More than 95% of the area below the erfc-function is covered by η in the range from 0 to 1.2. We can therefore use that $\int_0^\infty \text{erfc}(\eta)d\eta = 1/\sqrt{\pi}$, and we get the subsidence (7.161) as a function of time.

7.14 Backstripping and tectonic subsidence

It is often of great interest to know the amount of crustal stretching the lithosphere has undergone. The amount of stretching can be obtained from a subsidence history, which gives the thickness, age and the lithology of the different layers that constitute a sedimentary basin. We have seen two basic mechanisms that create the accumulation space for the sediments. The first is the subsidence caused by deposition into a water-filled basin, and the other is the subsidence caused by stretching and thinning of the crust. The latter subsidence is called the *tectonic subsidence* because it is the subsidence from tectonic processes. We will now see how the sediment load gives the tectonic subsidence.

The (basement) subsidence $z(t)$ of a sedimentary basin is made of three components at any time t; the water depth $w(t)$, the basin thickness $s(t)$ and the global sea level change $\Delta(t)$

$$z(t) = s(t) + w(t) - \Delta(t), \tag{7.168}$$

and it is often denoted *total subsidence*. Global sea level changes, which are also referred to as *eustatic* sea level changes, are distinguished from water depth changes by not being related to changes in the subsidence. The sea level has changed through geohistory due to global phenomena like changes in volume of the ocean basins or melting of the polar ice caps. A positive $\Delta(t)$ is a sea level rise.

The tectonic subsidence through the geohistory is found using the principle of isostasy, assuming that the sediment thickness $s(t)$, the average basin density $\bar{\varrho}_b(t)$, the water depth $w(t)$ and the (eustatic) sea level change $\Delta(t)$ are all known. It is then possible to replace the isostatic subsidence of the sediments with the corresponding amount of water. A sedimentary basin and its corresponding tectonic subsidence is shown in Figure 7.30. Isostasy gives that pressures at the same position in the ductile mantle beneath the two columns are the same

$$\varrho_w w(t) + \bar{\varrho}_b(t)s(t) + \varrho_c c + \varrho_m a_1 = \varrho_w y(t) + \varrho_c c + \varrho_m a_2 \tag{7.169}$$

where c is the crustal thickness, and where a_1 and a_2 are the thicknesses of a part of the asthenospheric mantle (see Figure 7.30). The reference level is the present-day sea level and we have

$$w(t) - \Delta(t) + s(t) + c + a_1 = y(t) + c + a_2. \tag{7.170}$$

These two equations give the tectonic subsidence

$$y(t) = w(t) + \left(\frac{\varrho_m - \bar{\varrho}_b(t)}{\varrho_m - \varrho_w}\right)s(t) - \frac{\varrho_m}{\varrho_m - \varrho_w}\Delta(t), \tag{7.171}$$

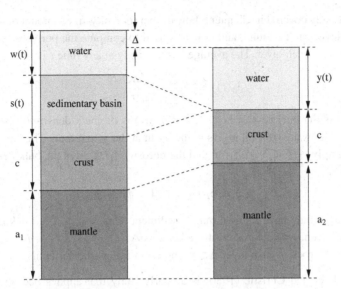

Figure 7.30. *The load of a sedimentary basin is replaced by the isostatically corresponding load of a water filled basin. The subsidence $y(t)$ of the water load is called the tectonic subsidence.*

where we identify the second term as the amount of water we get by removing the sediments (see equation (7.3)). The tectonic subsidence $y(t)$ can be compared with the subsidence of a tectonic model, like the McKenzie model, when only a water load is used. This is the conventional way to estimate lithospheric β (stretching) factors. For instance, there has been no stretching at all if $y(t)$ does not change through the geohistory. On the other hand, if $y(t)$ shows short time intervals of rapid subsidence and long intervals of slow subsidence we might be able to estimate the amount of stretching. The short intervals of rapid subsidence may be interpreted as periods of extension, and the long intervals of slow subsidence as periods of thermal subsidence following the extension.

We have so far said little about the paleo-water depth $w(t)$ and the global sea level changes $\Delta(t)$ through the geohistory. There are published data for the latter, but the paleo-water depth is often poorly constrained. There is no unique way to compute the basin thickness, or the average basin density either (see Chapter 5), which means that there are uncertainties in most quantities that enter the calculation of the tectonic subsidence. Nevertheless, the tectonic subsidence is an important basis for the determination of lithospheric stretching factors for extensional basins.

The equation for the tectonic subsidence is often referred to as the *backstripping equation*. Backstripping is the process of computing the paleo-layer thicknesses backwards through the geohistory by stripping off layer after layer, and by decompacting the remaining layers at each step. Backstripping is normally done with an Athy porosity function for each lithology in the basin. (See Chapter 5 for how the paleo-porosity and the paleo-thickness can be computed for different types of porosity functions.)

We have already covered in Chapter 5 how the net (porosity-free) amount of rock in each sedimentary layer can be found, and how it is used to compute the porosity and the (real) paleo-thickness of each layer. The average basin density at any time t is

$$\bar{\varrho}_b(t) = \frac{\sum_i \varrho_{b,i}(t) \, \Delta z_i(t)}{\sum_i \Delta z_i(t)} \tag{7.172}$$

where $\Delta z_i(t)$ is the thickness of layer i, and $\varrho_{b,i}(t)$ is the bulk density of the sediments in layer i. The sums are over all layers in the basin at the actual time. The basin thickness at time t is simply $s(t) = \sum_i \Delta z_i(t)$, and the porosity $\phi_i(t)$ gives the bulk density of the layer at time t as

$$\varrho_{b,i}(t) = \phi_i(t)\varrho_w + (1 - \phi_i(t))\,\varrho_{s,i} \tag{7.173}$$

where each layer may have its own matrix sediment density $\varrho_{s,i}$. The thickness of each layer i is also given by the porosity of the layer as $\Delta z(t)_i = \Delta \zeta_i / (1 - \phi_i(t))$, where the net amount of the rock in each layer $\Delta \zeta_i$ is constant through the burial history.

Note 7.14 Sclater and Christie (1980) is an early study that applies backstripping as a procedure for obtaining the tectonic subsidence. They estimated the tectonic subsidence for several North Sea wells, and compared it with the subsidence of the McKenzie model in an attempt to estimate the amount of lithospheric stretching. The work of Sclater and Christie (1980) is an instructive study of tectonic modeling.

Exercise 7.28 Derive the backstripping equation (7.171).

Exercise 7.29
(a) Show that the basin thickness at any time t can be written as

$$s(t) = \sum_i (1 + e_i(t))\,d\zeta_i \tag{7.174}$$

where $e_i(t)$ is the void ratio in layer i, and where $d\zeta_i$ is the net (porosity-free) thickness of the layer.
(b) Show that the average basin density is

$$\bar{\varrho}_b(t) = \frac{\sum_i \left(e_i(t)\varrho_w + \varrho_{s,i}\right)d\zeta_i}{\sum_i (1 + e_i(t))\,d\zeta_i}. \tag{7.175}$$

The only time-dependent property that is needed is the paleo-porosity, because both the net layer thickness $d\zeta_i$ and the sediment matrix density $\varrho_{s,i}$ of each layer are constants (independent of time).

Exercise 7.30 Assume that the porosity in a basin can be represented as a function of depth by the Athy function $\phi = \phi_0 \exp(-z/z_0)$.
(a) Show that the average basin porosity is

$$\bar{\phi} = \frac{1}{s} \int_0^s \phi(z)dz = \frac{\phi_0 z_0}{s}\left(1 - e^{-s/z_0}\right) \tag{7.176}$$

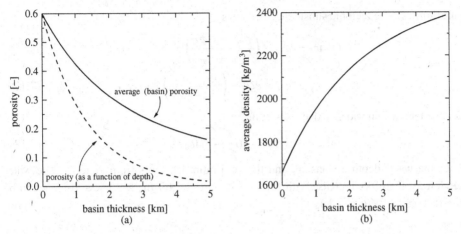

Figure 7.31. *(a) The Athy porosity function and the average basin porosity as a function of basin thickness. (b) The average basin (sediment) density as a function of basin thickness.*

when the basin is s thick. Figure 7.31a shows the Athy porosity function and the corresponding average basin porosity for $\phi_0 = 0.6$ and $z_0 = 1350$ m.

(b) Show that the average bulk basin density is

$$\bar{\varrho}_b = \bar{\phi}\varrho_w + (1 - \bar{\phi})\varrho_s, \qquad (7.177)$$

when the matrix density ϱ_s is constant (independent of depth). Figure 7.31b shows the average basin sediment density for the average basin porosity in Figure 7.31a. The water density and the sediment matrix density are $\varrho_w = 1000$ kg m^{-3} and $\varrho_s = 2650$ kg m^{-3}, respectively.

(c) The principle of isostasy (7.3) implies that sediments of thickness s,

$$s = \frac{(\varrho_m - \varrho_w)}{(\varrho_m - \bar{\varrho}_b)} w, \qquad (7.178)$$

are needed to replace water of depth w, when the sediments have the density $\bar{\varrho}_b$. What is the factor $(\varrho_m - \varrho_w)/(\varrho_m - \bar{\varrho}_b)$ that relates the subsidence s to the water depth w when the basin is shallow and when the basin is deep? Use the average basin density plotted in Figure 7.31b and the mantle density is 3300 kg m^{-3}. (Answer: 1.8 and 2.3)

Exercise 7.31
(a) What is the tectonic subsidence as a function of the average basin porosity ϕ, when the average sediment density is $\varrho(\phi) = \phi\varrho_w + (1 - \phi)\varrho_s$, where ϱ_w is the water density and ϱ_s is the matrix density? Use that the basin thickness is $S = \zeta_0/(1 - \phi) = \zeta_0(1 + e)$, where ζ_0 is the net (porosity-free) amount of sediments in the basin and $e = \phi/(1 - \phi)$ is the void ratio.
(b) Chemical processes makes a basin compact as a function of time although there is no sediment infill. What is the tectonic subsidence from a decreasing average basin porosity when the water depth is kept constant?

Solution: (a) The tectonic subsidence becomes

$$w_t = w_0 + \left(\frac{\varrho_m - \varrho(\phi)}{\varrho_m - \varrho_w} \right) S$$

$$= w_0 + e\zeta_0 - \left(\frac{\varrho_m - \varrho_s}{\varrho_m - \varrho_w} \right) \zeta_0. \tag{7.179}$$

(b) The tectonic subsidence deceases with

$$\Delta w_t = (e_2 - e_1)\zeta_0 \tag{7.180}$$

when the water depth is constant and the void ratio decreases from e_1 to e_2. The water depth must therefore increase with $(e_1 - e_2)\zeta_0$ in order for the tectonic subsidence to be constant. (See Exercise 7.1.)

7.15 Subsidence of the Vøring margin, NE Atlantic

In this section we will apply some of the models we have been through so far and compute the subsidence history of a profile across the Vøring margin (west of mid-Norway). The Vøring profile is shown in Figure 7.32a, and it spans the distance from the continental–oceanic boundary to mainland Norway. The direction of extension has been along the profile. There are four sub-basins, Hel Graben, Naagrind syncline, Træna basin and Helgeland basin, from the west to the east. The Vøring margin has developed by regional subsidence from several episodes of lithospheric extension and crustal thinning, since the end of the Caledonian orogeny. It began with a post-Caledonian extension, which controlled the formation and sedimentation of Devonian sedimentary basins. Then followed a Permian extension, a Late Jurassic extension and finally a Late Cretaceous to Paleocene rift phase, which led to continental break-up and the formation of the North Atlantic ocean. We use the following four rift phases in the modeling of the Vøring margin:

Devonian	-400 Ma to -360 Ma
Permian	-310 Ma to -260 Ma
Late Jurassic	-160 Ma to -140 Ma
Late Cretaceous and Paleocene	-80 Ma to -56 Ma

in line with earlier work (see for instance Lundin and Doré, 1997, Mjelde *et al.*, 1997, Doré *et al.*, 1999, Reemst and Cloetingh, 2000, Ren *et al.*, 2003, Gernigon *et al.*, 2006, Wangen *et al.*, 2007). We notice from Figure 7.32a that the Paleozoic, Triassic and Jurassic are present across the entire profile, and that thick Cretaceous formations are deposited mainly to the west of Nordland ridge. The upper part of the basin has Pliocene and Pleistocene sediments that covers most of the profile.

Figure 7.32a shows the crust at the present time, which also includes a lower crustal body (LCB). The origin of the LCB is debated – it has been taken to be magmatic underplating during the time of continental break-up, alternatively it could be Caledonian crust, or a mixture of magmatic underplating and Caledonian crust. See Gernigon *et al.* (2003) for

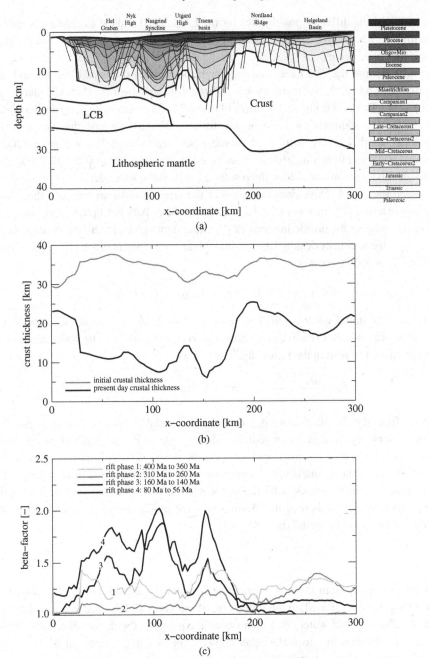

Figure 7.32. *(a) The sedimentary basin and crust for a profile at the Vøring margin. LCB is a lower crustal body. (b) The thickness of the present day crust and an estimate for the initial crustal thickness. (c) The β-factors along the profile for each rift phase.*

a discussion of the different possibilities for the LCB. The simplest alternative is used in the following, where the LCB is Caledonian crust. It has therefore been a part of the crust through the entire geohistory.

We also assume that we have modeled (or backstripped) the burial history, and that we have the load of the sedimentary basin through the geohistory. (See Chapter 5 and Section 7.14.) The load is known by means of average sediment density $\varrho_s(t)$ and the thickness $s(t)$ of the sedimentary basin at any time t. We also assume that we have the paleo-water depth, and we will use the same water depths as in Wangen *et al.* (2007). The paleo-water depth is a highly uncertain piece of input, and Exercise 7.32 gives some simple sensitivity estimates of how the results depend on the water depth.

We need the initial (Devonian) thickness of the crust in order to compute the crustal stretching in terms of β-factors and subsidence. The use of Airy isostasy, knowledge of the present-day state of the profile in terms of the water depth (w_{N+1}), thickness of the basin (s_{N+1}), average sediment density (ϱ_{N+1}), and thickness of the crust (c_{N+1}), gives that the initial thickness of the crust is

$$c_0 = c_{N+1} + f_w(w_{N+1} - w_0) + f_{N+1}s_{N+1} \qquad (7.181)$$

where w_0 is the initial water depth. The number of rift phases is N and the index $N+1$ denotes the present day. (This indexing turns out to be convenient.) The water depth w_0 is for simplicity set to zero in the following. The factors

$$f_w = \frac{\varrho_m - \varrho_w}{\varrho_m - \varrho_c} \quad \text{and} \quad f_{N+1} = \frac{\varrho_m - \varrho_{N+1}}{\varrho_m - \varrho_c} \qquad (7.182)$$

are made from the mantle density ϱ_m, the average crustal density ϱ_c, water density ϱ_w and the present-day average basin sediment density ϱ_{N+1}. Figure 7.32b shows both the present day thickness of the crust and its initial thickness. The present day crust is thin underneath Træna basin, the deepest depo-center. Other local minima of the present day thickness of the crust coincide with the depocenters of the other sub-basins. The initial crustal thickness (7.181) is roughly 35 km across the profile, and it gives right away the maximum (present day) crustal thinning

$$\beta_{\max} = \frac{c_0}{c_{N+1}} = 1 + f_w\frac{w_{N+1}}{c_{N+1}} + f_{N+1}\frac{s_{N+1}}{c_{N+1}} \qquad (7.183)$$

when we look away from eventual thermal uplift from the last rift phase. The modeling of the rifting phases is based on the initial crustal thickness (7.181) because it is the crust (without sediments and water) that is in isostatic equilibrium with the present day lithosphere. It is also possible to make other choices, for example a constant 35 km initial thickness, since we know very little about the upper lithosphere at the beginning.

The lithosphere along the 2D profile is now represented by a row of columns, where each column becomes stretched with its own β-factors (one for each rift phase). The β-factors are therefore piecewise constant along the profile, and the columns have vertical sides that remain vertical during the extension process. The columns have had periods of extension and the β-factor for rift phase i in a column is denoted β_i. The product of the β-factors for each rift phase is the total (or cumulative) β-factor

$$\beta_1 \beta_2 \beta_3 \beta_4 = \beta_{\max}. \tag{7.184}$$

This is the previous equation (7.89). The initial thickness of the crust (7.181) can be rewritten as an expression for the thickness of the crust at any time t

$$c(t) = c_0 - f_w w(t) - f(t) s(t) \tag{7.185}$$

where $w(t)$ is the paleo-water depth, $s(t)$ is the sediment thickness, $\varrho_s(t)$ is the average sediment density. The coefficient $f(t) = (\varrho_m - \varrho_s(t))/(\varrho_m - \varrho_c)$ is the factor f_{N+1} as a function of time. The knowledge of the crustal thickness $c(t)$ gives the cumulative amount of crustal stretching during the geohistory as

$$\beta_{\max}(t) = \frac{c_0}{c(t)} = \frac{c_0}{c_0 - f_w w(t) - f_s(t) s(t)}. \tag{7.186}$$

Alternatively, the cumulative amount of stretching (7.186) can be written

$$\beta_{\max}(t) = \frac{c_0}{c_0 - f_w s_T(t)} \tag{7.187}$$

when expressed by the tectonic subsidence

$$s_T(t) = w(t) + f_T(t) s(t) \tag{7.188}$$

where f_T is the factor $f_T(t) = (\varrho_m - \varrho_s(t))/(\varrho_m - \varrho_w)$. The time of the beginning of rift phase number i is denoted t_i (and the number of rift phases is N). The time t_{N+1} denotes the present time although no rifting is necessarily beginning at this time. The cumulative amount of crustal stretching at the beginning of each rift phase i becomes

$$\beta_{\max,i} = \beta_{\max}(t_i) \tag{7.189}$$

which makes the β-factor for the ith rift phase

$$\beta_i = \frac{\beta_{\max,i+1}}{\beta_{\max,i}} \quad \text{for} \quad i = 1, \ldots, N. \tag{7.190}$$

The cumulative stretching is initially $\beta_{\max,1} = 1$, because there is no crustal thinning before the first rift phase, and $\beta_{\max,N+1}$ is the present-day (maximal) β-factor. The β-factors (7.190) obviously give the product (7.184) when multiplied together. We notice that expression (7.190) is based only on information for the times for the beginning of the rift phases and present time. If we have some knowledge about the water depth $w(t_i)$, then we get the remaining data, the average basin density $\varrho_s(t_i)$ and the basin thickness $s(t_i)$ from the burial-history modeling (or backstripping) of the basin. We can use a constant mantle density at the times t_i when the time interval between the rift phases is sufficiently long for the thermal uplift to have died out. The periods between the rift phases are often quite long (more than 50 Ma) and the modeling of the burial history therefore gives us important estimates for the β-factors. Figure 7.32c shows such estimates for the Vøring profile. These β-factors shows that the depocenters have undergone the most stretching. They also show that the Helgeland basin (in the interval 200 km to 300 km) formed primarily from the Devonian and Permian rift phases. The next two rift phases (Late Jurassic and the Late Cretaceous) are mainly responsible for the sub-basins to the west of the Nordland ridge.

It is not straightforward to use relation (7.190) to compute factor β_i when the thermal uplift is important. The average mantle density at the beginning of rift phase $i + 1$ then depends on the transient temperature at time t_{i+1}, which again depends on β_i. The only way to deal with this situation is to model each time interval between the rift phases iteratively until we match the tectonic subsidence at time t_{i+1}. A simple strategy is to try out all β-factors in the interval from 1 to 5 with a step size 0.2. One can then continue with a finer search in the interval ± 0.2 around the β-factor that gives the best match, and then yet a finer search, and so on. This scheme is not computationally demanding in 1D, although better options exist. Figure 7.33a shows the results of the optimization of tectonic subsidence for

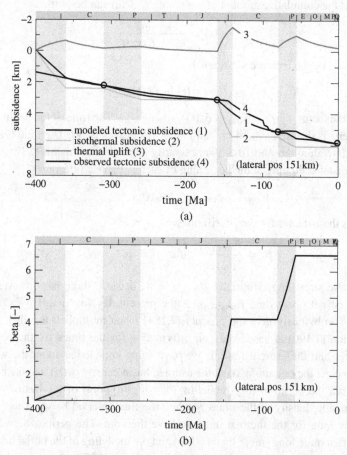

Figure 7.33. *(a) The tectonic subsidence of the position 151 km. (b) Lithospheric stretching at position 151 km.*

Table 7.2. *An example of computation of β-factors.*

i	s_i	ϱ_i	w_i	f_i	c_i	$\beta_{\max,i}$	β_i
1	1	2200	0	2.2	35028	1.00	1.18
2	1800	2200	300	2.2	29690	1.18	1.07
3	2300	2250	550	2.1	27670	1.27	1.77
4	10000	2550	950	1.5	15660	2.24	1.96
5	15000	2560	1050	1.5	8000	4.38	-

each of the four rift-phases. The modeling is for the present-day position 151 km along the profile in Figure 7.32a, which is for the quite deep Træna basin. This is also a place where the present day crust has a local minimum, and therefore undergoes most stretching. The black circles show the match of the modeled tectonic subsidence against the tectonic sub-sidence for the burial history modeling. The modeled tectonic subsidence is decomposed into isothermal subsidence and thermal uplift, and we see that the thermal uplift is only noticeable for the third and Late Jurassic rift phase. Figure 7.33b shows how the β-factor increases during rifting, when each rift phase has a constant strain-rate. These β-factors are similar to the β-factors in figure 7.32c, which were computed using a constant mantle density, assuming that thermal uplift has died out at the beginning of each rift phase.

Note 7.15 *Example:* The estimation of the crustal thickness c_i at time t_i using expression (7.185) is an instructive example that shows how the crust is thinned through the rift phases. The phases of extension start at time t_i and there are $N = 4$ phases of rifting. The time t_5 (for $i = N + 1$) is the present time. It is once again assumed that the time between periods of extension is sufficiently long for thermal transients to have died out. The computation is based on the thicknesses of the sedimentary basin s_i, the average sed-iment density ϱ_i and the water depth w_i at each time t_i. These input data are the first three columns of Table 7.2. The present day crustal thickness is in this case $c_5 = 8$ km, and the densities of water, crust and mantle are $\varrho_w = 1000$ kg m^{-3}, $\varrho_c = 2800$ kg m^{-3}, $\varrho_m = 3300$ kg m^{-3}, respectively. The data in Table 7.2 are taken from the Naagrind syn-cline at position 98.5 km, which is close to its deepest position. The first step is to make the factors

$$f_i = \frac{(\varrho_m - \varrho_i)}{(\varrho_m - \varrho_c)} \tag{7.191}$$

from the densities, and the next step is to estimate the initial crustal thickness

$$c_0 = c_5 + f_w(w_5 - w_0) + f_5 s_5 \tag{7.192}$$

from the present-day data (last line of the table). Once we have c_0 we can derive the thickness of the crust at the beginning of each rift phase

$$c_i = c_0 - f_w w_i - f_i s_i \tag{7.193}$$

which gives the cumulative stretching factors

$$\beta_{max,i} = \frac{c_0}{c_i} \tag{7.194}$$

and finally the β-factors

$$\beta_i = \frac{\beta_{max,i+1}}{\beta_{max,i}}. \tag{7.195}$$

These results are collected in the four last columns in Table 7.2. The estimation of the β-factors by this algorithm suggests that the deep Naagrid syncline was mainly created by the last two rift phases, each with $\beta \approx 2$. The crust has been stretched from 35 km thickness to the present-day 8 km, which gives that the total (accumulative) β-factor is $\beta_{max,5} = 4.38$.

Exercise 7.32 The extension factor β_i depends on the water depths $w_i = w(t_i)$ and $w_{i+1} = w(t_{i+1})$.

(a) Show that

$$\Delta\beta_i = -f_w \, \beta_{max,i+1} \frac{\Delta w_i}{c_0}. \tag{7.196}$$

(b) How large is the change $\Delta\beta_i$ when the water depth is increased by $\Delta w_i = 500$ m? Assume that $\varrho_w = 1000 \text{ kg m}^{-3}$, $\varrho_c = 2800 \text{ kg m}^{-3}$, $\varrho_c = 3500 \text{ kg m}^{-3}$, $c_0 = 35$ km and $\beta_{max,i+1} = 3$.

(c) Show that

$$\frac{\partial \beta_i}{\partial w_{i+1}} = -\frac{1}{c_0} f_w \beta_i \beta_{max,i+1}. \tag{7.197}$$

(d) Estimate the change $\Delta\beta_{i+1}$ when $\Delta w_i = 500$ m, $\beta_i = 1.5$ and the other parameters are as in (b).

7.16 Stretching and thinning of the sediments

Section 7.15 showed how the β-factors for each rift phase could be obtained from the cumulative amount of crustal stretching, which again is found from knowledge of the basin weight and the water depth through the geohistory. It is also possible to use a constant mantle density if the time spans between the rift phases are "long." We will now look at one problem with this approach, which is that the stretching and thinning of the sediments is not accounted for. Only the crust is stretched and thinned. It is reasonable to assume that the sedimentary basin undergoes the same extension and thinning as the crust. If not we may have a mass conservation problem, unless we can explain how the crust is stretched without affecting the basin. In this section we will take the basin into account too, and make a formulation that is mass-conservative for the entire lithosphere during stretching and thinning. Before we do that we need to review some notation – we have that s_1 to s_5 are the thicknesses the sedimentary basin would have had, at the respective times t_1 to t_5, according to the present-day amount of net (porosity-free) rock in each formation. These are the basin thickness from burial history modeling or back-stripping, and they do not involve stretching of the sedimentary basin. We will also assume four rift phases in the following. A generalization to more or less rift phases is straightforward. The cumulative β-factor at the beginning of the first rift phase is

$$\beta_{max,1} = 1 \tag{7.198}$$

because the crust is unstretched until then. At the beginning of the second rift phase (at time t_2) the cumulative β-factor becomes

$$\beta_{max,2} = \frac{c_0}{c_0 - f_w w_2 - \beta_2 \beta_3 \beta_4 f_2 s_2} \tag{7.199}$$

where $f_i = (\varrho_m - \varrho_i)/(\varrho_m - \varrho_c)$ and ϱ_i is the average basin density at time t_i. The thickness of the basin is now increased by the product $\beta_2 \beta_3 \beta_4$. The sediments deposited to the beginning of the second rift phase, with the thickness s_2, must have been a factor $\beta_2 \beta_3 \beta_4$ thicker, because they have experienced rift phases 2, 3, and 4. Note 7.18 shows how porous sediments can be stretched by a β-factor in a mass-conservative manner. Similarly, the cumulative β-factor for the third rift phase is

$$\beta_{max,3} = \frac{c_0}{c_0 - f_w w_3 - \beta_3 \beta_4 f_3 s_3} \tag{7.200}$$

because the sediments deposited until the beginning of the third rift phase, with the thickness s_3, have gone through rift phases 3 and 4. Therefore, this thickness was a factor $\beta_3 \beta_4$ thicker at the beginning of rift phase 3. At the beginning of the fourth rift phase we have

$$\beta_{max,4} = \frac{c_0}{c_0 - f_w w_4 - \beta_4 f_4 s_4} \tag{7.201}$$

and at the present time

$$\beta_{max,5} = \frac{c_0}{c_0 - f_w w_5 - f_5 s_5}. \tag{7.202}$$

At the same time the β-factor for each rift phase is given by the cumulative β-factor by expression (7.190). These equations for the β-factors can be solved by the following procedure, as shown in Note 7.17, where the last β-factor is found first, which is then used to obtain the next last β-factor and so forth. The β-factors in reverse order become

$$\beta_4 = \frac{c_0 - f_w w_4}{c_0 - f_w w_5 - (f_5 s_5 - f_4 s_4)} \tag{7.203}$$

$$\beta_3 = \frac{c_0 - f_w w_3}{c_0 - f_w w_4 - \beta_4 (f_4 s_4 - f_3 s_3)} \tag{7.204}$$

$$\beta_2 = \frac{c_0 - f_w w_2}{c_0 - f_w w_3 - \beta_4 \beta_3 (f_3 s_3 - f_2 s_2)} \tag{7.205}$$

$$\beta_1 = \frac{c_0 - f_w w_1}{c_0 - f_w w_2 - \beta_4 \beta_3 \beta_2 (f_2 s_2 - f_1 s_1)}. \tag{7.206}$$

The last β-factor depends on the difference $f_5 s_5 - f_4 s_4$, which is close to being proportional to the difference $s_5 - s_4$. This difference is the accumulation space needed for sediments that are filled in after the last rift phase. The preceding β-factors are similarly dependent on the accumulation space needed for the sediments that follow the rift phase, and they are inflated by β-factors of the rift phases that follow. The β-factor for the first rift phase is therefore dependent on the β-factors of the second, third and fourth rift phase. How sensitive these β-factors are for changes in the paleo-water depth is studied in Exercise 7.33.

Figure 7.34a shows the total cumulative β-factors (7.183) across the profile 7.32, where stretching and thinning of the sediments are not accounted for. Figure 7.34b shows the β-factors (7.203) to (7.206), which includes the thinning of the sedimentary load during each rifting interval. We notice that the final stretching is the same for both β-factors, but the amount of stretching during each rift phase is different. The first rift phase now becomes much more important, which is reasonable because the early (Paleozoic) sediments have gone through three rift phases and therefore were much thicker before they

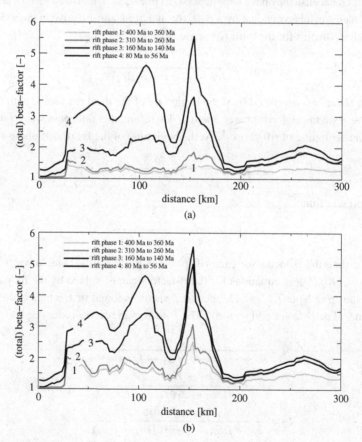

Figure 7.34. *(a) β-factors when the sediments are not stretched. (b) β-factors that account for stretching.*

Table 7.3. *An example of computation of β-factors where thickness of the sediments are taken into account. This is an alternative version of the case in Table 7.2.*

i	s_i	ϱ_i	w_i	f_i	c_i	$\beta_{max,i}$	β_i
1	1	2200	0	2.2	35030	1.00	1.56
2	1800	2200	300	2.2	22515	1.56	1.11
3	2300	2250	550	2.1	20271	1.73	1.90
4	10000	2550	950	1.5	10667	3.28	1.33
5	15000	2560	1050	1.5	8002	4.38	–

became stretched and thinned. A larger factor β_1 is therefore needed to make the extra accommodation space for these early sediments, see note 7.16. The β-factor for the last rift phase has become less when accounting for the sediment thickness, but the late Jurassic rifting has a larger β-factor. It is also reasonable that the accommodation spaces for the thick Cretaceous formations are made by the Jurassic rifting. If the thick Cretaceous formations were to have experienced the late Cretaceous and Paleocene rift phase with a β-factor nearly as large as 2, then they would have been unrealistically thick prior to rifting.

It is possible to calculate how the lithosphere develops through the geohistory once the β-factors have been estimated. Figure 7.36 shows the lithosphere at the beginning and end of the three last rift phases and at the present time. The β-factors are computed by equations (7.203) to (7.206) for a series of columns along the transect. We can now see how the crust and the basin are stretched and thinned, and how extension creates a thermal transient in the lithosphere. Especially, the Late Jurassic rifting creates a noticeable thermal transient. Figure 7.36 also indicates that the thermal transients for rifting are considerably reduced in the time interval between the rift phases. The sedimentary basin though the geohistory is shown in Figure 7.37. It shows the deposition, stretching and thinning of the sediments. This modeling is a mass-conservative handling of the basin through the rift phases (see Wangen and Faleide, 2008). The state of the basin is also shown at the beginning of each rift phase and at the present time.

Note 7.16 *Example*: The use of equations (7.203) to (7.206) compared to the example in Note 7.15 demonstrates what is different with this algorithm. The two first steps are the same – the computation of the factors f_i and the estimate (7.192) for the initial crustal thickness. The equations (7.203) to (7.206) are then used to make the β-factors for $i = 1$ to 4. These β-factors are then multiplied together to make the products $\beta_{max,i}$, which finally gives the crustal thickness $c_i = c_0 / \beta_{max,i}$. There are differences in the results for these two versions of computing the β-factors. Table 7.3 shows that the inclusion of the sediments makes the first rift phase more important, and the last rift phase less important. The early sediments must have been thicker because they have gone through several rift phases. The first rift phase must therefore have been stronger to make the extra accommodation space.

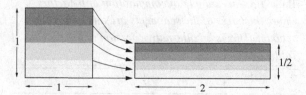

Figure 7.35. *The porosity during extension.*

The third (Late Jurassic) rift phase, which makes space for the thick Cretaceous formations, is now stronger. The last rift phase cannot have been strong, with $\beta \approx 2$, because it implies that all the sediments deposited prior to the last rift phase would have been twice as thick prior to rifting.

Note 7.17 *The β-factors*: The last β-factor is given as

$$\beta_4 = \frac{\beta_{\max,5}}{\beta_{\max,4}} = \frac{c_0 - f_w w_4 - \beta_4 f_4 s_4}{c_0 - f_w w_5 - f_5 s_5} \tag{7.207}$$

which gives β_4 as expressed by (7.203). The next β-factor is found in the same way from

$$\beta_3 = \frac{\beta_{\max,4}}{\beta_{\max,3}} = \frac{c_0 - f_w w_3 - \beta_3 \beta_4 f_3 s_3}{c_0 - f_w w_4 - \beta_4 f_4 s_4} \tag{7.208}$$

where we solve for β_3 without inserting β_4, and we get β_3 as given by (7.204). The two remaining β-factors are found by a straightforward continuation.

Note 7.18 *Mass-conservative stretching of the sedimentary basin*: The mass of a sedimentary basin of thickness s and width l before and after extension is

$$M = \varrho_{\text{av}} s l = \varrho_{\text{av}} \left(\frac{s}{\beta}\right) (l\beta) \tag{7.209}$$

and it is conserved as long as the average basin density ϱ_{av} remains unchanged by the extension. The average basin density can be related to the average basin porosity ϕ_{av} by

$$\varrho_{\text{av}} = \phi_{\text{av}} \varrho_w + (1 - \phi_{\text{av}}) \varrho_s \tag{7.210}$$

where ϱ_w and ϱ_s are the water and the matrix densities, respectively. The extension becomes mass-conservative if the average porosity is unchanged. The average porosity is not constant in general, but it remains constant for porosity functions of the form $\phi(z) = f(z/z_0)$, when the characteristic length z_0 is also stretched. See Figure 7.35. The new characteristic length after a rift phase becomes z_0/β. We have

$$\phi_{\text{av}} = \frac{1}{s} \int_0^s f(z/z_0) \, dz = \frac{1}{s'} \int_0^{s'} f(z'/z_0') \, dz' \tag{7.211}$$

when both the integration variable $z' = z/\beta$ and the characteristic depth $z_0' = z_0/\beta$ are scaled with the β-factor.

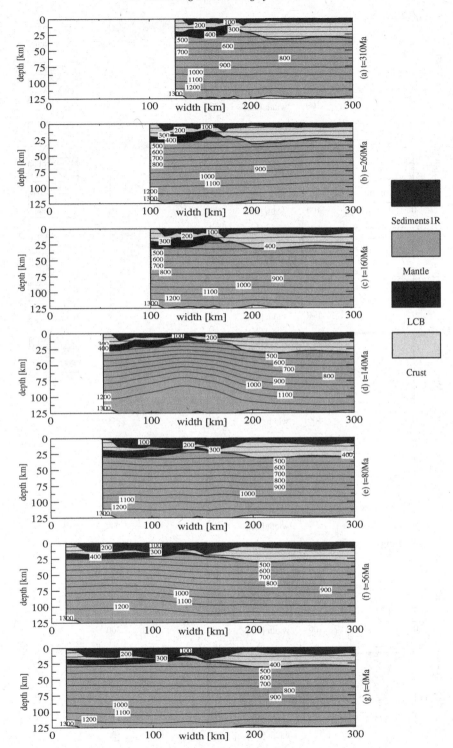

Figure 7.36. *The lithosphere at the beginning and end of the three last rift phases.*

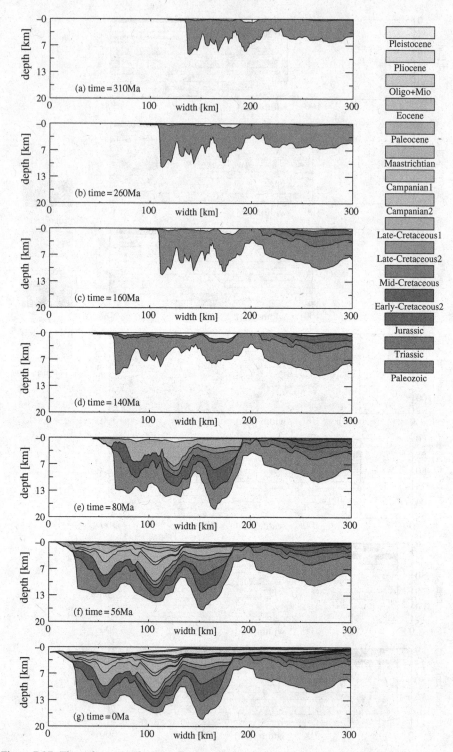

Figure 7.37. *The sedimentary basin at the beginning and end of the three last rift phases.*

Exercise 7.33 The stretching factors β_i, given by equations (7.203)–(7.206) for four rifting phases, depend on the water depth w_i at the beginning of the rift phases.
(a) Show that the relative change in the β-factor is

$$\frac{\Delta\beta_i}{\beta_i} = -\frac{\Delta w_i}{c_0 - f_w w_i}. \tag{7.212}$$

(b) What is the change in the β-factor when the water depth increases by $\Delta w_i = 500$ m? Assume that $\varrho_w = 1000$ kg m^{-3}, $\varrho_c = 2800$ kg m^{-3}, $\varrho_c = 3500$ kg m^{-3}, $c_0 = 35$ km $w_i = 1$ km and $\beta_i = 2$.

7.17 Further reading

A sedimentary basin will in general develop very complex geometries through its geohistory. A discussion of these structures are beyond the scope of this book. Allen and Allen (1990) give a broad view on basin forming tectonics, the related sedimentology and petroleum geology. Davis and Reynolds (1996) is a comprehensive text on structural geology. Turcotte and Schubert (1982) has an introduction to, and applications of, isostasy. The two essential references to the McKenzie model are McKenzie (1978) and Jarvis and McKenzie (1980). An early application of the McKenzie model to study the extension and subsidence of the North Sea is Sclater and Christie (1980). Inverse modelling of extensional sedimentary basins has been developed and applied by White (1994); Bellingham and White (2000); Poplavskii et al. (2001); Jones et al. (2004). Jones et al. (2004) were the first to carry out inverse strain rate modelling on the Vøring transect in Sections 7.15 and 7.16

8

Rheology: fracture and flow

Faulting and folding are important processes that shape the Earth. These different types of deformations reflect two basic types of rheological behavior of rocks – *brittle* and *ductile*. *Rheology* is the study of the deformation and flow of rock under the influence of stress. In the brittle regime the deformation is by fracture and by sliding along fault planes, and in the ductile regime the rock yields. Whether a rock is brittle or ductile depends on temperature and stress. The rock behaves as brittle when the stress needed for brittle deformation (fracture or frictional sliding) is less than for yielding, and vice versa. This chapter presents simple empirical laws that quantify the brittle and the ductile behavior of rocks.

8.1 Faults

A fault is a planar fracture where the two sides of the fracture have been displaced relative to each other. The direction of the fault block movement defines three different types of faults, as shown by Figure 8.1. A *strike-slip* fault is a steeply dipping (near vertical) fault where a horizontal slip has occurred along the strike, as illustrated in Figure 8.1a. Most large strike-slip faults in continental settings are plate boundaries. The two other fault types are dominated by translation directly up or down the dip of the fault plane. The two sides of a non-vertical fault are called the footwall and the hanging wall, where the block above the fault is the *hanging wall*, and the block below is the *footwall*. A *normal fault* results from extension, where the hanging wall moves down relative to the footwall (see Figure 8.1b). The opposite case, compression, creates a *reverse (thrust) fault*, where the hanging wall moves up relative to the footwall, as shown in Figure 8.1c. These three types of faults can also be understood in terms of the orientation of the principal stress.

8.2 Friction

Fault blocks move by sliding along fault planes, where the sliding is resisted by friction, see Figure 8.2. The friction is proportional to the normal force pressing the fault planes together. The normal stress σ_n therefore gives the shear stress τ_f in the fault plane as

$$\tau_f = \mu \sigma_n \tag{8.1}$$

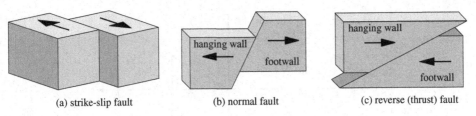

(a) strike-slip fault   (b) normal fault   (c) reverse (thrust) fault

Figure 8.1. *Three types of faults.*

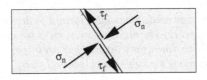

Figure 8.2. *The friction stress τ_f acts in the fault plane as a result of the (compressive) normal stress σ_n when the fault blocks slide against each other.*

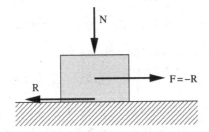

Figure 8.3. *The friction force R that resists sliding is proportional to the normal force N, with a coefficient that does not depend on the area between the block and the surface.*

where μ is the coefficient of friction. This relationship is *Amontons' law*. It is the same law as for the friction of a block sliding along a surface, which says that the friction is proportional to the normal force (the weight of the block), see Figure 8.3. The coefficient of proportionality between the friction force and the normal force does not depend on the area of contact, because both forces relate to the same area. Dividing the forces by the area of contact then gives Amontons' law.

Byerlee (1977, 1978) found that the crust is characterized by a coefficient of friction $\mu \approx 0.85$ for normal stress less than 200 MPa. At larger depths the friction stress is estimated as $\tau_f \approx 0.6\sigma_n + 60$ MPa. This specific version of Amontons' law

$$\tau_f = \begin{cases} 0.85\sigma_n & \sigma_n \leq 200\,\text{MPa} \\ 0.60\sigma_n + 60\,\text{MPa} & \sigma_n > 200\,\text{MPa} \end{cases} \qquad (8.2)$$

is called *Byerlee's law*, and it applies to a wide range of rocks.

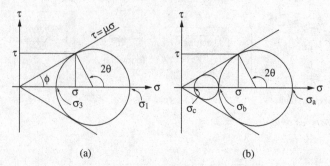

Figure 8.4. (a) *The principal stress needed for sliding along a fault plane becomes a Mohr's circle that touches the failure envelope. (b) The principal stress σ_b is the overburden $\varrho g z$. The smallest Mohr's circle represents extension, where the least principal stress is $\sigma_c = 0.2\sigma_b$. The largest Mohr's circle represents compression, where the largest principal stress is $\sigma_a = 5\sigma_b$.*

The normal stress and shear stress on any plane can be expressed by principal stresses using Mohr's circles as shown in Section 3.8. The normal stress on a fault plane with a normal vector that makes an angle θ with the largest principal stress is

$$\sigma_n = \frac{1}{2}(\sigma_1 + \sigma_3) + \frac{1}{2}(\sigma_1 - \sigma_3)\cos 2\theta \tag{8.3}$$

and the corresponding shear stress has the absolute value

$$\tau = \frac{1}{2}(\sigma_1 - \sigma_3)\sin 2\theta. \tag{8.4}$$

There is no sliding along the fault blocks unless the shear stress τ in the fault plane reaches the friction stress, $\tau = \tau_f = \mu\sigma_n$, or surpasses it. There is normally a large number of faults over a range of angles θ, and sliding takes place along the faults with an angle where the friction is least. These faults have a Mohr's circle that touches the failure envelope, as illustrated by the example in Figure 8.4a. The ratio of principal stresses is then

$$\frac{\sigma_1}{\sigma_3} = \left(\sqrt{1 + \mu^2} + \mu\right)^2 \tag{8.5}$$

as shown in Note 8.1. A coefficient of friction $\mu = 0.85$ gives that $\sigma_1/\sigma_3 \approx 5$, and it means that a substantial difference between the least and the largest principal stress is needed for sliding. One of the principal stresses is the (vertical) overburden stress. The other principal stress (in 2D) is therefore in the horizontal plane. The overburden may either be the largest or the least principal stress. In the first case $\sigma_1 = \varrho g z$ and $\sigma_3 = \sigma_1/5$, which is the condition for sliding under extension. The other case has $\sigma_3 = \varrho g z$ and $\sigma_1 = 5\sigma_3$, which is the stress state needed for sliding under compression. Figure 8.4b shows the Mohr's circles for these two stress states.

Note 8.1 We want to derive the ratio (8.5) between the largest and the least principal stress for a Mohr's circle that touches the failure envelope. The failure envelope makes an angle θ with the σ-axis, and $\tan\phi = \mu$. The normal and shear stresses are given by the angle 2θ in the Mohr's circle. An inspection of Figure 8.4a shows that $2\theta = \pi/2 + \phi$, and we therefore get that $\cos 2\theta = \cos(\pi/2+\phi) = -\sin\phi$ and $\sin 2\theta = \sin(\pi/2+\phi) = \cos\phi$. The trigonometric relations $\sin\phi = \tan\phi/(1 + \tan^2\phi)^{1/2}$ and $\cos\phi = 1/(1 + \tan^2\phi)^{1/2}$ together with $\tan\phi = \mu$ then gives

$$\cos 2\theta = -\frac{\mu}{\sqrt{1+\mu^2}} \quad \text{and} \quad \sin 2\theta = \frac{1}{\sqrt{1+\mu^2}}. \tag{8.6}$$

The condition for failure $\tau = \mu\sigma$ is

$$\frac{1}{2}(\sigma_1 - \sigma_3)\sin 2\theta = \frac{\mu}{2}(\sigma_1 + \sigma_3) + \frac{\mu}{2}(\sigma_1 - \sigma_3)\cos 2\theta \tag{8.7}$$

and when $\sin 2\theta$ and $\cos 2\theta$ are replaced by expressions (8.6), we get the ratio (8.5) after a little algebra.

8.3 Stick–slip faulting

Fault movements are rarely by a steady sliding or creep. Instead faults accumulate stress and elastic energy over long time, which is suddenly released in earthquakes. The faults are therefore locked, except for the rare events of earthquakes. The accumulated elastic energy is released as seismic waves that shake the ground and as heat from friction along the fault.

Figure 8.5 shows a simple model for the stick–slip behavior of a fault, where the fault is initially in a relaxed situation, without any shear stress in the fault plane. The fault is then subjected to shear strain at a constant rate, which accumulates until the corresponding shear stress exceeds the *static friction* of the fault. The fault then slips, and the fault is opposed by *dynamic friction* during motion. The static friction is the shear stress that opposes the initialization of movement, and the dynamic friction is the shear stress that opposes the motion. When the fault slips, and the shear stress along the fault equals the static friction τ_s, we have

$$\tau_s = \tau_d + G\epsilon \tag{8.8}$$

where τ_d is the dynamic friction, G is the shear modulus and ϵ is the amount of the shear strain necessary for the fault to slip. Relation (8.8) between static and dynamic friction

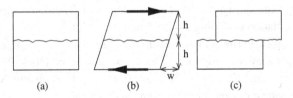

(a)   (b)   (c)

Figure 8.5. *Three stages in stick–slip faulting: (a) The fault is relaxed (right after fault slip). (b) The fault accumulates stress. (c) The fault has slipped and the elastic strain is converted to a permanent displacement.*

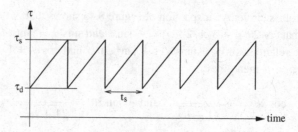

Figure 8.6. *The shear stress along a stick–slip fault as a function of time.*

shows that the static friction is larger than the dynamic friction (because the elastic shear stress $G\epsilon$ is larger than zero). The shear strain of the fault is

$$\epsilon = \frac{\tau_s - \tau_d}{G} \tag{8.9}$$

which is proportional to the difference between static and dynamic friction. If the sheared zone has the width h on both sides of the fault the displacement during fault slip is $w = 2h\epsilon$. The fault in Figure 8.5 is in a state of simple shear as shown by the example in Figure 3.2. If the shear strain accumulates linearly in time as $\epsilon = \dot{\epsilon}t$, at a constant strain rate $\dot{\epsilon}$, the time between each fault slip is

$$t_s = \frac{\tau_s - \tau_d}{G\dot{\epsilon}} \tag{8.10}$$

which is inversely proportional to the strain rate. The fault therefore accumulates stress over a long period of time if the strain rate is very small. Figure 8.6 shows the shear stress increasing linearly with time, from τ_d to τ_s, until the fault slips.

Exercise 8.1 What is the displacement w along a fault when $\tau_s - \tau_d = 100$ MPa, $G = 30$ GPa and $h = 100$ m?

8.4 The slider-block model of stick–slip motion

Another simple model for this stick–slip motion of faults is the movement of a block drawn by a spring over a surface as shown in Figure 8.7. The block has the weight m and the free tip of the spring moves with a constant velocity v. The block does not begin to move until the force from the spring exceeds the *static* friction. The static friction is the force that opposes the initiation of sliding, while *dynamic* friction is the force that opposes the movement of the block during motion. Both the static and the dynamic friction follow Amontons' law, and they are proportional to the weight of the block. The coefficient of static friction and dynamic friction are denoted μ_s and μ_d, respectively. Newton's second law gives the following equation for the movement of the block:

$$m\frac{d^2x}{dt^2} = k(x_s + vt - x) - \mu_d mg \tag{8.11}$$

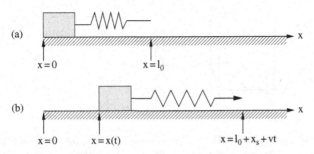

Figure 8.7. (a) *The length of the block with the spring is l_0, when the spring is relaxed. (b) The (rightmost) tip of the spring moves at a constant velocity v, but the block does not move until the spring is stretched a distance x_s. Time $t = 0$ is when the block starts to move, and $x(t)$ is the distance the block has moved.*

where the left-hand side is the mass times the acceleration, and the right-hand side is the sum of the forces acting on the block. The sum is the force from the spring minus the dynamic friction. The spring force is the spring constant k times its elongation $x_s + vt - x$, and $t = 0$ is the time when the force from the spring overcomes the static friction and the block begins to move. The spring is therefore stretched a length $x_s = \mu_s mg/k$ at $t = 0$. The elongation of the spring at time t is the position of its tip $l_0 + x_s + vt$ minus the position the tip would have had if it had been unstretched, $l_0 + x(t)$ (see Figure 8.7). The solution of equation (8.11) gives the position of the block as a function of time. In order to solve the second-order equation we need two boundary conditions. The first is that the block is at position $x = 0$ at $t = 0$, and the second is that the block has initially zero velocity, $dx/dt = 0$ at $t = 0$. The solution for the block position is

$$x(t) = vt + x_s - x_d - \frac{v}{\lambda}\sin(\lambda t) - (x_s - x_d)\cos(\lambda t) \tag{8.12}$$

as shown in Note 8.2, where $\lambda = \sqrt{k/m}$ and $x_d = \mu_d mg/k$. The velocity of the block is therefore

$$\frac{dx}{dt} = v - v\cos(\lambda t) + \lambda(x_s - x_d)\sin(\lambda t). \tag{8.13}$$

The block starts out with zero velocity and it comes to rest after a time span t_1. The time t_1, for the duration of the movement, is a solution of equation $dx/dt = 0$, which is

$$t_1 = \frac{2}{\lambda}\left(\pi - \tan^{-1}\left(\frac{\lambda(x_s - x_d)}{v}\right)\right) \tag{8.14}$$

where Note 8.3 shows the details of the solution. The block is then at position

$$x_1 = x(t_1) = vt_1 + 2(x_s - x_d) \tag{8.15}$$

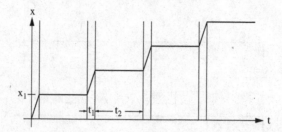

Figure 8.8. *Stick–slip fault behavior. The fault moves during a short time interval t_1 and rests during a long time interval t_2. The fault accumulates sufficient elastic stress during the time interval t_2 to overcome the static friction of the fault and to initiate slip.*

as shown in Note 8.4. Once the block is at rest it stays at rest until the force of the spring once more reaches $F_s = kx_s$ and overcomes static friction. That happens after a time span t_2 given by

$$F_s = kx_s = k\left(x_s + v(t_1 + t_2) - x_1\right) \tag{8.16}$$

or

$$t_2 = \frac{x_1 - vt_1}{v} = \frac{2(x_s - x_d)}{v} \tag{8.17}$$

where the last equality follows from equation (8.15). The time spans t_1 and t_2 are related by

$$\frac{1}{2}\lambda t_1 = \pi - \tan^{-1}\left(\frac{1}{2}\lambda t_2\right). \tag{8.18}$$

The movement of the block is periodic – it moves for a time interval t_1 and then rests a time interval t_2, before the cycle repeats, as shown in Figure 8.8. The time t_1 is much less than the time t_2 in the case of faulting. The time of fault movement may be less than a second, while the fault is at rest for several years. The large difference between the time spans t_1 and t_2 is caused by a very small velocity v. The distance (8.15) the block moves during slip can therefore be approximated as

$$x_1 \approx 2(x_s - x_d) = 2(\mu_s - \mu_d)\frac{mg}{k} \tag{8.19}$$

which is proportional to the difference between the coefficients of static and dynamic friction.

Equation (8.17) shows that the time the block is at rest becomes zero if the static and dynamic friction are equal. Stick–slip behavior of faults therefore requires that the coefficient of static friction is larger than the coefficient of dynamic friction, $\mu_s > \mu_d$. (See Exercise 8.2.)

Even if this block–spring model of the fault-behavior is simple, it nevertheless captures the main features of the stick–slip behavior observed in earthquakes. This kind of model has been the basis for more realistic models, consisting of several blocks connected with springs, which can reproduce the chaotic and fractal behavior of earthquakes.

Exercise 8.2 This exercise treats the special case when $\mu_s = \mu_d$.
(a) Show that $x(t) = vt - (v/\lambda)\sin(\lambda t)$.
(b) Show that $dx/dt = 0$ for $t_n = 2n\pi/\lambda$ for any integer $n > 0$.
(c) Show that the block does not stick, because the force is $F_s = kx_s$ at the times t_n.
The block therefore oscillates around position vt with an amplitude $A_0 = v/\lambda$ and a period $t_0 = 2\pi/\lambda$.

Exercise 8.3 Let $2\lambda(x_s - x_d)/v \gg \pi$.
(a) Show that $t_1 \ll t_2$.
(b) Show that $t_1 \approx \pi/\lambda$.
(c) Show that $t_2/t_1 \approx 2\lambda(x_s - x_d)/(\pi v)$.

Note 8.2 The solution of equation

$$\frac{d^2x}{dt^2} + \lambda^2 x = f(t) \quad \text{where} \quad f(t) = \lambda^2(vt + x_s - x_d) \tag{8.20}$$

is the sum of the homogeneous solution and a particular solution. The homogeneous solution solves the equation for $f(t) = 0$, and the particular solution is any solution of the equation with the given $f(t) \neq 0$. The homogeneous solution is $u(t) = A\sin(\lambda t) + B\cos(\lambda t)$, and a particular solution is $v(t) = vt + x_s - x_d$, and both solutions are verified by inserting them into equation (8.20). The solution is therefore $x(t) = u(t) + v(t)$, and it has two coefficients A and B that are given by the boundary conditions. The first boundary condition at $x = 0$ implies that $B = -(x_s - x_d)$, and the second boundary condition $dx/dt = 0$ at $x = 0$ gives that $A = -v/\lambda$, which together give the position (8.12).

Note 8.3 The block comes to rest after the time span t_1, which is a solution of $dx/dt = 0$. We have

$$\frac{dx}{dt} = v(1 - \cos(\lambda t)) + \lambda(x_s - x_d)\sin(\lambda t). \tag{8.21}$$

In order to solve equation $dx/dt = 0$ it is advantageous to rewrite expression (8.21) using the trigonometric formulas $1 - \cos(\lambda t) = 2\sin^2\left(\frac{1}{2}\lambda t\right)$ and $\sin(\lambda t) = 2\sin\left(\frac{1}{2}\lambda t\right)\cos\left(\frac{1}{2}\lambda t\right)$. It then follows that

$$\tan\left(\frac{1}{2}\lambda t_1\right) = -\frac{\lambda(x_s - x_d)}{v} \tag{8.22}$$

or

$$\frac{1}{2}\lambda t_1 = N\pi - \tan^{-1}\left(\frac{\lambda(x_s - x_d)}{v}\right) \tag{8.23}$$

where N is any integer. Adding (or subtracting) any number of π's to an angle does not change the tan-value. We therefore add π to the solution ($N = 1$) since we want the smallest positive solution for the time span.

Note 8.4 To arrive at expression (8.15) for $x(t_1)$ it is beneficial to rewrite $x(t)$ using the trigonometric relationships

$$\sin(\lambda t) = \frac{2 \tan\left(\frac{1}{2}\lambda t\right)}{1 + \tan^2\left(\frac{1}{2}\lambda t\right)} \quad \text{and} \quad \cos(\lambda t) = \frac{1 - \tan^2\left(\frac{1}{2}\lambda t\right)}{1 + \tan^2\left(\frac{1}{2}\lambda t\right)}. \tag{8.24}$$

The next step is to replace $\frac{1}{2}\lambda t_1$ with $-\tan^{-1}\left(\lambda(x_s - x_d)/v\right)$, and the x_1-position (8.15) follows after a little algebra.

8.5 Fracture

There is a maximum stress a rock (or any other material) can support before it fails. Much of what is known about the limits of the brittle strength of rocks is found from triaxial tests, where cylindrical rock specimens are subjected to a constant confining pressure and a variable axial stress, as shown in Figure 8.9. The confining pressure is normally the least principal stress and the axial stress is the largest principal stress. The stress states that lead to fracture can be mapped by increasing the axial stress until failure, and analysis of the results from such experiments have lead to empirical failure criteria. It is often seen that failure under compressive pressure is by means of a shear fracture, along a shear plane inclined an angle more than 45° relative to the direction of the least principal stress. These observations are the basis for the Coulomb fracture criterion, which relates the shear stress and the normal stress on a fracture plane. The fracture criterion is

$$\tau = S_0 + \mu_0 \sigma \tag{8.25}$$

and it becomes a linear envelope, as shown in Figure 8.11, where shear stress τ is plotted as a function of normal stress σ. The parameters in the Coulomb relationship are the *cohesive strength* S_0 and the coefficient of *internal friction* μ_0. The coefficient of internal friction may be also given by the *angle of internal friction*, ϕ, as $\mu = \tan\phi$. The cohesive strength is the shear stress necessary for failure at zero normal stress, and the coefficient of internal friction gives the increase in shear strength with increasing normal stress. Although the

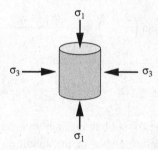

Figure 8.9. *In a triaxial test a cylindrical rock specimen is subjected to an axial (largest principal) stress σ_1 and a compressive pressure σ_3.*

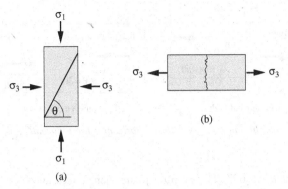

Figure 8.10. *(a) Shear fracture making an angle $\theta \approx 60°$ produced by a triaxial test. (b) Tension fracture that is normal to the least principal stress (σ_3).*

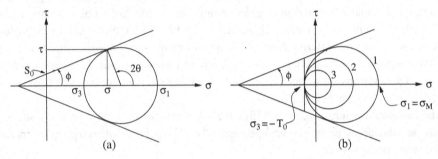

Figure 8.11. *(a) The Coulomb fracture criterion gives the stress states for fracture by a linear envelope. Stress states represented by Mohr's circles that touch the envelope lead to failure. (b) Stress states with $\sigma_3 = -T_0$ lead to tension failure. The largest possible σ_1 for tensional fracture is $\sigma_1 = \sigma_M$.*

Coulomb fracture criterion is similar to Amontons' law, they represent different phenomena. Shear fracture is the creation of a new fracture plane, while frictional sliding is along an already existing plane. The material parameters are also different, even if μ and μ_0 are of the same order of magnitude.

It is often convenient to give a stress state in terms of principal stress, and the shear stress and the normal stress on any plane then follow from relations (8.3) and (8.4). Stress states that touch the fracture envelope lead to failure (along a plane with orientation θ) as shown in Figure 8.11. The least principal stress and the largest principal stress are related at failure, where we have

$$\left(\sqrt{1 + \mu_0^2} - \mu_0\right)\sigma_1 = \left(\sqrt{1 + \mu_0^2} + \mu_0\right)\sigma_3 + 2S_0 \tag{8.26}$$

as shown in Note 8.1. Figure 8.11a shows that failure is then along a plane oriented at an angle $\theta = \pi/4 + \phi/2$, and Figure 3.11 shows that the θ-angle is relative to the least principal stress. A typical coefficient of internal friction is $\mu_0 \approx 0.6$ or alternatively $\phi \approx 30°$,

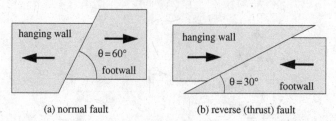

Figure 8.12. *The angles the fault makes with the horizontal for a normal fault (a) and a reverse (thrust) fault (b).*

which implies that $\theta \approx 60°$. The fracture plane then makes an angle $\sim 60°$ with the horizontal when the overburden is the largest principal stress. Such fractures are normal faults as shown by Figures 8.1 and 8.12a. The opposite situation is when the overburden is the least principal stress and the lateral compressive stress is the largest principal stress. Assuming the same $\mu_0 \approx 0.6$ then gives an angle $\theta \approx 30°$. Such fractures are reverse faults or thrust faults as shown by Figures 8.1 and 8.12b. This use of a Coulomb fracture envelope to explain the different types of faults is referred to as Anderson's theory of faulting. Although Anderson's theory is not exact, it nevertheless explains the main features of faulting. Observations show that normal faults are often steeper than 60°, and that reverse faults are often less steep than 30°.

The Coulomb fracture envelope (8.26) relates the least and the largest principal stress at failure, and the largest principal stress necessary for fracture is the following linear function of the least principal stress:

$$\sigma_1 = \left(\sqrt{1 + \mu_0^2} + \mu_0 \right)^2 \sigma_3 + C_0 \tag{8.27}$$

where

$$C_0 = 2S_0 \left(\sqrt{1 + \mu_0^2} + \mu_0 \right). \tag{8.28}$$

The largest principal stress necessary for fracture at zero compressive stress (σ_3) is $\sigma_1 = C_0$, which is called the uniaxial compressive strength. The uniaxial compressive strength is measured in triaxial tests with zero confining pressure. Figure 8.13 shows the largest principal stress as a function of the least principal stress, with the uniaxial compressive strength and the tension cutoff.

There is a lower limit for the least principal stress in the linear relation (8.27). The case of tension (negative confining pressure) is different from compression. The rock normally fails for a negative stress $\sigma_3 = -T_0$, and it fails along a plane that is normal to σ_3. The fracture envelope therefore stops at $\sigma_3 = -T_0$ as shown in Figure 8.11, which is the tension cutoff. The largest stress σ_1 that is possible with $\sigma_3 = -T_0$ is $\sigma_1 = \sigma_M$, where

$$\sigma_M = C_0 \left(1 - \frac{C_0 T_0}{4S_0^2} \right). \tag{8.29}$$

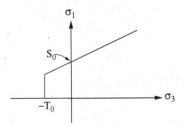

Figure 8.13. *The largest principal stress* (σ_1) *as a function of the least principal stress* (σ_3). *The* σ_1-*line cuts the* σ_3-*axis at* C_0, *which is the uniaxial compressive strength. The lowest* σ_3-*value is the tension cutoff* $-T_0$.

(This expression for σ_M is derived in Exercise 8.4.) A fracture criterion that distinguishes between shear fracture and tension fracture is summarized as follows:

$$\left(\sqrt{1+\mu_0^2}-\mu_0\right)\sigma_1-\left(\sqrt{1+\mu_0^2}+\mu_0\right)\sigma_3=2S_0 \quad \text{for} \quad \sigma_1 > \sigma_M \qquad (8.30)$$

$$\sigma_3 = -T_0 \qquad\qquad\qquad\qquad\qquad\qquad\qquad \text{for} \quad \sigma_1 < \sigma_M.$$

Exercise 8.4 Show that expression (8.29) follows from equations (8.27) and (8.28) when $\sigma_3 = 0$.

Exercise 8.5
(a) Show that the mean stress $p = \frac{1}{2}(\sigma_1+\sigma_3)$ and the shear stress $q = \frac{1}{2}(\sigma_1-\sigma_3)$ at fracture are related as

$$q = p\sin\phi + S_0\cot\phi. \qquad (8.31)$$

(b) Show that equation (8.27) for the largest principal stress as a function of the least principal stress can be written as

$$\sigma_1 = \left(\frac{1+\sin\phi}{1-\sin\phi}\right)\sigma_3 + 2S_0\frac{\cos\phi}{1-\sin\phi}. \qquad (8.32)$$

Solution: Figure 8.11 shows a Mohr's circle that touches the fracture envelope. The center of the circle is at $\frac{1}{2}(\sigma_1+\sigma_3)$ and its radius is $\frac{1}{2}(\sigma_1-\sigma_3)$. For the right-angled triangle in Figure 8.11 we have

$$\frac{1}{2}(\sigma_1-\sigma_3) = \left(S_0\cot\phi + \frac{1}{2}(\sigma_1+\sigma_3)\right)\sin\phi. \qquad (8.33)$$

This is equation (8.31), and we get equation (8.32) with the help of a little algebra.

8.6 Hydrofracturing

Fluid pressure changes the stress state of a saturated porous rock, and we have already seen (Section 3.15) that it is the effective stress that controls the deformations of the rock. The normal stress is reduced by the fluid pressure, while the shear stress is unaffected by the

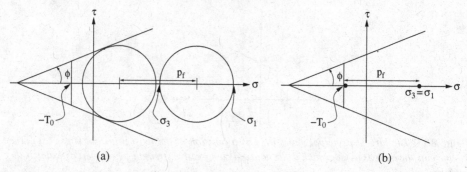

Figure 8.14. (a) A (positive) fluid pressure moves the Mohr's circle to the right. (b) An isotropic stress state can be moved leftwards until tensional fracture.

fluid pressure. Stress states represented by a Mohr's circle are therefore shifted towards the left by the value of the fluid pressure, because both the least and the largest principal stress are normal stresses. Figure 8.14a shows how a stress state becomes translated to the left. An increasing fluid pressure can therefore cause failure. The fluid pressure can increase until the Mohr's circle touches the fracture envelope, as shown in Figure 8.14a, which is the condition for fracture. Fracturing caused by the fluid pressure is hydrofracturing, and it leads to enough pathways for the fluid to prevent further increase in the fluid pressure. Figure 8.14b also shows how increasing fluid pressure leads to tensile fracture for an isotropic stress state. The effect of the fluid pressure is the same for sliding (Amontons' law) as well as for fracturing.

The principal stresses σ_1 and σ_3 in the Coulomb fracture criterion (8.30) are therefore replaced by the effective principal stress

$$\sigma_1' = \sigma_1 - p_f \quad \text{and} \quad \sigma_3' = \sigma_3 - p_f \tag{8.34}$$

and the fracture criterion (8.30) becomes

$$\left(\sqrt{1+\mu_0^2} - \mu_0\right)\sigma_1' - \left(\sqrt{1+\mu_0^2} + \mu_0\right)\sigma_3' = 2S_0 \quad \text{for} \quad \sigma_1' > \sigma_M$$
$$\sigma_3' = -T_0 \qquad\qquad\qquad\qquad\qquad \text{for} \quad \sigma_1' < \sigma_M. \tag{8.35}$$

This fracture criterion can also be rewritten as an expression for the fluid pressure necessary for fracturing. Some more notation is now introduced to simplify this task. We let $a = (1 + \mu_0^2)^{1/2} - \mu_0$, $b = (1+\mu_0^2)^{1/2} + \mu_0$ and $f = \sigma_3/\sigma_1$. The latter coefficient express the amount of stress anisotropy. The coefficient f is less than 1 (since σ_1 is the largest principal stress), but sufficiently large for $bf - a > 0$, or alternatively $f > a/b = ((1+\mu_0^2)^{1/2} - \mu_0)^2$. The condition (8.35) for shear fracture is then

$$a(\sigma_1 - p_f) = b(\sigma_3 - p_f) + 2S_0 \tag{8.36}$$

which gives that the fluid pressure necessary for fracturing is

$$p_f = \left(\frac{bf - a}{b - a}\right)\sigma_1 + \frac{2S_0}{b - a}.$$ (8.37)

The condition (8.35) for tension fracture is

$$-T_0 = \sigma_3 - p_f$$ (8.38)

which gives that the fluid pressure necessary for fracturing is

$$p_f = f\sigma_1 + T_0.$$ (8.39)

An increasing fluid pressure leads to tension fracture if the necessary fluid pressure for tension fracture is less than for shear fracture, and vice versa. A near isotropic stress state leads to tension fracture, while a sufficient amount of anisotropy leads to shear fracture. The condition (8.39) for tension fracture simplifies to

$$p_f \approx f\sigma_1$$ (8.40)

for depths where $\sigma_1 \gg T_0$. Observations from overpressured oil and gas reservoirs show that the fluid pressure is normally bounded by 85% of the overburden, which might indicate that the amount of anisotropy is $f \sim 0.85$.

8.7 Ductile flow and yield strength envelopes

Rock behaves more ductile than brittle at higher temperatures, and its mode of deformation becomes flow rather than fracturing and faulting. The ductile flow of rock is measured in the laboratory by a strain rate, when the rock is subjected to differential stress. It could for example be the strain rate of a cylindrical specimen compressed with a constant axial stress σ_1 and a surrounding stress $\sigma_2 = \sigma_3$. The stresses σ_1 and σ_3 are the largest and the least principal stresses, respectively. Such an experiment resembles the triaxial test shown in Figure 8.9, but the stresses σ_1 and σ_3 are now kept constant, while a strain rate is measured in the direction of σ_1. Experiments show that the strain rate $\dot{\epsilon}$ for a rock sample under a stress difference $\sigma_1 - \sigma_3$ is

$$\dot{\epsilon} = (\sigma_1 - \sigma_3)^n A \exp\left(-\frac{E}{RT}\right)$$ (8.41)

where n is an empirical exponent. The strain rate is temperature dependent by means of an Arrhenius factor $A\exp(-E/RT)$, where A is the Arrhenius prefactor, E is the activation energy, R is the gas constant and T is the temperature in kelvin. The stress–strain-rate relationship (8.41) is called *power law creep*, because the strain rate is proportional to the differential stress to the power of n. We notice that the strain rate becomes zero in the case of zero differential stress, $\sigma_1 - \sigma_3 = 0$. A stress anisotropy is therefore necessary for rocks to deform by ductile flow. The stress anisotropy is often given by means of the deviatoric stress, and for this special situation ($\sigma_2 = \sigma_3$) the deviatoric stress is proportional to the differential stress (see Section 3.10). The Arrhenius factor makes the rocks ability to flow strongly temperature dependent. Experimental data shows that the Arrhenius factor may increase by several orders of magnitude for a temperature increase of 100°C (see Exercise 8.6). The formulation (8.41) of power law creep applies only in the principal stress system. Note 8.5 shows how power law creep can be written with respect to any reference frame.

The strain rate (8.41) can be rewritten as an expression for the stress difference as a function of strain rate and temperature:

$$\sigma_1 - \sigma_3 = \left(\frac{\dot{\epsilon}}{A}\right)^{1/n} \exp\left(\frac{E}{nRT}\right). \tag{8.42}$$

This stress difference can be combined with the stress difference necessary for brittle failure as expressed by Byerlee's law (8.2) to decide the type of deformation – brittle failure or flow by creep. If the stress difference for brittle failure is less than for ductile flow the mode of deformation is brittle failure, otherwise it is by ductile flow. Figure 8.15 shows examples of such plots, which show the yield strength as an "envelope." This type of plot is therefore called a *yield strength envelope* or simply *YSE*. Byerlee's law applies for a large range of rocks and it is also independent of strain rate and temperature. On the other hand, the stress difference for ductile flow is dependent on lithology in addition to strain rate and temperature. For example, silica rich crustal rocks are less strong than mantle rocks.

Figure 8.15 shows how YSE depends on the temperature, strain rate and crustal thickness. An increasing temperature makes both the lower part of the crust and the lithospheric mantle more ductile. Figure 8.15a shows the YSE for the four different geotherms in Figure 8.15b, which differ by the amount of the radioactive heat generation in the crust. Table 8.1 gives the remaining parameters. The Moho is at the depth 35 km in Figures 8.15a and 8.15b, and these figures show that both the upper part of the crust and the mantle have differential stress by frictional sliding according to Byerlee's law. This is the linear part of the YSE. The point where Byerlee's law ends is where the rock becomes ductile, and this point is called the *brittle–ductile transition*. The figures show one such point for the crust and one for the mantle. The nature of the brittle–ductile transition is not point-like as in Figure 8.15, but it is a depth interval where the deformation gradually goes from brittle to ductile. If the crust is subdivided into several layers of different lithologies it is possible to have a brittle–ductile transition zone in each layer. Figure 8.15c shows how an increasing strain rate increases the differential stress, and Figure 8.15d shows that decreasing the thickness of the crust increases the differential stress. Mantle rock is stronger than crustal rocks, and thinning of the crust therefore leads to a stronger lithosphere, assuming that the geotherm remains the same. It should be noted that the YSE shown in Figure 8.15 applies for extension, where σ_1 is in the direction of extension.

Table 8.1. *Parameters used in Figures 8.15 and 8.17. The rheological parameters for the crust and the mantle are taken from Stüwe (2002), Table 5.3.*

Parameter	Value	Units
λ_c (heat conductivity crust)	2.5	$[\mathrm{W\,m^{-1}\,K^{-1}}]$
λ_m (heat conductivity mantle)	3.0	$[\mathrm{W\,m^{-1}\,K^{-1}}]$
T_0 (surface temperature)	0	$[°C]$
T_a (temperature base lithosphere)	1300	$[°C]$
z_m (thickness of crust)	35	$[\mathrm{km}]$
z_a (thickness of the asthenosphere)	120	$[\mathrm{km}]$
S_0 (radioactive heat production)	$1 \cdot 10^{-6}$	$[\mathrm{W\,m^{-3}}]$
μ (coefficient of friction)	0.85	$[\text{-}]$
$\dot{\epsilon}$ (strain rate)	$1 \cdot 10^{-15}$	$[\mathrm{s^{-1}}]$
A_c (prefactor crust)	$5 \cdot 10^{-6}$	$[\mathrm{MPa^{-n}s^{-1}}]$
A_m (prefactor mantle)	$7 \cdot 10^{4}$	$[\mathrm{MPa^{-n}s^{-1}}]$
E_c (activation energy crust)	190	$[\mathrm{kJ\,mole^{-1}}]$
E_m (activation energy mantle)	520	$[\mathrm{kJ\,mole^{-1}}]$
n_c (exponent crust)	3	$[\text{-}]$
n_m (exponent mantle)	3	$[\text{-}]$

The creep laws measured in the laboratory are for strain rates that are many orders of magnitude larger than the strain rates in nature. Section 7.9 shows that the strain rate during pure shear stretching of the lithosphere is

$$\dot{\epsilon} = \frac{\ln\beta_s}{t_s} \tag{8.43}$$

where t_s is the time interval of stretching and β_s is the stretching factor. A lithosphere being stretched a factor $\beta_s = 2$ during a time interval $t_s = 20$ Ma has a strain rate $\dot{\epsilon} \approx 1 \cdot 10^{-15}$ s^{-1}. A pure shear experiment carried out in the laboratory, where a rock sample undergoes a relative shortening $\Delta l/l = 10^{-3}$ during a time span of $\Delta t = 10$ days, gives the strain rate $\dot{\epsilon} = (\Delta l/\Delta t)/l = 1.1 \cdot 10^{-9}$ s^{-1}. The strain rate from the laboratory is 6 orders of magnitude larger than the strain rate for the stretching of the lithosphere. The flow laws measured in the laboratory should therefore be applied with some care, because of this large difference in strain rate. Figure 8.16 shows the strain rates for the example of rifting from Section 7.16. The β-factors along the profile are given in Figure 8.16a for each rift phase. The corresponding strain rates (8.43) are shown in Figure 8.16b, where the duration of the rift phases is shown on the figures. We recall that the strain rate (8.43) is constant through the rift phase at positions that follow the stretching. It can therefore be considered as an average strain rate. The real stretching is more likely to be intermittent due to seismic events rather than a continuous process.

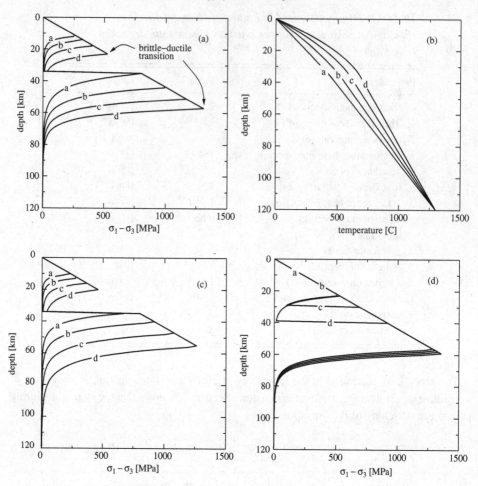

Figure 8.15. *(a) YSE for the four different geotherms shown in (b). (b) Geotherms for different amounts of crustal heat production. a: 0 Wm^{-3}; b: $0.5 \cdot 10^{-6}$ Wm^{-3}; c: $1 \cdot 10^{-6}$ Wm^{-3}; d: $1.5 \cdot 10^{-6}$ Wm^{-3}; (c) YSE for different strain rates. a: 10^{-18} s^{-1}; b: 10^{-16} s^{-1}; c: 10^{-14} s^{-1}; d: 10^{-12} s^{-1}; (d) YSE for different crustal thicknesses. a: 10 km; b: 20 km; c: 30 km; d: 40 km.*

The average strain rates combined with temperature can be used to estimate the stress state over the profile. The YSE integrated over the entire depth interval of the lithosphere is an expression for the strength of the lithosphere. The integral

$$S_l = \int_0^{z_a} (\sigma_1 - \sigma_3) \, dz \qquad (8.44)$$

is therefore called the *lithospheric strength*, where z_a is the depth to the asthenosphere. During rift phases the largest principal stress can be assumed to be in the direction of stretching. In a 2D model the least principal stress is then in the vertical direction. The vertical stress is now assumed to be due to gravity, $\sigma_3 = \varrho_b g z$,

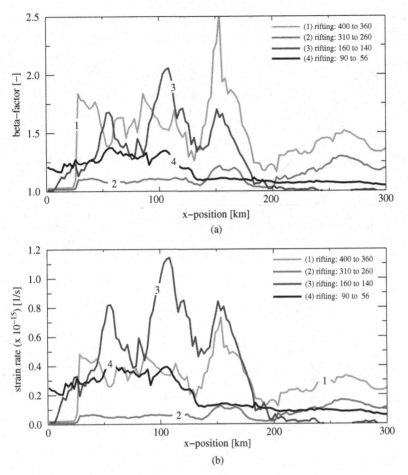

Figure 8.16. *(a) β-factors for four rift phases. (b) The strain rates of the rift phases. All rift phases are plotted with the present day lateral positions.*

where ϱ_b is the average bulk density of the lithosphere. The force needed to stretch the lithosphere is

$$F = \int_0^{z_a} \sigma_1 \, dz = S_l + \frac{1}{2}\varrho_b g z_a^2 \tag{8.45}$$

which is the lithospheric strength S_l added to a gravity term. Variations in the gravity term are mostly due to the density difference between the crust and mantle. Unless the thickness of the crust is changing or the thermal state of the lithosphere is altered the gravity term will not change much, and it can often be taken as a constant offset.

How the lithospheric strength (8.44) depends on radioactive heat production in the crust, crustal thickness and strain rate is shown in Figure 8.17. Figure 8.17a shows how the lithosphere gets weaker with increasing amounts of crustal heat generation. The geotherms for four values of crustal heat generation are shown in Figure 8.15b. Figure 8.17b shows

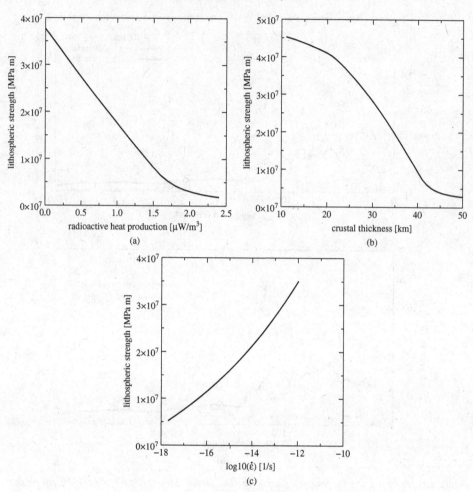

Figure 8.17. *Lithospheric strength as a function of (a) radioactive heat production, (b) crustal thickness and (c) strain rate. Table 8.1 gives the remaining parameters.*

the decrease of lithospheric strength with increasing crustal thickness, and Figure 8.17c shows the increase of lithospheric strength with increasing strain rate.

There are two opposing effects during stretching. The thinning of the crust makes the mantle fraction of the lithosphere larger and the lithosphere becomes stronger. On the other hand, stretching makes the lithosphere hotter and therefore weaker. Which of these two processes is strongest depends on how fast the stretching is. Figure 8.18 shows an example of YSE for stretching with a constant strain rate, where the lithosphere is stretched a β-factor 2 during a time span of 10 Ma. The crust has initially a 40 km thickness, which becomes reduced to 28 km at 5 Ma and finally reduced to 20 km at the end of rifting at 10 Ma. (Recall that the β-factor increases exponentially with time during rifting with a constant strain rate, $\beta = \exp(Gt)$, where G is the strain rate.) Although the crust is stretched

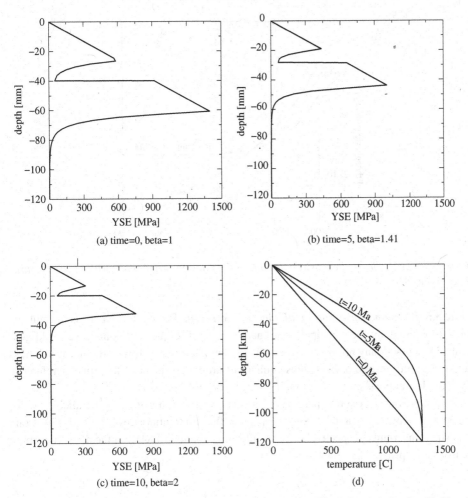

Figure 8.18. *The YSE during rifting with a constant strain rate. The β-factor is* 2 *after* 10 *Ma of rifting. Table 8.1 gives the rheological data used. (a) The YSE at time* 0 *Ma. (b) The YSE at time* 5 *Ma. (c) The YSE at time* 10 *Ma. (d) The temperature of the lithosphere at the times* 0 *Ma,* 5 *Ma and* 10 *Ma.*

the lithosphere becomes gradually weaker during the stretching because of the hot mantle that moves upwards. The lithospheric strength as a function of the $β$-factor is shown in Figure 8.19 for two different durations of the rift phase. One lasts for 10 Ma and the other for 100 Ma. Rifting over a "short " time interval (10 Ma) leads to continuously decreasing lithospheric strength, while the "slow" stretching (over 100 Ma) has a lithospheric strength that reaches a minimum before halfway through the rift phase, and then stays at the minimum for the rest of the rifting. Notice that the strain rate becomes zero when stretching has ceased, which implies that the stress state in the ductile parts of the lithosphere becomes isotropic.

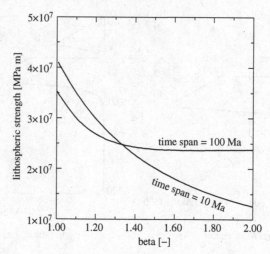

Figure 8.19. *Lithospheric strength as a function of the β-factor during rifting with a constant strain rate for the intervals 10 Ma and 100 Ma.*

Note 8.5 *Invariant formulation of power law creep.* Power law creep as formulated in (8.41) applies in the principal coordinate system. The question is how we could generalize this law to any reference system. The answer has to be expressed in terms of stress and the strain tensors, because these quantities are independent of a particular reference system. The expression we are looking for should become equation (8.41) in the special case of a principal system where $\sigma_2 = \sigma_3$. The first step towards a generalized law for creep is to replace the axial strain rate ($\dot{\epsilon}$) and the differential stress ($\sigma_1 - \sigma_3$) by strain and stress invariants, respectively. We recall from equation (3.84) that the deviatoric stress tensor in the principal system is

$$\boldsymbol{\sigma}' = \begin{pmatrix} \frac{2}{3} & 0 & 0 \\ 0 & -\frac{1}{3} & 0 \\ 0 & 0 & -\frac{1}{3} \end{pmatrix} (\sigma_1 - \sigma_3) \tag{8.46}$$

when $\sigma_2 = \sigma_3$. The strain rate tensor in the principal system is

$$\dot{\boldsymbol{\epsilon}} = \begin{pmatrix} 1 & 0 & 0 \\ 0 & -\frac{1}{2} & 0 \\ 0 & 0 & -\frac{1}{2} \end{pmatrix} \dot{\epsilon} \tag{8.47}$$

where the axial strain rate is $\dot{\epsilon} = \dot{\epsilon}_1$ and components ϵ_2 and ϵ_3 are $\dot{\epsilon}_2 = \dot{\epsilon}_3 = -\frac{1}{2}\dot{\epsilon}$ because of incompressibility. (See Section 3.4 for strain and volume change.) The first invariants (the sum of the diagonal elements), of these two tensors cannot be used to formulate a creep law because they are both zero. (See Section 3.7 and Exercises 3.6 and 3.7 for the definition of invariants.) A formulation is therefore based on the second invariants, which in this case are

$$J_2 = \frac{1}{2}\sigma'_{ij}\sigma'_{ij} \quad \text{and} \quad \dot{E}_2 = \frac{1}{2}\dot{\epsilon}_{ij}\dot{\epsilon}_{ij}. \tag{8.48}$$

An evaluation of J_2 and $\dot{E}_2$ using representations (8.46) and (8.47) in the principal system, respectively, give that

$$J_2 = \frac{1}{3}(\sigma_1 - \sigma_3)^2 \quad \text{and} \quad \dot{E}_2 = \frac{3}{4}\dot{\epsilon}^2. \tag{8.49}$$

The next step is to use the invariants to postulate the following effective stress and effective strain rate, respectively,

$$\sigma'_E = \left(\frac{1}{2}\sigma'_{ij}\sigma'_{ij}\right)^{1/2} \quad \text{and} \quad \epsilon_E = \left(\frac{1}{2}\dot{\epsilon}_{ij}\dot{\epsilon}_{ij}\right)^{1/2} \tag{8.50}$$

where we have

$$\sigma_1 - \sigma_3 = \sqrt{3}\sigma'_E \quad \text{and} \quad \dot{\epsilon} = \frac{2}{\sqrt{3}}\dot{\epsilon}_E. \tag{8.51}$$

Power law creep (8.41) in the principal system is therefore

$$\dot{\epsilon}_E = (\sigma'_E)^n A' \exp\left(-\frac{E}{RT}\right) \quad \text{where} \quad A' = \frac{3^{(n+1)/2}}{2} A \tag{8.52}$$

in terms of the effective and invariant properties σ'_E and $\dot{\epsilon}_E$. It is also possible to go one step further by postulating the following tensorial relationship between the strain rate and deviatoric stress:

$$\dot{\epsilon}_{ij} = (\sigma'_E)^{(n-1)}\sigma'_{ij} A' \exp\left(-\frac{E}{RT}\right). \tag{8.53}$$

Making the second invariant of the tensor relationship (8.53) gives back the invariant scalar law (8.52), which again is the same as the formulation (8.41) for the special case in the principal system.

Exercise 8.6

(a) Show that the strain rate increases with a factor $\exp(T_E/T_1 - T_E/T_2)$ when the temperature in kelvin increases from T_1 to T_2, and where $T_E = E/R$.

(b) What is the increase from $100\,°C$ to $1000\,°C$ when $E = 200\ \text{kJ mole}^{-1}$ and $R = 8.134\ \text{J K}^{-1}\text{mole}^{-1}$?

8.8 Further reading

Fundamentals and applications of rheological properties of rocks are treated in Jaeger, Cook and Zimmerman (2007), Ranalli (1995) and Turcotte and Schubert (1982). Rheology is an important basis for structural geology, see for example Davis and Reynolds (1996) and Parks (1989).

9

Flexure of the lithosphere

The lithosphere acts as a rigid plate that floats on a ductile mantle (asthenosphere), and we have seen how (Airy) isostasy can be used to predict how deep the lithosphere floats under the weight of a load. Isostasy is a simplification that ignores the possible bending of the lithospheric plate due to lateral variations in the surface load. Isostasy applies for loads of large lateral extent, like for instance a mountain range or a continent. On the other hand, "small" scale features like a valley or a mountain peak do not lead to any isostatic uplift or subsidence because they are completely supported by the elastic strength of the lithosphere. There is a length scale in between where surface loads are partly supported by the lithosphere, and partly supported by the buoyancy of the displaced mantle. This length scale can be estimated assuming that the lithosphere acts like a thin linear elastic plate floating on the mantle. The Hawaiian islands, which are piles of volcanic rock on the oceanic plate, are good examples of loads that bend the lithosphere. The flexure is in this case reproduced by simple solutions of the equation for the deflection of a thin elastic plate. The solutions also predict uplift in the form of a flexural bulge, a feature that isostasy alone cannot predict. Furthermore, observation of flexure allows for estimations of the thickness of the elastic part of the lithosphere.

In this chapter we will first derive the equation for the deflection of a thin elastic plate under loads that are only x-dependent. Some simple solutions for the deflection of a plate are presented and discussed. The elastic plate model is extended to a viscoelastic model of flexure, and both the viscoelastic and the viscous behavior of a plate are studied by simple solutions.

9.1 Equation for flexure of a plate

Loads on the lithospheric plate, like for instance mountain ranges or volcanic islands, may bend the lithosphere. An equation for the flexure of an elastic plate is needed to study how such loads may deflect the lithosphere. It turns out that the torque (or the bending moment) caused by the load is an important quantity that controls the flexure. The terms torque and bending moment mean the same thing here, and they are used interchangeably. A torque is the cause of rotation. If an object can rotate around an axis its rotation accelerates as

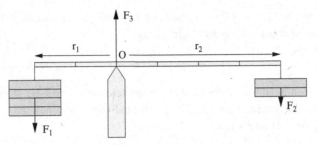

Figure 9.1. *A horizontal lever arm is an example of force and torque balance. The forces* $\mathbf{F}_i$ *add up to zero. The sum of the torques* $\mathbf{M}_i = \mathbf{r}_i \times \mathbf{F}_i$ *is also zero, regardless of where the origin is placed. The torque balance is simplified with respect to the point O, because the torque from the force* $\mathbf{F}_3$ *is then zero.*

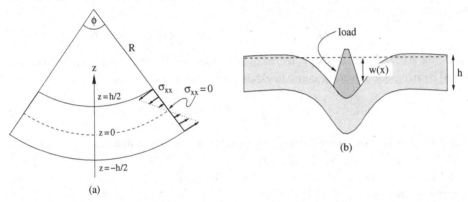

Figure 9.2. *(a) The lateral stress* σ_{xx} *in a bent plate (or beam). (b) The deflection* $w(x)$ *as a function of distance along a 2D plate.*

long as there is a non-zero torque. See Figure 9.1, which shows a lever arm with torques in balance. The effectiveness of the force increases with the perpendicular distance from the rotation point to the point of action. The torque is therefore defined as the force times the perpendicular distance to the force, which in general is the vector product $\mathbf{r} \times \mathbf{F}$. Since a deflected plate does not rotate there must be torque balance. The sum of all torques on the plate must add up to zero. A torque is always defined with respect to an axis, and all torques must total zero regardless of where the axis is placed. We will begin by finding the torque along a vertical cross-section of the plate, because it has to match a part of the torque caused the surface loads (as we will soon see).

Let us for a moment assume that the plate is bent slightly into a circular arc of radius R. Figure 9.2a shows a vertical cross-section of plate where the bending is exaggerated. It is seen from Figure 9.2a that the upper part of the plate undergoes contraction and that the lower part undergoes extension. The center arc of the plate has its length unchanged by the bending, and it is therefore called the *neutral surface*. Let us look at the arc (or fiber) in the plate that is a distance z from the neutral surface (see Figure 9.2a).

The length of the fiber is $l = (R - z)\phi$ when ϕ is the angle. The strain in the fiber is the elongation of the fiber divided by its original length,

$$\epsilon_{xx} = -\frac{dl}{l} = \frac{z}{R} \tag{9.1}$$

where a minus sign is added to ensure that compression gives positive stress. If we for a moment assume that we are bending a beam instead of a plate then the fiber stress σ_{xx} follows directly from Hooke's law,

$$\sigma_{xx} = E\epsilon_{xx} = E\frac{z}{R} \tag{9.2}$$

where E is Young's modulus. We will now compute the bending moment along a vertical cross-section at an arbitrary lateral position in a beam. The contribution to the bending moment (or torque) from a fiber of thickness dz and a distance z from the neutral surface is then

$$dM = z\,dF = z\,\sigma_{xx}\,dz = (E/R)z^2 dz \tag{9.3}$$

where the force in the 2D fiber is $dF = \sigma_{xx}\,dz$. The total bending moment in the beam is found by integration over the entire thickness of the plate, and we get

$$M = \frac{E}{R}\int_{-h/2}^{h/2} z^2 dz = \frac{E\,h^3}{12\,R}. \tag{9.4}$$

The deflection is in the general case represented by a function of distance, $w(x)$, which is not exactly a circular arc (see Figure 9.2b). Nevertheless, the local curvature of a function $w(x)$ at a point x is found from the circle that fits the curve locally. The radius of this circle is approximated by

$$\frac{1}{R} \approx \frac{d^2 w}{dx^2} \tag{9.5}$$

as shown in Note 9.1, and the bending moment for the beam is

$$M = \frac{E\,h^3}{12}\frac{d^2 w}{dx^2}. \tag{9.6}$$

We have so far assumed that bending a 2D plate is similar to bending a beam. The stress state of a slightly bent beam can be represented by just one non-zero stress component, σ_{xx}, and we can apply Hooke's law (9.2). The stress in a bent plate is a little different. It is first assumed that there is no stress in the plate in the y-direction, $\sigma_{yy} = 0$. Instead of assuming that σ_{zz} is zero too, which is the assumption for a beam, it is instead assumed that the strain in the vertical direction (ϵ_{zz}) is zero. Then it follows from equation (3.90) that

$$\epsilon_{zz} = \frac{1}{E}\left(\sigma_{zz} - v\sigma_{xx} - v\sigma_{yy}\right) = 0. \tag{9.7}$$

Using that $\sigma_{yy} = 0$ we get that $\sigma_{zz} = v\sigma_{xx}$, and when $\sigma_{yy} = 0$ and $\sigma_{zz} = v\sigma_{xx}$ are inserted into equation (3.88) for lateral strain we get

$$\epsilon_{xx} = \frac{1}{E}\left(\sigma_{xx} - v\sigma_{yy} - v\sigma_{zz}\right) = \frac{1}{E}\left(1 - v^2\right)\sigma_{xx}. \tag{9.8}$$

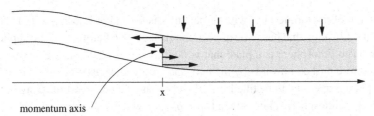

Figure 9.3. *The stresses acting along the surface of the shaded part of the plate. The torque balance is made with respect to the axis shown.*

With this relation between lateral stress and lateral strain, rather than Hooke's law (9.2), we have the following equation for the bending moment:

$$M = \frac{E h^3}{12(1 - v^2)} \frac{d^2 w}{dx^2}. \tag{9.9}$$

The difference between the moments in an equally bent beam and a plate is the factor $1/(1-v^2)$. A typical value for Poisson's ratio like $v = 0.25$ makes this factor 1.07, which is close to 1. A beam is therefore a good approximation of a thin 2D plate as long as $v < 0.25$.

For a body to be at rest the laws of mechanics require both a force balance and a torque balance. The body accelerates if there is no force balance, and its rotation accelerates if there is no torque balance. It should be emphasized that a torque is about a given axis. The torques are evaluated here with the axis at the center of a vertical cross-section (see Figure 9.3). The torque balance for the part of the plate beginning at position x is

$$M(x) = \tau(x), \tag{9.10}$$

where $M(x)$ denotes the torque (9.9) along the internal vertical cross-section at x, and $\tau(x)$ denotes the torque along the external boundaries from position x and to the right. The torque balance is made for the part of the plate that is shown as gray in Figure 9.3. Notice that M and τ denote torque with opposite sign conventions. The torque from the surface load $q(x)$ to the right of position x is

$$\tau(x) = \int_x^\infty (x' - x) q(x') dx' \tag{9.11}$$

where $(x' - x)$ is the lever arm and $q(x')dx'$ is the force on the line element dx' (and $q(x)$ is force per unit length). The torque balance (9.10) can be written using the internal bending moment (9.9) and the external load (9.11) as

$$\frac{E h^3}{12(1 - v^2)} \frac{d^2 w}{dx^2} = \int_x^\infty (x' - x) q(x') dx' \tag{9.12}$$

and a differential equation for the deflection is found after differentiating two times

$$\frac{E h^3}{12(1 - v^2)} \frac{d^4 w}{dx^4} = q(x). \tag{9.13}$$

The derivation of the torque $\tau(x)$ with respect to x is treated in Exercise 9.1. There is one important modification that has to be done to the surface load before it can be applied to the lithosphere. The lithosphere is a plate that is floating on a ductile mantle, and a deflected plate is therefore pushed upwards by the buoyancy of the displaced mantle. The deflection above the plate is assumed to be filled with the load, and the net pressure acting on the base of the deflected lithospheric plate is then $(\varrho_m - \varrho_c)\, w\, g$ where ϱ_m and ϱ_c are the densities of mantle rock and the load, respectively. The buoyancy pressure acts upwards and the surface load $q(x)$ acts downwards, and the total vertical stress on the plate is $q(x) - \Delta\varrho g w(x)$, where $\Delta\varrho = \varrho_m - \varrho_c$. The equation for deflection caused by a surface load on an elastic plate buoyed by the mantle is therefore

$$D \frac{d^4 w}{dx^4} + \Delta\varrho\, g\, w = q(x) \tag{9.14}$$

where the parameter

$$D = \frac{E\, h^3}{12(1 - v^2)} \tag{9.15}$$

is called the *flexural rigidity*. We see right away that a plate with zero flexural rigidity has the deflection

$$w(x) = \frac{q(x)}{\Delta\varrho\, g} \tag{9.16}$$

which is simply Airy isostatic equilibrium. The isostatic subsidence (9.16) can be rewritten as a force balance $q + \varrho_c g w = \varrho_m g w$, which shows that q is the pressure from the load above $z = 0$. The remaining part of the load, which is below $z = 0$, is the rock with density ϱ_c that fills the deflection. In other words, the solution for the deflection tells us how much rock fills the deflection for a given surface load. This depth was obtained earlier with equation (7.14) for the crustal root of a mountain range in isostatic equilibrium.

Note 9.1 *Radius of curvature.* A circle that fits a curve at a position x is found by first drawing the tangent at position x and at a neighbor position $x + dx$. The point where the normals to both these tangents meet is the center of a circle that fits the curve, see Figure 9.4. The optimal circle that fits the curve at the point x is made by letting dx approach zero. The radius of a circle is related to the arc length of the curve ds and the angle ϕ by

$$ds = R\,\phi. \tag{9.17}$$

The angle ϕ is $\phi = \alpha_2 - \alpha_1$ where the angles α_1 and α_2 are given by the steepness of the curve dz/dx at the two points, respectively, using that $\alpha = \arctan(dz/dx)$. The radius of curvature can therefore be written as

$$\frac{1}{R} \approx \frac{1}{ds}\left(\arctan\left(\frac{dz}{dx}(x + dx)\right) - \arctan\left(\frac{dz}{dx}(x)\right)\right) \tag{9.18}$$

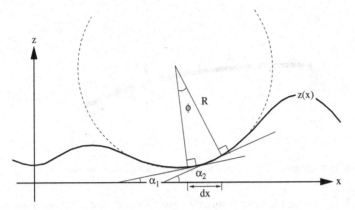

Figure 9.4. *The radius of curvature for a point along an arbitrary curve.*

or in terms of derivation as

$$\frac{1}{R} = \frac{\frac{d}{du}(\arctan u)\,\frac{du}{dx}}{\frac{ds}{dx}} \tag{9.19}$$

where $u = dz/dx$. From the length of a line element $ds^2 = dx^2 + dz^2$ it follows that

$$\frac{ds}{dx} = \sqrt{1 + (dz/dx)^2} \tag{9.20}$$

and we have

$$\frac{d}{du}(\arctan u) = \frac{1}{1 + u^2}. \tag{9.21}$$

When (9.20) and (9.21) are inserted into expression (9.19) for the radius of curvature we get

$$\frac{1}{R} = \frac{d^2z/dx^2}{\left(1 + (dz/dx)^2\right)^{3/2}}. \tag{9.22}$$

A slightly bent curve has $dz/dx \ll 1$ and we can make the approximation

$$\frac{1}{R} \approx \frac{d^2z}{dx^2} \tag{9.23}$$

which is the expression used above.

Exercise 9.1 Show that $d\tau/dx = -\int_x^\infty q(x')dx'$, where τ is the external torque (9.11). Hint: make $\Delta\tau = \tau(x + \Delta x) - \tau(x)$.

Exercise 9.2 Show that $V = \int_x^\infty q(x')dx'$ can be interpreted as the average shear force along a vertical cross-section at $x = 0$. Notice that it then follows that $dV/dx = q(x)$.

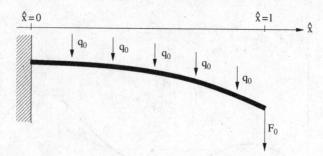

Figure 9.5. *The deflection of a loaded horizontal beam clamped at one end.*

Exercise 9.3 Show that the neutral surface passes through the center of mass of the plate.

Exercise 9.4 Find the deflection of a horizontal beam of length l that is clamped at one end and loaded at the free end with a force F_0 (see Figure 9.5). The beam is also loaded uniformly along its length by a load q_0, which could be the weight of the beam.
Solution: The total torque at the position x is

$$\tau(x) = F_0(l - x) + \int_x^l q_0(x' - x)dx' = F_0(l - x) + \frac{1}{2}q_0(l - x)^2 \qquad (9.24)$$

where the first term is the torque from the force F_0, and the second term is the torque from the uniform load q_0. The internal bending moment of the beam must balance the external moment $\tau(x)$, and we have the equation

$$D\frac{d^2w}{dx^2} = F_0(l - x) + \frac{1}{2}q_0(l - x)^2 \qquad (9.25)$$

where $D = Eh^3/12$ is the flexural rigidity of the beam. The equation for the deflection can be rewritten as

$$\frac{d^2w}{d\hat{x}^2} = \frac{F_0 l^3}{D}(1 - \hat{x}) + \frac{q_0 l^4}{2D}(1 - \hat{x})^2 \qquad (9.26)$$

using the dimensionless coordinate $\hat{x} = x/l$. The deflection $w(\hat{x})$ is obtained after integrating twice

$$w(\hat{x}) = w_1(1 - \hat{x})^4 + w_2(1 - \hat{x})^3 + c_1\hat{x} + c_2 \qquad (9.27)$$

where $w_1 = q_0 l^4/(24D)$ and $w_2 = F_0 l^3/(6D)$, and where c_1 and c_2 are integration constants. The two boundary conditions $w = 0$ and $dw/d\hat{x} = 0$ at the clamped end ($\hat{x} = 0$) give the two integration constants. The first boundary condition $w = 0$ at $\hat{x} = 0$ gives that $c_2 = -w_1 - w_2$ and the second boundary condition $dw/d\hat{x} = 0$ gives that $c_1 = 4w_1 + 3w_2$. The wanted deflection of the beam is then

$$w(\hat{x}) = w_1\Big((1 - \hat{x})^4 + 4\hat{x} - 1\Big) + w_2\Big((1 - \hat{x})^3 + 3\hat{x} - 1\Big) \qquad (9.28)$$

where it is written as the sum of the deflection from the weight of the beam and the deflection from the end load. The solution can be used to find the maximum deflection at the free end, which is

$$w(\hat{x}{=}1) = 3w_1 + 2w_2, \tag{9.29}$$

where the maximum deflection due to the weight of the beam is $3w_1$ and the maximum deflection due to the end load is $2w_2$.

The surface load becomes $q_0 = W_0 g / l$ when the beam has the weight W_0, and the end load becomes $F_0 = W_b g$ when it is due to a weight W_b. It is then seen that the weight of the beam and the end load contribute equally to the deflection when the beam has 3/8 of the weight of the end load.

The force F_0 can also be represented as a load per unit length by use of the Dirac delta function $\delta(x)$ as $F_0 \delta(x - l)$, which makes the total surface load per unit length $q(x) = q_0 + F_0 \delta(x - l)$. The Dirac delta function is defined by the integral

$$\int_{-\infty}^{\infty} f(x')\delta(x - x')\,dx' = f(x) \tag{9.30}$$

and it is seen to be zero everywhere, except at $x = 0$ where it is infinite.

9.2 Flexure from a point load

The flexure of a 2D elastic plate under a point load has a solution that turns out to be simple and useful. It is useful because any load $q(x)$ can be discretized into a series of discrete point loads, and the flexure from the full load $q(x)$ can then be approximated by superposition of the solutions for the discrete point loads (see Note 9.3). A point load in 2D is actually a line load in 3D because it extends along a line (normal to the cross-section) through the point where it is applied. The point load V_0 is placed at $x = 0$ and it is assumed that the plate is unaffected at large distances away from the load, which gives $w = 0$ for $x \to \pm\infty$. Equation (9.14) for the deflection of the plate becomes

$$D\frac{d^4 w}{dx^4} + \Delta\varrho g w = 0 \tag{9.31}$$

where $\Delta\varrho = \varrho_m - \varrho_c$, because the load is zero everywhere except at $x = 0$. It is a fourth-order equation and we therefore need four boundary conditions. We already have two and the third follows from symmetry around $x = 0$, which implies that $dw/dx = 0$ for $x = 0$. The fourth condition, which involves the point load, is the force balance in the vertical direction. The load on the plate must be balanced by the buoyancy of the displaced mantle, which is written as

$$\frac{V_0}{2} + \varrho_c g \int_0^{\infty} w(x)dx = \varrho_m g \int_0^{\infty} w(x)dx. \tag{9.32}$$

The load is divided by 2 because the integral is over only half the plate. The integral (9.32) shows that V_0 is the surface part of the load, which is the part that is above $z = 0$. The rest of the load, which is below $z = 0$, is the part that is filling the deflection. The solution for the deflection tells us how deep the deflection becomes when it is filled with rock of density ϱ_c. It is shown in Note 9.2 that the solution to equation (9.31), with these four boundary conditions, is

$$w(\hat{x}) = w_{\max} e^{-\hat{x}}(\cos \hat{x} + \sin \hat{x}) \tag{9.33}$$

in terms of the dimensionless coordinate $\hat{x} = x/\alpha$, where α is the characteristic length

$$\alpha = \left(\frac{4D}{\Delta \varrho g}\right)^{1/4}, \tag{9.34}$$

and the maximum deflection at $x = 0$ is

$$w_{\max} = \frac{V_0}{2\Delta \varrho\, g\alpha}. \tag{9.35}$$

Notice that the length scale α is independent of the load. The deflection (9.33) is plotted in Figure 9.7. The half-width $\hat{x}_0$ of the depression is given by $w(\hat{x}_0) = 0$, which has the solution

$$\hat{x}_0 = \frac{3\pi}{4}. \tag{9.36}$$

Figure 9.7 shows that the point load also creates a forebulge at the distance x_1 given by $dw/d\hat{x} = 0$. We have

$$\frac{dw}{d\hat{x}}(\hat{x}_1) = -w_{\max} e^{-\hat{x}_1} \sin \hat{x}_1 = 0 \tag{9.37}$$

which has the solution $\hat{x}_1 = \pi$, and the height of the forebulge then becomes $w_1 = w_{\max} e^{-\pi}$.

The distance to the forebulge is found to be roughly $x_1 \approx 250$ km for the Hawaiian island chain, see Figure 9.6, a distance that can be used to estimate the elastic thickness of the lithosphere below the islands. The characteristic length scale becomes $\alpha = x_1/\pi = 80$ km. The parameters $E = 100$ GPa, $v = 0.25$, $\Delta \varrho = 600$ kg m^{-3} give that the elastic thickness of the lithosphere is $h = 30$ km. Such an estimate is denoted the *effective elastic thickness*. It is the thickness an homogeneous plate would have, a plate with the same linear elastic properties everywhere. The effective thickness is not a property of a plate that can be directly observed, and it is normally used as a means of comparing elastic properties of different plates.

Observations of the maximum deflection $w_{\max}$ can in combination with the characteristic length scale α be used to estimate the size of the point load. It follows from (9.35) that the point load is

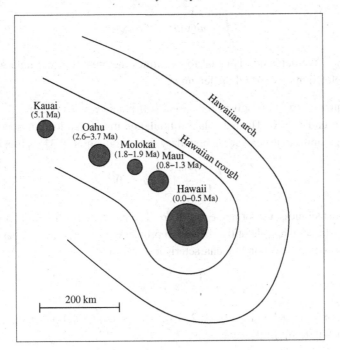

Figure 9.6. *A sketch of the Hawaiian island chain with the flexural trough and the flexural arch.*

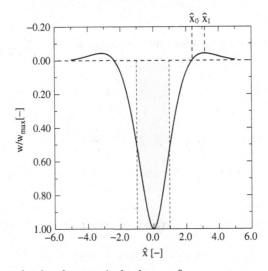

Figure 9.7. *The flexure of a plate from a point load at x = 0.*

$$V_0 = 2\Delta\varrho \, g\alpha w_{max} \tag{9.38}$$

which says that the point load corresponds to the weight of the material in the area $2\alpha w_{max}$ when it has the density $\Delta\varrho$. It also follows from the force balance (9.32) and the maximum deflection (9.35) that

$$\int_{-\infty}^{\infty} w(x)dx = 2w_{\max}\alpha. \tag{9.39}$$

The net area of the deflection is equal to two times the area of a rectangle with sides of maximum deflection ($w_{\max}$) and the length α.

Note 9.2 Equation (9.31) is solved for only a half-plate ($x > 0$) because the problem is symmetric around $x = 0$. The half-plate formulation makes the load disappear from the equation and it appears instead at the boundary $x = 0$. Equation (9.31) is first rewritten as

$$\left(\frac{1}{4}\frac{d^4}{d\hat{x}^4} + 1\right) w = 0 \tag{9.40}$$

using the dimensionless x-coordinate $\hat{x} = x/\alpha$. The parameter α has unit length and it is the characteristic length scale of the deflection problem. The solution of equation (9.40) is done by solving the corresponding characteristic equation

$$\frac{1}{4}r^4 + 1 = 0 \tag{9.41}$$

which has four roots r_k with $k = 1, 2, 3, 4$. The general solution of the fourth-order equation (9.40) in terms of the roots r_k is

$$w(\hat{x}) = \sum_{k=1}^{4} c_k e^{r_k \hat{x}} \tag{9.42}$$

which is seen by inserting the solution into equation (9.40). The characteristic equation is

$$\left(\frac{r}{\sqrt{2}}\right)^4 = -1 = e^{i\pi + i2\pi k} \tag{9.43}$$

which gives

$$r_k = \sqrt{2}\, e^{i(1/4+k/2)\pi} \tag{9.44}$$

for $k = 1, \ldots, 4$. The four roots of the characteristic equation are $r_1 = e^{i\pi/4} = 1 + i$, $r_2 = e^{i3\pi/4} = -1 + i$, $r_3 = e^{i5\pi/4} = -1 - i$ and $r_4 = e^{i7\pi/4} = 1 - i$. We recall that $e^{a+ib} = e^a(\cos b + i \sin b)$, and since we are solving the flexure problem for $x > 0$ we can only use r_2 and r_3, because r_1 and r_4 have positive real parts. Solution (9.42) is

$$w(\hat{x}) = e^{-\hat{x}}\left((c_2 + c_3)\cos\hat{x} + i(c_2 - c_3)\sin\hat{x}\right) \tag{9.45}$$

when r_2 and r_3 are inserted. We want a solution with only real numbers, which is achieved by replacing the coefficients c_2 and c_3 by real coefficients $d_1 = c_2 + c_3$ and $d_2 = i(c_2 - c_3)$. The solution in terms of the real coefficients d_1 and d_2 is then

$$w(\hat{x}) = e^{-\hat{x}}\left(d_1 \cos\hat{x} + d_2 \sin\hat{x}\right). \tag{9.46}$$

From the symmetry condition $dw/dx = 0$ at $x = 0$ it follows that $d_1 = d_2$. Finally, the force balance (9.32) gives

$$\frac{V_0}{2} = \Delta\varrho g d_1 \alpha \int_0^\infty w(\hat{x})d\hat{x} = \Delta\varrho g d_1 \alpha \qquad (9.47)$$

which implies that $d_1 = V_0/(2\Delta\varrho g \alpha)$, where the following integral is used:

$$\int e^{-\hat{x}}(\cos\hat{x} + \sin\hat{x})d\hat{x} = -e^{-\hat{x}}\cos\hat{x}. \qquad (9.48)$$

The coefficient d_1 is sometimes rewritten using that $\Delta\varrho g = 4D/\alpha^4$, which follows from definition (9.34) of α. The alternative version d_1 then becomes $d_1 = V_0\alpha^3/(8D)$.

Note 9.3 *Semi-numerical solution.* It has already been mentioned that a general load can be represented by a series of point loads and that the solution for the deflection then becomes the sum of the solutions for each point load. Such a superposition of solutions is possible because the partial differential equation for the deflection is linear (see Exercise 9.5). A general surface load $q(x)$ can be turned into a series of point loads by representing the surface by discrete intervals. The weight of the load in the interval i from x_i to x_{i+1} becomes the point load

$$V_i = \int_{x_i}^{x_{i+1}} q(x)dx \approx \frac{1}{2}\Big(q(x_i) + q(x_{i+1})\Big)(x_{i+1} - x_i). \qquad (9.49)$$

We let each point load V_i act at the center of interval i, which has the position $x_{c,i} = (x_i + x_{i+1})/2$. The solution for the deflection from N point loads can then be written as the sum

$$w(x) = \sum_{i=1}^N V_i\, f(|x - x_{c,i}|/\alpha) \qquad (9.50)$$

where f is the function $f(u) = e^{-u}(\cos u + \sin u)$. Figure 9.8 shows an example of a wide load that is represented by three point loads, and where the deflection is given as the superposition of the three solutions for the point loads. The solution is compared with an analytical solution based on Fourier series for the same load, and they are almost equal. How the deflection can be represented by a Fourier series is shown in Note 9.5. A finer discretization than only three point loads implies an even better approximation of the exact solution. The figure also shows the solution when the entire load is represented by just one point, which is inaccurate for the shown load. The superposition of point loads is a fast and accurate algorithm that is simple to implement.

Exercise 9.5 Show that if $w_1(x)$ and $w_2(x)$ are both solutions of equation (9.14) for the deflection, then it follows that $c_1 w_1(x) + c_2 w_2(x)$ is also a solution for any constant coefficients c_1 and c_2.

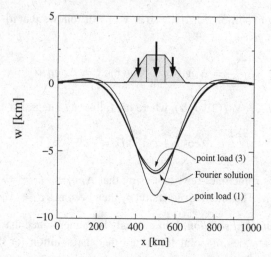

Figure 9.8. *A load is represented as three discrete point loads, and the deflection is the superposition of the deflections from each point load. The three-point-load solution (3) is compared with the deflection caused by having all the mass at the center (1). A Fourier series solution for the rectangular point load is also shown.*

9.3 Flexure from a point load on a broken plate

The plate beneath the Hawaiian island chain may be weak due to the volcanism that created the islands. The weakness of the plate along the chain can be modeled assuming a point load that acts on the edge of a broken plate. The assumption of a broken plate changes the point-load model of the previous section only with respect to the boundary condition at the end of the plate (at $x = 0$). For the infinite plate we had the symmetry condition $dw/dx = 0$ at $x = 0$. This boundary condition is now replaced by a zero bending moment, and it follows from equation (9.9) that a zero bending moment implies $d^2w/dx^2 = 0$. The solution for the deflection then becomes

$$w(\hat{x}) = w_{\text{max}} e^{-\hat{x}} \cos \hat{x} \tag{9.51}$$

where the dimensionless coordinate is $\hat{x} = x/\alpha$, and where the maximum deflection is

$$w_{\text{max}} = \frac{V_0}{\Delta \varrho \, g \alpha}. \tag{9.52}$$

The characteristic length scale α is the same as for the unbroken plate, but the maximum deflection of the broken plate is two times the deflection of the infinite plate (9.35).

The half-width of the deflection, which is given as the solution at $w = 0$, is $\hat{x}_0 = \pi/2$. It is seen to be $\pi/4$ shorter than the half- width for the unbroken plate. The distance to the forebulge, which is given by $dw/dx = 0$, is $\hat{x}_1 = 3\pi/4$. The distance to the forebulge is also $\pi/4$ less than the distance to the forebulge for the unbroken plate.

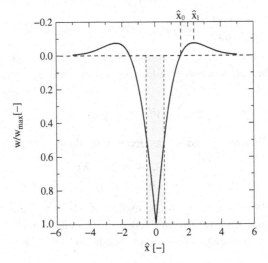

Figure 9.9. *The flexure of a broken plate from a point load at the free end* $(x = 0)$.

Note 9.4 The solution of equation (9.31) is shown in the previous section to be

$$w(\hat{x}) = e^{-\hat{x}}\left(d_1 \cos \hat{x} + d_2 \sin \hat{x}\right). \qquad (9.53)$$

The condition $d^2 w/d\hat{x}^2 = 0$ at $\hat{x} = 0$ implies that $d_2 = 0$. The coefficient d_1 is once more found from the buoyancy condition (9.32), which says that the weight of the load is equal to the weight of the displaced mantle. It then follows from the integral

$$\int e^{-\hat{x}} \cos \hat{x} \, d\hat{x} = \frac{e^{-\hat{x}}}{2}\left(\sin \hat{x} - \cos \hat{x}\right) \qquad (9.54)$$

that d_1 is the maximum deflection (9.52). We also have $\int_{-\infty}^{\infty} w(x)dx \doteq w_{max}\,\alpha$, which corresponds to the rectangle of area 1 in Figure 9.9.

9.4 Flexure and lateral variations of the load

The lithosphere deflects under the load of structures with a large lateral extent like a mountain range, but it is sufficiently strong to support the load of a single mountain peak. How large the lateral extent of a load needs to be in order to deflect the lithosphere can be determined by looking at the deflection of a plate under a periodic load (with an infinite extent)

$$q(x) = q_0\left(1 + \cos(2\pi x/\lambda)\right) \qquad (9.55)$$

where λ is the wavelength and q_0 is the maximum load. We will drop the constant part q_0 in the surface load, because it gives a constant subsidence $w = q_0/\Delta\varrho g$. Equation (9.14) for the deflection is then

$$D \frac{d^4 w}{dx^4} + (\varrho_m - \varrho_c)\, g\, w = q_0 \cos(2\pi x/\lambda) \tag{9.56}$$

which can be solved by trying a solution of the same form as the load, $w(x) = w_0 \cos(2\pi x/\lambda)$. When an expression of this form is inserted into equation (9.56) for the deflection we get

$$\left(D(2\pi/\lambda)^4 + \Delta\varrho g \right) w_0 = q_0 \tag{9.57}$$

which gives the coefficient w_0, and the deflection from the periodic load is

$$w(x) = \frac{q_0 \cos(2\pi x/\lambda)}{D(2\pi/\lambda)^4 + \Delta\varrho g}. \tag{9.58}$$

Note 9.5 shows that solution (9.58) can be used to construct a Fourier series solution for the deflection of nearly any load. An inspection of the deflection (9.58) shows that loads with sufficiently long wavelengths to fulfill the condition

$$D(2\pi/\lambda)^4 \ll \Delta\varrho g \tag{9.59}$$

will experience a deflection that is close to isostatic subsidence

$$w_{\text{iso}}(x) \approx \frac{q_0 \cos(2\pi x/\lambda)}{\Delta\varrho g}. \tag{9.60}$$

The wavelength of the load is then sufficiently large for the support from the elastic strength of the plate to be unimportant, and it is the buoyancy of the plate that controls the deflection. The condition (9.59) can be rewritten as $\lambda \gg \lambda_c$, where the critical wavelength λ_c is

$$\lambda_c = 2\pi \left(\frac{D}{\Delta\varrho g} \right)^{1/4}. \tag{9.61}$$

The deflection (9.58) is almost zero in the opposite regime, $\lambda \ll \lambda_c$, where the elastic strength of the plate is supporting the load. The critical wavelength λ_c therefore defines a length scale where the periodic load is partly supported by the elastic strength of the plate and partly by the buoyancy of the displaced mantle. The degree of isostatic equilibrium can be measured by the ratio of deflection over isostatic subsidence, $w(x)/w_{\text{iso}}(x)$, which is

$$c = \frac{w(x)}{w_{\text{iso}}(x)} = \frac{1}{(\lambda_c/\lambda)^4 + 1}. \tag{9.62}$$

Figure 9.10a shows a plot of c as a function of the scaled wavelength λ/λ_c. An important parameter in the critical wavelength is the flexural rigidity of the plate, $D = E h^3/(12(1 - v^2))$, which is controlled by the effective elastic thickness. For example, the parameters $h = 30$ km, $E = 70$ GPa, $v = 0.25$, $\Delta\varrho = 600$ kg m^{-3} give that $\lambda_c = 46$ km. Figure 9.10b also shows this, which plots the critical wavelength λ_c as a function of the plate thickness h. The distance from the center of the Hawaiian island chain to the flexural arch is estimated to be $\sim$250 km, which is less than $\lambda_c = 400$ km. These islands are therefore to a large degree supported by the elastic strength of the plate. On the other hand, large-scale structures

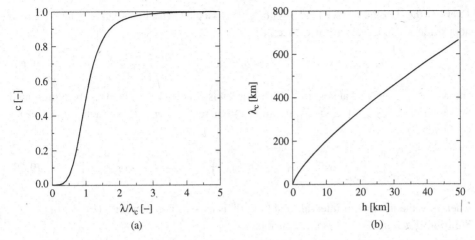

Figure 9.10. (a) The degree of isostatic equilibrium plotted as a function of the scaled wavelength. (b) The critical wavelength λ_c is plotted as a function of the plate thickness.

with a lateral extent of several thousand km like mountain ranges or entire continents have a subsidence that can be estimated using isostasy, even if they are supported by thicker plates.

Note 9.5 *Fourier series solution.* The solution (9.58) for the deflection from a periodic load can be used to construct a Fourier series solution. We have immediately that a periodic load

$$q(x) = \sum_{n=1}^{\infty} a_n \sin(k_n x), \qquad (9.63)$$

where $k_n = \pi n / L$ and L is the lateral length of the system, gives the deflection

$$w(x) = \sum_{n=1}^{\infty} \frac{a_n \sin(k_n x)}{D k_n^4 + \Delta \varrho g}. \qquad (9.64)$$

The Fourier coefficient a_n is

$$a_n = \frac{2}{L} \int_0^L q(x) \sin(k_n x) dx, \qquad (9.65)$$

which follows from the orthogonality property of the sine functions

$$\int_0^L \sin(k_n x) \sin(k_m x) dx = \begin{cases} 0 & n \neq m \\ L/2 & n = m. \end{cases} \qquad (9.66)$$

(The latter property can be viewed as a scalar product of $\sin(k_n x)$ and $\sin(k_m x)$, and we say that the $\sin(k_n x)$-functions are orthogonal with respect to the scalar product defined by integration.) Figure 9.8 shows an example of a Fourier series solution.

Exercise 9.6 Let the surface load $q(x)$ be given as a piecewise linear curve, where the load at position x in interval i from x_i to x_{i+1} is

$$q(x) = c_i(x - x_i) + q_i \quad \text{where} \quad c_i = \frac{q_{i+1} - q_i}{x_{i+1} - x_i} \tag{9.67}$$

and where q_i is the load at position x_i. The piecewise linear surface load can be represented as a Fourier series $q(x) = \sum_{n=1}^{\infty} a_n \sin(k_n x)$ where $k_n = \pi n/L$. Show that the Fourier coefficients are

$$a_n = \frac{2}{L} \sum_{i=1}^{N} \left[c_i \left(\frac{x}{k_n} \cos(k_n x) + \frac{1}{k_n^2} \sin(k_n x) \right) - \frac{(q_i - c_i x_i)}{k_n} \cos(k_n x) \right]_{x_i}^{x_{i+1}} \tag{9.68}$$

when N is the number of intervals and $[f(x)]_a^b$ is used as notation for $f(b) - f(a)$.

Solution: Each interval i has the contribution

$$I_{n,i} = \frac{2}{L} \int_{x_i}^{x_{i+1}} q(x) \sin(k_n x) dx \tag{9.69}$$

to each Fourier coefficient a_n. Useful integral: $\int x \sin(ax) dx = -(x/a) \cos(ax) + (1/a^2) \sin(ax)$.

Comments: The basis function $\sin(k_n x)$ imposes the boundary condition $w = 0$ at the boundaries $x = 0$ and $x = L$. The Fourier solution is in this respect different from the superposition of point load solutions (see Note 9.3), because the point load solutions have $w = 0$ at $x = \pm\infty$. The Fourier series solution is more computationally demanding than the corresponding point load solution, because of the sum over n in addition to the sum over intervals i. The infinite sum over n can be truncated at a sufficiently large number, depending on the wanted accuracy, and could typically be 100.

Exercise 9.7 Show that coefficients in the Fourier series (9.65) are zero for every even n, ($n = 2, 4, \ldots$) when the load $q(x)$ is symmetric around $L/2$.

9.5 The deflection of a plate under compression

An already bent lithospheric plate will buckle further under the action of an external compressive horizontal force (see Figure 9.11). We therefore start by extending the equation for the deflection of a plate by adding the bending moment (torque) from the horizontal force. The bending moment from a horizontal force F becomes Fw when the vertical deflection is w. The torque balance, where the internal bending moment is equal to the external bending moment, is

$$D \frac{d^2 w}{dx^2} = \int_x^{\infty} (x' - x) \Big(q(x') - \Delta \varrho g w(x') \Big) dx' - Fw \tag{9.70}$$

where the internal bending moment is the left-hand side, and the external bending moment is the right-hand side. The pressure on the plate in the vertical direction is the surface load

Figure 9.11. *The buckling of a horizontal plate by a horizontal compressive force F.*

$q(x)$ minus the counteracting buoyancy pressure $\Delta \varrho g w$. A differential equation for the deflection is found after differentiating twice, and we get

$$D\frac{d^4 w}{dx^4} + F\frac{d^2 w}{dx^2} + \Delta \varrho g w = q(x) \tag{9.71}$$

(see Exercise 9.1 for how to carry out the first of the two differentiations). Equation (9.71) gives the deflection from the combined action of a surface load $q(x)$ and a horizontal force F, when the plate is supported by a buoyancy pressure $\Delta \varrho w$. We have already seen in Section 9.4 that a periodic surface load produces a periodic deflection, and we will see that the deflection from horizontal compression depends on the wavelength. The impact of the horizontal force F is therefore studied in the case of the same periodic surface load as in Section 9.4,

$$q(x) = q_0 \cos(2\pi x/\lambda), \tag{9.72}$$

where λ is the wavelength of the load. A constant surface load q_0 could have been added to the periodic load, in order to make it oscillate between 0 and q_0, but it is left out since it only adds a constant subsidence $q_0/(\Delta \varrho g)$ to the deflection. A periodic surface load can also be expressed using the wave number $k = 2\pi/\lambda$, which simplifies the expressions. The deflection in the case of the surface load (9.72) is found by guessing that the solution has the same form as the surface load

$$w(x) = w_0 \cos(kx). \tag{9.73}$$

By inserting solution (9.73) into equation (9.71) we get

$$w_0\left(Dk^4 - Fk^2 + \Delta \varrho g\right) = q_0 \tag{9.74}$$

or

$$w_0 = \frac{q_0}{k^2\left((Dk^2 + \Delta \varrho g/k^2) - F\right)}. \tag{9.75}$$

We first notice that zero compression ($F = 0$) leads to the same deflection as in Section 9.4 for a periodic load. The next thing we notice is that the deflection increases with an increasing compressive force, but the compressive force cannot increase beyond the limiting force

$$F_c = Dk^2 + \Delta \varrho g/k^2, \tag{9.76}$$

which would imply an infinite deflection. The upper bound for the compressive force F_c depends on the wave number (or alternatively on the wavelength). The critical force F_c is seen to increase with decreasing wave numbers (increasing wavelengths), which means that a compressive force will not have much impact on nearly plane plates. The same applies for a plate deflected with large wave numbers (small wavelengths). The critical force F_c reaches a minimum for a wave number between the two limiting regimes of very short or very long wave numbers. The wave number for the minimum critical force is found by solving $dF_c/k = 0$, which gives

$$k_{min} = \left(\frac{\Delta \varrho g}{D}\right)^{1/4} \quad \text{or} \quad \lambda_{min} = 2\pi \left(\frac{D}{\Delta \varrho g}\right)^{1/4}. \tag{9.77}$$

The least possible critical force is

$$F_{c,min} = F_c(k_{min}) = 2(D\Delta \varrho g)^{1/2}. \tag{9.78}$$

We see that both the wavelength λ_{min} and the critical force $F_{c,min}$ depend on the coefficient of flexural rigidity $D = Eh^3/(12(1 - \nu^2))$, which again depends on the plate thickness h. The force $F_{c,min}$ can be related to the corresponding compressive stress by

$$\sigma_{c,min} = F_{c,min}/h. \tag{9.79}$$

The compressive stress $\sigma_{c,min}$ can be compared with lithostatic pressure in order to get an idea of how large it is. It should be noted that lithostatic pressure increases linearly towards the bottom of the plate, unlike the stress $\sigma_{c,min}$, which is distributed uniformly across the plate just as the force F. Both the stress $\sigma_{c,min}$ and the lithostatic pressure at the base of of the plate are shown in Figure 9.12 as functions of the plate thickness. The compressive stress $\sigma_{c,min}$ is seen to increase as $h^{1/2}$ while the lithostatic pressure increases as h. We can conclude that the deflection of the lithosphere is caused by surface loads, unless it is compressed by horizontal forces that are much larger than the lithostatic pressure at the base of the elastic part of the plate.

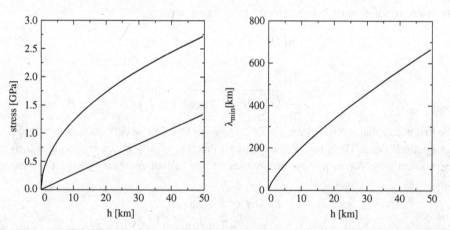

Figure 9.12. (a) The critical stress $\sigma_{c,min}$ is plotted as a function of the plate thickness. The lithostatic pressure $\varrho_c gh$ at the depth h is shown for reference (where $\varrho_c = 2700\ kg\ m^{-3}$ is used for the density of the crust). (b) The wavelength λ_{min} as a function of the plate thickness h.

Exercise 9.8 Show that the compressive force F needed to increase the deflection by a factor f from an uncompressed state is $F = (f - 1)F_c/f$. How large must F be for the deflection to be increased by 1% when $k = k_{min}$?

9.6 Damped flexure of a plate above a viscous mantle

An elastic plate that is rapidly loaded does not instantly deflect under the load because of the viscous damping of the ductile mantle. For example if the lithosphere is rapidly loaded by a volcanic island then some time will pass before the full elastic deflection is reached. A similar situation is the rapid deglaciation of a continent where the postglacial rebound will continue for some time after the deglaciation. We would like to know how long the lithosphere needs to adjust to changes in the surface load. A simple approach to this problem is to represent the viscous damping from the mantle by a dashpot. The dashpot is similar to the shock absorber in a car and it produces a resisting stress that is proportional to the velocity.

$$\sigma = \eta \frac{dw}{dt} = \left(\frac{\mu}{l}\right) \frac{dw}{dt} \tag{9.80}$$

where the parameter η is the viscosity μ divided by the mantle length scale l. The last equality (9.80) is the viscosity μ multiplied with the strain rate approximated as $(1/l)$ dw/dt. The stress σ acts on the base of the lithosphere and it is in addition to the restoring pressure from buoyancy. The equation for the deflection of the lithosphere is therefore

$$\frac{\partial^4 w}{\partial x^4} + \Delta \varrho g w + \eta \frac{\partial w}{\partial t} = q(x). \tag{9.81}$$

We can use the equation to study rapid loading and unloading. Let us first study rapid unloading when the plate has been deflected for a very long time by a periodic load $q(x) = q_0 \cos(kx)$, where $k = 2\pi/\lambda$ is the wave number. We then have that $\partial w/\partial t = 0$ before the plate is unloaded and that the initial deflection is

$$w_e(x) = \frac{q_0 \cos(kx)}{Dk^4 + \Delta \varrho g} \tag{9.82}$$

as shown in Section 9.4. When the load is suddenly removed we have to solve equation (9.81) for $q(x) = 0$ with $w_e(x)$ as the initial condition. The deflection as a function of time is then

$$w(x, t) = w_e(x)e^{-t/t_0(k)}, \tag{9.83}$$

where the characteristic time t_0 is dependent on the wave number,

$$t_0(k) = \frac{\eta}{Dk^4 + \Delta \varrho g}. \tag{9.84}$$

It is seen that the deflections of the plate decay to zero as time goes to infinity. The time $\ln 2 \, t_0(k)$ is the half-life for the rebound at the given wave number, and most of the deflection is lost after a time span $t \gg t_0(k)$. We see that the rebound is slowest for a periodic load with very long wavelength ($k \approx 0$) where we have

$$t_0 \approx \frac{\eta}{\Delta \varrho g}. \tag{9.85}$$

A very long wavelength corresponds to the case of isostatic subsidence where there is almost no flexure of the plate. A plate with very short wavelengths (large wave numbers) will rebound faster since there are stronger elastic forces to recover the shape of the plate to what it was before it was loaded. If the initial deflection can be decomposed into a superposition of deflections for a large range of wave numbers, then the time needed for rebound will be dominated by the shortest wave number (longest wavelength). The dashpot parameter η can be estimated using uplift data for a deglaciation. For example, $\eta = 5.7 \cdot 10^{15}$ Pa s m^{-1} if the time period for post glacial rebound is $t_0 = 3 \cdot 10^4$ years and $\Delta \varrho = 600$ kg m^{-3}. The mantle length scale l is a difficult parameter, because it is not a precise number. However, assuming that it is in the range of 10^2 km to 10^3 km gives a corresponding viscosity in the range of $5.7 \cdot 10^{20}$ Pa s to $5.7 \cdot 10^{21}$ Pa s.

Note 9.6 Equation (9.81) is solved by separation of variables, where it is taken as a starting point that the solution has the form

$$w(x,t) = Y(t)\cos(kx). \tag{9.86}$$

This solution is inserted into equation (9.81) when $q(x) = 0$, and we get

$$(Dk^4 Y + \Delta \varrho g Y + \eta \dot{Y})w_0 \cos(kx) = 0 \tag{9.87}$$

where the dot above Y denotes time differentiation. We then get the following equation for $Y(t)$:

$$\dot{Y} = -\frac{1}{\eta}(Dk^4 + \Delta \varrho g)\, Y = -\frac{1}{t_0(k)} Y. \tag{9.88}$$

An integration gives that $Y(t) = Y(0)e^{-t/t_0}$, where the value of Y at $t = 0$ is found from the initial condition

$$w(x, t{=}0) = Y(0)\cos(kx) = w_e(x). \tag{9.89}$$

We then get solution (9.83).

9.7 The equation for viscoelastic flexure of a plate

A lithospheric plate does not only behave elastically, but it may also deform viscously. The latter property implies that the lithosphere becomes deformed permanently as opposed to the elastic deformations that are fully recoverable. The strain ϵ of a viscoelastic plate therefore has to be decomposed into elastic (recoverable) strain ϵ_e and viscous (permanent) strain ϵ_v. The total deflection w is decomposed in the same way into an elastic (recoverable) deflection w_e and a viscous (permanent) deflection w_v, and we have

$$\epsilon = \epsilon_e + \epsilon_v \quad \text{and} \quad w = w_e + w_v. \tag{9.90}$$

Hooke's law relates elastic strain to the stress σ:

$$\epsilon_e = \frac{\sigma}{E} \tag{9.91}$$

where E is Young's modulus. The similar law for viscous strain says that the strain rate is proportional to the stress

$$\dot{\epsilon}_v = \frac{\sigma}{\mu} \tag{9.92}$$

where μ is the viscosity, and where a dot above a symbol denotes time differentiation. The elastic and the viscous strain can be related to the elastic and viscous deformations, respectively, using the radius of curvature as shown by equations (9.2) and (9.5). We therefore have

$$\epsilon_e = z \frac{\partial^2 w_e}{\partial x^2} \quad \text{and} \quad \epsilon_v = z \frac{\partial^2 w_v}{\partial x^2}. \tag{9.93}$$

The rate of total strain in the lateral direction is therefore

$$\dot{\epsilon} = \dot{\epsilon}_e + \dot{\epsilon}_v = z \frac{\partial^2 \dot{w}_e}{\partial x^2} + z \frac{\partial^2 \dot{w}_v}{\partial x^2} = z \frac{\partial^2 \dot{w}}{\partial x^2}, \tag{9.94}$$

but we also have

$$\dot{\epsilon} = \dot{\epsilon}_e + \dot{\epsilon}_v = \frac{1}{E} \left(\dot{\sigma} + \frac{\sigma}{t_e} \right) \tag{9.95}$$

where t_e is the time constant $t_e = \mu/E$. It turns out that viscous deformations become similar in size to the elastic deformations after time span t_e (see Exercise 9.10). From equations (9.94) and (9.95) we get

$$z \frac{\partial^2 \dot{w}}{\partial x^2} = \frac{1}{E} \left(\dot{\sigma} + \frac{\sigma}{t_e} \right) \tag{9.96}$$

where σ is the lateral stress in the plate. We can now form the internal bending moment along a vertical cross-section, just as in Section 9.1, by integrating across the plate:

$$M = \int_{-h/2}^{h/2} z \sigma \, dz. \tag{9.97}$$

The force in a 2D fiber in the plate is $\sigma \, dz$ and the moment arm is z. (In 3D we would have to use the area dA of the fiber to obtain the force $\sigma \, dA$ in the fiber.) The multiplication of z and the integration over z from $-h/2$ to $h/2$ is done on both sides of equation (9.96), and we get

$$D \frac{\partial^2 \dot{w}}{\partial x^2} = \dot{M} + \frac{M}{t_e} \tag{9.98}$$

where D is the flexural rigidity $D = Eh^3/12$. Recall from Section 9.1 that we could use an alternative expression for the flexural rigidity based on zero vertical strain rather than zero lateral stress. The same argument cannot be applied here because the deflection w

is not an elastic deflection, but it is the sum of both elastic and viscous deflections. Equation (9.98), which relates the total deflection of a viscoelastic plate or a beam to the internal bending moment M, is the fundamental equation for viscoelastic deformations. It is combined with the momentum balance (9.10), which says that the internal bending moment M must be equal to the external moment. A vertical surface load (stress), as shown by equation (9.11), gives the external moment, and it replaces the internal bending moment in equation (9.98). We can differentiate equation (9.98) two times with respect to x in the same way as equation (9.12), and we obtain

$$D\frac{\partial^4 \dot{w}}{\partial x^4} = \dot{q} + \frac{q}{t_e}. \tag{9.99}$$

This is the equation for the deflection of a viscoelastic beam (or plate) by a surface load q. The vertical load will be $q - \Delta\varrho g w$ when the plate is supported from below by a buoyancy pressure $\Delta\varrho g w$, and we then get

$$D\frac{\partial^4 \dot{w}}{\partial x^4} + \Delta\varrho g \dot{w} + \Delta\varrho g \frac{1}{t_e} w = \dot{q} + \frac{q}{t_e}. \tag{9.100}$$

Equation (9.100) for viscoelastic deformations has an elastic limit when $t_e \to \infty$, in which case $1/t_e \approx 0$. A time integration then recovers equation (9.13) for the flexure of an elastic plate. The other limit $t_e \to 0$, which is the viscous limit, is obtained by multiplying equation (9.100) with t_e and then approximating the two terms $t_e\Delta\varrho g \dot{w}$ and $t_e\dot{q}$ by zero. We then get the equation for the deformation of a purely viscous plate,

$$D_v\frac{\partial^4 \dot{w}}{\partial x^4} + \Delta\varrho g w = q \tag{9.101}$$

where the coefficient $D_v = \mu h^3/12$ is the flexural rigidity for a viscous plate.

9.8 Elastic and viscous deformations

Once we have solved equation (9.100) for the total deflection of a viscoelastic plate it would be interesting to know how much of the total deflection is viscous (permanent) and how much is elastic (recoverable). We must therefore return to the basic relationships (9.91) and (9.92) between stress and strain, which give that $\dot{\epsilon}_v = \epsilon_e/t_e$. The rate of change of total strain can therefore be expressed by only elastic strain as

$$\dot{\epsilon}_e + \frac{\epsilon_e}{t_e} = \dot{\epsilon} \tag{9.102}$$

which is an equation that can be integrated to a relationship between total strain and elastic strain (as shown in Note 9.7):

$$\epsilon_e(t) = e^{-t/t_e}\epsilon_e(0) + e^{-t/t_e}\int_0^t \dot{\epsilon}(t')\, e^{t'/t_e} dt'$$

$$= \epsilon(t) - \frac{e^{-t/t_e}}{t_e}\int_0^t \dot{\epsilon}(t')\, e^{t'/t_e} dt'. \tag{9.103}$$

Subtracting the elastic strain (9.103) from the total strain $\epsilon(t)$ gives the viscous strain

$$\epsilon_v(t) = \epsilon(t) - \epsilon_e(t) = \frac{e^{-t/t_e}}{t_e} \int_0^t \dot{\epsilon}(t') e^{t'/t_e} dt'. \tag{9.104}$$

These expressions for the elastic and viscous strain can be used to make similar expressions for the elastic and viscous deflections. The strains are related to the deflections by equations (9.93) and (9.94), which can be inserted into equations (9.103) to (9.104). Integration with respect to x two times then gives

$$w_e(t) = w(t) - \frac{e^{-t/t_e}}{t_e} \int_0^t \dot{w}(t') e^{t'/t_e} dt' \tag{9.105}$$

and

$$w_v(t) = \frac{e^{-t/t_e}}{t_e} \int_0^t \dot{w}(t') e^{t'/t_e} dt'. \tag{9.106}$$

We see that the elastic and viscous strain and deflections are related to the total strain and deflections, respectively, by similar relationships.

Note 9.7 Equation (9.102) is multiplied by e^{t/t_e}, and we get

$$\dot{\epsilon}_e e^{t/t_e} + \frac{1}{t_e} \epsilon_v e^{t/t_e} = \dot{\epsilon} e^{t/t_e}. \tag{9.107}$$

The left-hand side of the equation can be rewritten as follows using differentiation with respect to time:

$$\frac{\partial}{\partial t}\left(\epsilon_e e^{t/t_e}\right) = \dot{\epsilon} e^{t/t_e} \tag{9.108}$$

which is then straightforward to integrate. The integral of the right-hand side with respect to time is done by integration by parts:

$$\int_0^t \dot{\epsilon}(t') e^{t'/t_e} dt = \left[\epsilon(t') e^{t'/t_e}\right]_0^t - \frac{1}{t_e} \int_0^t \dot{\epsilon}(t') e^{t'/t_e} dt'. \tag{9.109}$$

An important point is that the initial total strain (or deformation) is equal to the initial elastic strain (or deformation), $\epsilon(t=0) = \epsilon_e(t=0)$, because it takes time to accumulate viscous (permanent) strain.

Exercise 9.9 Show that $w_v(t) = w_v(0) + (1/t_e) \int_0^t w_e(t') dt'$.
Hint: use that $\dot{\epsilon}_v = \epsilon_e/t_e$.

Exercise 9.10 (a) Find the deflection of a viscoelastic beam of length l that is clamped at one end and loaded at the free end with a time dependent force $F(t)$. (See Figure 9.5, and also Exercise 9.4.)
(b) What are the elastic and viscous parts of the deflection?
(c) At what time is the viscous (permanent) deflection equal to the elastic deflection when the load is constant $F(t) = F_0$?

Solution: (a) A starting point for obtaining the deflection is equation (9.98),

$$D\frac{\partial^2 \dot{w}}{\partial x^2} = \dot{\tau} + \frac{1}{t_e}\tau \tag{9.110}$$

where the torque from force F is

$$\tau(x, t) = (l - x)F(t). \tag{9.111}$$

An integration with respect to time gives

$$\frac{\partial^2 w}{\partial x^2} = \left(F(t) + \frac{1}{t_e}\int_0^t F(t')dt'\right)\frac{1}{D}(l - x). \tag{9.112}$$

This equation can be rewritten in the dimensionless (unit) coordinate $\hat{x} = x/l$ as

$$\frac{\partial^2 w}{\partial \hat{x}^2} = 6w_2(t)(1 - \hat{x}) \tag{9.113}$$

where

$$w_2(t) = \left(F(t) + \frac{1}{t_e}\int_0^t F(t')dt'\right)\frac{l^3}{6D}. \tag{9.114}$$

The integration with respect to $\hat{x}$ then leads to

$$w(\hat{x}, t) = w_2(t)(1 - \hat{x})^3 + c_1\hat{x} + c_2 \tag{9.115}$$

where c_1 and c_2 are two time-dependent integration constants. The clamped end ($\hat{x} = 0$) has the two boundary conditions $w = 0$ and $\partial w/\partial \hat{x} = 0$. From the first boundary condition ($w = 0$) we get that $c_2 = -w_2$, and from the second boundary condition ($\partial w/\partial \hat{x} = 0$) we get that $c_1 = 3w_2$. The deflection then becomes

$$w(\hat{x}, t) = w_2(t)\left((1 - \hat{x})^3 + 3\hat{x} - 1\right), \tag{9.116}$$

which is the same solution as in Exercise 9.4, except for the time-dependence in w_2.
(b) The solution can be decomposed into an elastic deflection $w_e(x)$ and a viscous deflection $w_v(x, t)$ where

$$w_e(\hat{x}, t) = F(t)\left(\frac{l^3}{6D}\right)\left((1 - \hat{x})^3 + 3\hat{x} - 1\right) \tag{9.117}$$

and

$$w_v(\hat{x}, t) = \frac{1}{t_e}\int_0^t F(t')dt'\left(\frac{l^3}{6D}\right)\left((1 - \hat{x})^3 + 3\hat{x} - 1\right). \tag{9.118}$$

The elastic deflection is proportional to the force $F(t)$ (see Exercise 9.4), and the viscous deflection is zero at time $t = 0$.
(c) When $F(t) = F_0$ (constant) we get that $w_e(x, t) = w_v(x, t)$ after the time span $t = t_e$. This justifies the interpretation that t_e is the time span needed for the viscous deflection to be of similar size to the elastic deflections.

9.9 Flexure of a viscoelastic plate

The flexure of an elastic plate under a periodic load

$$q(x) = q_0 \cos(2\pi x/\lambda) \tag{9.119}$$

was studied in Section 9.4, where λ is the wavelength. We found a critical wavelength $\lambda_c = 2\pi(D/\Delta\varrho g)^{1/4}$ that defines two regimes with respect to the size of the wavelength λ. Loads with $\lambda \gg \lambda_c$ were almost fully supported by buoyancy, while loads with $\lambda \ll \lambda_c$ were almost fully supported by the elastic strength of the plate. Loads that are fully buoyed by the displaced mantle are stable (because they cannot subside further), but loads that are partly supported by the elastic strength of the plate may subside further by viscous flow. We will now find out how long we can expect the elastic support of a load to last.

We will start out with a flat (unloaded) plate where the periodic load (9.119) is suddenly applied at $t = 0$. Equation (9.100) for viscoelastic deflections then becomes

$$D\frac{\partial^4 \dot{w}}{\partial x^4} + \Delta\varrho g \dot{w} + \frac{\Delta\varrho g}{t_e} w = \frac{q_0}{t_e} \cos(kx) \tag{9.120}$$

for $t \geq 0$, where $k = 2\pi/\lambda$ is the wave number. This equation for the deflection is solved by guessing that the solution has the same form as the load,

$$w(x, t) = \frac{q_0}{\Delta\varrho g} Y(t) \cos(kx) \tag{9.121}$$

where the unknown function $Y(t)$ is found by inserting the solution into equation (9.120) for the deflection. We then get the flexure

$$w(x, t) = \frac{q_0}{\Delta\varrho g} \left(1 + \left(\frac{t_e}{t_0(k)} - 1\right) e^{-t/t_0(k)}\right) \cos(kx) \tag{9.122}$$

as Note 9.8 shows, where the characteristic time t_0 is dependent on the wave number,

$$t_0(k) = \left(\frac{Dk^4}{\Delta\varrho g} + 1\right) t_e. \tag{9.123}$$

The first thing we notice is that the solution at $t = 0$ is equal to the flexure (9.58) for a purely elastic plate. (This property of the solution is in fact an initial condition that has been imposed on it, as shown in Note 9.8.) The next thing we look for is the deflection after infinite time, and we see that the deflection becomes $w \approx (q_0/\Delta\varrho g) \cos(kx)$ for $t \gg t_0$. This is precisely isostatic subsidence, and all loads will become almost fully supported by buoyancy after a time span $t \gg t_0$. The characteristic time t_0 can be rewritten in terms of the critical wavelength λ_c as

$$t_0(\lambda) = \left(\left(\frac{\lambda_c}{\lambda}\right)^4 + 1\right) t_e. \tag{9.124}$$

The regime of large wavelengths, $\lambda \gg \lambda_c$, has a characteristic time $t_0 \approx t_e$, and the deflection (9.122) is weakly dependent on time (see Exercise 9.11). The opposite regime of short wavelengths, $\lambda \ll \lambda_c$, has a characteristic time $t_0 \gg t_e$, and the characteristic time

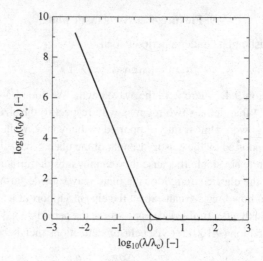

Figure 9.13. *The scaled time constant t_0/t_e as a function of the scaled wavelength λ/λ_c.*

increases with decreasing wavelength, see Figure 9.13. For example, a wavelength that is a factor $1/10$ less than λ_c will last 10^4 longer than t_e.

The deflection (9.122) can be decomposed into an elastic part (w_e) and a viscous part (w_v) as shown in Section 9.8, and we get

$$w_e(x,t) = \frac{q_0}{\Delta \varrho g} \left(\frac{t_e}{t_0(k)} \right) e^{-t/t_0(k)} \cos(kx) \qquad (9.125)$$

$$w_v(x,t) = \frac{q_0}{\Delta \varrho g} \left(1 - e^{-t/t_0(k)} \right) \cos(kx). \qquad (9.126)$$

The elastic part of the deflection is initially equal to the flexure of a purely elastic plate, and it then decays to zero with the half-life $t_0 \ln 2$. The viscous part is initially zero, but it increases with time and approaches isostatic subsidence as time goes far beyond t_0. The following table summarizes some of the wave-dependent properties of the solution:

Wavelength	Characteristic time	Initial deflection
$\lambda \ll \lambda_c$	$t_0 \approx (\lambda_c/\lambda)^4 t_e$	$w_e \ll w_{\text{iso}}$
$\lambda \gg \lambda_c$	$t_0 \approx t_e$	$w_e \approx w_{\text{iso}}$

where isostatic subsidence is $w_{\text{iso}} = (q_0/\Delta \varrho g) \cos(kx)$.

A simple estimate for the time constant t_e is 0.3 Ma, using that $E = 100$ GPa and $\mu = 10^{24}$ Pa s. This implies that loads with a typical wavelength similar to the critical wavelength (like volcanic islands) should not remain partly supported by the elastic forces of the plate for several tens of Ma. But there are loads, like for example the Emperor Seamounts, that seems to have been supported by the plate for more than 60 Ma. Such observations cannot be easily explained by the viscoelastic model. One possible solution to the problem is that the load itself should be included in the plate, which would make the plate thicker and more rigid. This problem also suggests that simple elastic and viscoelastic models may be too simple to account for certain observations of flexure.

The viscoelastic model shows that elastic support of surface loads decays to zero with increasing time. It is therefore possible that old loads are supported mainly by buoyancy, which may explain why simple isostatic equilibrium calculations often are so successful.

Note 9.8 The deflection (9.122) is obtained by inserting the guess for a solution (9.121) into equation (9.120) for the deflection. We then get the following equation for the unknown function $Y(t)$:

$$Dk^4 \dot{Y} + \dot{Y} + \frac{1}{t_e} Y = \frac{1}{t_e}, \tag{9.127}$$

which can be rewritten as

$$t_0 \dot{Y} + Y = 1. \tag{9.128}$$

This equation has the solution

$$Y(t) = 1 + Ce^{-t/t_0}, \tag{9.129}$$

where C is an integration constant that is obtained from the initial condition, which is the elastic deflection at $t = 0$. We have already found the elastic deflection for the same periodic load, equation (9.58), which can be rewritten as

$$w(x) = \frac{q_0}{\Delta \varrho g} \frac{t_e}{t_0} \cos(kx). \tag{9.130}$$

The viscoelastic deflection at $t = 0$ is

$$w(x) = \frac{q_0}{\Delta \varrho g} \left(1 + C\right) \cos(kx) \tag{9.131}$$

and we therefore get that $C = (t_e/t_0) - 1$.

Exercise 9.11 Show that the deflection (9.122) can be approximated by

$$w(x, t) \approx w_{\text{iso}} \left(1 - \left(\frac{\lambda_c}{\lambda}\right)^4 e^{-t/t_e}\right) \approx w_{\text{iso}} \tag{9.132}$$

in the regime $\lambda \gg \lambda_c$. Hint: use that $1/(1 + x) \approx 1 - x$ for small x.

Exercise 9.12 Show that the deflection can be approximated by

$$w(x, t) = w_{\text{iso}} \left(1 - e^{-t/t_0}\right) \tag{9.133}$$

in the regime $\lambda \ll \lambda_c$.

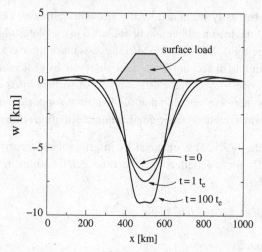

Figure 9.14. *The time-dependent deflection when the shown surface load is kept constant. Notice that keeping the surface load constant implies that the deflection is continuously filled as it gets deeper.*

Exercise 9.13 Use the result of Exercise (9.12) to show that the viscous subsidence for a periodic load at $t = 0$ fulfills

$$\frac{\partial w}{\partial t} \approx \frac{w}{t_0}. \tag{9.134}$$

Exercise 9.14 Derive equations (9.126) and (9.126) using expressions from Section 9.8.

Exercise 9.15 Show that a load that is written as a Fourier series

$$q = \sum_{n=1}^{\infty} a_n \sin(\pi n x / L) \tag{9.135}$$

has the viscoelastic (time-dependent) deflection

$$w(x, t) = \sum_{n=1}^{\infty} \frac{a_n}{\Delta \varrho g} \left(1 + \left(\frac{t_e}{t_n} - 1 \right) e^{-t/t_n} \right) \sin(k_n x) \tag{9.136}$$

where $k_n = \pi n / L$ and $t_n = t_0(k_n)$. Exercise 9.6 shows how the Fourier coefficients can be obtained for a rectangular load, and Figure 9.14 shows an example of the solution (9.136). (The deflection at $t = 0$ is the elastic deflection, which is also shown in Figure 9.8.)

9.10 Buckling of a viscous plate

An elastic plate with buckles, due to for example a periodic surface load, will buckle more when compressed by a horizontal force. We saw in Section 9.5 that there is an upper limit for how large a compressive force an elastic plate can sustain. If the force approaches this limiting value the plate will most likely fracture. We will now look once more at a plate under compression, but this time at a viscous plate. The starting point is equation (9.98) for the deflection of a viscoelastic plate,

$$D\frac{\partial^2 \dot{w}}{\partial x^2} = \dot{M} + \frac{M}{t_e}. \tag{9.137}$$

This equation relates the deflection at position x of a viscoelastic plate to the torque M vertically through the plate at the same position. The torque balance requires that M has to be equal to the torque acting on the right side of position x. Section 9.5 shows that this torque is

$$\tau(x) = \int_x^\infty (x' - x)\Big(q(x') - \Delta\varrho g w(x')\Big)dx' - Fw \tag{9.138}$$

when it is subjected to both a vertical (net) pressure $q(x') - \Delta\varrho g w(x')$ and a compressive force F. Two times differentiation of equation $M = \tau$ with respect to x gives the equation for the deflection:

$$D\frac{\partial^4 \dot{w}}{\partial x^4} + \left(\dot{F} + \frac{F}{t_e}\right)\frac{\partial^2 w}{\partial x^2} + F\frac{\partial^2 \dot{w}}{\partial x^2} + \Delta\varrho g\left(\dot{w} + \frac{w}{t_e}\right) = \dot{q} + \frac{q}{t_e}. \tag{9.139}$$

Equation (9.139) gives the deflection of a viscoelastic plate under both a surface load and a compressive force, when the plate is floating on a substratum. A model for a pure viscous plate is obtained in the limit where Young's modulus goes to infinity, which is when the elastic deflections become negligible compared with viscous deformations. The time constant $t_e = \mu/E$ approaches zero in this limit. Equation (9.139) is therefore multiplied with t_e, and all terms that contain t_e are dropped. The only exception is the factor $t_e D$, which becomes $D_v = \mu h^3/12$, because Young's modulus factors out. We are then left with the equation for a pure viscous plate,

$$D_v\frac{\partial^4 \dot{w}}{\partial x^4} + F\frac{\partial^2 w}{\partial x^2} + \Delta\varrho g w = q. \tag{9.140}$$

The impact of the horizontal force F is once more studied for a plate that initially has a periodic deflection

$$w(x) = w_0 \cos(kx) \tag{9.141}$$

where $k = 2\pi/\lambda$ is the wave number and where λ is the wavelength. It is shown in Exercise 9.142, for a plate that has no surface load ($q = 0$), that this periodic deformation develops through time as

$$w(x, t) = w_0 e^{t/t_0(k)} \cos(kx) \tag{9.142}$$

where t_0 is a characteristic time that depends on the wave number k as

$$t_0(k) = \frac{D_v k^4}{Fk^2 - \Delta\varrho g}. \tag{9.143}$$

We notice that t_0 may be either larger than zero or less than zero depending on whether Fk^2 is larger than or less than $\Delta\varrho g$, and that t_0 becomes infinite when $Fk^2 = \Delta\varrho g$. The characteristic force that makes t_0 infinite is

$$F_s = \frac{\Delta \varrho g}{k^2} = \frac{\Delta \varrho g \lambda^2}{(2\pi)^2} \qquad (9.144)$$

when the wavelength is fixed. The characteristic wavelength that gives an infinite t_0 when the force is fixed is

$$\lambda_s = 2\pi \sqrt{\frac{F}{\Delta \varrho g}}. \qquad (9.145)$$

How the characteristic time depends on the force and the wavelength is summarized as follows:

$$
\begin{array}{lll}
t_0 < 0, & F < F_s & \text{or} \quad \lambda > \lambda_s \\
t_0 = \infty, & F = F_s & \text{or} \quad \lambda = \lambda_s \\
t_0 > 0, & F > F_s & \text{or} \quad \lambda < \lambda_s.
\end{array}
$$

The initial deformations will decrease with time when $t_0 < 0$, and they will increase with time when $t_0 > 0$. An infinite characteristic time t_0 means that the periodic buckling of the plate is stable under the compressive force F_s. The initial buckling becomes preserved. The deformations will decrease with time for a compressive force less than F_s, and increase for a force larger than F_s.

Wavelengths longer than λ_s will die out, and the longer the wavelengths are the faster will they die out. Wavelengths that are $\sim \lambda_s$ will be preserved, and wave lengths that are less than λ_s will increase with time. The wavelengths that can be observed in fold belts should therefore be the wavelengths that are less than λ_s. It is seen that the limiting wavelength (9.145) does not depend on the viscosity of the layer, and that it is only dependent on the compressive force F.

A periodic deflection of a lithospheric plate can be written as a Fourier series

$$w(x) = \sum_n a_n \sin(k_n x), \quad \text{with} \quad k_n = \pi n / L \qquad (9.146)$$

where the sum is a superposition of deflections of different wavelengths. The development of this deflection through time is the superposition of the solution (9.142):

$$w(x, t) = \sum_n a_n e^{t/t_0(k_n)} \sin(k_n x). \qquad (9.147)$$

Compressing the plate with a force F implies that wavelengths $\lambda_n = 2\pi/k_n = 2L/n$ larger than λ_s in the Fourier decomposition will die out, and that wavelengths less than λ_s will increase. Figure 9.15 shows the time development of the deflection from Figures 9.8 and 9.14 in the case of a viscous plate. The initial state is just the deflection filled with rock of density ϱ_c when the surface load is absent. The compressive force is measured relative to the average force from the isostatic pressure through the plate,

$$F_u = \int_0^h \varrho_c g z' dz' = \frac{1}{2} \varrho_c g h^2. \qquad (9.148)$$

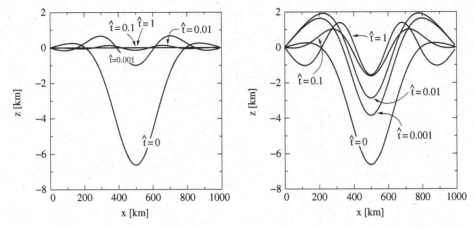

Figure 9.15. *(a) A viscous plate where the deflection dies out during compression. (b) A plate where the compressive force is large enough for short wavelengths to grow at the same rate as the longer wavelengths decay.*

The plate in Figure 9.15a is compressed with a force that is $F = 0.25 \times F_u$, and time is measured relative to the unit time $t_u = 4D_v \Delta \varrho g / F^2$ (see Exercise 9.17). The deflection through time is independent of the viscosity when measured with unit time t_u. The unit force is $F_u = 1.2 \cdot 10^9$ N for a plate that has the thickness $h = 30$ km, and the unit time is $t_u = 180$ Ma when the viscosity is $\mu = 10^{24}$ Pa s and force is $F = 0.25 \times F_u$. The figure shows the deformation of the plate at time steps $\hat{t} = 0, 0.001, 0.01, 0.1$ and 1, where $\hat{t} = t/t_u$ is dimensionless time. Figure 9.15a shows that the force is not large enough to preserve the deformations, and the deformations have almost died out at $\hat{t} = 1$. A closer look at the Fourier coefficients shows that they are decreasing with increasing n, and that the first unstable wavelength is $n = 15$. The characteristic wavelength is $\lambda_s = 1.4 \cdot 10^5$ m for the given force F.

n	a_n [m]	λ_n [m]	Unstable
1	3.5e+03	2.0e+06	–
3	−2.3e+03	6.7e+05	–
5	6.6e+02	4.0e+05	–
7	−1.2e+02	2.9e+05	–
9	1.2e+01	2.2e+05	–
11	3.6e+00	1.8e+05	–
13	−3.2e+00	1.5e+05	–
15	1.3e+00	1.3e+05	Unstable
17	−3.2e–01	1.2e+05	Unstable

The table shows the Fourier coefficients at $x = 0$, and the coefficients are zero for even n because the load is symmetric around $x = L/2$ (see Exercise 9.7). (Any other

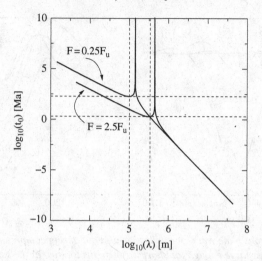

Figure 9.16. *The characteristic time t_0 plotted as a function of the wavelength for two different compressive forces. (See the text for details.)*

position than $x = 0$ shows a similar behavior of the Fourier coefficients.) The first Fourier coefficient is $a_1 \approx 3500$ m, but the first unstable coefficient is only $a_{15} = 1.3$ m. The initial deflection is almost dying out because the unstable Fourier coefficients are small, and they need a long time to grow large.

If the compressive force is multiplied by 10 we get the deflections shown in Figure 9.15b. The first unstable Fourier coefficient is now $a_5 = 660$ m. The deflection is not dying out in this case, because of the large unstable Fourier coefficients that are growing. The unit time is $t_u = 1.8$ Ma because the force is now ten times larger than in Figure 9.15a.

In nature there are always small deflections with small wavelengths. These deflections will grow large under compression after a sufficiently long time, assuming that the plate behaves viscously. The condition for the deflections to grow large is $t \gg t_0$, where $t_0 = t_0(\lambda)$ is the characteristic time for the wavelength λ. Figure 9.16 shows the characteristic time $|t_0|$ as a function of the wavelength λ for the two cases in Figure 9.15. The minimum of t_0 in the regime $\lambda < \lambda_s$ is marked with dashed lines (see Exercise 9.17).

Exercise 9.16 Show that the time-dependent deflection (9.142) is a solution of equation (9.140) for a viscous plate when the initial deflection of the plate is $w(x) = w_0 \cos(kx)$.

Solution: The equation is solved by trying a solution of the same form as the initial deflection,

$$w(x, t) = Y(t) \cos(kx). \tag{9.149}$$

This method of solving a partial differential equation is called separation of variables because the solution is written as a product of a function of only t and another function of only x. When function (9.149) is inserted into equation (9.140) we get

$$\left(D_v k^4 \dot{Y} - F k^2 Y + \Delta \varrho g Y\right) \cos(kx) = 0 \tag{9.150}$$

which can be written as

$$\frac{\dot{Y}}{Y} = \frac{F k^2 - \Delta \varrho g}{D_v k^4} = \frac{1}{t_0(k)}. \tag{9.151}$$

This expression is slightly rewritten as

$$\frac{dY}{Y} = \frac{dt}{t_0} \tag{9.152}$$

and integrated, which gives the function $Y(t) = Y_0 e^{t/t_0}$. The integration factor Y_0 is given by the initial condition (at $t = 0$), and becomes $Y_0 = w_0$.

Exercise 9.17 The characteristic time t_0 has a minimum in the regime $\lambda < \lambda_s$ (see Figure 9.16). Show that the minimum is

$$t_{0,\min} = \frac{4 D_v \Delta \varrho g}{F^2} \quad \text{for} \quad \lambda_{0,\min} = \frac{\lambda_s}{\sqrt{2}}. \tag{9.153}$$

Exercise 9.18 Show that the characteristic time $t_0(k)$ as function of the wave number can be approximated as follows:

$$t_0(k) \approx \begin{cases} -\dfrac{D_v k^4}{\Delta \varrho g}, & F \ll \Delta \varrho g \\[2mm] \dfrac{D_v k^2}{F}, & F \gg \Delta \varrho g. \end{cases} \tag{9.154}$$

Figure 9.16 shows this approximation by the straight lines at each side of wavelength λ_s.

9.11 Further reading

Turcotte and Schubert (1982) has a chapter that covers the theory of plates and flexure, where the equation for flexure of a plate is derived and solved. The solutions are applied to a variety of cases.

Watts (2001) is dedicated to isostasy, flexure and gravity. The book begins with two chapters on the history of how these concepts developed, before the theory is expanded, applied and discussed in the next chapters. The book has a rich set of case studies from all continents, and it is equipped with an extensive list of references to the work done in the field.

Chapters 8 and 9 in Nadai (1963) are devoted to the mathematical treatment of flexure of both elastic plates and viscoelastic plates. The mathematical treatment is complete, with a large set of examples and applications. A number of the applications are geological, like folding of viscoelastic layers under compression, and the viscoelastic rebound of Fennoscandia after the last deglaciation.

10

Gravity and gravity anomalies

The gravitational acceleration is often taken to be the constant $g = 9.8 \text{ m s}^{-2}$, although this is not quite accurate. The value of g depends on where on the planet it is measured, because the Earth is not a perfect sphere and also because the Earth is rotating. There are also small regional and local variations in g due to density variations in the subsurface. These small variations in g can be measured with a great deal of precision, and they are an important source of information about the distribution of mass (or density) in the subsurface. Figure 10.1 shows an example of a gravity anomaly measured along the sea surface (free-air gravity). The sea bed is smooth and cannot explain the observed gravity. The increased gravity turns out to be caused by a ridge of crustal rocks with a density larger than the surrounding sedimentary rocks.

Differences in g over mountains are closely linked to the concept of isostasy. Measurements show that g is less in high mountain areas than close to sea level. One might easily have guessed the opposite – that gravity would be higher in the mountains since they are large masses of rock. We have seen that isostatic equilibrium means that there is the same mass in all columns down to the same depth in the ductile mantle. Each column is simply floating on the ductile mantle, and a column in the mountains therefore needs deep crustal roots of "low" density to float high compared to areas close to sea level. The mass of the mountains is compensated by a mass deficiency in the crust below. These deep crustal roots are precisely what is measured by the lowered values for g in the mountains compared with areas close to sea level.

We have seen that mountains or peaks with a small lateral extent are only partly supported by buoyancy of the ductile mantle, because they are also supported by the elastic strength of the lithosphere. In this chapter will see how the degree of isostatic equilibrium is reflected by variations in g.

10.1 Newton's law of gravity

The gravitational attraction between two masses is given by Newton's law. It says that the attractive force between the masses m and M a distance r apart is

$$F = G \frac{m M}{r^2}$$

(10.1)

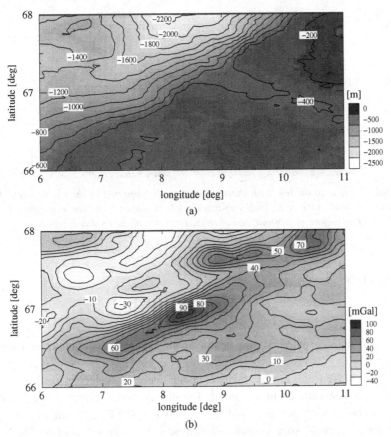

Figure 10.1. *(a) Water depth over a part of the Lofoten ridge, Norway. (b) Free-air gravity over the area in (a) reveals the ridge as a gravity anomaly. The ridge is basement rock that pierces through sediments with a substantially lower density.*

where G is the *universal gravitational constant* $G = 6.67 \cdot 10^{-11}$ m³ kg⁻¹ s⁻². The gravitational acceleration of an object, for example a piece of rock, with mass m falling towards the Earth follows from Newton's second law of mechanics. It says that the acceleration is the force divided by the mass:

$$g = \frac{F}{m} = G\,\frac{M}{r^2} \qquad (10.2)$$

where M is the mass of the Earth. Force and acceleration are vectors, because they are quantities with a direction. The gravity of a point mass M on the point mass m is directed from m to M. The gravitational force is written with direction using the unit vector $\mathbf{n}_r = \mathbf{r}/r$, where $\mathbf{r}$ is the vector from M to m, and where $r = |\mathbf{r}|$. Newton's law is then

$$\mathbf{F} = -G\,\frac{Mm}{r^2}\,\mathbf{n}_r \qquad (10.3)$$

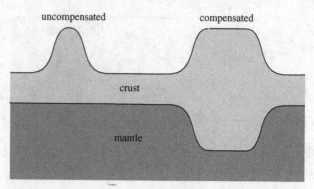

Figure 10.2. *The load to the left, with short wavelength, is supported by the elastic strength of the lithosphere. The load to the right, with long wavelength, is supported by buoyancy from the ductile mantle, and it is therefore in isostatic equilibrium. The mass of the load is compensated by a mass deficiency from low-density crustal roots.*

and the acceleration is the vector

$$\mathbf{g} = -G\,\frac{M}{r^2}\,\mathbf{n}_r.\qquad(10.4)$$

There is also a force of equal size acting in the opposite direction from mass m to mass M.

The masses in Newton's law of gravity are point masses, which is an idealization where all the mass is concentrated in one point. The distance r in Newton's law is therefore between two points with masses m and M, respectively. The Earth is almost a point mass at the length scale of the solar system and Newton's law of gravity applies. The situation might be different on the surface of the Earth, because the Earth is then a mass with a large lateral extent. The question is now how we can apply Newton's law when the mass occupies a very large volume. The acceleration of gravity from a body like the Earth can be obtained by dividing it into a large (infinite) number of small (infinitesimal) cells and adding the contribution from each cell. The contribution to the gravity from one single (small) cell a distance r away is

$$\mathbf{d}g = G\,\frac{\varrho\,dV}{r^2}\,\mathbf{n}_r\qquad(10.5)$$

where the mass of the cell is $dm = \varrho\,dV$. The acceleration of gravity from the entire body of volume V is then the sum of the contributions from all cells in the volume, which is the integral over the entire volume V:

$$\mathbf{g} = G\int_V \frac{\varrho}{r^2}\,\mathbf{n}_r\,dV.\qquad(10.6)$$

An interesting example of this integral is for a spherical object like the Earth, when the density is only dependent on the distance r from the center of the sphere, as shown

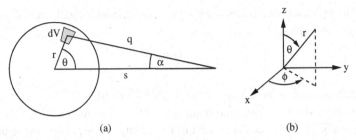

Figure 10.3. *(a) The gravity from a small volume element dV of a sphere. (b) Spherical coordinates. See Exercise 10.2.*

in Figure 10.3. The integral is carried out using spherical coordinates, as shown in Exercise 10.2, and it is

$$g = G\frac{M}{r^2}, \quad \text{where} \quad M = \int_V \varrho\, dV. \tag{10.7}$$

The gravity from a sphere with radially dependent density, $\varrho = \varrho(r)$, is simply the same as if all the mass is placed at the center of the sphere. In other words, the gravity from such a spherical body acts as if it was a point mass. Newton's law (10.2) can be applied to the Earth as if it is a point mass (with all its mass placed at the center), under the assumption that it has a spherically symmetric density distribution. As shown in Exercise 10.2, the gravity is also independent of a particular radial density distribution as long the total mass of the sphere is constant.

The mass and the average density of the Earth can be found from equation (10.2) using the gravitational constant G, the gravitational acceleration at the surface g and the radius of the Earth r. The average density of the Earth is then

$$\varrho = \frac{3g}{4\pi Gr} = 5500 \text{ kg m}^{-3}. \tag{10.8}$$

This is a much larger density than what is found for rocks at the Earth's surface, which normally have densities in the range 2000 kg m^{-3} to 3000 kg m^{-3}. The interior of the Earth must therefore be made of quite dense-material.

Exercise 10.1 Let ϱ be the average density of the Earth and a the radius of the Earth. Show that the relative increase in the gravitational acceleration is

$$\frac{\Delta g}{g} = \left(\frac{r}{a}\right)^3 \frac{\Delta \varrho}{\varrho} \tag{10.9}$$

when the inner part r ($r < a$) of the Earth gets its density increased by $\Delta \varrho$.

Exercise 10.2 Show that gravity outside a sphere with a radial density distribution $\varrho(r)$ is the same as if all the mass is placed at the center of the sphere.

Solution: The gravitational attraction dF on a mass m from a small volume element dV of the sphere with mass $dM = \varrho\, dV$ is

$$dF' = \frac{Gm\, dM}{q^2} \tag{10.10}$$

where q is the distance to the volume element, see Figure 10.3. The distance to the center of the sphere is s. The force from the small mass dF' is directed parallel to the side q of the triangle in Figure 10.3. The component of the gravity that is parallel to the line s is the projection of dF' on line s:

$$dF = \frac{Gm\, dM}{q^2} \cos\alpha. \tag{10.11}$$

The gravity F from the sphere is found by integration of equation (10.11) over the entire sphere. Spherical coordinates are the natural choice for the task, where the volume element is then $dV = r^2 \sin\theta\, d\theta\, d\phi\, dr$. The angle α can be expressed by the length of the sides q, s and r using the law of cosines,

$$r^2 = q^2 + s^2 - 2sq\cos\alpha \quad \text{or} \quad \cos\alpha = \frac{q^2 + s^2 - r^2}{2sq}. \tag{10.12}$$

The integral then becomes

$$F = 2\pi Gm \int_0^a \int_0^\pi \frac{q^2 + s^2 - r^2}{2sq^3} r^2 \sin\theta\, d\theta\, dr \tag{10.13}$$

where the radius of the sphere is a. The factor 2π comes from integration over ϕ. The angle θ can also be expressed by the sides q, s and r using the law of cosines, since we have

$$q^2 = r^2 + s^2 - 2sr\cos\theta \quad \text{or} \quad \cos\theta = \frac{r^2 + s^2 - q^2}{2sr}. \tag{10.14}$$

The integration over θ can be replaced by an integration over q, where the second version of equation (10.14) gives the differential

$$\sin\theta\, d\theta = \frac{q\, dq}{sr} \tag{10.15}$$

when s and r are kept constant. The integration limits in the new integration variable q are $q = r - s$ and $q = r + s$ for $\theta = 0$ and $\theta = \pi$, respectively. The integral (10.13) is

$$F = \frac{\pi Gm}{s^2} \int_0^a \varrho(r)r \int_{s-r}^{s+r} \left(1 + \frac{s^2 - r^2}{q^2}\right) dq\, dr \tag{10.16}$$

where the integral over q becomes $4r$, and the force F is therefore

$$F = \frac{Gm}{s^2} \int_0^a 4\pi \varrho(r)r^2 dr \tag{10.17}$$

where the integral over r becomes the total mass of the sphere. A constant density gives the mass $M = 4\pi a^3 \varrho/3$, which is a check of the integral. We see that the gravity from the sphere is constant and independent of the density distribution $\varrho(r)$ as long as the total mass of the sphere is constant.

10.2 Potential energy and the potential

We will first now show that g is the gradient of a potential, which implies that gravity is what is called a conservative field. We start by calculating the work done by moving a mass m in the gravitational field from a (point) mass M. Work is only done by moving in the radial direction, and the total work done by moving from the radial position r_1 to the radial position r_2 is

$$W = \int_{r_1}^{r_2} F\, dr = \int_{r_1}^{r_2} \frac{GmM}{r^2} dr = -\frac{GmM}{r_2} + \frac{GmM}{r_1}. \tag{10.18}$$

The work done on the mass m is a difference in the potential energy

$$E = -\frac{GmM}{r} \tag{10.19}$$

between the radii r_2 and r_1, respectively. The potential is defined as energy per mass, $U = E/m$, and it is

$$U = -\frac{GM}{r}. \tag{10.20}$$

The value of the potential cannot be measured. It is only differences in the potential that are observed. We are therefore free to choose any reference value for the potential by adding a constant to the left side of (10.20). A convenient choice is to keep the potential as it is in equation (10.20), where $U = 0$ at $r = \infty$. The potential is then a negative value that increases with increasing r towards 0 at an infinite distance away from the mass M.

Equation (10.20) gives the potential from a point mass. The potential from any mass distribution is obtained by dividing the mass into a large (infinite) number of small (infinitesimal) cells. The mass of a cell with volume dV is $dm = \varrho\, dV$ when ϱ is the density. The cells become point masses in the limit where $dV \to 0$. The potential in position $\mathbf{r}_0$ from one cell in position $\mathbf{r}$ is therefore

$$dU(\mathbf{r}_0) = -G \frac{\varrho(\mathbf{r})\, dV}{|\mathbf{r} - \mathbf{r}_0|}. \tag{10.21}$$

The potential from the entire mass is the sum of the potentials of all point masses, which is an integration over the entire volume

$$U(\mathbf{r}_0) = -G \int_V \frac{\varrho(\mathbf{r})}{|\mathbf{r} - \mathbf{r}_0|}\, dV(\mathbf{r}) \tag{10.22}$$

where $dV(\mathbf{r})$ denotes the volume of the cell in position $\mathbf{r}$. The calculation of the potential is often more straightforward than the corresponding calculation of the gravitational acceleration (10.6), because the potential is a scalar while the acceleration is a vector. Furthermore, the acceleration is found from the gradient of the potential, $\mathbf{g} = -\nabla U$, as we will soon see.

Exercise 10.3 Show that the potential from a spherical density distribution is the same as if all the mass is placed at the center.

Solution: The sphere can be divided into a large (infinite) number of small (infinitesimal) cells, and the potential for the entire sphere is the sum of the potentials from each cell. The mass of a small cell is $dm = \varrho \, dV$, and the potential from the mass dm is

$$dU = G \frac{\varrho(r)}{q} dV \tag{10.23}$$

where q is the distance to the cell and where r is the distance from the center of the sphere to the cell, as shown in Figure 10.3. Integration over the sphere gives the potential from the entire mass, and the integration is simplified when using spherical coordinates, where $dV = r^2 \sin\theta \, d\phi \, dr \, d\theta$. The potential from the entire sphere is then

$$U = 2\pi G \int_0^a \int_0^\pi \frac{\varrho}{q} r^2 \sin\theta \, d\theta \, dr. \tag{10.24}$$

The distance q can be expressed by the radius r, the distance to the center of the sphere s and the angle θ using the law of cosines,

$$q^2 = r^2 + s^2 - 2sr \cos\theta. \tag{10.25}$$

A change of integration variable from θ to $u = \cos\theta$ gives

$$U = 2\pi G \int_0^a \int_{-1}^1 \frac{\varrho(r)}{(r^2 + s^2 - 2rsu)^{1/2}} r^2 \, du \, dr. \tag{10.26}$$

The integration over u is carried out using that $\int dx/\sqrt{ax+b} = (2/a)\sqrt{ax+b}$, and the potential becomes

$$U = \frac{2\pi G}{s} \int_0^a \varrho(r) \left((s+r) - (s-r) \right) r \, dr$$

$$= \frac{G}{s} \int_0^a 4\pi \varrho(r) r^2 \, dr \tag{10.27}$$

where the total mass of the sphere is $M = \int_0^a 4\pi \varrho(r) r^2 \, dr$. The potential from a sphere with a radial density distribution is the same as from the corresponding point mass. The calculation of the potential from a sphere is more straightforward than for the gravity as shown in Exercise 10.2, because the potential is a scalar quantity. This is an alternative and simpler way to show that Newton's law for gravity (10.1) applies for a spherically symmetric mass distributions.

10.3 Conservative fields

We notice that a difference in potential energy between two positions is only dependent on the radii of the two positions. Any path connecting the same end-points is therefore separated by the same potential energy. Such a force field, where the difference in potential energy is independent of the chosen path, is called a *conservative* field. An important property of a conservative field, like gravity, is that the force **F** is proportional to the gradient of the potential energy E:

$$\mathbf{F} = -\nabla E. \tag{10.28}$$

This property is shown by looking at the difference in potential energy between two nearby points $\mathbf{r}_1$ and $\mathbf{r}_2$, separated by a distance $\mathbf{ds} = \mathbf{r}_2 - \mathbf{r}_1$. The work done by going from $\mathbf{r}_1$ to $\mathbf{r}_2$ for a conservative field is

$$-\mathbf{F} \cdot \mathbf{ds} = E(\mathbf{r}_2) - E(\mathbf{r}_1) \tag{10.29}$$

where a minus sign has been added to the left-hand side because **F** and **ds** have opposite positive directions. The difference in potential energy can also be written

$$E(\mathbf{r}_2) - E(\mathbf{r}_1) = \nabla E \cdot \mathbf{ds} \tag{10.30}$$

because **ds** is small. The difference between the two expressions (10.29) and (10.30) is

$$(\mathbf{F} + \nabla E) \cdot \mathbf{ds} = 0 \tag{10.31}$$

for any small distance **ds**, which implies that $\mathbf{F} = -\nabla E$. Furthermore, a conservative force field has zero curl, $\nabla \times \mathbf{F} = 0$, because we have from calculus that $\nabla \times \nabla E = 0$ for any well-behaved function E.

The mass m has the gravitational acceleration $\mathbf{g} = \mathbf{F}/m$ and the potential is $U = E/m$. Since we have that $\mathbf{F} = -\nabla E$ it follows that

$$\mathbf{g} = -\nabla U. \tag{10.32}$$

The quantities **g** and U are simply the quantities **F** and E divided by m, respectively.

10.4 Gauss's law

This section shows a useful relationship called Gauss's law. It says that the integral of $\mathbf{g} \cdot \mathbf{n}$ over a closed surface S yields the mass inside the surface times $-4\pi G$,

$$\oint_S \mathbf{g} \cdot \mathbf{n} \, dS = -4\pi G M \tag{10.33}$$

where the scalar dS is a small area of the surface and the vector **n** is the outward unit vector on the area dS. (The sign $\oint$ denotes integration over a closed surface.) Gauss's law is first validated for a spherical surface around a point mass M as shown in Figure 10.4.

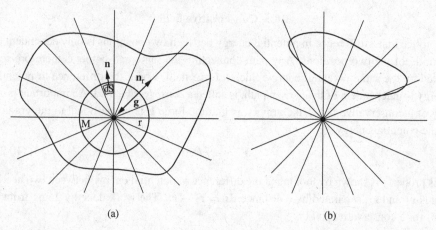

Figure 10.4. *(a) The figure shows the field lines that radiate from a point mass. Any closed surface around the point mass is intersected by the same number of field lines. (b) The volume does not contain the point mass. The same number of field lines that go into the volume also come out of the volume.*

The acceleration g is $\mathbf{g} = -(M/r^2)\mathbf{n}_r$ along a spherical surface with radius r and $\mathbf{g}$ is everywhere normal to the surface. We therefore have

$$\oint_S \mathbf{g} \cdot \mathbf{n} \, dS = -\frac{GM}{r^2} 4\pi r^2 = -4\pi GM \tag{10.34}$$

where we have used that $4\pi r^2$ is the surface area of a sphere with radius r. Gauss's law is also valid for any closed surface around a point mass M (not just a sphere), because the integral

$$\oint_S \frac{\mathbf{n}_r \cdot \mathbf{n}}{r^2} \, dS = \oint d\Omega = 4\pi \tag{10.35}$$

is simply the integration of the solid angle $d\Omega = \mathbf{n}_r \cdot \mathbf{n} \, dS/r^2$ over the closed surface. A solid angle is a generalization of the normal angle from 2D to 3D (see Exercise 10.4). The solid angle $d\Omega$ is equal to the surface area of the unit sphere ($r = 1$) that is bounded by the same field lines as the area dS (see Figure 10.4). The integral (10.35) for a closed surface therefore becomes the surface area of the unit sphere, which is 4π. Another example of a surface element dS and the corresponding solid angle $d\Omega$ is shown in Figure 10.5. The unit used for a solid angle is *steradian*.

Gauss's law therefore applies to one point mass. The generalization to any mass (or density) distribution is by means of superposition. Any mass distribution can be thought of as a large number of small point masses, and the contributions of each point mass are added together:

$$\sum_i \oint_S \mathbf{dg}_i \cdot \mathbf{n} \, dS = -4\pi G \sum_i dM_i. \tag{10.36}$$

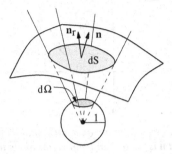

Figure 10.5. *The surface area dS a distance r away from the origin corresponds to the surface area $d\Omega$ on the unit sphere. The surface area $d\Omega$ is the solid angle.*

The gravitational acceleration of the point mass dM_i is $\mathbf{dg}_i$. The acceleration of all the point masses is $\mathbf{g} = \sum_i \mathbf{dg}_i$, and the total mass inside the surface is $M = \sum_i dM_i$. Gauss's law is shown by bringing the summation on the left-hand side through the integration sign.

It is now straightforward to show that the gravitational acceleration from a sphere with a radial mass distribution is the same as the acceleration from the corresponding point mass. The spherical symmetric mass distribution implies that the acceleration is only dependent on r and since equation (10.34) applies for any mass distribution, not only a point mass, it follows that $g(r) = GM/r^2$. This simple application of Gauss's law could have saved us the trouble of calculating the gravitational force and potential outside a sphere with radial density distribution (as done in Exercises 10.2 and 10.3), to show that Newton's law of gravitation applies as if the sphere is a point mass.

Exercise 10.4 Show that $\oint_s \mathbf{n}_r \cdot \mathbf{n}/r\, ds = 2\pi$ for any closed surface in 2D, where r is the distance from the origin, $\mathbf{n}_r$ is the radial unit vector and $\mathbf{n}$ is the outward unit vector to the curve.

10.5 Bouguer's formula for gravity due to a horizontal layer

A useful application of Gauss's law is the derivation of Bouguer's formula for the gravitational acceleration from a horizontal layer of infinite extent. Figure 10.6 shows a sketch of a layer and the gravitational acceleration due to it. The acceleration is normal to the horizontal sides because the layer has an infinite extent. Gauss's law says that the scalar product $\mathbf{g} \cdot \mathbf{n}$ integrated over a closed surface S is $-4\pi GM$, where M is the mass inside the surface S. Figure 10.6 shows how the closed surface S follows the layer, and the surface integral is

$$\oint_s \Delta\mathbf{g} \cdot \mathbf{n}\, ds = -\Delta g A - \Delta g A = -2\Delta g A \tag{10.37}$$

where Δg is the acceleration due to the layer. The acceleration $\Delta\mathbf{g}$ points inwards on both horizontal sides, and the normal vector $\mathbf{n}$ points outwards. The scalar product $\Delta\mathbf{g} \cdot \mathbf{n}$ is therefore $-\Delta g$ along both sides (where $\Delta g > 0$). The acceleration is parallel to the vertical

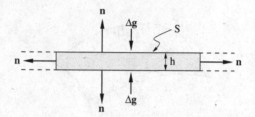

Figure 10.6. *The gravitational acceleration around a horizontal plate of infinite extent. The surface S encloses the shaded mass, and* **n** *is the outward unit vector along the surface S.*

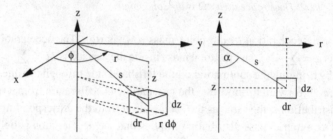

Figure 10.7. *Cylinder coordinates.*

sides of S where $\Delta \mathbf{g} \cdot \mathbf{n} = 0$. The mass inside the surface S is $M = \varrho A h$, and Gauss's law gives

$$- 2\Delta g A = -4\pi G \varrho A h \tag{10.38}$$

or

$$\Delta g = 2\pi G \varrho h \tag{10.39}$$

which is Bouguer's formula. It applies regardless of the thickness h. As an example, let us find the gravitational acceleration from a layer of thickness $h = 1$ km and density $\varrho = 2500 \text{ kg m}^{-3}$. The acceleration is $\Delta g = 2\pi G \varrho h = 0.001 \text{ m s}^{-2}$. This is roughly 10^{-4} of the gravity from the Earth. Small values of the gravitational acceleration are normally measured in the unit mGal (milligal in honor of Galileo), where 1 mGal is 10^{-5} m s^{-2}. The acceleration from the plate in this example is therefore 100 mGal. Bouguer's formula will later be used to correct gravity measurements for the mass between the observation and a reference height.

Note 10.1 Bouguer's formula is also straightforward to find by integration using cylinder coordinates (r, ϕ, z), see Figure 10.7. The origin is placed at the surface of the layer, and the gravitational acceleration from a small cell a distance $s = (r^2 + z^2)^{1/2}$ away is

$$dg = \frac{G\,dm}{s^2} \quad \text{and} \quad dm = \varrho\,dr\,(r\,d\phi)\,dz, \tag{10.40}$$

see Figure 10.7, where the volume of the cell is $dV = dr\,(r\,d\phi)\,dz$. The component of the gravitational acceleration in the vertical direction is

$$dg_z = dg\cos\alpha \quad \text{where} \quad \cos\alpha = \frac{z}{s}. \tag{10.41}$$

The vertical gravitational acceleration is the contribution for all mass elements in the infinite plate, which is the integral

$$g_z = G\varrho \int_0^{2\pi} \int_{-h}^{0} \int_0^{\infty} \frac{zr}{(r^2+z^2)^{3/2}}\,dr\,dz\,d\phi. \tag{10.42}$$

The z-axis is pointing upwards and the plate is in the interval $-h$ to 0. The integration over ϕ is just the factor 2π, and the integration over r becomes

$$\int_0^{\infty} \frac{zr}{(r^2+z^2)^{3/2}}\,dr = \left[-\frac{z}{(r^2+z^2)^{1/2}}\right]_0^{\infty} = 0 - \left(-\frac{z}{z}\right) = 1. \tag{10.43}$$

We are then left with

$$g_z = 2\pi G\varrho \int_{-h}^{0} 1\,dz = 2\pi G\varrho h \tag{10.44}$$

which is Bouguer's formula.

10.6 Laplace's and Poisson's equations for the potential

Gauss's law (10.33) can be rewritten using volume integrals. The surface integral in Gauss's law is converted to a volume integral by the divergence theorem, which gives

$$\oint_S \mathbf{g}\cdot\mathbf{n}\,dS = \int_V \nabla\cdot\mathbf{g}\,dV \tag{10.45}$$

for any vector field $\mathbf{g}$. The mass on the right-hand side of equation (10.33) can also be expressed as a volume integral by means of the density $M = \int \varrho\,dV$, and Gauss's law becomes

$$\int_V \nabla\cdot\mathbf{g}\,dV = \int \varrho\,dV \quad \text{or} \quad \nabla\cdot\mathbf{g} = -4\pi G\varrho. \tag{10.46}$$

The gravitational acceleration is minus the gradient of the potential, $\mathbf{g} = -\nabla U$, and we therefore get that $\nabla\cdot\nabla U = 4\pi G\varrho$ or

$$\nabla^2 U = 4\pi G\varrho \tag{10.47}$$

which is Poisson's equation. The gravitational field is the solution of Poisson's equation with relevant boundary conditions. Poisson's equation becomes Laplace's equation

$$\nabla^2 U = 0 \tag{10.48}$$

when the density is zero. We have already seen that the integral (10.22) gives the potential U, and Exercise 10.6 shows that it is a solution of Poisson's equation. One might wonder why we need Poisson's equation for the potential when we already have a solution. The reason is that we don't always know the density distribution, which is required in the solution (10.22), but instead we might for example know the potential along the xy-plane. The potential above the plane is then found as a solution of Laplace's equation using the potential at the xy-plane as a boundary condition.

Note 10.2 Newton's law of gravitation applies to point masses, and a point mass is a finite mass placed at a point, which by definition has no volume. The normal definition of density doesn't make sense when something has a zero volume. Yet it is still possible to work with a point mass as if it has a density, by using the Dirac δ-function. The δ-function has two properties: $\delta(\mathbf{r})$ is zero everywhere except at $\mathbf{r} = 0$, and the integral over any region V that includes $\mathbf{r} = 0$ is $\int_V \delta(\mathbf{r}) \, dV = 1$. The "density" of a point mass M placed in position $\mathbf{r}_0$ is therefore

$$\varrho(\mathbf{r}) = M \, \delta(\mathbf{r} - \mathbf{r}_0). \tag{10.49}$$

The integral (10.22) gives the potential in position $\mathbf{r}$, and we have

$$U(\mathbf{r}) = -GM \int_V \frac{\delta(\mathbf{r}' - \mathbf{r}_0)}{|\mathbf{r}' - \mathbf{r}|} \, dV(\mathbf{r}') = -\frac{GM}{|\mathbf{r} - \mathbf{r}_0|} \tag{10.50}$$

where V is any volume sufficiently large to include both positions $\mathbf{r}$ and $\mathbf{r}_0$. We now have that the density (10.49) gives the potential (10.50). The potential is also a solution of Poisson's equation (10.47) with the "density" of a point mass, and we therefore have

$$\nabla^2 \left(\frac{1}{|\mathbf{r} - \mathbf{r}_0|} \right) = -4\pi \delta(\mathbf{r} - \mathbf{r}_0). \tag{10.51}$$

That this really is the case is checked in Exercise 10.5.

Exercise 10.5
(a) Show that

$$\nabla^2 \left(\frac{1}{r} \right) = 0 \tag{10.52}$$

for $r \neq 0$ by differentiation.
(b) Show that

$$\nabla^2 \left(\frac{1}{r} \right) = -4\pi \tag{10.53}$$

for $r = 0$ using a volume integral that covers $r = 0$.
Solution: (a) A direct calculation shows that $\nabla^2(1/r) = 0$ for $r \neq 0$, where the distance from the origin is $r = (x^2 + y^2 + z^2)^{1/2}$. We get that $dr/dx = x/r$ and that $d^2r/dx^2 = -(r^2 - 3x^2)/r^5$, and similar calculations of d^2r/dy^2 and d^2r/dz^2 then give that $\nabla^2(1/r) = 0$.

(b) The case $r = 0$ is shown by integration over a volume that includes the origin. The left-hand side then becomes

$$\int_V \nabla^2 \left(\frac{1}{r}\right) dV = \oint_S \nabla \left(\frac{1}{r}\right) \cdot \mathbf{n}\, dS = -\oint_S \frac{\mathbf{n}_r \cdot \mathbf{n}}{r^2} dS = -\oint d\Omega = -4\pi \quad (10.54)$$

where the volume integral was converted to a surface integral using the divergence theorem. We have used that $\nabla r = \mathbf{n}_r$ (the unit vector in the direction of $\mathbf{r}$), and that Ω is a solid angle, see equation (10.35). Figure 10.4 shows the difference between $r = 0$ inside the volume and $r = 0$ on the outside of the volume.

Exercise 10.6 Show that the potential

$$U(\mathbf{r}) = -G \int_V \frac{\varrho(\mathbf{r}')}{|\mathbf{r}' - \mathbf{r}|} dV(\mathbf{r}') \quad (10.55)$$

is a solution of Poisson's equation. Hint: use equation (10.51).

10.7 Gravity from a buried sphere

A buried object with a different density than the surrounding rock leads to a weak perturbation of the gravitational acceleration at the surface. This perturbation is now calculated for some simple objects. It is important to keep in mind that the gravity from different bodies can be superposed. We can therefore compute the gravity for one body at a time and add the contribution from all bodies to obtain the total effect. Another point is that perturbation of gravity from a buried body is caused by the density difference between the body and the surrounding rock. The buried body can be viewed as two overlapping bodies, one with the density of the surrounding rock and another with the density difference. It is the latter object that perturbates the gravity.

We have already shown that the gravitational acceleration from a spherical object with a constant density is the same as for a point mass

$$g = \frac{GM}{r^2} \quad (10.56)$$

where $M = \Delta\varrho\,(4/3)\pi a^3$ is the excess mass of a sphere with radius a. The difference in density between the sphere and the surrounding rock is $\Delta\varrho$. The vertical component of the acceleration is $g_z = g \cos\alpha$ where $\cos\alpha = h/r$ (see Figure 10.8a). The vertical component g_z in position (x, y) in the xy-plane from a sphere at depth h is therefore

$$g_z(x, y) = \frac{GMh}{(x^2 + y^2 + h^2)^{3/2}} \quad (10.57)$$

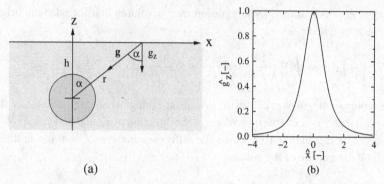

Figure 10.8. *(a) The gravity from a buried sphere. (b) The vertical component of the gravitational acceleration as a function of the horizontal distance from the sphere.*

because the radius is $r = (x^2 + y^2 + h^2)^{1/2}$. The gravity anomaly from the sphere is at its maximum at the origin, where it is

$$g_{z,\max} = \frac{GM}{h^2}. \tag{10.58}$$

Figure 10.8b shows how the gravity depends on distance x. The acceleration is scaled by its maximum value, and the distance along the x-axis is scaled by the depth h,

$$\hat{g}_z(\hat{x}) = \frac{g_z}{g_{z,\max}} = \frac{1}{(\hat{x}^2 + 1)^{3/2}}. \tag{10.59}$$

The expression for $\hat{g}_z(\hat{x})$ shows that g_z is reduced by a factor $2^{-3/2} \approx 35\%$ a distance h away from the origin, and that it is reduced to less than 10% more than $2h$ away from the origin (see Figure 10.8).

If a gravity anomaly similar to the one shown in Figure 10.8 is observed then we know that it may be a buried sphere, but can we say anything more? If it is a sphere, it tells us how deep the sphere is buried, which is the distance between the position of the maximum and where the acceleration is reduced by $\sim 35\%$. The maximum tells us how large the excess mass M is, but there is no information that gives us the radius of the sphere. It is impossible to measure the radius of the sphere from its gravitational field, because it behaves like a point mass.

Exercise 10.7 At what distance from the origin is the gravitational acceleration reduced to one-half?

10.8 Gravity from a horizontal cylinder

A gravitational acceleration from a buried horizontal and infinitely long cylinder is found in a similar way to the gravity from a buried sphere. Gauss's law

$$\oint \mathbf{g} \cdot \mathbf{n} \, dS = -4\pi GM \tag{10.60}$$

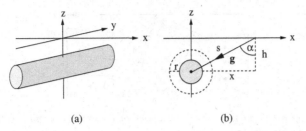

(a)   (b)

Figure 10.9. *A buried horizontal cylinder.*

is a simple tool to find the gravity around a cylinder as well as a sphere. It is assumed that the density is only dependent on the radius of the cylinder. The vector field **g** is then radial towards the center of the cylinder. We obtain the gravitational acceleration, a distance r away from the center axis, by considering a cylindrical surface. For a distance l along the axis of the cylinder we have

$$\oint \mathbf{g} \cdot \mathbf{n}\,dS = -g2\pi rl \tag{10.61}$$

where g is the absolute value of the gravity and r is the radius of the cylinder. The vector **g** is everywhere normal to the surface and the surface area is $2\pi rl$. Gauss's law then gives that the acceleration a distance r away from the center is

$$g2\pi rl = 4\pi GM \quad \text{or} \quad g = \frac{2Gm}{r} \tag{10.62}$$

where m is the mass per length along the cylinder, $m = M/l$. We notice that the gravity from a cylinder acts like a line load, where all the mass is placed along the line at the center of the cylinder. There is no information in the expression that tells us anything about the size of the cylindrical mass, only its mass per length. The vertical component of the gravitational acceleration at position x is

$$g_z = \frac{2Gm}{s}\cos\alpha = \frac{2Gmh}{s^2} \tag{10.63}$$

because $\cos\alpha = h/s$, where s is the distance to the center of the cylinder and h is the (vertical) depth of the cylinder (see Figure 10.9). The gravitational acceleration as a function of x is

$$g_z(x) = \frac{2Gmh}{(x^2 + h^2)}. \tag{10.64}$$

The maximum of g_z, which is at $x = 0$, is

$$g_{z,\text{max}} = \frac{2Gm}{h} \tag{10.65}$$

and the scaled version of the gravitational acceleration is

$$\hat{g}_z(\hat{x}) = \frac{g_z}{g_{z,\text{max}}} = \frac{1}{\hat{x}^2 + 1} \tag{10.66}$$

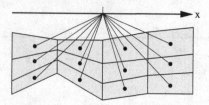

Figure 10.10. *The gravitational acceleration at a point on the surface is approximated by the sum of the line loads through the cells.*

where $\hat{x} = x/h$. We see right away that the acceleration is reduced to one-half at position $x = h$ (or $\hat{x} = 1$) on the surface.

The line mass is useful when computing the 3D gravity for a grid of 2D vertical cross-sections. The simplest approach is to assume that each cell in the cross-section has a line load through its center, where the mass per length of the line load through a cell is $m = \varrho A$, and where A is the area of the cell, see Figure 10.10. A more elaborate approach is to integrate the contribution from a large number of (small) line loads that fill the cells of the cross-section. An example of this approach is shown in the next section.

Exercise 10.8 Show that a line load needs a line density m that is $m > M/2h$ to have a stronger maximum than a spherical load with mass M at the same depth.

10.9 Gravity from a prism with rectangular cross-section

An application of the line load from Section 10.8 is the calculation of the gravitational acceleration from a prism with rectangular cross-section. We will calculate the gravity at the origin from a prism that has a cross-section restricted to the rectangle defined by the point (x_0, z_0) (see Figure 10.11a). Later we will see that we can superpose several such prisms to obtain the gravity from any rectangular prism (with horizontal sides). The rectangular cross-section of the prism is considered to be made of small cells with area $dx\,dz$, and the gravity from each cell is represented by a line load (see Figure 10.11a). Equation (10.63) gives the gravitational acceleration at the origin due to a line load at position (x, z), which is

$$dg_z = \frac{2Gz\varrho\,dx\,dz}{x^2 + z^2} \tag{10.67}$$

where $\varrho\,dx\,dz$ is the mass per unit length (in the y-direction). The integral over the line loads that fill the rectangle is the total gravity

$$g_z = 2G\varrho \int_0^{x_0} \int_0^{z_0} \frac{z}{x^2 + z^2}\,dx\,dz \tag{10.68}$$

which becomes (as shown in Note 10.3)

$$g_z = G\varrho \left(x_0 \ln\left(1 + \left(\frac{z_0}{x_0}\right)^2\right) + 2z_0 \tan^{-1}\left(\frac{x_0}{z_0}\right) \right). \tag{10.69}$$

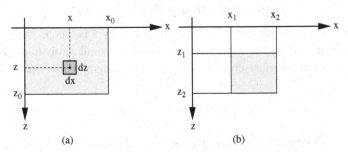

Figure 10.11. (a) The rectangular prism extending from the origin to (x_0, z_0) is made of a (large) number of (small) line loads with mass per length $\varrho\, dx\, dz$. (b) The gravity from the rectangular cross-section (x_1, z_1) to (x_2, z_2) is found by superposition.

The gravity from any rectangular cross-section is obtained by superposition of cross-sections as the one found (with the origin as one corner). The gravitational accelera-tion (10.69) is now written

$$g_z = f(x_0, z_0) \tag{10.70}$$

and the gravity from a rectangular cross-section from (x_1, z_1) to (x_2, z_2) is then

$$g_z = f(x_2, z_2) - f(x_1, z_2) - f(x_2, z_1) + f(x_1, z_1), \tag{10.71}$$

see Figure 10.11b. The gravity from several rectangular cross-sections are the superposition of the gravity from each rectangular cross-section.

The gravity from a rectangular cross-section that extends laterally to infinity is $g_z = G\varrho\pi z_0$ (using that $\ln(1 + (z_0/x_0)^2) \to 0$ and that $\tan^{-1}(x_0/z_0) \to \pi/2$ when $x_0 \to \infty$ in equation (10.69)). Bouguer's formula is then obtained by adding a factor 2, which accounts for the cross-section along the negative x-axis.

Note 10.3 The integration over z in equation (10.68) gives

$$g_z = 2G\varrho \int_0^{x_0} \left(\int_0^{z_0} \frac{du}{2u} \right) dx = G\varrho \int_0^{x_0} \left(\ln(x^2 + z_0^2) - \ln(x^2) \right) dx \tag{10.72}$$

using the substitution $u = x^2 + z^2$, and the integration over x is done using the integrals

$$\int \ln(x^2 + a^2)\, dx = x \ln(x^2 + a^2) + 2a \tan^{-1}\left(\frac{x}{a}\right) - 2x \tag{10.73}$$

$$\int \ln(x^2)\, dx = x \ln(x^2) - 2x. \tag{10.74}$$

The term $x \ln(x^2) = 2x \ln x$ is evaluated at $x = 0$ using l'Hopital's rule, where $\ln x/(1/x)$ is an ∞/∞ expression that approaches 0 when $x \to 0$.

10.10 Gravity from a 2D polygonal body

Most 2D bodies can be quite accurately approximated by a polygon. The gravity at the origin $(0, 0)$ from a polygonal cross-section is found in a similar way to the gravity from a 2D body with a rectangular cross-section. It is the integral of line loads that fills the polygon

$$g = 2G\varrho \iint_A \frac{z}{x^2 + z^2}\, dx\, dz \tag{10.75}$$

where $dm = \varrho\, dx\, dz$ is the mass per unit length in the y-direction, and where the density ϱ is constant inside the area A. Note 10.4 shows that the gravity at the origin then becomes

$$g = 2G\varrho \sum_{n=1}^{N} \frac{b_n}{a_n^2 + 1} \left[\ln\frac{r_{n+1}}{r_n} + a_n \left(\tan^{-1}\frac{x_{n+1}}{z_{n+1}} - \tan^{-1}\frac{x_n}{z_n} \right) \right] \tag{10.76}$$

when the polygon has N corners with coordinates (x_n, z_n) anticlockwise around the perimeter, and where $r_n = (x_n^2 + z_n^2)^{1/2}$ is the distance from the origin to the point n. The coefficients a_n and b_n define the line $X_n(z) = a_n z + b_n$ that interpolates the points (x_n, z_n) and (x_{n+1}, z_{n+1}), as shown in Figure 10.12, and they are therefore

$$a_n = \frac{x_{n+1} - x_n}{z_{n+1} - z_n} \tag{10.77}$$

$$b_n = -a_n z_n + x_n. \tag{10.78}$$

The summation in equation (10.76) assumes that point $N + 1$ is the same as point 1. Other assumptions are that the polygon cannot have horizontal lines, and that the z-coordinate cannot be 0. Expression (10.76) for the gravity from a polygon is often called *Talwani's formula* in honor of Talwani, who together with Worzel and Landisman (Talwani *et al.*, 1959) first presented a version of the formula for computer applications, although the same method was introduced earlier by Hubbert (1948).

Figure 10.13 shows Talwani's formula (10.76) tested on a 2D load with a square cross-section. The result is compared with the gravity from a line load. The discrepancy between the two is reduced to almost zero when both loads represent the same mass per unit length (normal to plot).

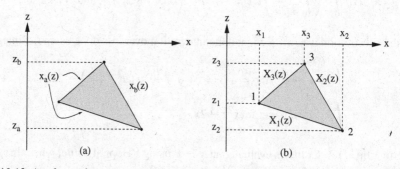

Figure 10.12. *A polygonal cross-section of a 2D body.*

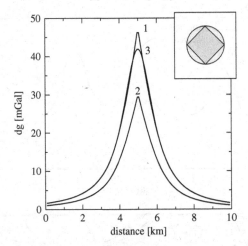

Figure 10.13. *Curve 1 is the gravity from a 2D load with a circular cross-section, and curve 2 is the gravity from a load with a square cross-section and the same density. Curve 3 shows the gravity from the square cross-section when it has the same mass per length (normal to the plane) as the load with a circular cross-section (which implies that its density is increased by a factor 1.57).*

Talwani's formula is the basis for most numerical computations of gravity anomalies in 2D. Figure 10.14 gives an example for the Lofoten ridge from mid-Norway. The geometry of the water, sediments, crust and the mantle are represented by polygons that each are assigned a density. The gravity is computed with Talwani's formula (10.76) for discrete positions along the water surface, by summing the contribution from each polygon. The example in Figure 10.14 shows that the computed gravity is in accordance with the observation. The density distribution is coarse in this case, and the match could have been improved by a refinement of the density model for the sediments and the crust. The usefulness of a simple density model is that it provides a consistency check of the mapped geometry against the observed gravity. It should be mentioned that 3D effects also play a role here, which are not correctly represented by a 2D model. Another point is the non-uniqueness of gravity models. Different density distributions may produce almost the same surface gravity.

There are a few points to note concerning the computation of the gravity, as shown in Figure 10.14. We are interested in variations of the gravity along the water surface. We can obtain these variations by using density differences in Talwani's formula or we can use the actual densities. In the first case we can for instance make density differences with respect to the water, which implies that the contribution from the polygons that represent water become zero. Another issue is boundary effects. In order to have well-behaved boundaries we can elongate the model beyond the vertical sides with laterally long rectangles that match the height of each lithology at the boundaries. These rectangles normally have to be as long as several hundred km depending on the size of the model. Extending the model by rectangles beyond the vertical sides is a simple way to compute a well-behaved gravity.

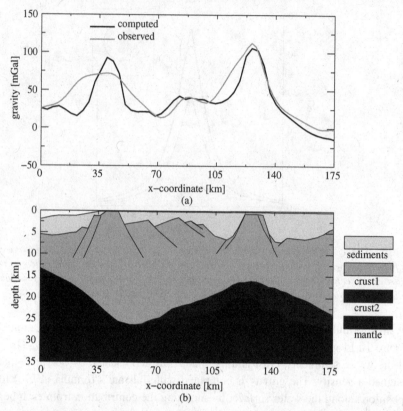

Figure 10.14. *(a) The observed and computed free-air gravity anomaly across the Lofoten ridge.* *(b) A cross-section of the ridge that shows the water, sediments, the crust, the lower crust and the mantle.*

It is also possible to compute the gravity from a 2D body using the fast Fourier transform (FFT) by means of Parker's formula (Parker, 1972). The Fourier transform of gravity can be written as

$$\mathcal{F}(\Delta g) = 2\pi G\varrho \, e^{-kz_0} \sum_{n=1}^{\infty} \frac{k^{n-1}}{n!} \mathcal{F}\left(h_2^n(x) - h_1^n(x)\right) \qquad (10.79)$$

where h_2 and h_1 are the vertical positions of the top and the base of a body with density ϱ. (See Note 10.5 for details.) Parker's formula refers to a coordinate system where the z-axis points upwards, and $\mathcal{F}(\Delta g)$ is the Fourier transform of gravity along a horizontal line a distance z_0 above the x-axis. Setting $z_0 = 0$ gives the Fourier transform of the gravity along the x-axis. We therefore have that the difference between the Fourier transforms of gravity along two different horizontal lines separated a vertical distance z_0 is simply the factor $\exp(-kz_0)$. We will meet this factor again in Section 10.12, where we once more look at how the gravity can be vertically extended from one horizontal line to another. The Fourier transform of $h_2^n(x) - h_1^n(x)$ is computed with FFT for the terms in the series expansion,

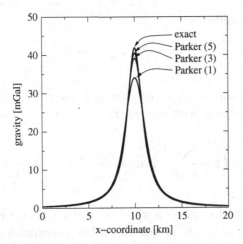

Figure 10.15. *The gravity computed with Parker's formula with 1, 3 and 5 terms are compared with the exact gravity for a line load.*

and the inverse FFT of $\mathcal{F}(\Delta g)$ gives the gravity. Using only the first term in the power series expansion gives often a quite accurate approximation of the gravity. Figure 10.15 compares the exact gravity for a line load with the gravity from Parker's formula with 1, 3 and 5 terms. The line load is represented as a cylinder in the y-direction with a 1 km radius and a density 1000 kg m^{-3}, which touches the xy-plane. It should be mentioned that Parker's formula not only applies for gravity along a line, but also for the gravity over an xy-plane. The derivation given in Note 10.5 is for the general case, where the gravity is found over a horizontal plane from a 3D body.

Note 10.4 The gravity at the origin from the polygon is proportional to the integral

$$I = \int_{z_a}^{z_a} \int_{x_a(z)}^{x_b(z)} \frac{z}{x^2 + z^2} \, dx \, dz \tag{10.80}$$

where the left side and right sides of the polygon are given by the curves $x_a(z)$ and $x_b(z)$, respectively. The base and the top of the polygon along the z-axis are z_a and z_b, respectively, as shown in Figure 10.12a. An integration over x then gives

$$I = \int_{z_b}^{z_a} \left(\tan^{-1} \frac{x_b(z)}{z} - \tan^{-1} \frac{x_a(z)}{z} \right) dz. \tag{10.81}$$

The points that define the polygon are (x_n, z_n) for $n = 1$ to N anticlockwise around the perimeter. The line that interpolates the points n and $n+1$ is $X_n(z) = a_n z + b_n$, where the coefficients a_n and b_n are given by (10.77) and (10.78), respectively. Figure 10.12b shows a polygon made of three points, for which integral (10.81) becomes

$$I = \int_{z_2}^{z_1} \left(\tan^{-1} \frac{X_2(z)}{z} - \tan^{-1} \frac{X_1(z)}{z} \right) dz + \int_{z_1}^{z_3} \left(\tan^{-1} \frac{X_2(z)}{z} - \tan^{-1} \frac{X_3(z)}{z} \right) dz. \tag{10.82}$$

This integral can be rewritten as

$$I = \sum_{n=1}^{N} \int_{z_{n+1}}^{z_n} \tan^{-1} \frac{X_n(z)}{z} dz \tag{10.83}$$

for $N = 3$, where point $N + 1$ is the same as point 1. The same reasoning applies equally well for a polygon with any number N of points. The remaining integration (10.83) is

$$\int \tan^{-1} \frac{az + b}{z} dz = \left(z + \frac{ab}{a^2 + 1}\right) \arctan \frac{az + b}{z} + \frac{b}{2(a^2 + 1)} \ln\left((az + b)^2 + z^2\right) \tag{10.84}$$

which can be verified by derivation. (It is also straightforward to carry out the integration, using the substitution $u = az + b$ as a first step, and then integration by parts as the second step, with $U' = -b/(u - a)^2$ and $V = \tan^{-1} u$ as the two factors.) The integral (10.83) then becomes

$$I = \sum_{n=1}^{N} \frac{b_n}{a_n^2 + 1} \left[\ln\frac{r_{n+1}}{r_n} + a_n \left(\tan^{-1} \frac{x_{n+1}}{z_{n+1}} - \tan^{-1} \frac{x_n}{z_n}\right)\right]$$

$$+ \sum_{n=1}^{N} \left(z_{n+1} \tan^{-1} \frac{x_{n+1}}{z_{n+1}} - z_n \tan^{-1} \frac{x_n}{z_n}\right) \tag{10.85}$$

which is almost Talwani's formula (10.76). The last summation (10.85) is zero because each term in the summation appears twice with opposite signs.

Note 10.5 *Parker's formula.* Parker's formula (Parker, 1972) gives the Fourier transform of the gravity along a xy-plane from a 3D body as

$$\mathcal{F}(\Delta g) = 2\pi G \exp(-|\mathbf{k}|z_0) \sum_{n=1}^{\infty} \frac{|\mathbf{k}|^{n-1}}{n!} \mathcal{F}(h_2^n - h_1^n) \tag{10.86}$$

where $h_1(x, y)$ and $h_2(x, y)$ are the vertical positions of the top and the base of the body, and ϱ is the body's density. The derivation of Parker's formula makes use of the 2D Fourier transform of a function in the xy-plane

$$\mathcal{F}(f) = \int_{\mathcal{R}^2} dx\, dy\, f(x, y)\, e^{i(k_x x + k_y y)} \tag{10.87}$$

where $i = \sqrt{-1}$ and $\mathbf{k} = (k_x, k_y)$ is the 2D wave vector in the horizontal plane. The gravity is derived from the potential, and the potential in position $\mathbf{r}_0$ is

$$U(\mathbf{r}_0) = -G\varrho \int_V \frac{dV}{|\mathbf{r} - \mathbf{r}_0|} = -G\varrho \int_D dS \int_0^{h(x,y)} \frac{dz}{|\mathbf{r} - \mathbf{r}_0|} \tag{10.88}$$

for a body of volume V and density ϱ, and where $dS = dx\, dy$ denotes integration over the xy-plane. The derivation follows Parker (1972) and we assume that body is between the xy-plane at $z = 0$ and a surface $z = h(x, y)$. Notice that the position vector is 3D, $\mathbf{r} = (x, y, z)$, but the wave vector is 2D, $\mathbf{k} = (k_x, k_y, 0)$, because we are using a 2D Fourier transform. We therefore have that $\mathbf{k} \cdot \mathbf{r} = k_x x + k_y y$. The Fourier transform of the potential becomes

$$\mathcal{F}(U(\mathbf{r}_0)) = \int_{\mathcal{R}^2} dS_0 \, U(\mathbf{r}_0) \, e^{i\mathbf{k}\cdot\mathbf{r}_0} = -G\varrho \int_{\mathcal{R}^2} dS \int_0^{h(x,y)} dz \int_{\mathcal{R}^2} dS_0 \frac{e^{i\mathbf{k}\cdot\mathbf{r}_0}}{|\mathbf{r} - \mathbf{r}_0|} \quad (10.89)$$

when the order of integration is interchanged. At this point we need the following integral:

$$\int_{\mathcal{R}^2} \frac{e^{i\mathbf{k}\cdot(\mathbf{r}_0-\mathbf{r})}}{|\mathbf{r}_0 - \mathbf{r}|} \, dx_0 \, dy_0 = \frac{2\pi}{|\mathbf{k}|} e^{-|\mathbf{k}|(z_0-z)} \quad (10.90)$$

which can be found in Bracewell (1965) according to Parker (1972). It is also possible to show the integral by use of two cosine-transformations (Halvorsen, 2003). Integration over z can then be carried out and we get

$$\mathcal{F}(U(\mathbf{r}_0)) = -2\pi G\varrho \frac{e^{-|\mathbf{k}|z_0}}{|\mathbf{k}|} \int dS \, e^{i\mathbf{k}\cdot\mathbf{r}} \frac{1}{|\mathbf{k}|} \left(e^{|\mathbf{k}|h} - 1 \right). \quad (10.91)$$

The exponential of $|\mathbf{k}|h$ can be expanded as a power series

$$e^{|\mathbf{k}|h} = \sum_{n=0}^{\infty} \frac{(|\mathbf{k}|h)^n}{n!} \quad (10.92)$$

which gives the Fourier transform of the potential as the power series

$$\mathcal{F}(U(\mathbf{r}_0)) = -2\pi G\varrho \, e^{-|\mathbf{k}|z_0} \sum_{n=1}^{\infty} \frac{|\mathbf{k}|^{n-2}}{n!} \mathcal{F}(h^n). \quad (10.93)$$

The gravity is the gradient of the potential and in the vertical direction $\Delta g = -dU/dz_0$, and the Fourier transform of the gravity becomes

$$\mathcal{F}(\Delta g) = \mathcal{F}\left(\frac{\partial U}{\partial z_0}\right) = \frac{\partial}{\partial z_0} \mathcal{F}(U) = -|\mathbf{k}|\mathcal{F}(U) \quad (10.94)$$

which gives the wanted expression for one surface $h(x, y)$. The Fourier transform for the gravity between the two surfaces is simply the difference of the Fourier transform applied to each of the two surfaces. An implementation of Parker's formula is shown below for Octave/MatLab. It has only been tested with Octave.

```
function gx = get_gravity_2D(L, h2, h1, drho, z0, Nterms)
    if (rows(h1) != rows(h2)) printf("Error!\n"); exit(1); endif
    N = rows(h2);
    prod_h2 = ones(N, 1);
    prod_h1 = ones(N, 1);
    kn = zeros(N, 1);
    Gk = zeros(N, 1);
    Hk = zeros(N, 1);
    N2 = floor(N / 2);
    factorial = 1;
```

```
        for n = 1:Nterms
            prod_h2 = prod_h2 .* h2;
            prod_h1 = prod_h1 .* h1;
            Hk = fft(prod_h2 - prod_h1);
            factorial *= n;
            for i = 1:N2
                i1 = i + 1;
                i2 = N - i + 1;
                kn(i) = 2 * pi * i / L;
                fac1 = exp(-kn(i) * z0) * (kn(i)^(n - 1)) / factorial;
                Gk(i1) += fac1 * Hk(i1);
                if (i1 == i2) continue; endif
                Gk(i2) += fac1 * conj(Hk(i1));
            endfor
        endfor
    G = 6.67e-11;
    mGal = 1.0e+5;
    gx = mGal * 2 * pi * G * drho * ifft(Gk);
endfunction
```

Exercise 10.9 (Technical) Use integrals (10.81) and (10.84) to derive equation (10.71) for the gravity from a 2D body with rectangular cross-section.

10.11 Excess mass causing gravity anomalies

Gauss's law

$$\oint \mathbf{g} \cdot \mathbf{n} \, dS = -4\pi G M \tag{10.95}$$

can be used to determine the total excess mass (or the mass deficiency) in the subsurface that causes a gravity anomaly g_z in the horizontal xy-plane. A surface that encloses the anomalous mass M can be made of two parts, where one part is in the xy-plane and the other part is a hemisphere underneath the xy-plane, see Figure 10.16. We let

$$\oint \mathbf{g} \cdot \mathbf{n} \, dS = I_1 + I_2 \tag{10.96}$$

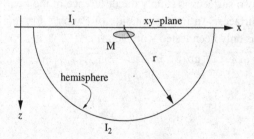

Figure 10.16. *The closed surface around the mass anomaly M is made of two parts, a (circular) area in the xy-plane and a hemisphere.*

where I_1 is the integral over the xy-surface and I_2 is over the hemisphere. The latter integral is approximated as

$$I_2 = \int \mathbf{g} \cdot \mathbf{n} \, dS \approx -2\pi GM \qquad (10.97)$$

because the anomalous mass M becomes almost a point mass when the radius of the hemisphere is large. The gravitational field along the hemisphere is then $g \approx GM/r^2$, according to Newton's law of gravitation. It is pointing in the opposite direction to the unit normal vector along the hemisphere, which causes the minus sign in front of $2\pi GM$ in equation (10.97). Gauss's law written as $I_1 + I_2 = -4\pi GM$ then becomes $I_1 = -2\pi GM$ or

$$\int g_z \, dx \, dy = 2\pi GM \qquad (10.98)$$

since $\mathbf{g} \cdot \mathbf{n} = -g_z$ along the xy-plane. Integral (10.98) can be rewritten as an expression for the anomalous mass M in terms of an integral over the corresponding anomalous gravity field g_z in the xy-plane:

$$M = \frac{1}{2\pi G} \int g_z \, dx \, dy. \qquad (10.99)$$

Exercise 10.10 gives an example and also a check of this expression.

Exercise 10.10 The vertical gravitational acceleration from a spherical mass M buried at a depth h is $g_z(x, y) = GMh/(x^2 + y^2 + h^2)^{3/2}$ at position (x, y) according to equation (10.57). Use expression (10.99) to show that the gravitational anomaly g_z really is caused by a mass M.

Solution: Equation (10.99) gives the total mass M that causes the field g_z. The integral over the plane in equation (10.99) is

$$\int g_z \, dx \, dy = GMh \int_0^{2\pi} \int_0^{\infty} \frac{d\phi \, r \, dr}{(r^2 + h^2)^{3/2}} \qquad (10.100)$$

when it is rewritten in cylinder coordinates, and using that $r^2 = x^2 + y^2$ and $dx \, dy = r \, d\phi$. A change of integration variable from r to $u = (r^2 + h^2)^{1/2}$ then gives

$$\int g_z \, dx \, dy = 2\pi GMh \int_h^{\infty} \frac{du}{u^2} = 2\pi GM \qquad (10.101)$$

and we are done.

10.12 Vertical continuation of gravity

There are situations where we have the gravity along a horizontal line (or plane), and we want to know the gravity above the line (or the plane). The question is how can we extend the known gravity from the line upwards above the line? We will see that the answer is a solution of Laplace's equation, where the known gravity along the line or (the plane) is used as a boundary condition. The potential U obeys Laplace's equation (in the absence of mass)

$$\frac{\partial^2 U}{\partial x^2} + \frac{\partial^2 U}{\partial z^2} = 0. \tag{10.102}$$

The gravity now extends in 2D from a line, assuming it does not depend on the y-coordinate. The vertical component of the gravitational acceleration is the gradient of the potential (with a minus sign)

$$g_z = -\frac{\partial U}{\partial z} \tag{10.103}$$

and Laplace's equation for g_z appears if we apply $-\partial/\partial z$ on Laplace's equation for the potential

$$-\frac{\partial}{\partial z}\left(\frac{\partial^2 U}{\partial x^2} + \frac{\partial^2 U}{\partial z^2}\right) = \frac{\partial^2 g_z}{\partial x^2} + \frac{\partial^2 g_z}{\partial z^2} = 0. \tag{10.104}$$

There is fortunately a simple solution to the Laplace equation for g_z, when the boundary condition is

$$g_z = g_1 \cos(kx) + g_2 \sin(kx) \tag{10.105}$$

along the x-axis. We solved this problem in Section 6.18 for the temperature into the subsurface when the surface temperature is a sine or a cosine function. The solution is

$$g_z(x, z) = (g_1 \cos(kx) + g_2 \sin(kx)) \exp(-kz) \tag{10.106}$$

which is verified by inserting the solution into Laplace's equation (10.104). Setting $z = 0$ gives the boundary condition (10.105). Nearly any function can be represented by a Fourier series

$$g_z(x, 0) = a_0 + \sum_n (a_n \sin(k_n x) + b_n \cos(k_n x)) \tag{10.107}$$

where $k_n = 2\pi n/L$ is the wave number, and L is the extent of the data set along the x-axis. The gravity along a horizontal line a vertical distance z away from the x-axis is the superposition of a series of solutions (10.106):

$$g_z(x, z) = a_0 + \sum_n (a_n \sin(k_n x) + b_n \cos(k_n x)) \exp(-k_n z). \tag{10.108}$$

We see right away that the Fourier components with small wavelengths (large wave numbers) die out much faster than components with long wavelengths (small wave numbers) when we move vertically away from the x-axis. The gravity signal therefore appears more smoothed with increasing z because the short wavelengths die out.

An important application of formula (10.108) is the gravity anomaly from the density contrast between the crust and mantle along the Moho. We can make an estimate of the gravity along the top of a rectangular band that contains the Moho and then extend the

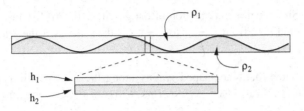

Figure 10.17. *A sine shaped boundary between rocks of density ϱ_1 and ϱ_2. The wavelength is sufficiently long for the boundary to appear as almost horizontal.*

gravity to the surface. The Moho is for simplicity taken to be a sine function as shown in Figure 10.17. The band that contains the Moho has thickness $2h_0$ and is at a depth z. It is assumed that the Moho has sufficiently long wavelengths to appear locally as almost horizontal. The gravity along the surface of the band can be approximated as

$$g = 2\pi G \varrho_c h_c(x) + 2\pi G \varrho_m h_m(x) \tag{10.109}$$

when the gravity from each layer is given by Bouguer's formula. The thicknesses $h_c(x) = h_0 - h_0 \sin(kx)$ and $h_m(x) = h_0 + h_0 \sin(kx)$ are sections of density ϱ_c and ϱ_m, respectively. The gravity along the layer is

$$g = 2\pi G(\varrho_m + \varrho_c)h_0 + 2\pi G(\varrho_m - \varrho_c)h_0 \sin(kx) \tag{10.110}$$

where the anomaly due to the sine-shape of the Moho is

$$\Delta g = 2\pi G(\varrho_m - \varrho_c)h_0 \sin(kx). \tag{10.111}$$

The gravity anomaly then becomes

$$\Delta g = 2\pi G(\varrho_m - \varrho_c)h_0 \sin(kx)\exp(-kz). \tag{10.112}$$

The general case where the Moho is given as a Fourier series

$$h(x) = \sum_n (a_n \sin(k_n x) + b_n \cos(k_n x)) \tag{10.113}$$

at the depth z gives the gravity anomaly at the surface:

$$\Delta g = 2\pi G(\varrho_m - \varrho_c) \sum_n (a_n \sin(k_n x) + b_n \cos(k_n x))\exp(-k_n z). \tag{10.114}$$

We will return to this topic in a discussion of gravity anomalies from the deflection of the crust by surface loads.

It should be mentioned that solution (10.106) corresponds to Parker's formula for the Fourier transform of the gravity as a function of the Fourier transform of the topography, when only the first term is included. The gravity from a sine-shaped layer is approximated with Bouguer's formula, which applies for flat layers. This approximation therefore assumes that the layer has "long" wavelengths. We also notice that the factor $\exp(-k_n z)$ reduces the short wavelengths (large wave number) much more with increasing z than the long wavelengths (small wave number).

Exercise 10.11 Show that a continuation of the gravity (10.106) from the horizontal line $z = 0$ to $z = z_1$ and then from $z = z_1$ to $z = z_2$ is the same as if the gravity is extended directly from $z = 0$ to $z = z_2$.

Exercise 10.12 Show that sine-shaped gravity along the x-axis with a wavelength λ is reduced to one-half at a distance $z_{1/2} = (\ln 2/2\pi)\lambda \approx 0.11\lambda$ vertically away from the x-axis.

10.13 Reduction of gravity data

The vertical component of the gravitational acceleration is measured along the surface of the Earth. These gravity measurements cannot easily be compared with each other when they are obtained at different heights. The process of transforming the measurements to a common reference height is called *reduction* of the gravity observations (see Figure 10.18). The common reference level for the Earth is a spheroid. A spheroid is an ellipsoid where two axes are equal, which in the case of the Earth is the two axes of the equatorial plane. The radius of the Earth is slightly larger at the equator than at the poles ($\sim$0.3%). The actual equipotential surface of the Earth is called the *geoid*, and it deviates slightly from the spheroid because of mass anomalies in the subsurface. The geoid coincides with the sea surface when the sea is not disturbed by winds and tides. The difference in height between the spheroid and the geoid is slightly more than 100 m at the maximum. The reference gravity is the value of the acceleration at the spheroid, assuming an Earth without any mass anomalies. Its value at the latitude ϕ is given by the *international gravity formula*

$$g_0(\phi) = 9.780318 \cdot (1 + 5.278895 \cdot 10^{-3} \sin^2 \phi + 2.3462 \cdot 10^{-5} \sin^4 \phi) \qquad (10.115)$$

where the gravitational acceleration along the equator is 9.780318. The reference gravity along the surface of the Earth depends on the latitude because the Earth has a slightly flattened shape due to its rotation. Setting $\sin \phi = 1$ in the formula gives the gravity at the poles, which is 5200 mGal more than at the equator. This is roughly 0.5 mGal/km, which becomes noticeable when the survey area is as long as 100 km in the north–south direction. Details about the international gravity formula are published by the National Imagery and Mapping Agency (1984).

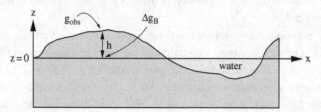

Figure 10.18. *Gravity measurements along the rock surface have to be transformed to a common reference height before they can be compared. The x-axis at sea level is the reference height.*

Gravity measurements are first corrected for latitude variations using the gravity formula. The next correction removes the difference in height of the gravity measurements, and it is therefore called the *free-air correction*. The gravitational acceleration depends on the radius of the Earth as $g = GM/r^2$, where M is the mass of the Earth. A change dr in the radius gives a change

$$dg = \frac{dg}{dr}\,dr = -\frac{2GM}{r^3}\,dr = -2g\frac{dr}{r} \tag{10.116}$$

in the acceleration. Using $2g/r \approx 0.306$ mGal/m gives that the free-air correction is an addition of $0.306\,h$ mGal, when h is the distance from the reference level (spheroid) up to the observation point in meters. In practice, one measures the height from the geoid rather than the spheroid, because it coincides with the sea level. Spheroid heights were also difficult to measure before the emergence of satellite technology. The geoid is smooth, and the difference between the geoid and the spheroid can be considered constant over reasonable study areas. The errors introduced by the use of the geoid instead of the spheroid as a reference level are therefore insignificant. The topics of spheroid, geoid and gravity are discussed in a tutorial by Li and Götze (2001). Note 10.6 shows how the potential difference between the geoid and spheroid can be related to their difference in height.

The third correction removes the contribution of the mass between the observation point and the reference level. This is normally done by assuming that the topography is sufficiently flat to allow for the use of Bouguer's formula, and this correction is therefore called the *Bouguer correction*. It is the subtraction of $2\pi G\varrho_c h$ from the observed gravity, when the observation point is a height h above the reference line. The density of the rocks are normally set to $\varrho_c = 2670$ kg m^{-3}, and the Bouguer correction becomes the subtraction of $0.112\,h$ mGals (when h is in unit of meters). Figure 10.18 illustrates the Bouguer transform.

When these three corrections are put together we get that the corrected (or reduced) gravity anomaly is

$$\Delta g_B = g_{obs} - g_0(\phi) + \frac{2g}{r}h - 2\pi G\varrho_c h \tag{10.117}$$

which is called the *Bouguer gravity anomaly on land*. When the Bouguer correction is not included we have the free-air gravity anomaly

$$\Delta g_{fa} = g_{obs} - g_0(\phi) + \frac{2g}{r}h. \tag{10.118}$$

The relationship between the Bouguer and the free-air anomalies on land is therefore

$$\Delta g_B = \Delta g_{fa} - 2\pi G\varrho_c h. \tag{10.119}$$

There is also a *Bouguer anomaly at sea*:

$$\Delta g_B = \Delta g_{fa} + 2\pi G(\varrho_c - \varrho_w)w \tag{10.120}$$

where w is the water depth, which gives the gravity when water is replaced by crust (see Exercise 10.14). Gravity observations are nearly always discussed in terms of Bouguer and

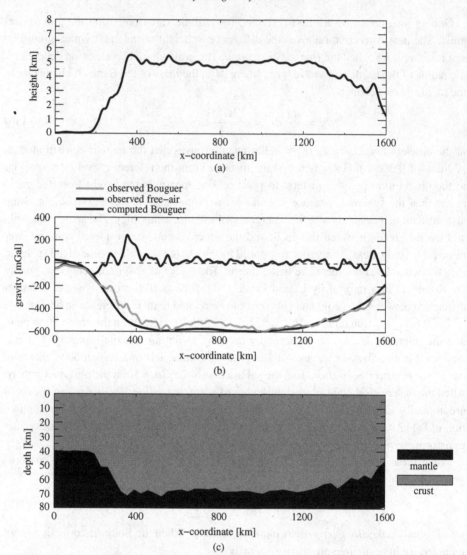

Figure 10.19. *(a) The topography of a profile from Tibet along the 90° longitude. (b) Observed and computed Bouguer anomaly for the profile. (c) Crustal roots along the profile.*

free-air anomalies. Figures 10.19a and 10.19b show the topography and the free-air gravity, respectively, for a profile from Tibet taken from the SAR Topex database (Sandwell
and Smith, 1997, 2009). The Bouguer correction (10.119) in Figure 10.19b is obtained
using the topography in Figure 10.19a. We notice that the short wavelength free-air gravity
correlates with the topography. The Bouguer correction therefore gives a smoother gravity
with smaller short wavelength oscillations. We also notice that the free-air gravity is close
to zero, except for at the South front, and we therefore expect these high mountains to

be in nearly isostatic equilibrium. The deviation from zero at the front could be related to active compression from the Indian continent. Figure 10.19c shows the crustal roots corresponding to the topography in Figure 10.19a assuming a density $\varrho_c = 2800$ kg m^{-3}. The Bouguer correction gives the gravity anomaly along $z = 0$, and using the crustal model we can also calculate the gravity along $z = 0$ to compare it with the Bouguer correction. The calculation is done with Talwani's formula as explained in Section 10.10. The Bouguer gravity and the calculated gravity from the crustal model are comparable, because the Bouguer anomaly represents free-air data close to zero (or isostatic equilibrium) and the calculated gravity is for crustal roots in isostatic equilibrium. We notice that the crust has a thickness ~ 40 km where the topography is almost at sea level. This reference depth, which is not obtained from the data set, has been taken from Rajesh and Mishra (2003). The topography and the assumption of isostasy gives the depth of the crustal roots relative to this reference depth. Rajesh and Mishra (2003) discuss gravity data for a similar transect across Tibet and present a more detailed model of the density distributions in the crust.

Note 10.6 *Brun's formula.* The difference between the gravitational potential for the geoid and the spheroid is called the disturbing potential. There is a linear relationship between the disturbing potential and the difference in height between the geoid and the spheroid. Differentiation of the potential gives to first order that

$$U \approx U_0 + \frac{\partial U}{\partial r}(r - r_0) \tag{10.121}$$

where U_0 is the reference potential. The disturbing potential is

$$V = U - U_0 = -g(r - r_0) \tag{10.122}$$

because $g = -\partial U/\partial r$, and the relationship between V and the height $N = r - r_0$ is therefore

$$V = -g\,N \tag{10.123}$$

which is called *Brun's formula*.

10.14 Gravity and isostasy over continents

This section shows how gravity anomalies may reflect the degree of isostatic equilibrium over continents. In Chapter 9 we saw how large loads on the lithosphere like mountain ranges are in isostatic equilibrium, by being supported by the buoyancy of deep crustal roots. Loads with a short lateral extent are supported by the elastic strength of the lithosphere. Since the Moho is depressed in the first case, but not in the latter, we expect a difference in gravity between these two cases, since the density difference between the mantle and the crust is ~ 500 kg m^{-3}.

We will first find the gravity anomaly from a sine-shaped surface load, because we already know the lithosphere flexure under such a load (see Section 9.4). Furthermore, any loads (with a reasonable shape) can be represented by a Fourier series, which is a superposition of cosine and sine loads of different wavelengths. The starting point is a sine-shaped surface topography at sea level

$$h(x) = h_0 \sin(kx) \tag{10.124}$$

where $k = 2\pi/\lambda$ is the wave number and λ is the wavelength. A constant offset may be added to $h(x)$ to make a topography above sea level (see Exercise 10.13). This topography gives the surface load

$$q(x) = \varrho_c g h_0 \sin(kx) \tag{10.125}$$

when it is made of a rock with density ϱ_c. It is shown in Section 9.4 that the deflection of the lithosphere from this surface load is

$$w(x) = \frac{\varrho_c}{(\varrho_m - \varrho_c)} \Phi(k) h_0 \sin(kx) \tag{10.126}$$

where the function

$$\Phi(k) = \frac{1}{1 + \dfrac{Dk^4}{(\varrho_m - \varrho_c)g}} \tag{10.127}$$

is the degree of isostatic equilibrium. The surface topography and the corresponding deflection of the Moho is shown in Figure 10.20. We recall that D is the flexural rigidity (9.15) of the lithosphere, and that the function $\Phi(k)$ can be written as

$$\Phi(k) = \frac{1}{1 + (k/k_c)^4} \tag{10.128}$$

where $k_c = ((\varrho_m - \varrho_c)g/D)^{1/4}$ is the critical wave number. The wave number k_c defines the transition between loads in isostatic equilibrium and loads that are fully supported by the elastic strength of the plate. (The wavelength that corresponds to the critical wave number is $\lambda_c = 2\pi/k_c$.) Loads with $k \ll k_c$ (or $\lambda \gg \lambda_c$) have $\Psi \approx 1$ and $w(x) \approx h(x)\varrho_c/(\varrho_m - \varrho_c)$, and are in isostatic equilibrium. The loads of the opposite regime, $k \gg k_c$ (or $\lambda \ll \lambda_c$), have $\Psi \approx 0$ and $w(x) \approx 0$, and they are therefore supported by the elastic strength of the plate. We must remember that this is a simple model of an elastic plate. The plate is assumed to be homogeneous with constant thickness, and the load has no elastic strength.

The gravity anomaly at the depth of the Moho caused by its shape $w(x)$ is found using Bouguer's formula with the density difference $\varrho_m - \varrho_c$. The gravity from the deflected Moho at its average depth $z = d$ is then

$$\Delta g(z = d) = -2\pi G (\varrho_m - \varrho_c) w(x)$$
$$= -2\pi G \varrho_c \Phi(k) h_0 \sin(kx) \tag{10.129}$$

when the deflection (10.126) is inserted. Bouguer's formula applies for a layer of infinite extent and constant thickness. The sine-shaped deflection of the Moho is assumed to have

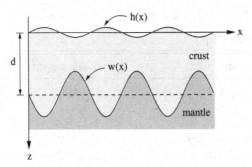

Figure 10.20. *A sine-shaped topography at sea-level with the Moho at an average depth d.*

a sufficiently large wavelength to appear locally as a "long" layer with almost the same thickness. A minus sign has been added to the gravity anomaly (10.129) because it is decreasing with increasing topography and deflection. The gravity anomaly (10.129) is at the average depth of the Moho, but we want the gravity anomaly along the x-axis (at $z = 0$), where it becomes the Bouguer anomaly. (See Figure 10.20.) A continuation of the sine-shaped gravity field from a horizontal line at the depth $z = d$ up to the depth $z = 0$ is as shown in Section 10.12. The sine-shaped gravity along the x-axis ($z = 0$) is then $\Delta g(0) = \Delta g(z) \exp(-kz)$ when it is $\Delta g(z)$ at the depth z. The gravity decreases exponentially upwards with the factor $\exp(-kz)$. The Bouguer gravity anomaly from the deflected Moho is therefore

$$\Delta g_B(x) = -2\pi G \varrho_c \, \Phi(k) \, e^{-kd} \, h_0 \sin(kx). \tag{10.130}$$

The free-air gravity anomaly is the Bouguer anomaly added to the gravity from the mass between the x-axis and the height of the topography. Bouguer's formula is used once more to obtain the gravity from the topography, assuming that it appears locally like a "long" plate. The free-air gravity is therefore

$$\Delta g_{fa}(x) = \Delta g_B(x) + 2\pi G \, h(x)$$
$$= 2\pi G \varrho_c \left(1 - \Phi(k) \, e^{-kd}\right) h_0 \sin(kx). \tag{10.131}$$

The anomalies (10.130) and (10.131) are the Bouguer and the free-air anomalies, respectively, caused by the flexure of the sine-shaped surface load. We will now take a closer look at these anomalies when the load has a sufficiently low wave number for $e^{-kd} \approx 1$. The Bouguer anomaly is then seen to be negative for sine-loads that are in isostatic equilibrium, and it is close to zero for sine-loads that are supported by the elastic strength of the plate. The free-air anomaly behaves oppositely, it is zero for sine-loads in isostatic equilibrium, and positive for sine-loads that are supported by the elastic stiffness. The relationships between isostasy and gravity for a sine-load are summarized in Table 10.1.

Table 10.1. *A summary of gravity anomalies and isostasy.*

Isostasy	Elastic support
$k \ll k_c$	$k \gg k_c$
$\lambda \gg \lambda_c$	$\lambda \ll \lambda_c$
$\Phi(k) \approx 1$	$\Phi(k) \approx 0$
$\Delta g_{\mathrm{B}} \approx -2\pi G \varrho_c h(x)$	$\Delta g_{\mathrm{B}} \approx 0$
$\Delta g_{\mathrm{fa}} \approx 0$	$\Delta g_{\mathrm{fa}} \approx 2\pi G \varrho_c h(x)$

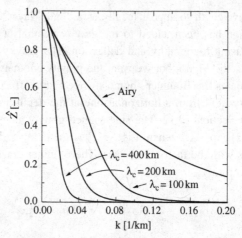

Figure 10.21. *The admittance function for different critical wavelengths when* $d = 10$ *km. See text for further details.*

The expressions (10.130) and (10.131) between the gravity anomaly and a sine-shaped topography show that the ratio

$$Z_{\mathrm{B}}(k) = -\frac{\Delta g_{\mathrm{B}}(x)}{h(x)} = 2\pi G \varrho_c \, \Phi(k) \, e^{-kd} \qquad (10.132)$$

is only dependent on the wave number (k). The ratio $Z_{\mathrm{B}}(k)$ is called the Bouguer *admittance* function. Admittance is a term adopted from circuit analysis, where it denotes the output or the system response (Δg_{B}) related to an input signal (h). The scaled Bouguer admittance function

$$\hat{Z}_{\mathrm{B}}(k) = \frac{Z_{\mathrm{B}}(k)}{2\pi G \varrho_c} = \Phi(k) \, e^{-kd} \qquad (10.133)$$

is shown in Figure 10.21 for $d = 10$ km and the different critical wavelengths $\lambda_c = 100$ km, 200 km and 400 km. (Recall that $\lambda_c = 2\pi/k_c$, $k_c = (\Delta \varrho g / D)^{1/4}$ and $D = E T_e^3 / (12(1 - \nu^2))$, where T_e is the effective elastic thickness and E is Young's modulus.)

These critical wavelengths correspond to the effective elastic thicknesses $T_e = 5.6$ km, 14.1 km and 35.5 km, when Young's modulus is $E = 70$ GPa and the crustal density is $\varrho_c = 2700 \text{ kg m}^{-3}$. The Airy curve shows the admittance function for isostatic equilibrium, which is $\Phi = 1$.

The topography of any load can be approximated by a Fourier series with N terms as

$$h(x) = a_0 + \sum_{n=1}^{N} (a_n \cos(k_n x) + b_n \sin(k_n x)) \tag{10.134}$$

where the wave number $k_n = 2\pi n/L$ is related to the size L of the load. (The appendix shows how a discrete function is approximated by a Fourier series.) The gravity anomaly from the load (10.134) becomes the superposition of the anomalies from each term in the Fourier series:

$$\Delta g(x) = -Z(k_n) (a_n \cos(k_n x) + b_n \sin(k_n x)). \tag{10.135}$$

If we have observations of both the topography and the Bouguer anomaly, where the gravity anomaly is represented by the Fourier series

$$\Delta g(x) = A_0 + \sum_{n=1}^{N} (A_n \cos(k_n x) + B_n \sin(k_n x)) \tag{10.136}$$

we might expect the ratio of the Fourier coefficients to be the admittance function, since we have

$$Z(k_n) = \left| \frac{A_n}{a_n} \right| = \left| \frac{B_n}{b_n} \right| \quad \text{for} \quad n = 1, 2, \dots. \tag{10.137}$$

Observations of topography and especially gravity include noise, and it turns out that it is not sufficient to use one data set and the simple ratios (10.137) to obtain the admittance function. In order to reduce the noise it is necessary to take the average of the Fourier coefficients for a number of transects. The Fourier coefficients are most easily obtained by the FFT, which returns the coefficients a_n and b_n as the real and imaginary parts of a complex Fourier coefficient c_n. (FFT and complex Fourier coefficients are explained in the appendix.) In order to average the Fourier coefficients we assume that we have data for M transects. The complex Fourier coefficients for the topography and the gravity along the mth transect are denoted $c(k_n)_m$ and $d(k_n)_m$, respectively, for wave number k_n. A straightforward average of the Fourier coefficients is not used, because Z is best estimated from the following averages:

$$G(k_n) = \frac{1}{M} \sum_{m=1}^{M} c(k_n)_m \, d(k_n)_m^* \tag{10.138}$$

and

$$H(k_n) = \frac{1}{M} \sum_{m=1}^{M} c(k_n)_m \, c(k_n)_m^* \tag{10.139}$$

where a star is the complex conjugate. The admittance function for wave number k_n is then

$$Z(k_n) = \frac{|G(k_n)|}{H(k_n)} \qquad (10.140)$$

where we notice that Z_n is $d(k_n)/c(k_n)$ for just one transect ($M = 1$). The product of a complex number with its conjugate, like $c(k_n)_m \, c(k_n)_m^*$, is always a real number, and the average $H(k_n)$ is therefore real. The terms in the function $G(k_n)$ are normally not real and the absolute value of this function is therefore needed in the admittance function (10.140). The averaging of the Fourier coefficients for wave number k_n over M transects, as with the functions $H(k_n)$ and $G(k_n)$, require that the coefficients really are for the same wave number k_n. A simple way to achieve this is to use the same transect length and the same number of sample points for each transect. The use of the averages $G(k_n)$ and $H(k_n)$ was suggested by McKenzie and Bowin (1976) and appears to be an effective way to reduce noise.

We have already noticed that the only tuning parameter in the admittance function is the critical wave number k_c, which again is a function of the elastic thickness of the lithosphere. A plot of the admittance function based on several transects is actually one of the most used methods to estimate the effective elastic thickness. Figure 10.22a shows an example of the free-air admittance function based on 19 north–south profiles across Tibet at longitudes 85.5° to 94.5° (0.5° separation), which go from latitude 25° to latitude 40°. The plot shows the admittance for $k_1 = 0.004$ km^{-1} to $k_{19} = 0.077$ km^{-1}, which corresponds to the wavelengths $\lambda_1 = 1557$ km to $\lambda_{19} = 82$ km, respectively. (The length of the profile is λ_1.) The free-air admittance function

$$Z_{\mathrm{fa}}(k) = 2\pi G \varrho_c \left(1 - \Phi(k)\, e^{-kd}\right) \qquad (10.141)$$

from equation (10.131) is plotted for elastic effective thicknesses $T_e = 0$ km, 40 km and 80 km. The corresponding critical wave numbers and wavelengths are $k_c = \infty$, $k_c \approx 0.01$ km^{-1}, and $k_c \approx 0.005$ km^{-1}, and $\lambda_c = 0$, $\lambda_c = 686$ km and $\lambda_c = 1154$ km, respectively. The depth to the Moho is $d = 70$ km and $\varrho_c = 2900$ kg m^{-3}. The admittance function becomes $Z \approx 2\pi G \varrho_c = 121$ mGal when $k \gg k_c$ (or $\lambda \ll \lambda_c$), which is the regime where the topograph is supported by the elastic strength of the plate. The observations plotted in Figure 10.22a suggest that elastic strength is low (almost zero). Free-air admittance studies of Tibet by McKenzie and Fairhead (1997) suggest that $T_e \approx 40$ km. The same authors estimated the effective elastic thickness for various continental areas to be less than 25 km.

We can use the Fourier coefficients of the gravity and topography to study the similarity in these two data sets with the tools of signal processing. The similarity can be measured with the *coherence function*

$$\gamma^2 = \frac{G^2(k_n)}{F(k_n)\, H(k_n)} \qquad (10.142)$$

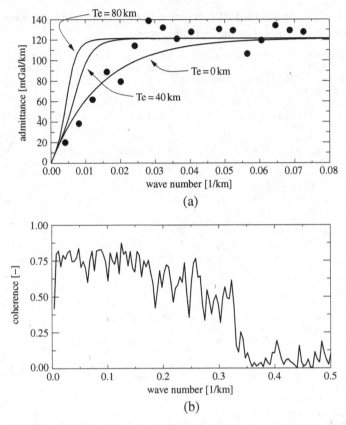

Figure 10.22. (a) The free-air admittance function based on 19 north–south lines across Tibet. (b) The coherence for the same data as in (a).

where $F(k_n)$ is the average

$$F(k_n) = \frac{1}{M} \sum_{m=1}^{M} d(k_n)_m d(k_n)_m^*.$$ (10.143)

The averages $F(k_n)$ and $H(k_n)$ are called *power spectral density* functions in signal analysis. The coherence function gives a number between 0 and 1 and the more correlated the two data sets are, the closer to 1 is the coherence function. Figure 10.22b shows the coherence for the same data set as the admittance plotted in Figure 10.22a. There is a certain degree of coherence for wave numbers less than 0.15 (wavelengths longer than 40 km), and the coherence is close to zero for short wavelengths.

Note 10.7 *Computation of admittance.* The Octave (MatLab) script below computes the admittance and coherence from a set of profiles with topography and gravity. There is one file for each profile, and they are numbered by their file names, for instance `line1.xy`, line2.xy, ... Notice that the FFT function in Octave returns the Fourier coefficients a_n and b_n in array position $n + 1$ as shown in the appendix.

```
function [lambda, kn, Z, coherence] = get_admittance(filename, M)
     for m = 1:M                          %% M is number of transects.
          file = [filename, num2str(m), ".xy"];
          input = load(file);
          xpos = input(:,1);              %% X-coordinates, [km]
          hx = input(:,2);                %% h(x), the topography, [m].
          gx = input(:,3);                %% g(x), the gravity, [mGal].
          N = rows(xpos);                 %% Number of points.
          L = xpos(N) - xpos(1);          %% Length of profile.
          if (m == 1)
               Lref = L;
               Nref = N;
               sumHH = zeros(N, 1);
               sumGH = zeros(N, 1);
               sumGG = zeros(N, 1);
          else
               if ((abs(L - Lref) > 1) || (N != Nref))
                    printf("Transects are incompatible!\n"); exit(1);
               end
          endif
          Hk = fft(hx);
          Gk = fft(gx);
          sumHH += Hk .* conj(Hk);        %% Is always real
          sumGH += Gk .* conj(Hk);
          sumGG += Gk .* conj(Gk);        %% Is always real
     endfor
     N2 = N / 2;                          %% Fourier coefs come twice
     lambda = zeros(N2, 1);               %% because the input is real numbers.
     kn = zeros(N2, 1);
     Z = zeros(N2, 1);
     coherence = zeros(N2, 1);
     %% The factor 1000 below is needed when topography is in meter
     %% and admittance is wanted in units mGal/km.
     for n = 1:N2
          kn(n) = 2 * pi * n / L;
          lambda(n) = 2 * pi / kn(n);
          p = n + 1;                      %% FFT returns a(n), b(n) in c(n+1).
          Z(n) = 1000 * abs(sumGH(p) / sumHH(p));
          coherence(n) = sumGH(p) * conj(sumGH(p)) / (sumHH(p) * sumGG(p));
     endfor
endfunction
```

Exercise 10.13

This exercise looks at gravity and isostasy for the mountain range shown in Figure 10.23.

(a) What is the Bouguer gravity from a mountain range of height $h = 2$ km and density $\varrho_c = 2700 \text{ kg m}^{-3}$?

(b) Show that the free-air anomaly is zero.

Solution: (a) The Bouguer gravity anomaly is

$$\Delta g_B = -2\pi G(\varrho_m - \varrho_c)d \qquad (10.144)$$

where d is the depth of the crustal root from the reference height $z = 0$ (the x-axis). Isostasy gives that the crustal root of a mountain range of height h is $d = h\varrho_c/(\varrho_m - \varrho_c)$

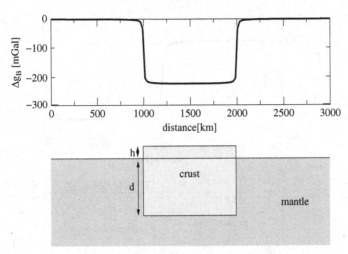

Figure 10.23. *The Bouguer gravity anomaly of a 2 km high mountain range. (See Exercise 10.13.)*

(see Section 7.2). We can use this expression for the depth d in the gravity (10.144), and the Bouguer anomaly becomes an expression of the topography h as

$$\Delta g_B = -2\pi G \varrho_c h. \tag{10.145}$$

The height $h = 2$ km and the density $\varrho_c = 2700$ kg m^{-3} give that $\Delta g_B = -226$ mGal.
(b) The free-air anomaly is the Bouguer gravity Δg_B added to the gravity from the mass of the topography $2\pi G \varrho_c h$, which gives $\Delta g_{fa} = \Delta g_B + 2\pi G \varrho_c h = 0$.

Exercise 10.14
(a) What is the Bouguer gravity anomaly at the surface of $w = 1$ km deep water when the free-air gravity anomaly is zero? Assume that the crustal density is $\varrho_c = 2700$ kg m^{-3}.
(b) What is the Bouguer gravity anomaly along a $h = 2$ km high mountain chain next to the water? (See Figure 10.24.)
Solution: (a) The Bouguer gravity anomaly over the water is

$$\Delta g_B = 2\pi G (\varrho_c - \varrho_w) w \approx 71 \text{ mGal} \tag{10.146}$$

which is the Bouguer correction for water with the depth w.
(b) The Bouguer gravity anomaly for the mountain range is due to two density differences in the subsurface. We have to compare the densities underneath the mountain with the densities to the left, in the ocean part. Firstly, we have the density difference between the crust and water, $\varrho_c - \varrho_w$, over the depth interval w, which acts as an excess mass. Secondly, we have the mass deficiency due to the density difference between the crust and mantle, $\varrho_c - \varrho_m$, over a depth interval d. The Bouguer gravity anomaly is therefore

$$\Delta g_B = 2\pi G (\varrho_c - \varrho_w) w - 2\pi G (\varrho_m - \varrho_c) d. \tag{10.147}$$

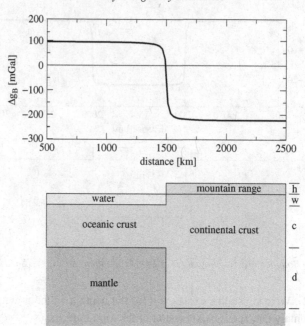

Figure 10.24. *The Bouguer gravity anomaly across the continental–ocean boundary, where the ocean is 1 km deep and the mountain range is 2 km high. The Bouguer anomaly is positive along the ocean and negative along the mountain when there is isostatic equilibrium.*

The mountain range is in isostatic equilibrium, which gives

$$d = h\frac{\varrho_c}{(\varrho_m - \varrho_c)} + w\frac{(\varrho_c - \varrho_w)}{(\varrho_m - \varrho_c)}. \tag{10.148}$$

When d is inserted into Bouguer gravity anomaly (10.147) we get

$$\Delta g_B = 2\pi G\varrho_c h \tag{10.149}$$

which gives that $\Delta g_B = -226$ mGal. We see right away, from the Bouguer anomaly (10.149), that the free-air anomaly is zero for the mountain range. This exercise shows that the Bouguer anomaly over the ocean is positive because of the Bouguer correction for the water, and the Bouguer anomaly over a mountain range is negative because of the deep crustal roots.

10.15 Gravity and sea bed topography

The effective elastic thickness of a continental plate was estimated in the previous section. The estimation was based on an admittance function that relates the topography and the gravity anomaly. We will now do the same for oceanic plates. Observations of free-air gravity over the oceans can be combined with the water depth (bathymetry) to obtain an

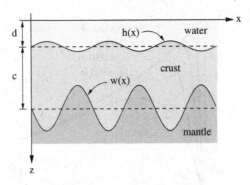

Figure 10.25. *A sine-shaped sea bed and the corresponding Moho.*

admittance, which is a function that tells us to what degree loads of various wavelengths are in isostatic equilibrium. The starting point is once more the gravity anomaly from the flexure from a sine-shaped density contrast. The gravity is now the free-air anomaly related to a sine-shaped sea bed. There are two density differences now, one between the water and the crust, and another between the crust and the mantle, as shown in Figure 10.25. The sine-shaped sea bed $h(x) = h_0 \sin(kx)$ creates a (linear elastic) deflection of the Moho:

$$w(x) = -\frac{(\varrho_c - \varrho_w)}{(\varrho_m - \varrho_c)} \Phi(k) \, h_0 \sin(kx) \tag{10.150}$$

where $\Phi(k)$ is the flexural response function (10.127). The gravity anomaly is obtained by superposing the gravity anomaly from the two density contrasts. The first is the gravity anomaly from the sine-shaped sea bed topography

$$\Delta g_1 = 2\pi G(\varrho_c - \varrho_w) h_0 \sin(kx) e^{-kd} \tag{10.151}$$

and the second is the anomaly from the deflected Moho

$$\Delta g_2 = -2\pi G(\varrho_m - \varrho_c) w(x) e^{-k(d+c)}. \tag{10.152}$$

The total gravity anomaly is

$$\Delta g_{fa} = \Delta g_1 + \Delta g_2$$
$$= 2\pi G(\varrho_c - \varrho_w) e^{-kd} \left(1 - \Phi(k) e^{-kc}\right) h_0 \sin(kx) \tag{10.153}$$

when the deflection (10.150) is inserted. We then get that the sine-shaped sea bed generates the free-air admittance function

$$Z_{fa}(k) = \frac{\Delta g(x)}{h_0 \sin(kx)} = 2\pi G(\varrho_c - \varrho_w) e^{-kd} \left(1 - \Phi(k) e^{-kc}\right), \tag{10.154}$$

a relationship between the gravity and the topograph. We notice that $Z_{fa}(k)$ is only dependent on the wave number. The scaled admittance function

$$\hat{Z}_{fa}(k) = \frac{Z_{fa}(k)}{2\pi G(\varrho_c - \varrho_w)} = e^{-kd} \left(1 - \Phi(k) e^{-kc}\right) \tag{10.155}$$

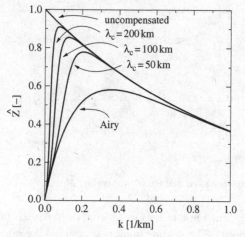

Figure 10.26. *The admittance function for different critical wavelengths. (See the text for details on the other parameters.)*

is plotted in Figure 10.26 for the different critical wavelengths $\lambda_c = 50$ km, 100 km and 200 km. These critical wavelengths correspond to the effective elastic thicknesses $T_e = 2.2$ km, 5.6 km and 14.1 km, respectively. The other parameters are the water depth $d = 1$ km, the thickness of the crust $c = 5$ km, Young's modulus $E = 70$ GPa and the density difference $\Delta\varrho = 1700$ kg m^{-3}. The Airy curve is the admittance function for isostatic equilibrium, which is given by $\Phi = 1$. The uncompensated curve has $\Phi = 0$.

In order to estimate the admittance function we need data for several transects, and we have to apply the averaging procedure from Section 10.14. Figure 10.27 shows the free-air admittance function estimated from nine transects across the Hawaiian–Emperor seamount chain. The bathymetry over the area, shown in Figure 10.27a, coincides with the free-air gravity shown in Figure 10.27b. Figures 10.27a and 10.27b also show the nine transects, which all have the same length 575 km. The seamounts rise from the depth of the ocean floor at $d \approx 4$ km to almost the sea surface. A look at Figure 10.27b shows that the gravity difference between the seafloor and the top of the seamounts is roughly 300 mGal. We can estimate the gravity from the seamounts assuming a flat plate by using the Bouguer formula (10.39), and we then get the gravity $\Delta g = 2\pi G(\varrho_c - \varrho_w)d \approx 300$ mGal, when $\varrho_c = 2800$ kg m^{-3}. Figure 10.27c shows observations of free-air admittance together with a plot of the admittance function (10.154). The parameters are the water depth $d = 4$ km, the crustal thickness $c = 10$ km and the crust density $\varrho_c = 2800$ kg m^{-3}. The effective elastic thickness appears to be in the interval from 20 km to 30 km. Watts (2001) also estimated the effective elastic thickness to be in the range 20–30 km, but with a data set that covered a larger part of the Hawaiian–Emperor seamount chain. The coherence function (10.142) for the gravity and the bathymetry is shown in Figure 10.27d. It is close to 1 for wave numbers less than $k = 0.1$ and then decreases with increasing k.

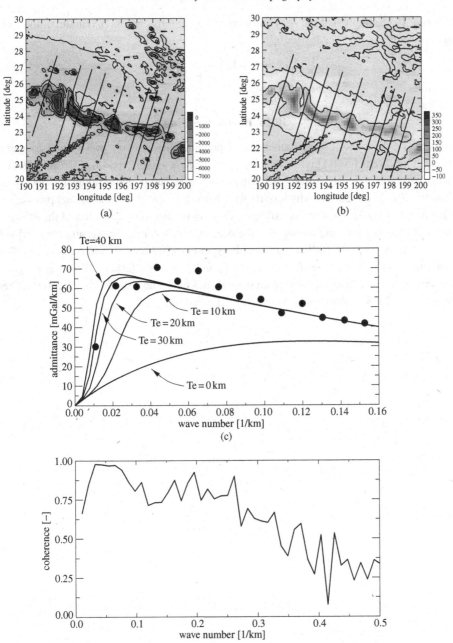

Figure 10.27. *(a) The bathymetry over a part of the Hawaiian–Emperor seamount chain. (b) The free-air gravity over the same area as in (a). (c) The admittance estimated from the nine transects shown in (a) and (b). (d) The coherence of free-air gravity and bathymetry.*

Exercise 10.15 Show that the Airy admittance function reaches its maximum at the wave number

$$k = -\frac{1}{c} \ln \left(\frac{d}{c+d} \right).$$ (10.156)

10.16 Further reading

Blakely (1996) covers the fundamentals of potential theory with gravity and magnetic applications. Watts (2001) gives a thorough presentation and discussion of isostasy, flexure and gravity, with extensive references to previous work. Turcotte and Schubert (1982) have a chapter devoted to gravity which starts with Newton's law of gravitation and proceeds to relationships between gravity and isostasy. There have been several studies of the effective elastic thickness of the continents and the oceans with a certain scatter in estimated values. A starting point for further studies is Forsyth (1985), McAdoo and Sandwell (1989), McKenzie and Fairhead (1997), Watts (2001) and McKenzie (2003). Global gravity and topography data are also available from several sites on the net (Sandwell and Smith, 2009, Pavlis *et al.*, 2008, International Association of Geodesy, 2009).

11

Quartz cementation of sandstones

Sediments are deposited as free or loose particles of a wide range of sizes: mud as fine-grained and sand as coarse-grained particles. When the sediments are buried they get transformed from loose sediments into solid rock by a variety of processes called diagenesis. Mud becomes mudrock or shale, and sand becomes sandstone. This chapter deals with diagenesis of quartzose sandstones, which is among the simplest diagenetic systems, because it basically involves just silica. Nevertheless, it is far from a trivial process and it indicates how complicated diagenesis can be when several minerals are involved.

11.1 Introduction

Well-cemented sandstones are really solid rock. There is little pore space left and the original pore space can be filled with so much quartz cement that it amounts to 25% of the bulk volume. There are several explanations for the source of the large amounts of quartz cement often seen. One is that fluid supersaturated with silica has been flowing through the rock, for instance, convective rolls where dissolution takes place in one end of the roll and precipitation at the other. However, this model is shown to be less likely because the number of pore volumes needed to fill the pore space with quartz cement is unrealistically high.

The alternative to silica imported with large-scale fluid flow is a local dissolution–precipitation model. One much-studied model with a local source for silica is the pressure-solution model, where the dissolution takes place in the fluid film between two quartz grains. The fluid pressure in the fluid film between two grains under large stress is larger than the pressure in the pore fluid. There is a larger equilibrium concentration in a fluid with a high pressure than in a fluid with low pressure, and there is therefore a concentration gradient from the fluid film under stress to the pore fluid. The concentration gradient drives dissolution at the grain contracts and precipitation in the pores. Although pressure solution at grain contacts is a physically appealing process, petrographic evidence does not support it as a mechanism for dissolution of quartz in sandstones. Cathodoluminescence, a technique that allows the detrital (precipitated) quartz to be distinguished from the original quartz grains, shows that the original

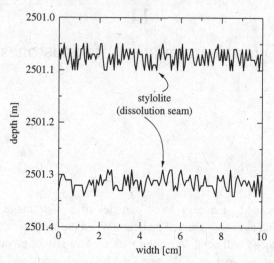

Figure 11.1. *Stylolites are highly irregular surfaces where dissolution takes place. They are rich in the clay minerals illite and mica, which play a key role in enhancing the quartz solubility along the dissolution surface.*

grains in many strongly cemented sandstones have not been dissolved along their grain contacts.

Another local dissolution–precipitation model, that fits the petrographic observations, is dissolution along stylolites seams within the sandstone as the source for silica. Stylolites are irregular (tooth-shaped) surfaces rich in the clay minerals illite and mica (see Figure 11.1), which are responsible for an enhanced quartz solubility. The solubility is estimated to be typically one percent. The exact mechanism responsible for enhanced quartz solubility at quartz–mica and quartz–clay contacts is still poorly understood. The silica dissolved along the stylolites is transported by diffusion into the region between the stylolites where it precipitates in the pore space. The precipitation process is normally the slowest part in the overall dissolution–precipitation process, and the silica concentration is therefore almost constant throughout the interstylolite region. This is seen from an evenly distributed amount of cement between the stylolites. The compaction of the rock takes place at the stylolites, where the quartz grains are dissolved along the seams, while there is no compaction in the interval between the stylolites. It is possible to model the cementation process because precipitation is the slowest step. The only requirement is the kinetics of quartz precipitation.

11.2 Quartz kinetics and precipitation rates

The rate of change of silica (H_4SiO_4) concentration in the pore water is given as the difference between a dissolution rate r_f and a precipitation rate r_r,

$$r_{\text{net}} = \frac{dm}{dt} = r_f - r_r \tag{11.1}$$

where m is the silica concentration (H_4SiO_4) in the pore fluid in units of molality (moles per kilogram). The unit for the rates r_f and r_r is therefore moles per kilogram fluid per second. We notice that the net rate is positive for dissolution and negative for precipitation. The dissolution rate is a zeroth-order law

$$r_f = k_f \frac{A_s}{M_w} \qquad (11.2)$$

where the rate is proportional to the dissolution rate k_f (mole $m^{-2}s^{-1}$) and the surface area A_s, and inversely proportional to the mass of water M_w. The precipitation rate is a first-order law

$$r_r = k_r \frac{A_s}{M_w} m \qquad (11.3)$$

where k_r (kg $m^{-2}s^{-1}$) is the precipitation rate. The net rate is zero when dissolution and precipitation take place at equal rates, and the pore fluid is in equilibrium with the quartz. The equilibrium concentration is then given by $r_r = r_f$, and it becomes

$$m_{eq} = \frac{k_f}{k_r}, \qquad (11.4)$$

which is the principle of detailed balancing. It says that the equilibrium concentration is equal to the ratio of the dissolution reaction and the precipitation reaction. The net rate (11.1) is

$$r_{net} = \frac{dm}{dt} = k_f \frac{A_s}{M_w} \left(1 - \frac{m}{m_{eq}} \right) \qquad (11.5)$$

where the precipitation rate k_r has been replaced by the equilibrium concentration m_{eq}. The net rate (11.5) is seen to be proportional to the degree of supersaturation $c = m/m_{eq} - 1$.

We are going to calculate porosity changes in sandstones, and we therefore need an expression for the rate of change of porosity in terms of r_{net}. The volume of quartz ΔV_{qz} precipitated during a time interval Δt can be expressed using the molar volume of quartz v (volume of quartz per number of moles of quartz). We then have that the volume of quartz precipitated from the mass M_w of water in the time interval Δt is

$$\Delta V_{qz} = r_{net} v M_w \Delta t. \qquad (11.6)$$

The quartz volume ΔV_{qz} becomes a volume fraction by dividing it by the bulk volume V_b of the sandstone, and we get

$$\frac{\Delta V_{qz}}{V_b} = k_f \frac{A_s}{V_b} c v \Delta t \qquad (11.7)$$

when expression (11.5) is inserted into equation (11.6). The volume fraction filled with quartz in the sandstone has to be equal to the porosity lost, $\Delta \phi = -\Delta V_{qz}/V_b$, because there is no compaction between the stylolites. The surface area A_s per bulk volume V_b is the specific surface $S(\phi)$. It is a function of the porosity, where the surface area available for quartz precipitation decreases with decreasing porosity, and we have

$$\frac{d\phi}{dt} = -k_f(T)S(\phi)\,c\,v. \qquad (11.8)$$

The reaction rate k_f is a function of temperature given by an Arrhenius law

$$k_f(T) = A_f e^{-E/RT} \qquad (11.9)$$

where A_f is the Arrhenius prefactor, E_f is the activation energy and R is the gas constant, and where the temperature is given in units of K. The minus sign in equation (11.8) says that the porosity decreases when the pore fluid is supersaturated with silica ($c > 0$). Equation (11.8) for the rate of change of porosity is a simple ordinary differential equation that can be integrated once we have a (reasonably simple) expression for the specific surface as a function of the porosity.

11.3 Surface area

Section 2.2 showed how the specific surface can be obtained from a correlation function of the porous medium. We also derived the specific surface as a function of the porosity from the correlation function for a medium of overlapping spherical grains of equal size (see Section 2.3). The specific surface area (surface area per bulk volume) of a sandstone is now expressed by the function

$$S(\phi) = S_0\,\hat{S}(\phi) \quad \text{with} \quad \hat{S}(\phi) = \left(\frac{\phi - \phi_c}{\phi_0 - \phi_c} \right)^n \qquad (11.10)$$

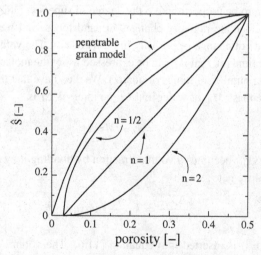

Figure 11.2. *The figure shows the porosity dependence of the specific surface.*

which is a function of the porosity. The specific surface is written as the product of the specific surface S_0 at the initial porosity ϕ_0 and a unit function $\hat{S}(\phi)$ of the porosity. The porosity-dependent part of the specific surface $\hat{S}(\phi)$ is a function that decreases from 1 to 0 as the porosity decreases from the surface porosity ϕ_0 towards the minimum porosity ϕ_c. The limit ϕ_c is the porosity at the percolation threshold, which is where the pore space becomes disconnected. The simple expression (11.10) for the specific surface can be used to approximate most experimental or theoretical relationships by tuning n, although it turns out that $n = 1$ is an often used exponent. Figure 11.2 shows the porosity dependent factor $\hat{S}(\phi)$ for $n = 1/2$, 1 and 2, together with the normalized specific surface of the penetrable grain model.

11.4 Isothermal quartz cementation

It is straightforward to integrate expression (11.8) for a constant temperature, when the specific surface is given by function (11.10). We then get

$$\phi(\hat{t}) = \begin{cases} \phi_c + (\phi_0 - \phi_c) \cdot \left(1 - (1-n)\,\hat{t}\right)^{1/(1-n)}, & n \neq 1 \\ \phi_c + (\phi_0 - \phi_c) \cdot \exp(-\hat{t}), & n = 1 \end{cases} \qquad (11.11)$$

where dimensionless time $\hat{t} = t/t_0$ is the time t scaled with characteristic time

$$t_0 = \frac{\phi_0 - \phi_c}{k_f S_0 \, v \, c}. \qquad (11.12)$$

The solution (11.11) shows that the porosity reaches the percolation threshold after the finite time $\hat{t} = 1/(1-n)$ when $n < 1$. This is seen in Figure 11.3 where the porosity

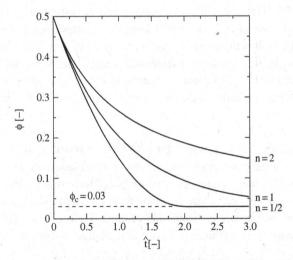

Figure 11.3. *The porosity is plotted as a function of time for isothermal conditions. Notice that the porosity reaches the lower limit ϕ_c after a finite time $\hat{t} = 2$ for the exponent $n = 1/2$.*

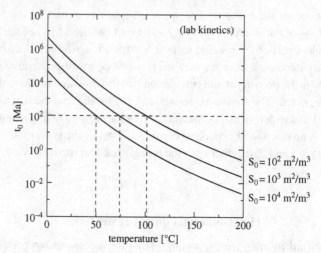

Figure 11.4. *The characteristic time (11.12) is plotted as a function of temperature for three different choices of the specific surface S_0 that are 10^2 m^2/m^3, 10^3 m^2/m^3 and 10^4 m^2/m^3.*

solution (11.11) is plotted as a function of time for several values of the exponent n. The porosity for the specific surface exponent $n = 1/2$ reaches the lower limit ϕ_c for $\hat{t} = 2$. For a specific surface exponent $n \geq 1$ the porosity will never reach the percolation threshold after finite time. The time needed for quartz cementation to reduce the porosity to one-half is found by solving for the time $\hat{t}_{1/2}$ when $\phi = (\phi_0 - \phi_c)/2$. We then get

$$\hat{t}_{1/2} = \begin{cases} \left(2^{n-1} - 1\right)/(n - 1), & n \neq 1 \\ \ln 2, & n = 1 \end{cases} \tag{11.13}$$

where $\hat{t}_{1/2} = 0.59$, 0.69 and 1 for $n = 1/2$, 1 and 2, respectively. The characteristic time t_0 is therefore a useful estimate of how long we can expect the porosity to last under isothermal conditions. It is therefore interesting to see what t_0 is for various temperatures. Figure 11.4 shows t_0 as a function of temperature using the data from Table 11.1. The characteristic time (11.12) shows that the parameters A_f, S_0, c and v all enter t_0 in the same product. The parameter that is the most difficult to constrain is the specific surface S_0. Figure 11.4 therefore shows the characteristic time for three different choices of S_0, which are 10^2 m^2/m^3, 10^3 m^2/m^3 and 10^4 m^2/m^3. We notice from Figure 11.4 that the specific surface needed to preserve the porosity for ~ 100 Ma at $100°C$ is as low as 100 m^2/m^3. Such a low specific surface corresponds to a coarse-grained sandstone with a grain diameter as large as ~ 1 cm. The quartz grains are often measured to be two orders of magnitude less in their diameters. This indicates that the laboratory-measured quartz kinetics may be too fast by as much as two orders of magnitude or more. Oelkers *et al.* (2000) discuss this important observation with reference to a large compilation of data for quartz kinetics.

Exercise 11.1 Derive expression (11.11) for the porosity as a function of time.
Solution: The rate of change of porosity, equation (11.8), can be written as

Table 11.1. *The data for quartz kinetics are taken from Tester et al. (1994) and the calibrated data are obtained by fitting the Arrhenius kinetics to the precipitation rates obtained by Walderhaug (1996). The specific surface 10^3 m^2/m^3 corresponds to a grain diameter of 1 mm.*

Parameter	Value	Units
ϕ_0	0.5	[–]
ϕ_c	0.03	[–]
R	8.314	$[\text{J K}^{-1}\text{mole}^{-1}]$
S_0	$1 \cdot 10^3$	$[\text{m}^2/\text{m}^3]$
c	0.01	[–]
v	$2.4 \cdot 10^{-5}$	$[\text{m}^3\text{mole}^{-1}]$
A_f (Tester)	24	$[\text{mole m}^{-2}\text{s}^{-1}]$
E_f (Tester)	90	$[\text{kJ mole}^{-1}]$
A_f (calibrated)	$8.37 \cdot 10^{-6}$	$[\text{mole m}^{-2}\text{s}^{-1}]$
E_f (calibrated)	60.1	$[\text{kJ mole}^{-1}]$

$$\frac{d\phi}{\left(\dfrac{\phi - \phi_c}{\phi_0 - \phi_c}\right)^n} = -k_f S_0 \, v \, c \, dt \tag{11.14}$$

when the specific surface function (11.10) is inserted. A change of variable from ϕ to $x = (\phi - \phi_c)/(\phi_0 - \phi_c)$ yields the integral

$$\frac{dx}{x^n} = -\frac{dt}{t_0} = -d\hat{t} \tag{11.15}$$

which is straightforward to integrate. The integration limits are time from 0 to t and the porosity from the initial porosity ϕ_0 to ϕ (or x from 1 to x).

11.5 Calibration of quartz kinetics

We saw in the preceding section that laboratory-measured quartz kinetics is too fast. Fortunately, there have been efforts to estimate the precipitation rates for quartz in sandstones. Walderhaug (1996) was able to express quartz cementation rates in the temperature window where quartz cementation operates (90°C to 120°C), with an exponential temperature dependence

$$r_w(T) = a \, 10^{b(T-T_0)} \tag{11.16}$$

where the rate r and the parameter a both have units $\text{mole m}^{-2}\text{s}^{-1}$ (just like A_f). The temperature is in kelvin and $T_0 = 273$ K, and the parameter b therefore has units °C^{-1}.

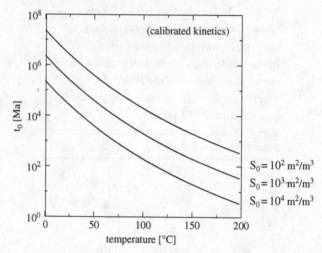

Figure 11.5. *The characteristic time (11.12) is plotted as a function of temperature for three different choices of the specific surface S_0 that are 10^2 m^2/m^3, 10^3 m^2/m^3 and 10^4 m^2/m^3.*

Walderhaug found that the parameter values $a = 1.98 \cdot 10^{-18}$ mole m^{-2}s^{-1} and $b = 0.022°$C^{-1} describe the quartz cementation process observed in many sandstone reservoirs. It is now possible to use Walderhaug's data to obtain the corresponding Arrhenius data. We can make two equations, $c\,k_f(T_1) = r_w(T_1)$ and $c\,k_f(T_2) = r_w(T_2)$, for the two unknowns A_f and E_f by requiring that the two rates are equal at the two temperatures T_1 and T_2. The solution of the two equations for the temperatures $T_1 = 90°$C and $T_2 = 120°$C yields the Arrhenius parameters $A_f = 8.37 \cdot 10^{-6}$ mole m^{-2}s^{-1} and $E_f = 60.1$ kJ mole^{-1}. We notice that these Arrhenius data are quite different from the laboratory-measured data in Table 11.1. Figure 11.5 shows the characteristic time once more, but this time using the calibrated Arrhenius data from the precipitation rate of Walderhaug. We see that a specific surface a little larger than 10^4 m^2/m^3 preserves the pore space for $\sim$100 Ma at $\sim$100°C. We also notice that t_0 is above 1 Ma for a temperature as high as 200°C, although it is uncertain how accurate the Arrhenius data are at 200°C. We could have applied Walderhaug's empirical precipitation rate (11.16) in t_0, but it is limited to the temperature window of quartz cementation. It is reasonable to expect Arrhenius kinetics ($Ae^{-E/RT}$) to be a better way to extrapolate the observed precipitation rates to temperatures outside the window of quartz cementation, instead of using the exponential temperature dependence ($a\,10^{b(T-T_0)}$) of the empirical precipitation rate.

Exercise 11.2 Solve the equations $c\,k_f(T_1) = r_w(T_1)$ and $c\,k_f(T_2) = r_w(T_2)$ for the unknowns A_f and E_f.
Solution: The two equations are

$$cA_f e^{-T_f/T_1} = a\,10^{b(T_1-T_0)} \tag{11.17}$$
$$cA_f e^{-T_f/T_2} = a\,10^{b(T_2-T_0)} \tag{11.18}$$

where $T_f = E_f/R$. Dividing the two left-hand sides and the two right-hand sides yields

$$e^{-T_f(1/T_1 - 1/T_2)} = 10^{b(T_1 - T_2)} \tag{11.19}$$

where we get

$$E_f = RT_f = R\ln(10)\, b\, T_1 T_2 \tag{11.20}$$

and

$$A_f = (a/c)\, 10^{b(T_1 - T_0)} e^{T_f/T_1}. \tag{11.21}$$

11.6 Cementation during constant burial

We have seen how we can integrate the rate of change of porosity $d\phi/dt$ when the temperature is constant. The temperature is normally not constant over long geological periods, and we are therefore going to integrate the same expression for the case of burial at a constant rate. We will then be able to predict the porosity for simple burial histories. Burial is assumed to take place along a constant thermal gradient, which implies that the heating rate is also constant, because the heating rate is the product of the thermal gradient and the burial rate. The temperature as a function of time is therefore

$$T(t) = T_0 + r_h t = T_0 + \frac{dT}{dz}\omega t \tag{11.22}$$

where T_0 is the surface temperature, r_h is the heating rate, dT/dz is the thermal gradient and ω is the (real) burial rate. We will follow a sample of quartzose sandstone from the surface, during burial, when only quartz cementation reduces the porosity. Mechanical compaction is therefore left out, although it could easily be added to the model. For example, one could let mechanical compaction operate from deposition (at the surface) until the depth at which only a small percentage of quartz has precipitated. A few percent of quartz cement will fix the grain framework and mechanical compaction stops. Quartz cementation takes over at this point during burial as the dominating process that reduces the pore space further. A nice feature of this simplified picture of the compaction process is that mechanical compaction and quartz cementation are considered as non-overlapping processes.

It is difficult to integrate equation (11.8) exactly when the temperature is increasing linearly because the temperature dependence is through an Arrhenius law (see Exercise 11.4). Fortunately, good approximations exist (see Exercise 11.3), and the porosity during burial can be written (as shown in Note 11.1)

$$\phi(T) = \begin{cases} \phi_c + (\phi_0 - \phi_c)\left\{1 - (1-n)N_B\Big(F(T) - F(T_0)\Big)\right\}^{1/(1-n)}, & n \neq 1 \\[2ex] \phi_c + (\phi_0 - \phi_c)\exp\left\{-N_B\Big(F(T) - F(T_0)\Big)\right\}, & n = 1 \end{cases} \tag{11.23}$$

where $F(T)$ is the function

$$F(T) = (T/T_f)^2 \exp(-T_f/T),$$ (11.24)

with $T_f = E_f/R$ and where the number N_B is

$$N_B = \frac{c v_q S_0 A_f E_f}{(\phi_0 - \phi_c) R \omega \, dT/dz}.$$ (11.25)

The porosity during the burial is written as a function of temperature, but it could easily have been written as a function of either time or depth. The temperature is related to time by expression (11.22), or to depth by the thermal gradient, $T = (dT/dz)\,z$. Notice that the parameters c, v_q, S_0, A_f, dT/dz and ω only enter the porosity–temperature expression (11.23) through the number N_B. The only parameters that appear in other places are n, ϕ_0, ϕ_c and E_f. The function F, which can be written $F(x) = x^2 e^{-1/x}$ for $x = T/T_f$, is an approximate expression of the integral $\int e^{-1/x} dx$ that for most practical purposes is sufficiently accurate. Exercises 11.3 and 11.4 give examples of better approximations than F. Figure 11.6a shows a comparison of the approximate porosity obtained with function F and the exact porosity. The approximation is quite accurate, especially considering the uncertainties normally associated with the input data, for instance the specific surface and the thermal gradient. Figure 11.6b shows how the porosity depends on the initial specific surface S_0. A large specific surface leads to more quartz precipitation than a smaller specific surface, as expected. Figure 11.6 shows a burial history with burial at a constant rate $\omega = 25$ m/Ma along a thermal gradient $dT/dz = 35°$C/km. (See Table 11.2 for the full set of parameters.) We see from Figure 11.6 (right) that cementation starts at the depth 2500 m for a sandstone with a typical specific surface $S_0 = 10^4$ m^2/m^3. The same sandstone is

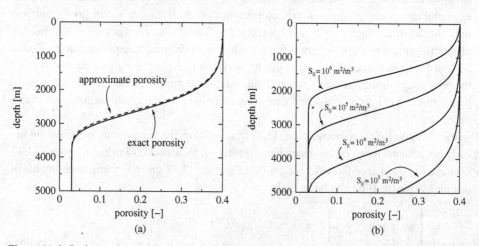

Figure 11.6. *Both panels are based on the data in Table 11.2, except for parameters added to the curves. (a) Comparison of the approximate porosity solution (11.23) with the exact porosity solution (see Exercise 11.4). (b) The porosity as a function of depth for different initial specific surfaces.*

Table 11.2. *The data used in the porosity–depth curves shown in Figures 11.6, 11.7 and 11.8 unless noted otherwise on the curves.*

Parameter	Value	Units
R	8.314	$[\text{J K}^{-1}\text{mole}^{-1}]$
S_0	$1 \cdot 10^5$	$[\text{m}^2/\text{m}^3]$
c	0.01	$[-]$
A_f	$8.36 \cdot 10^{-6}$	$[\text{mole m}^{-2}\text{s}^{-1}]$
E_f	60.1	$[\text{kJ mole}^{-1}]$
$T_f = E_f/R$	7227	$[\text{K}]$
v	$2.4 \cdot 10^{-5}$	$[\text{m}^3\text{mole}^{-1}]$
ϕ_0	0.4	$[-]$
ϕ_c	0.03	$[-]$
n	1.001	$[-]$
dT/dz	0.035	$[^\circ\text{C/m}]$
ω	25	$[\text{m Ma}^{-1}]$

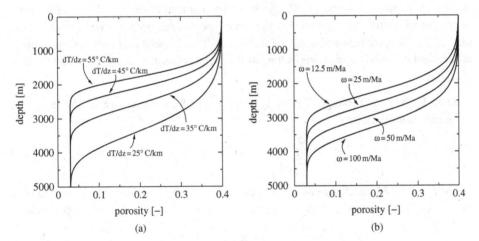

Figure 11.7. *These plots are based on the data in Table 11.2, except for parameters added to the curves. (a) The porosity as a function of depth for different thermal gradients. (b) The porosity as a function of depth for different burial rates.*

nearly filled with cement at the depth 4500 m. The porosity becomes quite sensitive to the thermal gradient, as seen from Figure 11.7 (left). A sandstone reservoir where the thermal gradient is as high as 50°C/km will easily lose most of its porosity before it is buried to the depth of 2500 m. The burial rate also has an important impact on the porosity reduction, as seen from Figure 11.7 (right). Slow burial means that a sandstone reservoir will experience a higher temperature–time impact than rapid burial before it reaches a certain depth.

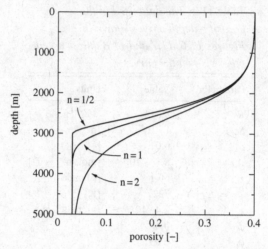

Figure 11.8. *The porosity as a function of depth for different exponents n in the specific surface function (11.10).*

Figures 11.6 and 11.7 show the porosity of the same (single) sample of sand/sandstone at different depths during burial. The figures represent also a porosity–depth relationship in the case of continuous deposition of the same type of sand, because every sample of the sand will experience the same temperature history. The temperature history of every piece of sediment deposited is the same as long as the burial is constant along a constant thermal gradient.

Note 11.1 The rate of change of porosity, equation (11.8), is once more the starting point, and it can be written as

$$\frac{d\phi}{\left(\frac{\phi - \phi_c}{\phi_0 - \phi_c}\right)^n} = -S_0 \, v \, c \, A_f e^{-T_f/T} dt \tag{11.26}$$

where $T_f = E_f/R$. A change of variable from t to $x = T/T_f$ where T is related to time by $T = (dT/dz)\,\omega\,t$, and a change of variable from ϕ to $y = (\phi - \phi_c)/(\phi_0 - \phi_c)$ yields

$$\int_1^{y_1} \frac{dy}{y^n} = -N_B \int_{x_0}^{x_1} e^{-1/x} dx. \tag{11.27}$$

The integration limits are $x_0 = x(T_0)$, $x_1 = x(T_1)$, $y_0 = y(\phi_0) = 1$ and $y_1 = y(\phi)$. Since the integration of $e^{-1/x}$ isn't a simple expression in well-known functions (as shown in Exercise 11.4) we will resort to approximations. A change of variable to $u = -1/x$ gives

$$\int e^{-1/x} dx = \int u^{-2} e^u du. \tag{11.28}$$

The latter integral can be rewritten using integration by parts as

$$\int u^{-2} e^u du = u^{-2} e^u + 2 \int u^{-3} e^u du \tag{11.29}$$

and after repeated use of integration by parts as

$$\int u^{-2}e^u du = u^{-2}e^u + 2u^{-3}e^u + 2\cdot 3u^{-4}e^u + \cdots + 2\cdot 3 \cdots (n-1)\int u^{-n}e^u du. \quad (11.30)$$

An inspection of the integration boundaries for x shows that $x \ll 1$, and therefore $u \gg 1$. The integral (11.30) is therefore accurately approximated by only the first term (see also Exercise 11.3). We can therefore write

$$\int_{x_0}^{x_1} e^{-1/x}dx \approx \left[x^2 e^{-1/x}\right]_{x_0}^{x_1}. \quad (11.31)$$

Inserting approximation (11.31) into the starting point (11.27) then gives the porosity (11.23) as a function of temperature during burial.

Exercise 11.3 Show that the first-order and second-order improvements to the approximation (11.31) are

$$\int_{x_0}^{x_1} e^{-1/x}dx \approx \left[x^2(1 - 2x)e^{-1/x}\right]_{x_0}^{x_1} \quad (11.32)$$

and

$$\int_{x_0}^{x_1} e^{-1/x}dx \approx \left[x^2(1 - 2x + 6x^2)e^{-1/x}\right]_{x_0}^{x_1}. \quad (11.33)$$

Calculate the definite integral $\int_{x_0}^{x_1} e^{-1/x}dx$ using typical integration limits as $x_0 = 0.045$ and $x_1 = 0.062$ for the lowest-order approximation (11.27) and then for the next two higher-order approximations. Are there any significant differences? (The integration limits $x_0 = 0.045$ and $x_1 = 0.062$ correspond to $T = 50°C$ and $T = 175°C$, respectively, when $E_f = 60.1 \text{ kJ mole}^{-1}$.)

Exercise 11.4 Show that

$$\int_{x_0}^{x_1} e^{-1/x}dx = \left[xe^{-1/x} - E_1(1/x)\right]_{x_0}^{x_1} \quad (11.34)$$

where the exponential integral function $E_1(x)$ is defined as

$$E_1(x) = \int_x^\infty \frac{e^{-u}}{u}du. \quad (11.35)$$

This is the exact solution to the right-hand side in equation (11.27), although it isn't very useful unless the function E_1 is available. The function E_1 can be replaced by a similar exponential integral function denoted E_i, where $E_1(x) = -E_i(-x)$.
Solution: A change of variable from x to $u = 1/x$ gives

$$\int e^{-1/x}dx = -\int \frac{e^{-u}}{u^2}du \quad (11.36)$$

and integration by parts then gives

$$-\int \frac{e^{-u}}{u^2}\,du = \frac{e^{-u}}{u} + \int \frac{e^{-u}}{u}\,du \tag{11.37}$$

and we are almost done.

11.7 Cementation for general burial histories

We are able to predict the porosity of a layer of quartzose sandstone that is heated at a constant rate during burial. The next step is to do the same for more general temperature histories. It turns out that a generalization of the results of the preceding section to a piecewise linear temperature history is straightforward. A piecewise linear temperature history can be specified by the temperature and the time of each interval in the history. We will let t_i and T_i denote the time and the temperature at the end interval i, respectively. The first interval is therefore from time t_0 to t_1 with initial temperature T_0 and final temperature T_1. The porosity for any temperature T in temperature interval k is (as shown in Note 11.2)

$$\phi(T) = \begin{cases} \phi_c + (\phi_0 - \phi_c)\left\{1 - (1-n)\sum_{i=1}^{k} N_{B,i}\Delta F_i\right\}^{1/(1-n)}, & n \neq 1 \\[2ex] \phi_c + (\phi_0 - \phi_c)\exp\left\{-\sum_{i=1}^{k} N_{B,i}\Delta F_i\right\}, & n = 1 \end{cases} \tag{11.38}$$

where the number $N_{B,i}$ during interval i is

$$N_{B,i} = \frac{cv_q S_0 A_f E_f}{(\phi_0 - \phi_c)R\,r_i} \tag{11.39}$$

and where r_i is the constant heating rate $r_i = (T_i - T_{i-1})/(t_i - t_{i-1})$ during the interval. The ΔF-function for interval i is

$$\Delta F_i = F(T_i) - F(T_{i-1}) \tag{11.40}$$

where F is function (11.24). There is an exception for the last interval, where $\Delta F_k = F(T) - F(T_{k-1})$, because T is any temperature between T_{k-1} and T_k.

Figure 11.9 shows an example of a porosity computation with equation (11.38). The temperature history in Figure 11.9a has three intervals, where the second interval shows a decreasing temperature with time, for instance caused by a major erosion process. Figure 11.9b shows the porosity corresponding to the temperature history in Figure 11.9a, and we see that the porosity is decreasing in all three intervals. The porosity reduction slows down with decreasing temperature in the second interval, before increasing once more with increasing temperature in the third interval. Porosity reduction by quartz cementation (or diagenesis in general) is different from mechanical compaction in this respect, because it does not stop until nearly all the porosity is lost. Mechanical compaction during uplifts

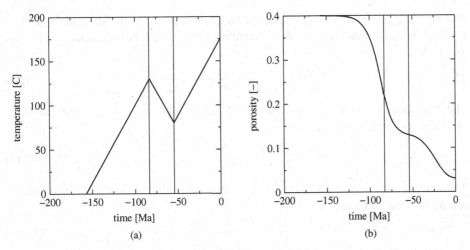

Figure 11.9. *(a) A piecewise linear temperature history made of three intervals. (b) The porosity as a function of time is shown for the temperature history in (a). The data used in the porosity plot, except for the burial history, are taken from Table 11.2.*

yields a slight porosity increase with decreasing effective stress. An example of such a slight recovery during unloading is shown in Figure 4.2.

Note 11.2 The porosity is found by integration of the rate of change of porosity, equation (11.8), and for a piecewise linear temperature history we have

$$\int_{\phi_0}^{\phi} \frac{d\phi}{\left(\frac{\phi - \phi_c}{\phi_0 - \phi_c}\right)^n} = -S_0 \, v \, c \, A_f \sum_{i=1}^{k} \int_{t_{i-1}}^{t_i} e^{-T_f/T} dt. \tag{11.41}$$

Each part of the sum is integrated by a change of variable from t to $x = T/T_f$ using the heating rate $dT/dt = r_i$. (We then get $dx = dT/T_f = (r_i/T_f)dt$.) A change of integration variable from ϕ to $y = (\phi - \phi_c)/(\phi_0 - \phi_c)$ then yields

$$\int_{1}^{y_1} \frac{dy}{y^n} = -\sum_{i} N_{B,i} \int_{x_{i-1}}^{x_i} e^{-1/x} dx \tag{11.42}$$

where the integration limits are $x_{i-1} = x(T_{i-1})$, $x_i = x(T_i)$, $y_0 = y(\phi_0) = 1$ and $y_1 = y(\phi)$. The remaining details are the same as in Note 11.1.

11.8 Strain rate

The rate of compaction can be measured as strain rate. Strain, in the case of compaction, is the relative shortening of a depth interval. The thickness l of a depth interval is conveniently expressed by the net amount of rock in the interval $(d\zeta)$ and the porosity (ϕ) as

$$l = \frac{d\zeta}{1 - \phi}, \tag{11.43}$$

because the net amount $d\zeta$ stays constant during compaction. The thickness of the interval is reduced by Δl when the interval is compacted and the porosity is decremented by $\Delta\phi$, and we have

$$\Delta l = \frac{d\zeta}{1-\phi} - \frac{d\zeta}{1-\phi-\Delta\phi}. \tag{11.44}$$

The difference Δl can be approximated as $\Delta l \approx l\Delta\phi/(1-\phi)$ assuming that $\Delta\phi \ll \phi$, and the strain can therefore be written as

$$\frac{\Delta l}{l} \approx \frac{\Delta\phi}{1-\phi}. \tag{11.45}$$

The strain (11.45) becomes the strain rate by dividing it with the time Δt of compaction, and we have the strain rate

$$\frac{1}{l}\frac{dl}{dt} = \frac{1}{(1-\phi)}\frac{d\phi}{dt}. \tag{11.46}$$

We already have expression (11.8) for $d\phi/dt$ and expression (11.23) for ϕ as a function of time during burial at a constant rate. These expressions can be used to make plots of the strain rate as a function of depth, as shown in Figure 11.10, where the strain rate is plotted for burial at a constant rate with the data from Table 11.1. The strain rates in Figure 11.10 show a clear peak in the depth interval where the rate of change of porosity is at its maximum. Note 11.3 shows that the temperature of maximum strain rate and maximum rate of change of porosity is accurately approximated by

$$T_{\max} \approx \frac{T_f}{a}\left(b - \ln N_B\right) - 273\,°\text{C} \tag{11.47}$$

in units $°\text{C}$, where $a = 444.2$ and $b = 48.4$ are constant parameters in a linear approximation (see below). The numbers N_B for the three curves shown in Figure 11.10

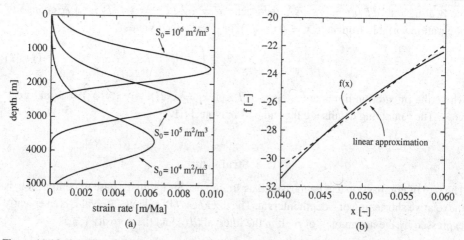

Figure 11.10. *(a) The strain rate for compaction as a function of depth. (b) The function $f(x) = 2\ln(x) - (1/x)$ and a linear approximation.*

are $N_B = 1.4 \cdot 10^{10}$, $1.4 \cdot 10^{11}$ and $1.4 \cdot 10^{12}$, for the initial specific surfaces areas $S_0 = 10^4 \text{ m}^2/\text{m}^3$, $10^5 \text{ m}^2/\text{m}^3$ and $10^6 \text{ m}^2/\text{m}^3$, respectively. Using these numbers for N_B we get that the depths where the strain rate is at its maximum are $z_{\max} = 3834$ m, 2763 m and 1693 m. Figure 11.10 shows that these depths are fairly accurate.

Note 11.3 The maximum rate of change of porosity during constant burial can be used to approximate the temperature of maximum strain rate. We notice that the numerator $d\phi/dt$ is a much more important term in the strain rate than ϕ is in the denominator. If we approximate ϕ in the denominator by a constant porosity like ~ 0.2 we get that the maximum strain rate is given by the point of maximum rate of change of porosity. Time in expression (11.8) for $d\phi/dt$ can be replaced by the dimensionless variable $x = (dT/dz)\omega t/T_f$, when burial is at a constant rate ω along a constant thermal gradient dT/dz. The porosity can be scaled to the unit interval as $\hat{\phi} = (\phi - \phi_c)/(\phi_0 - \phi_c)$, and we get that the rate of change of porosity is

$$\frac{d\hat{\phi}}{dx} = -N_B \hat{S}(\hat{\phi}) e^{-1/x}. \tag{11.48}$$

The point of maximum rate of change of porosity is then found by solving

$$\frac{d^2\hat{\phi}}{dx^2} = -N_B \left(\hat{S}'(\hat{\phi}) \frac{d\hat{\phi}}{dx} e^{-1/x} + \hat{S}(\hat{\phi}) \frac{e^{-1/x}}{x^2} \right) = 0 \tag{11.49}$$

for x, where $\hat{S}' = d\hat{S}/d\hat{\phi}$. The factor $d\phi/dx$ is replaced by expression (11.48) and we get that

$$\frac{d^2\hat{\phi}}{dx^2} = -N_B \hat{S}(\hat{\phi}) \left(-\hat{S}'(\hat{\phi}) N_B e^{-1/x} + \frac{1}{x^2} \right) e^{-1/x} = 0. \tag{11.50}$$

The differentiation $\hat{S}'$ is 1 for the specific surface exponent $n = 1$, and $d^2\hat{\phi}/dx^2 = 0$ then implies that

$$x^2 e^{-1/x} = 1/N_B. \tag{11.51}$$

This equation is not easily solved, but it can be written as $2\ln x - (1/x) = -\ln N_B$, where the left-hand side, $f(x) = 2\ln x - (1/x)$, can be replaced by a linear approximation $ax - b$. Figure 11.10b shows that the function $f(x)$ is close to a straight line in the interval $x = 0.04$ to $x = 0.06$, which corresponds to the temperature interval $T = 16°C$ to $160°C$ for T_f from Table 11.1. The straight line is equal to the curve $f(x)$ at the points $x = 0.045$ and 0.055, which yields the parameters $a = 444.2$ and $b = 48.4$. These numbers for a and b give equation (11.47) for the temperature the maximum rate of porosity change.

11.9 A reaction–diffusion equation for silica

We have so far taken for granted that the silica concentration stays constant between the stylolites regardless of their separation and the temperature. This is not necessarily the case, and to see why we have to solve a reaction–diffusion equation for the silica concentration in the interval between the stylolites. A reaction–diffusion equation is derived following the same line of reasoning as for mass conservation in Section 3.18. We consider silica concentration in a small volume element as shown in Figure 3.18. The flux F of silica in solution at any point along the x-axis is the superposition of two transport processes, advection and diffusion,

$$F(x) = mv - \frac{\phi D}{\tau} \frac{\partial m}{\partial x} \tag{11.52}$$

where m is the silica concentration in units of moles per m^3. The first term mv in the F-function is the silica flux due to advection (fluid flow), where v is the Darcy flux. The second term is the silica flux due to diffusion, where the flux is proportional to the concentration gradient $\partial m/\partial x$ and the diffusivity D of the fluid. This term is Fick's law for diffusion in porous media, and it differs from Fick's law for diffusion in a fluid by the two factors ϕ and τ. The porosity factor ϕ says that diffusion takes place only through this fraction of the cross-section area. This is precisely what we get from a model of porous media made of parallel tubes. The denominator τ is the tortuosity, and it accounts for the fact that pathways through a porous medium rarely are straight lines. The tortuosity is only slightly larger than 1 for porous media with a high porosity, and it may increase to above 2 at low porosities. The number of moles of silica M_{in} that is transported into a small volume with cross-section A during the time interval Δt, and the number of moles of silica M_{out} that is transported out of the same volume at the other side, are

$$M_{in} = F(x) A \Delta t \tag{11.53}$$
$$M_{out} = F(x + \Delta x) A \Delta t \tag{11.54}$$

respectively. (Such an elementary volume is shown in Figure 3.18.) The difference between M_{in} and M_{out} is

$$M_{in} - M_{out} \approx -\frac{\partial F}{\partial x} \Delta x \, A \, \Delta t. \tag{11.55}$$

The number of moles of silica in the pore volume at the beginning and the end of the time interval is

$$M_1 = m(x_c, t) \phi \Delta V \tag{11.56}$$
$$M_2 = m(x_c, t + \Delta t) \phi \Delta V, \tag{11.57}$$

respectively, where x_c is the x-coordinate of the center of the volume. The bulk volume of the box is $\Delta V = A \Delta x$ (and Δx is its length). The difference between M_2 and M_1 is

$$M_2 - M_1 \approx \frac{\partial(\phi m)}{\partial t} \Delta t \Delta V. \tag{11.58}$$

Conservation of silica requires that the difference in mass due to transport through the boundaries ($M_{in} - M_{out}$) plus the mass produced (or consumed) is equal to the increase in silica mass ($M_2 - M_1$) during the time interval (Δt). The mass balance of silica in the volume is therefore

$$M_2 - M_1 = M_{in} - M_{out} + q\Delta V \Delta t \qquad (11.59)$$

where q is a source term for silica that accounts for the rate of silica production in moles per bulk volume rock and per time due to dissolution or precipitation. (Notice that a negative source term is a sink term.) A division of equation (11.59) by $\Delta V \Delta t$ yields the reaction–transport equation for silica

$$\frac{\partial(\phi m)}{\partial t} + \frac{\partial(mv)}{\partial x} - \frac{\partial}{\partial x}\left(\frac{\phi D}{\tau}\frac{\partial m}{\partial x}\right) = q. \qquad (11.60)$$

Equation (11.8) gives the source/sink term for dissolution and precipitation

$$q = Sk_f\left(1 - \frac{m}{m_{eq}}\right). \qquad (11.61)$$

The following expression for the mass balance of the fluid simplifies equation (11.60) for reaction and transport:

$$\frac{\partial\phi}{\partial t} + \frac{\partial v}{\partial x} = 0, \qquad (11.62)$$

and we get

$$\phi\frac{\partial m}{\partial t} + v\frac{\partial m}{\partial x} - \frac{\partial}{\partial x}\left(\frac{\phi D}{\tau}\frac{\partial m}{\partial x}\right) = Sk_f\left(1 - \frac{m}{m_{eq}}\right). \qquad (11.63)$$

This reaction–transport equation is by no means restricted to silica only, although it was derived for silica. The same equation applies to reaction and transport of other species in the fluid phase (of a porous medium) with the only difference that different species and different processes may have different source terms.

Exercise 11.5 Show that equation (11.62) is an expression for mass conservation of a fluid with constant density. (We will return to this topic in Chapter 12.)

11.10 The silica concentration between stylolites

We have from the introduction (to this chapter) that dissolution and compaction take place along stylolites where the degree of supersaturation is estimated to be ~ 0.01. The dissolved silica is transported away from the stylolites by diffusion into the inter stylolite region where it precipitates as silica. The silica concentration between the stylolites is found by solving the reaction–transport equation (11.63). Before we do that we will first simplify it a little bit by setting $v = 0$. There is no fluid flow in the reservoir. Other simplifications are a constant porosity ϕ and a constant diffusivity D between the stylolites. The tortuosity is simply set to 1, because the uncertainty in the diffusivity is probably much larger than a factor 2. Next we will introduce the degree of supersaturation

$$c = \frac{m}{m_{eq}} - 1 \tag{11.64}$$

as the main variable instead of the concentration m. The reaction–transport equation (11.63) is then

$$\frac{\partial c}{\partial t} - D\frac{\partial^2 c}{\partial z^2} = -\frac{Sk_f}{\phi m_{eq}}c. \tag{11.65}$$

The stylolite seams are normally horizontal and the reaction–transport equation is therefore along the z-axis. Equation (11.65) is now rewritten using two characteristic times as

$$\frac{\partial c}{\partial t} - \frac{1}{t_D}\frac{\partial^2 c}{\partial \hat{z}^2} = -\frac{1}{t_p}c \tag{11.66}$$

where $\hat{z} = z/l_0$ is a dimensionless unit coordinate, and where l_0 is the distance between two stylolites. There are two characteristic times in the equation, where the first,

$$t_D = \frac{l_0^2}{D}, \tag{11.67}$$

is a time characteristic for the diffusion process, and the second one,

$$t_p = \frac{\phi m_{eq}}{Sk_f}, \tag{11.68}$$

is a characteristic time for the precipitation process. To see that t_D is a characteristic time for the diffusion process we let $t_p = \infty$. No reaction takes place and we then have that the concentration c is the solution of the diffusion equation. The solution of a diffusion equation tells us that a concentration gradient will be reduced to half after a time interval that is $\sim t_D$. We see that t_p characterizes the reaction (precipitation) process by a similar argument. There is no transport (diffusion) only reaction (precipitation) in a system where $t_D = \infty$, and the reaction–transport equation is

$$\frac{\partial c}{\partial t} = -\frac{1}{t_p}c \tag{11.69}$$

which has the solution $c(t) = c_0 e^{-t/t_p}$. An initial concentration c_0 will then decay with a half-life that is $t_{1/2} = \ln 2\, t_p$. A completely dimensionless reaction–transport formulation is made by scaling time with the characteristic time for diffusion, and we get

$$\frac{\partial c}{\partial \hat{t}} - \frac{\partial^2 c}{\partial \hat{z}^2} = -\mathrm{Da}\, c \tag{11.70}$$

where

$$\mathrm{Da} = \frac{t_D}{t_p} = \frac{l_0^2 Sk_f}{\phi m_{eq} D} \tag{11.71}$$

is the Damköhler number. The Da-number measures the time scale of transport (diffusion) relative to the time scale of reaction (precipitation). We will soon see how the reaction–transport process is characterized by the Da-number. Notice that the Da-number is the only parameter in the equation for the silica concentration. Two boundary conditions are needed before we can solve equation (11.70) for the degree of supersaturation; there is a constant degree of supersaturation c_0 at both stylolites

$$c(x{=}0) = c_0 \quad \text{and} \quad c(x{=}1) = c_0. \tag{11.72}$$

We are interested in the solution of equation (11.70) after long time spans, and in order to simplify the task we assume that the silica concentration is stationary. A stationary state is reached for $t \gg t_D$ as we have seen for various solutions of the time-dependent temperature equation (which is exactly the same type of equation). The characteristic time t_D is ~ 1 year for a stylolite separation $l_0 = 0.2$ m and a diffusivity $D = 10^{-9}$ m^2 s^{-1}. This is a very short time compared to the cementation process that takes millions of years. In other words, we can assume that $\partial c / \partial \hat{t} \approx 0$. It is shown below that the solution of the stationary reaction–transport equation

$$\frac{d^2 c}{d\hat{z}^2} - \mathrm{Da}\, c = 0 \tag{11.73}$$

with boundary conditions (11.72) is

$$c(\hat{z}) = c_0 \frac{\cosh\left(\sqrt{\mathrm{Da}}\,(x - \tfrac{1}{2})\right)}{\cosh\left(\tfrac{1}{2}\sqrt{\mathrm{Da}}\right)}. \tag{11.74}$$

Figure 11.11 shows the solution (11.74) for different Da-numbers. We see that there is an almost constant degree of supersaturation between the stylolites for Da-numbers in the regime Da $\ll 1$, where the characteristic time for diffusion is smaller than the characteristic time for precipitation. In the other regime, Da $\gg 1$, where the characteristic time for diffusion is much larger than the characteristic time for precipitation, the degree of supersaturation approaches zero between the stylolites. The diffusion process does not keep pace with the precipitation process when Da $\gg 1$.

The porosity of a sandstone was computed in previous sections assuming that silica concentration was even between two stylolites. We have now seen that this assumption corresponds to the regime Da $\ll 1$ for the reaction–diffusion equation. It remains to check that this assumption is good. The Damköhler number (11.71) is a function of temperature, because both the rate for quartz kinetics (k_f) and the equilibrium concentration of silica (m_{eq}) are temperature dependent. The Arrhenius law (11.9) with parameters from Table 11.1 gives the rate k_f. The following function of temperature (in kelvin) taken from Gunnarsson and Arnórsson (2000) gives the equilibrium concentration m_{eq} (in units mole m^{-3}):

$$\log_{10} m_{eq} = -31.188 + \frac{197.47}{T} - 5.851 \times 10^{-6} T^2 + 12.245 \times \log_{10} T. \tag{11.75}$$

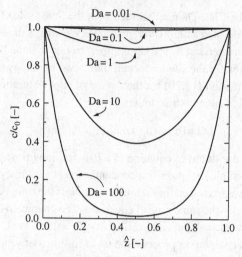

Figure 11.11. *The degree of supersaturation between two stylolites.*

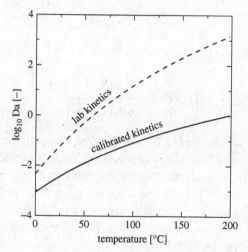

Figure 11.12. *The Da-number is plotted as a function of temperature for $S = 10^4 \, m^2/m^3$.*

Figure 11.12 shows the Da-number for both laboratory-measured data and calibrated data. We see that the cementation process during burial of a sandstone with $S = 10^4 \, m^2/m^3$ enters the regime Da $\gg 1$ already for temperatures at ~80°C with laboratory measured kinetics for quartz. The regime Da $\gg 1$ does not appear until the temperature has passed beyond 200°C when calibrated data for quartz kinetics is used. This is once more an indication that the laboratory-measured kinetics for quartz is too fast for geological applications. The rate k_f for quartz kinetics is always multiplied with the specific surface when applied, and an alternative explanation for the fast kinetics is that the specific surface of sandstones is orders of magnitude less than normally assumed.

Note 11.4 The stationary equation (11.70) is solved by factorization

$$\left(\frac{d}{d\hat{z}} - \sqrt{Da}\right)\left(\frac{d}{d\hat{z}} + \sqrt{Da}\right)c = 0 \tag{11.76}$$

where the function $c = c(\hat{z})$ becomes zero when acted on by one of the factors in the product. The equations for each factor are

$$\left(\frac{d}{d\hat{z}} \pm \sqrt{Da}\right)c = 0, \tag{11.77}$$

and they have the functions $\exp(\pm\sqrt{Da}\,\hat{z})$ as solutions. Equation (11.70) has as a solution any linear combination of these functions:

$$c(\hat{z}) = c_1 e^{\sqrt{Da}\,\hat{z}} + c_2 e^{-\sqrt{Da}\,\hat{z}}. \tag{11.78}$$

The two coefficients c_1 and c_2 are found from the two boundary conditions (11.72), which yield

$$c_1 = \frac{c_0\left(1 - e^{-\sqrt{Da}}\right)}{e^{\sqrt{Da}} - e^{-\sqrt{Da}}} \tag{11.79}$$

$$c_2 = \frac{c_0\left(e^{\sqrt{Da}} - 1\right)}{e^{\sqrt{Da}} - e^{-\sqrt{Da}}}. \tag{11.80}$$

When these coefficients are inserted into equation (11.78) we get the solution

$$c(\hat{z}) = \frac{c_0}{\sinh\sqrt{Da}}\left(\sinh(\sqrt{Da}\,\hat{z}) + \sinh(\sqrt{Da}\,(1 - \hat{z}))\right). \tag{11.81}$$

(We recall that $\sinh x = (e^x - e^{-x})/2$.) The denominator and numerator can be simplified using the formulas

$$\sinh(2x) = 2\cosh x\,\sinh x \tag{11.82}$$

$$\sinh x \pm \sinh y = 2\cosh\left(\frac{x \pm y}{2}\right)\cosh\left(\frac{x \mp y}{2}\right), \tag{11.83}$$

respectively, and we finally get the solution (11.74).

Note 11.5 A simplified system where quartz is the dominant mineral is assumed,

$$SiO_2 + 2H_2O \Leftrightarrow H_4SiO_4 \tag{11.84}$$

where the equilibrium constant for the reaction is

$$K = \frac{a_{H_4SiO_4}}{a_{SiO_2}\,a_{H_2O}} \tag{11.85}$$

and where $a_{H_4SiO_4}$, a_{SiO_2} and a_{H_2O} are the activity coefficients of silica, quartz and water, respectively. Furthermore, it is assumed that the activity coefficients of quartz and water can be approximated by 1, and that the solution is sufficiently dilute for the activity coefficient of silica also to be approximated by 1. The equilibrium concentration in the fluid is then $m_{eq} \approx K$.

Exercise 11.6 How large must the specific surface area be for the Damköhler number to become one at 100°C?

11.11 Further reading

There are several models for porosity reduction of (quartzose) sandstones by quartz cementation. The one presented here has dissolution of quartz at mica interfaces and stylolites as the source for silica. This model is presented in a series of papers of Bjørkum, Oelkers, and Walderhaug: Bjørkum (1994, 1996), Oelkers *et al.* (1992, 1993, 1996), Walderhaug (1994b, a, 1996). The degree of supersaturation created by stylolites is estimated by a work of Aase *et al.* (1996). Stylolites as the source for silica was suggested early by Heald (1955). Bjørkum (1996) gives petrographic observations that show precipitation to be the rate-limiting step in the cementation process. Oelkers *et al.* (1996) show that supersaturation of silica becomes almost uniform throughout the interstylolite region when the characteristic time for the diffusion process is much shorter than the characteristic time for the precipitation process. Porosity predictions based on this model have been made by several authors: Oelkers *et al.* (1996), Walderhaug (1996), Bjørkum *et al.* (1998) and Wangen (1999). Oelkers *et al.* (1996) solved the diffusion-reaction equation for the silica supersaturation between stylolites numerically to obtain the precipitation rates of quartz. Bjørkum *et al.* (1998) extended this model to porosity predictions in the presence of hydrocarbons, and compared the porosity predictions to a set of observations. Oelkers *et al.* (2000) give a thorough discussion of the problem that laboratory-measured quartz kinetics becomes too fast when applied to geological problems.

The intergranular pressure solution, where quartz is assumed to dissolve at grain contacts under high pressure, is another model that has been applied to quartz cementation of sandstones. The dissolved silica is then transported by diffusion in a thin fluid film between the grain contacts out to the pore space, where the silica precipitates as cement. Application of the pressure solution in basin modeling was carried out by Angevine and Turcotte (1993), Birchwood and Turcotte (1994), Ramm (1992), Dewers and Ortoleva (1990), Lemée and Guéguen (1996) and Schneider *et al.* (1996).

A third model that has been suggested for quartz cementation of sandstones is precipitation from a flow of supersaturated aqueous fluids. But it has been shown that unreasonably large volumes of fluids are needed to fill large fractions of a pore space compared to possible sources of aqueous fluids at the depths where quartz cementation operates (deeper than ≈ 1.5 km), see Bjørlykke (1993).

12

Overpressure and compaction: exact solutions

This chapter looks at overpressure build-up caused by porosity reduction during deposition and burial of sediments. Sediments compact when they get buried, either mechanically from the increasing weight above, or by chemical processes. Mechanical compaction takes place in the upper part of a basin, which is too cold for chemical (diagenetic) reactions to be sufficiently effective. The porosity reduction and compaction happens by chemical process further down in the sedimentary basin, where mechanical compaction is arrested by mineral cement. The rates of the chemical processes increase rapidly with increasing temperature. Porosity reduction implies expulsion of fluids from the pore space, and a pressure gradient is needed (in Darcy's law) to drive the expulsion. Fluid expulsion therefore leads to overpressure build-up during deposition and compaction. This chapter shows the basic behavior of overpressure build-up with simple analytical models. One common simplification of these models is that they consider deposition of just one lithology at a constant deposition rate.

Thanks to the oil industry a large number of wells have been drilled in offshore areas where deposition is currently taking place. The mud weight used to balance the formation pressure in wells is a valuable indication of the overpressure, although it is not a precise pressure measurement. It is normally larger than the formation pressure in order to ensure that there is no outflow from the formations. Figure 12.1a shows the mud density of an overpressured well offshore mid-Norway and Figure 12.1b shows where the corresponding fluid pressure is between the hydrostatic water pressure and the lithostatic pressure. The two latter pressures are normally the lower and the upper bounds for the fluid pressure. The upper part of the well is hydrostatic down to the depth 1.5 km – it is then overpressured further down, and from 4.5 km the overpressure approaches the lithostatic pressure. These overpressures could be associated with the rapid deposition of sediments during the Late Pliocene and Pleistocene, roughly 1000 m during the time interval from -3.2 Ma to -2 Ma. There has also been further deposition of 300 m until today.

Exercise 12.1 Assume that the mud weight balances the fluid pressure exactly when the mud density is ϱ_m and the mud column has the height z.
(a) What is the fluid pressure at z?
(b) What is the overpressure gradient at z?

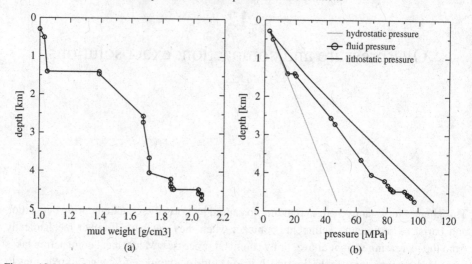

Figure 12.1. *Mud density data for the overpressured well 6406/2-7 offshore mid-Norway. The sea surface is at the depth 0 and the seafloor is at the depth 273 m. (a) Mud weight. (b) The fluid pressure that corresponds to the mud weight in (a).*

Solution: (a) The fluid pressure is $p_f = \varrho_m g z$.
(b) The overpressure is $p_e = p_f - p_h$ and the gradient is $dp_e/dz = (\varrho_m - \varrho_w)g$ where ϱ_w is the water density.

12.1 The pressure equation in 1D

The equation for overpressure build-up is derived from the mass conservation of fluid. It is written in the z-direction for a volume of rock that contains the same sedimentary grains through time. The volume is a box with the height $dz(t) = z_2(t) - z_1(t)$ and an area A, as shown in Figure 12.2. Mass conservation of fluid in the volume is

$$A\frac{d}{dt}\int_{z_1}^{z_2} \phi\varrho_f \, dz = A(\varrho_f v)\big|_{z_1} - A(\varrho_f v)\big|_{z_2} \qquad (12.1)$$

where v is the vertical Darcy flux, ϕ is the porosity and ϱ_f is the fluid density. Equation (12.1) says that the rate of change of fluid mass in the volume is equal to the mass flux of fluid into the volume minus the mass flux of fluid out of the volume. We see that the area A can be left out in equation (12.1). The fluid density is now assumed constant and it drops out too. The time-differentiation on the left-hand side does not go through the integration sign because compaction makes the height of the box $dz(t)$ dependent on time. This problem can be avoided by a change of integration variable from z to the Lagrange coordinate ζ, using that $dz = d\zeta/(1 - \phi)$ (see Section 3.17). Furthermore, along the ζ-axis the box is between the fixed positions ζ_1 and

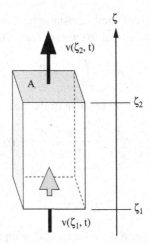

Figure 12.2. *Mass conservation in the vertical direction. The difference between the mass that flows in and the mass that flows out has to accumulate inside the box.*

ζ_2 through time. (The ζ-coordinates ζ_1 and ζ_2 correspond to z_1 and z_2.) The left-hand side becomes

$$\frac{d}{dt}\int_{z_1}^{z_2}\phi\,dz = \frac{d}{dt}\int_{\zeta_1}^{\zeta_2}\frac{\phi}{1-\phi}\,d\zeta = \int_{\zeta_1}^{\zeta_2}\frac{\partial}{\partial t}\left(\frac{\phi}{1-\phi}\right)d\zeta \qquad (12.2)$$

because ζ is a constant through time for the box. The right-hand side of equation (12.1) can also be rewritten in terms of the ζ-coordinate:

$$v(\zeta_1) - v(\zeta_2) = -\int_{\zeta_1}^{\zeta_2}\frac{\partial v}{\partial \zeta}d\zeta'. \qquad (12.3)$$

Both the left-hand side and the right-hand side of equation (12.1) are now expressed as integrals over the Lagrange coordinate ζ, by equation (12.2) and equation (12.3), respectively. We can drop the integration and get the following expression for fluid conservation in the vertical direction:

$$\frac{\partial}{\partial t}\left(\frac{\phi}{1-\phi}\right) = -\frac{\partial v}{\partial \zeta}. \qquad (12.4)$$

We recall that the void ratio is $e = \phi/(1-\phi)$ and conservation of fluid mass in the vertical direction is simply

$$\frac{\partial e}{\partial t} + \frac{\partial v}{\partial \zeta} = 0. \qquad (12.5)$$

The usefulness of the Lagrange coordinate ζ is that the time-differentiation is by (normal) partial derivative rather than by the material derivative.

The Darcy flux in the vertical direction is (see Section 2.5)

$$v = -\frac{k}{\mu}\left(\frac{\partial p_f}{\partial z} - \varrho_f g\right). \qquad (12.6)$$

The fluid pressure p_f is the sum of the hydrostatic pressure p_h and the overpressure p. The hydrostatic pressure is the weight of the water column up to the basin surface,

$$p_h = \int_0^z \varrho_f g \, dz, \qquad (12.7)$$

when the basin surface is at $z = 0$. The gradient of the water pressure is

$$\frac{\partial p_f}{\partial z} = \frac{\partial p}{\partial z} + \frac{\partial p_h}{\partial z} = \frac{\partial p}{\partial z} + \varrho_f g \qquad (12.8)$$

and Darcy's law is therefore

$$v = -\frac{k}{\mu} \frac{\partial p}{\partial z} \qquad (12.9)$$

in terms of the overpressure p. Since we want to work with the Lagrange coordinate ζ we need to convert the differentiation in z to a differentiation in ζ. We get

$$\frac{\partial p}{\partial z} = (1 - \phi) \frac{\partial p}{\partial \zeta} \qquad (12.10)$$

because $dz = d\zeta/(1 - \phi)$. We also have that the factor $1 - \phi$ can be written as $1/(1 + e)$ using the void ratio. Inserting the pressure gradient (12.10) into Darcy's law (12.9), and then Darcy's law into expression (12.5) for mass conservation gives the pressure equation

$$\frac{\partial e}{\partial t} - \frac{\partial}{\partial \zeta} \left(\frac{k}{(1 + e)\mu} \frac{\partial p}{\partial \zeta} \right) = 0. \qquad (12.11)$$

This is the basic equation for overpressure build-up that we are going to solve for two different models of the void ratio. In the first model the void ratio decreases exponentially with the depth, similar to the porosity function in Section 5.1, and in the second model the void ratio is a linear function of effective vertical stress.

12.2 The Darcy flux caused by compaction

The basic equation for fluid mass conservation can be used to derive the Darcy flux caused by compaction (porosity reduction) during deposition of sediments. We have

$$\frac{\partial v}{\partial \zeta} = -\frac{\partial e}{\partial t} \qquad (12.12)$$

which implies that

$$v = -\int_0^\zeta \frac{\partial e}{\partial t} d\zeta \qquad (12.13)$$

when there is no fluid flow into the base of the basin (at $\zeta = 0$). This integration can be carried out for any porosity (or void ratio) function that is only a function of the depth from the basin surface measured as net (porosity-free) rock. For such porosity functions we have that the void ratio is

$$e = e(\zeta^* - \zeta) \qquad (12.14)$$

where the net depth is $\zeta^* - \zeta$. ζ^* is the net thickness of the basin measured as net rock. (See Section 3.17.) Let the basin be in a state of constant deposition, $\zeta^* = \omega t$, where ω is the net deposition rate. We can now write the void ratio as a function of the argument $u = \omega t - \zeta$. The integration (12.13) is carried out by a change of integration variable from ζ to u, and we get

$$v(\zeta) = -\int_0^\zeta \frac{\partial e}{\partial t} d\zeta = -\int_0^\zeta \frac{de}{du} \frac{\partial u}{\partial t} d\zeta = \omega \int_{u(0)}^{u(\zeta)} \frac{de}{du} du = \omega \left(e(\zeta) - e_{\text{bot}} \right) \quad (12.15)$$

where e_{bot} is the void ratio at the base of the basin (at $\zeta = 0$). The Darcy flux can be approximated by $v \approx \omega e$, when the basin has become sufficiently deep for $e_{\text{bot}} \approx 0$. This turns out to be a good estimate for the Darcy flux in 1D models even if the porosity is not a function of the net sediment depth.

The results above can be used to answer how fast the fluid is flowing upwards relative to the subsidence of the sediments. The real (average) velocity of the fluid is $v_f = v/\phi$. At the base of the basin $v = 0$, and the fluid is therefore buried with the sediments during deposition towards the base of the basin. The fluid velocity increases upwards through the sedimentary column, and the maximum velocity is found at the basin surface. We have that v_f is a little less than $\omega/(1 - \phi_0)$ at the surface, because e_{bot} is not exactly equal to zero. (The variable ϕ_0 is the surface porosity.) The subsidence of the sediments at the surface is exactly $\omega/(1 - \phi_0)$, which implies that the fluid is buried with the sediments even at the surface. The conclusion that the fluid is buried with the sediments is based on the assumption of a 1D model with porosity as a function of the net (porosity-free) depth. Burial of fluid is not always the case in basins because of focusing of the fluid flow, for instance along faults.

12.3 Void ratio as a function of depth

The pressure equation (12.11) turns out to be difficult to solve for the case of deposition of sediments, because the solution domain then grows with time. However, there is an exception, and that is when the void ratio is given as an explicit function of depth. We have from the previous section that the Darcy velocity is given by the simple and exact relationship (12.15). It is an integration of this expression for the Darcy flux that allows us to obtain the overpressure in a basin during deposition. We are therefore going to assume once more an Athy-type of porosity function. Instead of the porosity we will let the void ratio decrease as a function of the depth measured as net (porosity-free) rock,

$$e = e_0 \exp\left(-\frac{\zeta^* - \zeta}{\zeta_0} \right), \qquad (12.16)$$

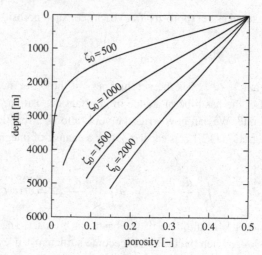

Figure 12.3. *The porosity as a function of z obtained from the void ratio function (12.16) and the z-coordinate function (12.18). The porosity function is plotted for different values of the parameter ζ_0.*

where $\zeta^* - \zeta$ is the net depth. The parameter ζ_0 controls how fast the void ratio approaches zero with increasing depth. Figure 12.3 shows the void ratio in terms of the porosity for different values of ζ_0. The variable e is now used exclusively for the void ratio, and exponentiation is denoted by the exp-function. This void ratio function allows us to calculate the real depth for a given ζ-coordinate. We have

$$z = \int_\zeta^{\zeta^*} \frac{d\zeta}{1 - \phi} = \int_\zeta^{\zeta^*} (1 + e)\, d\zeta. \tag{12.17}$$

Using the void ratio function (12.16) we get

$$z(\zeta) = (\zeta^* - \zeta) + \zeta_0(e_0 - e) \tag{12.18}$$

where e is the void ratio at the given ζ-position. (See Exercise 12.2.) We now have both the void ratio and the z-coordinate as a function of the ζ-coordinate. It is then possible to plot the porosity as a function of the z-coordinate. (Recall that $\phi = e/(1 + e)$.)

We will first recapture some pressure definitions before we proceed with the overpressure. The hydrostatic pressure is the pressure from the weight of the water column

$$p_h = \int_0^z \varrho_f g\, dz \tag{12.19}$$

where ϱ_f is the fluid density and g is the constant of gravity. The lithostatic pressure is from the weight of the fluid saturated sedimentary column

$$p_b = \int_0^z \left(\phi \varrho_f + (1 - \phi)\varrho_s\right) g\, dz \tag{12.20}$$

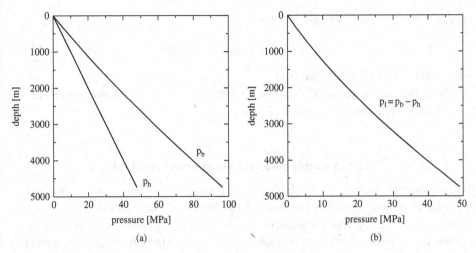

Figure 12.4. (a) The fluid pressure is normally bounded below by the hydrostatic pressure and above by the lithostatic pressure. (b) The same bounds on the overpressure are the zero pressure and the excess lithostatic pressure. These plots are based on the porosity–depth relationship (12.16).

where ϱ_s is the density of the solid matrix. The fluid pressure is normally bounded below by the hydrostatic pressure and above by the lithostatic pressure. The overpressure, which is $p_e = p_f - p_h$, is therefore between 0 and the so-called excess lithostatic pressure $p_l = p_b - p_h$. The excess lithostatic pressure can also be written

$$p_l = \int_0^z (\varrho_s - \varrho_f)(1 - \phi)g \, dz \qquad (12.21)$$

$$= (\varrho_s - \varrho_f)(\zeta^* - \zeta)g \qquad (12.22)$$

when both ϱ_s and ϱ_f are constant through the sedimentary column. We see that the excess lithostatic pressure does not depend on the porosity when it is expressed with the net depth from the basin surface. Figure 12.4 shows the hydrostatic pressure and the lithostatic pressure, where the latter pressure is based on the porosity function (12.16) and the z-coordinate (12.18).

Exercise 12.2 Show that the void ratio function (12.16) leads to equation (12.18) for the z-coordinate as a function of the ζ-coordinate.

Exercise 12.3 Show that the porosity is half the surface porosity at the net (porosity-free) depth $\zeta^* - \zeta = \zeta_0 \ln(2 + e_0)$.
Hint: show that $\phi_{1/2} = \phi_0/2 = e_0/(2 + 2e_0)$ and that $e_{1/2} = e_0/(2 + e_0)$.

Exercise 12.4
(a) Show that the lithostatic pressure can be written

$$p_b = \int_\zeta^{\zeta^*} (\varrho_s + e\varrho_f)g \, d\zeta. \qquad (12.23)$$

(b) Show that lithostatic pressure is

$$p_b = \varrho_s g (\zeta^* - \zeta) g + \varrho_f g \zeta_0 (e_0 - e) \tag{12.24}$$

when function (12.16) is the void ratio.

(c) Check that the answer in (b) is consistent with the z-coordinate from equation (12.18) and the excess lithostatic pressure (12.22).

12.4 A simple model for overpressure build-up

We are going to compute the overpressure for a basin in a state of constant deposition, when the void ratio is given by function (12.16). We already have an exact expression for the Darcy flux (12.15), which applies for any void ratio (or porosity) function as long as it is a function of the net depth from the basin surface. Darcy's law can therefore be written as

$$\frac{k}{\mu(1+e)} \frac{\partial p}{\partial \zeta} = \omega \cdot (e - e_{\text{bot}}) \tag{12.25}$$

whenever the void ratio is a function of the net depth, where it is used that $\partial p / \partial z = (\partial p / \partial \zeta)/(1 + e)$. It is possible to obtain the overpressure by integration of the gradient $\partial p / \partial \zeta$ when the permeability is the following function of the void ratio:

$$k(e) = k_0 \left(\frac{1+e}{1+e_0} \right) \left(\frac{e}{e_0} \right)^n, \tag{12.26}$$

where k_0 is the surface permeability. The permeability function is a kind of Kozeny–Carman permeability function, see equation (2.48), where the permeability decreases as e^n with decreasing void ratio e. The scaling factor $(1 + e)/(1 + e_0)$ is introduced because it simplifies the calculation considerably. The factor is a number in the range 0.5 to 1, and it is therefore considered insignificant compared to e^n which may span several orders of magnitude for $n \gg 1$. The overpressure then becomes (see Note 12.1)

$$p(\zeta, t) = \frac{\omega \mu \zeta_0 e_0 (1 + e_0)}{k_0} \left\{ \frac{1}{(1-n)} \left[1 - \exp\left(-\frac{(1-n)(\zeta^* - \zeta)}{\zeta_0} \right) \right] \right.$$
$$\left. + \frac{1}{n} \left[1 - \exp\left(\frac{n(\zeta^* - \zeta)}{\zeta_0} \right) \exp\left(-\frac{\zeta^*}{\zeta_0} \right) \right] \right\}. \tag{12.27}$$

An analysis of the overpressure solution (12.27) becomes simplified in dimensionless (unit) coordinates. The net deposition rate ω is a characteristic velocity of the problem. From the void ratio function ζ_0 is a characteristic length scale of the problem. Exercise 12.3 shows that the surface porosity is typically reduced to half at the net depth $\sim \zeta_0$. From the characteristic length ζ_0 and the characteristic velocity ω it follows that the characteristic time is $t_0 = \zeta_0 / \omega$. Dimensionless time is therefore $\tau = t / t_0$. The net height of the basin grows as $\zeta^* = \omega t$ and we therefore introduce the dimensionless vertical unit coordinate

$x = \zeta/(\omega t)$, which ensures that the basin always has a unit height. The void ratio function in dimensionless quantities is

$$e(x, \tau) = e_0 \exp\left((1 - x)\,\tau\right) \tag{12.28}$$

because

$$\frac{1}{\zeta_0}\left(\zeta^* - \zeta\right) = \frac{\zeta^*}{\zeta_0}\left(\frac{\zeta^*}{\zeta^*} - \frac{\zeta}{\zeta^*}\right) = \tau\,(1 - x). \tag{12.29}$$

Notice that $\tau = t/t_0 = \omega t/\zeta_0 = \zeta^*/\zeta_0$. The real depth of a unit position x at dimensionless time τ follows from expression (12.18), and it is

$$\frac{z}{\zeta^*} = (1 - x) + \frac{1}{\tau}\left(e_0 - e(x, \tau)\right). \tag{12.30}$$

The dimensionless Darcy flux is

$$\hat{v}(x, \tau) = \frac{v}{\omega} = e_0 \exp\left((1 - x)\,\tau\right) - e_0 \exp(-\tau). \tag{12.31}$$

The overpressure is normally less than the excess lithostatic pressure, and the maximum excess lithostatic pressure, $p_{l,\max}$, is found at the base of the sedimentary column. It follows from excess lithostatic pressure (12.22) that $p_{l,\max} = \Delta\varrho\, g\omega t$, where $\Delta\varrho = (\varrho_s - \varrho_f)$. A dimensionless unit overpressure $u = p/p_{l,\max}$ can be defined using $p_{l,\max}$. From the overpressure solution (12.27) it follows that the unit overpressure is

$$u(x, \tau) = \frac{e_0(1 + e_0)}{N_g}\left\{\frac{1}{(n-1)\,\tau}\left[\exp\left((n-1)(1-x)\,\tau\right) - 1\right]\right.$$
$$\left. - \frac{1}{n\,\tau}\exp(-\tau)\left[\exp\left(n\,(1-x)\,\tau\right) - 1\right]\right\} \tag{12.32}$$

where the (dimensionless) gravity number N_g is

$$N_g = \frac{k_0\Delta\varrho\, g}{\mu\omega}. \tag{12.33}$$

Solution (12.32) shows that two parameters, the permeability exponent n and the gravity number N_g, control the overpressure. The unit overpressure is simply inversely proportional to the gravity number. The two most important factors in the gravity number are the surface permeability and the deposition rate. The overpressure is proportional to the deposition rate and inversely proportional to the surface permeability. Both the deposition rate and the surface permeability are parameters that vary over several orders of magnitude from case to case.

The overpressure solution (12.32) is plotted in Figure 12.5 (left) at different times when the gravity number is $N_g = 1$ and the permeability exponent is $n = 3$. The dashed line

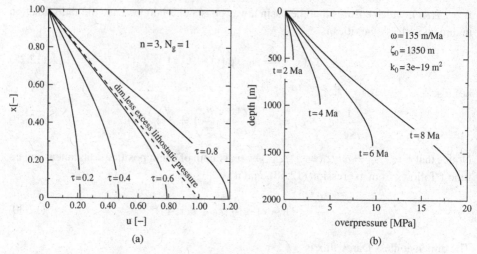

Figure 12.5. *(a) The unit overpressure when the gravity number is $N_g = 1$ and the permeability exponent is $n = 3$. (b) The overpressure plot to the left is plotted with units to the right.*

on the figure is the unit excess lithostatic pressure, which is obtained by scaling with the maximum excess lithostatic pressure, $p_{l,\max}$

$$\hat{p}_l = \frac{(\zeta^* - \zeta)\Delta \varrho g}{p_{l,\max}} = 1 - x. \tag{12.34}$$

The lithostatic pressure therefore corresponds to the straight line $\hat{p}_l = 1 - x$. Figure 12.5a shows that the overpressure at the time $\tau = 0.8$ is beyond the lithostatic pressure. In this simple model, where the porosity is given as a function of the depth, there is nothing that prevents the overpressure from exceeding the lithostatic pressure. It is therefore necessary to introduce hydrofracturing as a valve to limit the fluid pressure by the lithostatic pressure. Section 8.6 treats criteria for hydrofracturing, and we return in the next section to what the permeability of hydrofractured rocks might be.

The gravity number N_g is an important number that tells us whether we can expect high overpressure or low overpressure. If $N_g \ll 1$ then the basin is in a state of high overpressure after only a short time (given that the permeability exponent is $n > 1$). On the other hand, for $N_g \gg 1$ the overpressure is most likely low even after a long period with steady deposition of sediments.

Figure 12.5b shows the real overpressure as a function of real depth and time for the corresponding unit overpressure in Figure 12.5a. The full set of parameters that yields $N_g = 1$ is given in the table below. We see that the characteristic time becomes $t_0 = 10$ Ma. The Darcy fluxes at the same time steps as the overpressure in Figure 12.5 are shown in Figure 12.6, which shows that the dimensionless Darcy flux approaches $e_0 = 1$ as $\tau \to \infty$.

Parameter	Value	Units
e_0	1	[-]
ζ_0	1350	[m]
μ	$1 \cdot 10^{-3}$	[Pa s]
ω	135	[m Ma^{-1}]
$\Delta\rho$	1500	[kg m^{-3}]
k_0	$2.85714 \cdot 10^{-19}$	[m^2]
g	10	[m s^{-2}]

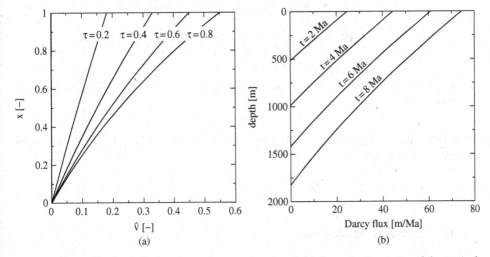

Figure 12.6. (a) The dimensionless Darcy flux as a function of the dimensionless time and the vertical unit coordinate. (b) The same as (a), but as a function of t and z.

How the permeability exponent n controls the overpressure can be analyzed by looking at the overpressure at the base of the basin (where the overpressure is at the maximum). The unit pressure at the base ($x = 0$) is

$$u_{\max}(\tau) = \frac{e_0(1 + e_0)}{N_g} \frac{1}{\tau} \left(\frac{\exp(-\tau)}{n} + \frac{\exp((n-1)\tau)}{n(n-1)} - \frac{1}{n-1} \right), \qquad (12.35)$$

and Figure 12.7 shows that there is strong increase in the overpressure with time for $n > 1$. On the other hand, $n < 1$ gives that the overpressure slowly decreases for $\tau \to \infty$. A closer inspection of $u_{\max}$ shows that

$$
\begin{aligned}
0 \le n < 1, \quad & u_{\max} \to 0, \quad && \tau \to \infty \\
n = 1, \quad & u_{\max} \to 1, \quad && \tau \to \infty \\
n > 1, \quad & u_{\max} \to \infty, \quad && \tau \to \infty,
\end{aligned}
$$

see Exercise 12.5.

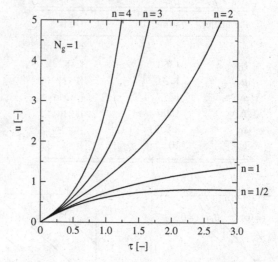

Figure 12.7. *The overpressure at the base of the basin as a function of time for different permeability exponents.*

Note 12.1 The exact pressure (12.27) is obtained from Darcy's law (12.25), which gives the overpressure gradient

$$\frac{\partial p}{\partial \zeta} = \frac{\mu\omega(1+e)}{k(e)}\,(e - e_{\text{bot}}).$$

(12.36)

Inserting the permeability function (12.26) leads to

$$\frac{\partial p}{\partial \zeta} = \frac{\mu\omega(1+e_0)\,e_0^n}{k_0}\left(e^{1-n} - e_{\text{bot}}e^{-n}\right),$$

(12.37)

where we see that the overpressure gradient is given by void ratio e (must not be mistaken for the base of the natural logarithm). Integration of the gradient (12.37) gives the overpressure solution

$$p = \frac{\mu\omega(1+e_0)\,e_0^n}{k_0}\left(F(1-n) - e_{\text{bot}}F(-n)\right),$$

(12.38)

where the function F is defined by

$$F(m) = \int_\zeta^{\zeta^*} e^m\,d\zeta.$$

(12.39)

(Notice that e is the void ratio and not the base of the natural logarithm.) We see that the pressure solution fulfills the boundary condition $p = 0$ for $\zeta = \zeta^*$. Notice that this is a general overpressure solution that applies for any void ratio function, as long as it is a function of net depth. For the specific void ratio function (12.16) we get

$$F(m) = e_0^m \int_{\zeta}^{\zeta^*} \exp\left(-\frac{m(\zeta^* - \zeta)}{\zeta_0}\right) d\zeta$$

$$= \frac{\zeta_0 e_0^m}{m} \left[1 - \exp\left(-\frac{m(\zeta^* - \zeta)}{\zeta_0}\right)\right]. \tag{12.40}$$

When this specific $F(m)$ is inserted into the general pressure solution (12.38) we finally get the overpressure solution (12.27).

Exercise 12.5 Show that u_{max} has a maximum at $\tau = 1.7$ when $n = 0$.

Exercise 12.6 Show that the function $F(m)$ given by equation (12.40) can be written as

$$F(m) = \frac{\zeta_0}{m}\left(e_0^m - e^m\right) \tag{12.41}$$

where e is the void ratio and not the base of the natural logarithm.

12.5 Hydrofracturing

The preceding section shows that the overpressure build-up during deposition of sediments can easily exceed the lithostatic pressure in a model where the porosity reduction is related to depth. Typical permeabilities for shales and typical deposition rates of sedimentary basins lead to fluid pressures far beyond the lithostatic pressure at depths below 2 km. Porosity reduction as a function of depth can be a reasonable model for chemical compaction, at least for quartz cementation where the dissolution process will operate for compressive stresses that are just a few percent of the overburden. However, overpressure above 95% of the lithostatic pressure is rarely observed. There are two alternative explanations for why the fluid pressure does not exceed the lithostatic pressure. The first is that the compaction process stops when the fluid pressure approaches the lithostatic pressure. The other is that the sediments hydrofracture when the fluid pressure exceeds the least horizontal stress, and thereby have their permeabilities enhanced.

We will look a little further at hydrofracturing, and possible implications of hydrofracturing on the permeability of sediments. It should be mentioned that natural hydrofracturing is a far from fully understood process. Nevertheless, it is often assumed that a fluid pressure slightly above the least horizontal stress will open vertical fractures or fault zones, see Figure 12.8. The least horizontal stress is normally less than the lithostatic pressure, and that is why a pressure lower than the lithostatic pressure limits the fluid pressure (see Section 8.6). As a starting point we will assume that we know the least horizontal stress, and that it is equal to the fluid pressure in the depth intervals of hydrofracturing, see Figure 12.9. The fact that we know the fluid pressure implies that we can find the

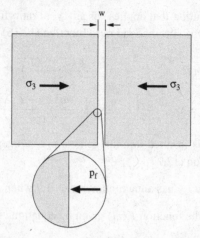

Figure 12.8. *The vertical fracture is opened by a fluid pressure p_f that is larger than the least horizontal stress σ_3.*

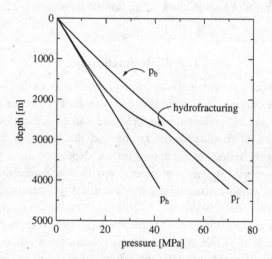

Figure 12.9. *The fluid pressure (p_f) given by the overpressure solution (12.27) is plotted from the basin surface until the point of hydrofracturing, and the fluid pressure is equal to the fracture pressure below the point of hydrofracturing. The hydrostatic pressure (p_h) and the lithostatic pressure (p_b) are also shown. The parameters used in the overpressure solution (12.27) are $\zeta^* = 3000\,m$ deposited during 10 Ma, $\zeta_0 = 1350\,m$, $k_0 = 5 \cdot 10^{-17}\,m^2$ and $n = 5$.*

permeability if we know the Darcy flux. We have already found an expression for the Darcy flux

$$v_z = \omega(e - e_{\text{bot}}) \qquad (12.42)$$

where ω is the net deposition rate, e is the void ratio and e_{bot} is the void ratio at the base of the sedimentary column. By inserting the Darcy flux into Darcy's law we get

$$\frac{k}{\mu}\frac{\partial p}{\partial z} = \omega\,(e - e_{\text{bot}}),$$
(12.43)

where p is the overpressure and μ is the viscosity. Using the fact that the fluid pressure is close to the lithostatic pressure implies that the overpressure gradient can be approximated as $\partial p/\partial z \approx \Delta\varrho g$, where $\Delta\varrho$ is the difference between the solid density and the fluid density. The permeability of hydrofractured rock is therefore

$$k_b \approx \frac{\mu\omega\,(e - e_{\text{bot}})}{\Delta\varrho g}.$$
(12.44)

We could also have written this as a gravity number

$$\frac{k_b \Delta\varrho g}{\mu\omega(e - e_{\text{bot}})} \approx 1,$$
(12.45)

and we get that the local gravity number of hydrofractured rock is close to one. Once we have an estimate for the permeability of hydrofractured rocks we can use it to estimate the aperture of vertical fractures, assuming that we have equally spaced fractures with equal apertures. The permeability k_b is the average permeability of the intact (unfractured) rock and the vertical fractures,

$$k_r l + \frac{w^2}{12}w = k_b(l + w),$$
(12.46)

where k_r is the permeability of the (unfractured) intact rock, $w^2/12$ is the permeability of a fracture with aperture w, and l is the average distance between the fractures. (Expression (12.46) is derived in Section 2.11 as the average permeability parallel to the bedding of a layered rock.) The fracture aperture is then

$$w \approx (12(k_b - k_r)l)^{1/3}$$
(12.47)

where we have used that the right-hand side in (12.46) can be approximated as $k_b l$ when the aperture w is much less than the fracture spacing ($w/l \ll 1$). We can take the estimate of w a step further when hydrofracturing enhances the permeability so much that $k_b \gg k_r$ or $k_b \approx w^3/12l$. The permeability is then dominated by flow in the fractures. The aperture is then

$$w \approx \left(\frac{12\mu\omega\,(e - e_{\text{bot}})l}{\Delta\varrho g}\right)^{1/3}.$$
(12.48)

The fracture spacing l could also have been replaced with the inverse fracture density, where the fracture density is the number of fractures per unit length, $\rho_l = 1/l$. We can get an idea of what the fracture aperture might be by using the parameters $l = 1$ m, $e = 0.5$, $e_{\text{bot}} = 0.05$, $\omega = 500$ m Ma^{-1}, $\Delta\varrho = 1400$ kg m^{-3} and $g = 10$ m s^{-2}, which gives that $w = 2\ \mu$m. The condition that most of the fluid passes through the fracture means that the rock has a permeability that is much less than $w^3/12l \approx 5 \cdot 10^{-19}$ m^2. An aperture as small as $\omega \sim 1\ \mu$m means almost closed fractures can drain sufficient amounts of fluid to prevent the pressure building up beyond the lithostatic pressure. It should be added that this reasoning is done in 2D, while it is likely that the dominant fluid flow in fractures could be in channels rather than evenly distributed in a plane.

The simple estimate (12.42) for a Darcy flux during burial and compaction can be used in an estimate of the bulk rock permeability, when it is combined with pressure observations. Figure 12.1 shows the fluid pressure for an overpressured well. Although these observations are based on mud weight they are sufficiently accurate to be used to estimate the permeability. It is not unusual that permeability measurements of heterogeneous rocks span an order of magnitude. Darcy's law with the flux (12.42)

$$v_D = \omega \, (e - e_{\text{bot}}) = \frac{k_b}{\mu} \frac{d p_e}{dz} \tag{12.49}$$

gives that the bulk permeability is

$$k_b = \frac{\mu \omega \, (e - e_{\text{bot}})}{d p_e / dz}. \tag{12.50}$$

The void ratio as a function of depth is obtained from Athy's porosity function with $\phi_0 = 0$ and $z_0 = 1200$. Using the overpressure gradient $d p_e / dz$ from the data in Figure 12.1 and the deposition rates $\omega = 50 \text{ m Ma}^{-1}$ and $\omega = 500 \text{ m Ma}^{-1}$ gives the permeability in Figure 12.10. Notice that the permeabilities are proportional to the burial rate. They therefore differ by one order of magnitude because that is the difference between the deposition rates. The pressure seal at 1.5 km depth in Figure 12.1 corresponds to an order of magnitude decrease in permeability. The permeability is then decreasing with two more orders of magnitude towards the base. The upper and lower limits for shale permeability from Section 2.7 are also shown. We see that the rapid deposition brings us beyond the maximum shale permeability. Hydrofracturing may therefore be a necessary process to enhance the bulk permeability.

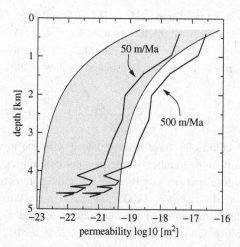

Figure 12.10. *The permeability that would give the overpressure in Figure 12.1 for the two deposition rates $\omega = 50 \text{ m Ma}^{-1}$ and $\omega = 500 \text{ m Ma}^{-1}$.*

The various ways hydrofracturing may operate are still poorly understood. It may be related to both large-scale faults and to micro-fractures, and it might be intermittent and continuous. The model presented above represents continuous flow in micro-fractures. The "dim zones" or "gas chimneys" seen with 2D and also 3D seismics may be an another important aspect of hydrofracturing as well as leakage from overpressured reservoirs (Hermanrud and Bolås, 2002, Judd and Hovland, 2007).

12.6 Gibson's solution for overpressure

There is probably only one exact and reasonable simple solution for overpressure build-up for a basin in a state of constant deposition of sediments, when the porosity is a function of the effective stress (and thereby the overpressure), and that is the Gibson solution (Gibson, 1958). It assumes a shallow basin where the porosity reduction is so little that it is negligible with respect to geometry. In other words, the starting point is mass conservation in a column where compaction is insignificant. The time-derivative in equation (12.1) (for conservation of fluid mass) then goes through the integration sign, and we have the following expression for mass conservation:

$$\frac{\partial(\phi \varrho_f)}{\partial t} = -\frac{\partial(\varrho_f v)}{\partial z}. \tag{12.51}$$

The fluid density ϱ_f is assumed constant and therefore drops out. The gradient of the overpressure gives the Darcy velocity in the vertical direction, and equation (12.51) for mass conservation then becomes

$$\frac{\partial \phi}{\partial t} = \frac{\partial}{\partial z}\left(\frac{k}{\mu}\frac{\partial p_e}{\partial z}\right). \tag{12.52}$$

The porosity is assumed to be a linear function of the vertical effective stress σ':

$$\phi = \phi_0 - \alpha\sigma' \tag{12.53}$$

where α is sediment compressibility. The difference between the lithostatic pressure (overburden) and the fluid pressure is the vertical effective stress (3.126).

The vertical effective stress is given by equation (3.126) as the difference between the lithostatic pressure (overburden) and the fluid pressure. For shallow depths from the basin surface we have that the lithostatic pressure and the fluid pressure are

$$p_b = \varrho_b g h \quad \text{and} \quad p_h = \varrho_f g h, \tag{12.54}$$

respectively, where ϱ_b is the bulk density and ϱ_f is the fluid density. The depth from the basin surface is h. The vertical effective stress is therefore

$$\sigma' = p_b - p_f = p_b - p_h - p_e = (\varrho_b - \varrho_f)gh - p_e. \tag{12.55}$$

By combining the porosity function (12.53) with the effective stress as a function of the overpressure we get

$$\frac{\partial \phi}{\partial t} = -\alpha \left((\varrho_b - \varrho_f)g\omega_b - \frac{\partial p_e}{\partial t} \right) \tag{12.56}$$

where the bulk deposition rate is $\omega_b = dh/dt$. This term is the left-hand side of equation (12.51) for conservation of fluid mass, which then leads to the following equation for the overpressure:

$$\frac{\partial p_e}{\partial t} - \frac{k}{\alpha \mu} \frac{\partial^2 p_e}{\partial z^2} = (\varrho_b - \varrho_f)g\omega_b \tag{12.57}$$

where both the permeability k and the viscosity μ are constant. Boundary conditions for the pressure equation are zero overpressure at the basin surface

$$p_e = 0, \quad \text{for} \quad z = 0 \tag{12.58}$$

and zero Darcy flux at the base of the basin. The latter boundary condition implies that the overpressure gradient is zero,

$$\frac{\partial p_e}{\partial z} = 0, \quad \text{for} \quad z = h = \omega_b t. \tag{12.59}$$

The solution to the pressure equation (12.57) with these boundary conditions for a domain that is growing with time is the solution found by Gibson (1958). The solution is in Gibson's notation using a vertical coordinate z_g, where $z_g = 0$ is the base of the sedimentary column at all times, and the top of the basin is $z_g = \omega_b t$. The solution for overpressure is

$$p_e(z_g, t) = \gamma \omega_b t - \gamma (\pi ct)^{-1/2} \exp\left(\frac{-z_g^2}{4ct} \right)$$

$$\times \int_0^\infty \xi \tanh\left(\frac{\omega_b \xi}{2c} \right) \cosh\left(\frac{z_g \xi}{2ct} \right) \exp\left(\frac{-\xi^2}{4ct} \right) d\xi \tag{12.60}$$

where $\gamma = (\varrho_b - \varrho_f)g$ and $c = k/(\alpha \mu)$. The integral has to be evaluated numerically, and we notice that the integration variable ξ has the unit of length (meter). Figure 12.11 shows a plot of the Gibson solution in the z-coordinate for the data given in Table 12.1. The plot uses the z-coordinate $z = \omega_b t - z_g$, where $z = 0$ is the basin surface.

Note 12.2 The pressure equation (12.57) is

$$\frac{\partial p_e}{\partial t} - c \frac{\partial^2 p_e}{\partial z_g^2} = \gamma \omega_b \tag{12.61}$$

with the boundary conditions

$$p_e(h, t) = 0 \quad \text{and} \quad \frac{\partial p_e}{\partial z_g}(0, t) = 0 \tag{12.62}$$

Table 12.1. *The parameters used in the overpressures shown in Figure 12.11.*

Parameter	Value	Units
k	1e–18	m^2
μ	0.001	Pa s
ϱ_f	1000	$kg\,m^{-3}$
ϱ_f	2100	$kg\,m^{-3}$
ω_b	3.1746e–11	$m\,s^{-1}$
ω_b	1000	$m\,Ma^{-1}$
α	1e–08	Pa^{-1}
g	10	$m\,s^{-2}$

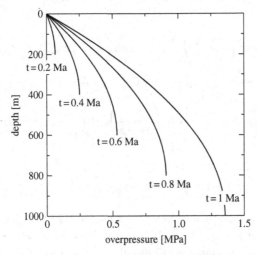

Figure 12.11. *The Gibson overpressure solution (12.60) for the data given in Table 12.1.*

and where the solution domain grows linearly with time, $h = \omega_b t$. It is possible to write down a solution to this problem, which is

$$p_e(z_g, t) = \gamma h - \frac{1}{\sqrt{t}} \int_0^\infty g(\xi) \left[\exp\left(\frac{-(z_g - \xi)^2}{4ct} \right) - \exp\left(\frac{-(z_g + \xi)^2}{4ct} \right) \right] d\xi$$

(12.63)

where the integration variable ξ has the unit meter (length). The above expression is a solution because the functions

$$f_{1,2} = \frac{1}{\sqrt{t}} \exp\left(-\frac{(z_g \pm u)^2}{4ct} \right)$$

(12.64)

solve the (homogeneous) pressure equation

$$\frac{\partial p_e}{\partial t} - c\frac{\partial^2 p_e}{\partial z_g^2} = 0 \qquad (12.65)$$

and the term γh yields the right-hand side (the source term). The solution (12.63) also fulfills the boundary condition at the base of the basin, $z_g = 0$, because we have

$$\frac{\partial p_e}{\partial z_g} = \frac{1}{\sqrt{t}}\int_0^\infty g(\xi)\left[2(z_g-\xi)\exp\left(\frac{-(z_g-\xi)^2}{4ct}\right) + 2(z_g+\xi)\exp\left(\frac{-(z_g+\xi)^2}{4ct}\right)\right]d\xi \qquad (12.66)$$

which yields $\partial p_e/\partial z_g = 0$ for $z_g = 0$. There is now only one thing left to do, and that is to find the function $g(\xi)$ that makes the solution fulfill the boundary condition $p_e = 0$ at the surface of the basin (at $z_g = h$). From the starting point, equation (12.63), we have that $p_e(h,t) = 0$ is

$$\frac{1}{2}\gamma h\sqrt{t} = \exp\left(\frac{-h^2}{4ct}\right)\int_0^\infty g(\xi)\cosh\left(\frac{2h\xi}{4ct}\right)\exp\left(\frac{-\xi^2}{4ct}\right)d\xi. \qquad (12.67)$$

This equation is now rewritten using new variables $s = \xi^2$ and $p = 1/(4ct)$ as

$$\frac{1}{8}\gamma\omega_b(pc)^{-2/3}\exp\left(\frac{\omega_b^2}{16c^2p}\right) = \int_0^\infty F(s)e^{-ps}ds \qquad (12.68)$$

where we have

$$F(s) = g(s^{1/2})\cosh\left(\frac{\omega_b s^{1/2}}{2c}\right)s^{-1/2}. \qquad (12.69)$$

The reason for rewriting the right-hand side of equation (12.68) is that we want to express it as a Laplace transform. If we can find a function $F(s)$, for which the Laplace transform is the left-hand side of the equation (12.68), then we will also have found the unknown function $g(\xi)$. From tables of Laplace transforms we find that

$$\int_0^\infty \sinh(\lambda\sqrt{s})e^{-ps}ds = \frac{\lambda\sqrt{\pi}}{2p^{3/2}}e^{\lambda^2/(4p)}, \qquad (12.70)$$

where $\sinh(\lambda\sqrt{s})$ is precisely the function $F(s)$ we are searching for. The parameter λ is seen to be $\lambda = \omega_b/(4c)$ and the function $F(s)$ becomes

$$F(s) = \frac{1}{2}\gamma\sqrt{\pi c}\sinh\left(\frac{\omega_b}{4c}\sqrt{s}\right). \qquad (12.71)$$

A combination of the equations (12.69) and (12.71) finally gives

$$g(\xi) = \frac{\gamma\xi}{2\sqrt{\pi c}}\tanh\left(\frac{\omega_b\xi}{2c}\right) \qquad (12.72)$$

and we have the solution (12.60).

Exercise 12.7 Show that the shallow depth condition $\phi \approx \phi_0$ can be written as

$$h \ll \frac{\phi_0}{\alpha(\varrho_b - \varrho_f)g}. \tag{12.73}$$

Exercise 12.8 Show that $f = t^{-1/2}\exp(-x^2/(4t))$ is a solution to the parabolic equation

$$\frac{\partial f}{\partial t} - \frac{\partial^2 f}{\partial x^2} = 0 \tag{12.74}$$

with the boundary conditions $f = 0$ for $x \to \pm\infty$.

12.7 Gibson's solution for porosity reduction

The Gibson solution for overpressure is based on an assumption that compaction is negligible, which implies that the solution has to be restricted to shallow depths. We will now show that the Gibson solution can be applied to a pressure equation in the vertical Lagrange coordinate ζ, where we don't have to impose this restriction. The starting point is equation (12.11) that accounts for compaction in the ζ-coordinate. We can make a similar pressure equation as the one solved by Gibson by choosing the void ratio function

$$e = e_0 - \alpha\sigma', \tag{12.75}$$

which is similar to the porosity function (12.53). The vertical effective stress (σ') is the difference between the lithostatic pressure (overburden) (p_b) and the fluid pressure (p_f), $\sigma' = p_b - p_f$. In the previous section it was assumed that the lithostatic pressure was related to a constant bulk density. Instead, we will let the porosity be

$$\varrho_b = (1 - \phi)\varrho_s + \phi\varrho_f, \tag{12.76}$$

which is the mean of the solid density ϱ_s and the fluid density ϱ_f weighted with the porosity. Both the solid density ϱ_s and the fluid density ϱ_f are constant. The bulk density (12.76) is not restricted to shallow depths because it accounts for the porosity. The effective vertical stress is then

$$\sigma' = (\zeta^* - \zeta)(\varrho_s - \varrho_f)g - p_e \tag{12.77}$$

which follows from the difference (12.22) between the lithostatic pressure and the fluid pressure. We therefore get

$$\frac{\partial e}{\partial t} = -\alpha\left((\varrho_s - \varrho_f)g\omega - \frac{\partial p_e}{\partial t}\right) \tag{12.78}$$

where $\omega = d\zeta^*/dt$ is the deposition rate given as net (porosity-free) rock. The permeability is taken to be the following function of the void ratio:

$$k(e) = k_0\frac{1 + e}{1 + e_0} \tag{12.79}$$

where k_0 is the surface permeability. This particular choice for a permeability function simplifies the pressure equation and we get

$$\frac{\partial p_e}{\partial t} - \frac{k_0}{(1+e_0)\alpha\mu} \frac{\partial^2 p_e}{\partial \zeta^2} = (\varrho_s - \varrho_f)\omega g. \qquad (12.80)$$

There are two boundary conditions for equation (12.80), where the first is zero overpressure at the basin surface ($\zeta = \zeta^*$), and the second is zero Darcy flux at the base of the basin ($\zeta = 0$). The latter boundary condition implies that $\partial p_e/\partial \zeta = 0$ at $\zeta = 0$. Equation (12.80) is precisely the same equation as the one solved by Gibson, where the basin thickness (as net rock) increases linearly with time instead of the real thickness. We only need to define the parameters

$$c = \frac{k_0}{(1+e_0)\alpha\mu} \quad \text{and} \quad \gamma = (\varrho_s - \varrho_f)\omega g \qquad (12.81)$$

and we can simply write down Gibson's solution for the overpressure in the ζ-coordinate:

$$p_e(\zeta, t) = \gamma\omega t - \gamma(\pi ct)^{-1/2} \exp\left(\frac{-\zeta^2}{4ct}\right)$$

$$\times \int_0^\infty \xi \tanh\left(\frac{\omega\xi}{2c}\right) \cosh\left(\frac{\zeta\xi}{2ct}\right) \exp\left(\frac{-\xi^2}{4ct}\right) d\xi. \qquad (12.82)$$

(Notice that the deposition rate is the net rate ω and not the bulk deposition rate ω_b.) The real z-coordinate for a given ζ-coordinate and a given time is as usual given by the integral

$$z = \int_\zeta^{\zeta^*} \frac{d\zeta}{1-\phi} = \int_\zeta^{\zeta^*} (1+e)d\zeta \qquad (12.83)$$

where the void ratio is a function of the effective stress and thereby the overpressure.

There is a possible depth limitation for this solution, which is related to the void ratio function (12.75). We see that the void ratio may become negative if the effective vertical stress is sufficiently large. An estimate for this depth is calculated in Exercise 12.9. The shortest depth occurs for hydrostatic compaction, because that is when the effective vertical stress is largest and the void ratio smallest.

Exercise 12.9 Show the void ratio is zero for a hydrostatic basin with net thickness

$$\zeta^* = \frac{e_0}{\alpha \Delta\varrho \, g} \qquad (12.84)$$

where $\Delta\varrho = \varrho_s - \varrho_f$.

Exercise 12.10 Show that the z-coordinate for a given ζ-coordinate under hydrostatic conditions is

$$z = (1+e_0)(\zeta^* - \zeta) - \frac{1}{2}\alpha\Delta\varrho \, g(\zeta^* - \zeta)^2 \qquad (12.85)$$

where $\Delta\varrho = \varrho_s - \varrho_f$.

12.8 Overpressure and mechanical compaction

Mechanical compaction is porosity reduction due to increasing effective stress, where the void ratio is related to the effective vertical stress through a function like (4.63). In general the void ratio is a function of the vertical effective stress

$$e = e(\sigma') \tag{12.86}$$

where the effective vertical stress is related to the overpressure by

$$\sigma' = \int_{\zeta}^{\zeta^*} (\varrho_s - \varrho_f)g \, d\zeta - p_e. \tag{12.87}$$

Equation (12.11) gives conservation of fluid mass in a compacting column of sediments. This equation can be turned into a general equation for overpressure build-up during deposition of sediments that compact mechanically. From the void ratio function (12.86) and the expression for the effective vertical stress (12.87) we get

$$\frac{\partial e}{\partial t} = \frac{de}{d\sigma'} \frac{\partial \sigma'}{\partial t} = \frac{de}{d\sigma'} \left((\varrho_s - \varrho_f)g \frac{d\zeta^*}{dt} - \frac{\partial p_e}{\partial t} \right) \tag{12.88}$$

and equation (12.11) becomes

$$\frac{\partial p_e}{\partial t} + \left(\frac{de}{d\sigma'} \right)^{-1} \frac{\partial}{\partial \zeta} \left(\frac{k(e)}{(1+e)\mu} \frac{\partial p_e}{\partial \zeta} \right) = (\varrho_s - \varrho_f)g\omega \tag{12.89}$$

which is the general equation for overpressure build-up and mechanical compaction. The deposition rate as net sediment is $\omega = d\zeta^*/dt$ and the permeability is represented by a function of the void ratio. We have already seen that the Gibson solution solves this equation for the choice of a linear void ratio function and a linear permeability function.

The right-hand side of equation (12.89) can be interpreted as a source term. It is the source term that is responsible for pressure build-up. If the source term is zero then there is nothing in the equation that can generate overpressure. That is the case when there is no deposition ($\omega = 0$). It is also the case for deposition of sediments with a solid density equal to the fluid density ($\varrho_s = \varrho_f$). We will look further at source terms in the next chapter, and then for various other mechanisms that can generate overpressure.

A dimensionless version of the general pressure equation (12.89) is now made in order to study the overpressure build-up when the compaction is mechanical. The basin will be in a state of constant deposition with the (net) rate ω, and we can immediately identify the deposition rate ω as a characteristic velocity of the problem. The ζ-coordinate will be replaced by the dimensionless unit coordinate

$$x = \frac{\zeta}{\omega t} \tag{12.90}$$

and pressure will be replaced by the dimensionless unit pressure

$$u = \frac{p_e}{\Delta \varrho \, g \omega t} \tag{12.91}$$

where $\Delta\varrho = \varrho_s - \varrho_f$. We recall that the overpressure is scaled with the maximum excess lithostatic pressure $p_{l,\max} = \Delta\varrho\, g\omega t$, see equation (12.32). The effective vertical stress is also scaled with $p_{l,\max}$ and we have

$$u_s = \frac{\sigma'}{\Delta\varrho\, g\omega t} = \frac{\Delta\varrho\, g(\omega t - \zeta) - p_e}{\Delta\varrho\, g\omega t} = 1 - x - u. \qquad (12.92)$$

The void ratio function is now restricted to functions of the form

$$e = e(\alpha\sigma') \qquad (12.93)$$

where the function e takes the dimensionless argument $\alpha\sigma'$. The parameter α thus defines a pressure $\sigma_0 = 1/\alpha$ that characterizes the compaction process. We expect compaction to become noticeable for effective vertical stress $\alpha\sigma' \sim 1$. The characteristic pressure σ_0 can in turn be used to define a characteristic depth ζ_0 using the excess lithostatic pressure $\Delta\varrho\, g\zeta_0$ such that $\alpha\Delta\varrho\, g\zeta_0 = 1$. Once we have a characteristic length ζ_0 and a characteristic velocity ω we also have a characteristic time $t_0 = \zeta_0/\omega$, which is

$$t_0 = \frac{1}{\alpha\Delta\varrho\, g\omega}. \qquad (12.94)$$

The dimensionless product $\alpha\sigma'$ is related to the dimensionless time τ and the dimensionless vertical effective stress u_s using equation (12.92), and we get

$$\alpha\sigma' = \alpha\Delta\varrho\, g\omega t (1 - x - u) = \tau\, (1 - x - u). \qquad (12.95)$$

The dimensionless version of the void ratio function is then

$$e = e\left(\tau\, (1 - x - u)\right). \qquad (12.96)$$

The characteristic pressure σ_0 can also be used to express dimensionless overpressure and we get from definition (12.91) that

$$\alpha p_e = \alpha\Delta\varrho\, g\omega t u = \tau\, u \qquad (12.97)$$

where the characteristic time (12.94) is used. The following is a summary of the dimensionless quantities introduced so far:

$$x = \frac{\zeta}{\omega t} \quad \text{(dimensionless vertical coordinate)} \qquad (12.98)$$

$$u = \frac{p_e}{\Delta\varrho\, g\omega t} \quad \text{(dimensionless overpressure)} \qquad (12.99)$$

$$\tau = \alpha\Delta\varrho\, g\omega\, t \quad \text{(dimensionless time)} \qquad (12.100)$$

$$u_s = 1 - x - u \quad \text{(dimensionless effective vertical stress)} \qquad (12.101)$$

$$e = e\left(\tau\, (1 - x - u)\right) \quad \text{(the void ratio function)}. \qquad (12.102)$$

When the pressure equation (12.89) is rewritten in terms of these dimensionless quantities we get the dimensionless pressure equation

$$\tau \frac{\partial u}{\partial \tau} + u - x \frac{\partial u}{\partial x} + \frac{N_g}{\tau e'} \frac{\partial}{\partial x} \left(\frac{\hat{k}(e)}{(1+e)} \frac{\partial u}{\partial x} \right) = 1 \qquad (12.103)$$

where e' is the derivative of e with respect to the dimensionless argument $\alpha \sigma'$. In the case of the linear void ratio function $e = e_0 - \alpha \sigma'$ we get that $e' = -1$, and in the general case e' is a negative number because the void ratio decreases with increasing effective stress. The parameter N_g is once more the gravity number

$$N_g = \frac{k_0 \Delta \varrho \, g}{\mu \omega}. \qquad (12.104)$$

The permeability function $k = k(e)$ is written as $k = k_0 \hat{k}(e)$, where k_0 is the surface permeability and where the dimensionless function $\hat{k}(e)$ takes care of how the permeability depends on the void ratio. It then follows that $\hat{k}(e_0) = 1$ for the void ratio e_0 at the basin surface. Note 12.3 gives the details of how this dimensionless pressure equation is obtained. An important result from the scaling of the pressure equation is that there is only one explicit parameter left in the model, and that is the gravity number N_g. There may by more parameters in the dimensionless formulation, parameters that are hidden in the permeability function, $\hat{k}(e)$, like a permeability exponent. The boundary conditions for the dimensionless pressure equation (12.103) are as usual zero overpressure at the basin surface,

$$u = 0, \quad \text{at} \quad x = 1 \qquad (12.105)$$

and zero Darcy flux into the base of the basin,

$$\frac{\partial u}{\partial x} = 0, \quad \text{at} \quad x = 0. \qquad (12.106)$$

The dimensionless Darcy flux is obtained by scaling with the burial rate (which is the characteristic velocity). When the dimensionless Darcy flux is expressed using the gradient of the dimensionless overpressure we get

$$\hat{v} = N_g \frac{\hat{k}(e)}{(1+e)} \frac{\partial u}{\partial x}. \qquad (12.107)$$

The scaled Darcy flux is simply proportional to the gravity number, and the gravity number can therefore be interpreted as a dimensionless permeability. The scaling of the Darcy flux is given as Exercise 12.12.

We have already seen in Section 12.4 that the gravity number defines three regimes of overpressure build-up:

1. $N_g \ll 1$ is the regime of high overpressure.
2. $N_g \sim 1$ is the intermediate regime between high and low overpressure.
3. $N_g \gg 1$ is the regime of low overpressure.

We notice that the right-hand side becomes 1 after the scaling. The gravity number there-fore decides when the unit source term will lead to overpressure. It is possible to expose the regimes defined by the gravity number by solving the overpressure equation for the end members $N_g = 0$ and $N_g \to \infty$. We expect a pressure solution to behave contin-uously as a function of the gravity number N_g, where a decreasing gravity number will lead to increasing overpressure, and an increasing gravity number will lead to decreasing overpressure. The extreme values for the overpressure are therefore found by inspecting the end members of the gravity number. The end member $N_g = 0$ leads to the first-order pressure equation

$$\tau \frac{\partial u}{\partial \tau} + u - x \frac{\partial u}{\partial x} = 1, \tag{12.108}$$

which has the simple (stationary) solution

$$u(x, \tau) = 1 - x. \tag{12.109}$$

The solution fulfills the boundary condition $u = 0$ at $x = 1$. We cannot use the other boundary condition because a first order equation needs only one boundary condition. From equation (12.34) we see that the stationary solution (12.109) corresponds to a fluid pressure equal to the lithostatic pressure. This is the maximum possible overpressure for mechanical compaction, because it implies that the effective vertical stress is zero. Zero effective vertical stress means that there is no compaction, and the fluid pressure is then carrying the entire sedimentary column above that point. The other end member $N_g \to \infty$ leads to the approximate pressure equation

$$\frac{\partial^2 u}{\partial x^2} = 0, \tag{12.110}$$

with the trivial solution $u(x, \tau) = 0$ that fulfills both boundary conditions. This is the solution for hydrostatic fluid pressure, which is the minimum overpressure possible. This final state of the overpressure is shown in Figure 12.12.

Overpressure build-up when compaction is mechanical is fundamentally different from the simple model of overpressure of Section 12.4. Compaction in the simple model was only related to depth and we had fluid expulsion as long as we had deposition. We saw that there was nothing that prevented the overpressure from exceeding the lithostatic pressure when there is no feedback from the fluid pressure on the compaction. With mechanical compaction there is feedback from the overpressure on the compaction. In the limit of zero permeability the fluid pressure becomes the lithostatic pressure, vertical effective stress becomes zero, and compaction stops.

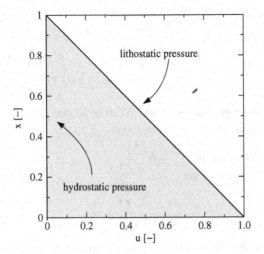

Figure 12.12. *The dimensionless formulation for overpressure build-up and mechanical compaction during deposition allows only overpressure in the lower (shaded) triangle.*

Note 12.3 We want to rewrite the pressure equation (12.89) in a dimensionless form. We start with the time derivative of the overpressure that is

$$\left(\frac{\partial p_e}{\partial t}\right)_\zeta = \left(\frac{\partial p_e}{\partial t}\right)_x + \frac{\partial p_e}{\partial x}\frac{\partial x}{\partial t}$$

$$= \left(\frac{\partial p_e}{\partial t}\right)_x - \frac{x}{t}\frac{\partial p_e}{\partial x}. \tag{12.111}$$

Notice that we are going from the Lagrangian coordinate ζ to the Eulerian coordinate x, and we therefore have to replace the partial time-derivative by a material derivative. How $\partial x/\partial t$ becomes $-x/t$ is shown in Exercise 12.11. When the overpressure is replaced by the dimensionless overpressure we get

$$\left(\frac{\partial p_e}{\partial t}\right)_\zeta = \frac{1}{t_0}\left(\frac{\partial p_e}{\partial \tau}\right)_\zeta$$

$$= \frac{1}{t_0}\frac{\partial}{\partial \tau}\left(\frac{1}{\alpha}\tau u\right)_\zeta$$

$$= \frac{1}{t_0\alpha}\left(\frac{\partial}{\partial \tau}(\tau u) - \frac{x}{\tau}\frac{\partial}{\partial x}(\tau u)\right)$$

$$= \frac{1}{t_0\alpha}\left(\tau\frac{\partial u}{\partial \tau} + u - x\frac{\partial u}{\partial x}\right). \tag{12.112}$$

It is assumed that the void ratio is a function of the dimensionless argument $\alpha\sigma'$, $e = e(\alpha\sigma')$. The derivative $de/d\sigma'$ then becomes $de/d\sigma' = e'\alpha$, where e' is the derivative of the dimensionless function e, and we get that the second term is

$$\left(\frac{de}{d\sigma'}\right)^{-1} \frac{\partial}{\partial \zeta} \left(\frac{k(e)}{(1+e)\mu} \frac{\partial p_e}{\partial \zeta}\right) = \frac{k_0}{\mu} \frac{1}{(\omega t)^2} \frac{\tau}{\alpha} \frac{1}{\alpha e'} \frac{\partial}{\partial x} \left(\frac{\hat{k}(e)}{(1+e)} \frac{\partial u}{\partial x}\right)$$

$$= \frac{1}{\alpha t_0} \frac{N_g}{\tau e'} \frac{\partial}{\partial x} \left(\frac{\hat{k}(e)}{(1+e)} \frac{\partial u}{\partial x}\right). \qquad (12.113)$$

The last term is now the right-hand side which can be written as $1/(\alpha t_0)$. Multiplication of all terms by αt_0 then gives the pressures equation (12.103).

Exercise 12.11 Show that $\partial x/\partial t = -x/t$.
Solution: From $x = \zeta/(\omega t)$ we get

$$\frac{\partial x}{\partial t} = -\frac{\zeta}{\omega t^2} = \frac{x}{t}. \qquad (12.114)$$

Exercise 12.12 Show that the scaled Darcy flux becomes expression (12.107).

12.9 The dimensionless Gibson solution

A solution of the general pressure equation (12.89) gives the overpressure in a basin during deposition of sediments when compaction is mechanical. We found that the Gibson solution solved the overpressure equation (12.89) when the void ratio is a linear function of the effective stress

$$e = e_0 - \alpha \sigma' \qquad (12.115)$$

and the permeability is a linear function of the void ratio

$$k(e) = k_0 \frac{1+e}{1+e_0}. \qquad (12.116)$$

A dimensionless version of the Gibson solution will therefore be a solution of the dimensionless overpressure equation (12.103). We will first write down the dimensionless pressure equation (12.103) for the Gibson assumptions, which are the void ratio and the permeability given by functions (12.115) and (12.116), respectively. The derivative e' in the dimensionless equation (12.103), which is with respect to the dimensionless argument $\alpha \sigma'$, becomes -1. When $e' = -1$ and the permeability function (12.116) are inserted into equation (12.103) we get that the dimensionless version of the Gibson overpressure equation (12.80) is

$$\tau \frac{\partial u}{\partial \tau} + u - x \frac{\partial u}{\partial x} - N \frac{1}{\tau} \frac{\partial^2 u}{\partial x^2} = 1. \qquad (12.117)$$

The boundary conditions for this equation are zero overpressure at the basin surface and zero Darcy flux into the base of the basin, as given by equations (12.105) and (12.106), respectively. The parameter $N = N_g/(1 + e_0)$ is the only parameter in the equation, and the gravity number therefore controls the overpressure build-up. The factor $1+e_0$ is of little importance since it is ~ 2, while the gravity number can span several orders of magnitude, typically 10^{-6} to 10^{+6}. The large span in the gravity number N_g is due to the large span in surface permeability k_0 and the deposition rate ω.

The dimensionless version of the Gibson solution (12.82) is obtained by the same scaling as the one carried out on the pressure equation, and we get as shown in Note 12.4 that

$$u(x, \tau) = 1 - \left(\frac{\tau}{\pi N}\right)^{1/2} \exp\left(\frac{-x^2\tau}{4N}\right)$$

$$\times \int_0^\infty \hat{\xi} \tanh\left(\frac{\tau\hat{\xi}}{2N}\right) \cosh\left(\frac{x\tau\hat{\xi}}{2N}\right) \exp\left(\frac{-\tau\hat{\xi}^2}{4N}\right) d\hat{\xi}. \qquad (12.118)$$

The scaled version of the Gibson solution (12.82) solves the scaled version of the pressure equation (12.80), because both the equation and its solution are scaled the same way. There is only one parameter in solution (12.118), which is as expected the (modified) gravity number N. We saw in the preceding section how the gravity number defines different overpressure regimes. These regimes can be illustrated by looking at the Gibson solutions in two regimes $N_g \ll 1$ and $N_g \gg 1$.

Figure 12.13 shows the dimensionless overpressure, porosity and dimensionless Darcy flux for the Gibson solution when $N_g = 0.1$ at the time steps $\tau = 0, 1/4, 1/2, 3/4$ and 1. We see that the overpressure is close to the lithostatic pressure in the upper half of the basin. This observation is consistent with the porosity plot that shows little compaction (porosity reduction) in this part of the basin. The Darcy flux is also constant in the part of the basin were there is not much compaction. The gravity number equal to 0.1 marks the beginning of the overpressure regime, and if we had selected a much smaller gravity number we would have had more extreme overpressure results.

The opposite regime for the Gibson solution, $N_g = 10$, is shown in Figure 12.14 for the same time steps, $\tau = 0, 1/4, 1/2, 3/4$ and 1, as in Figure 12.13. The dimensionless overpressures are small compared to those for $N_g = 0.1$ shown in Figure 12.13. This is clearly similar to the hydrostatic case. The porosity plot shows much compaction and the dimensionless Darcy flux plot shows an increasing Darcy flux up through the entire column. The gravity number $N_g = 10$ is the beginning of the hydrostatic regime, where much larger gravity numbers would have led to almost zero overpressure. See Exercise 12.13 for possible cases (with units) that have $N_g = 0.1$ and 10.

Note 12.4 The Gibson solution using the ζ-coordinate can be rewritten as

$$p_e(\zeta, t) = \gamma \zeta^* - \frac{\gamma(\zeta^*)^2}{\sqrt{\pi c t}} \exp\left(-\frac{(\zeta^*)^2(\zeta/\zeta^*)^2}{4ct}\right)$$

$$\times \int_0^\infty \hat{\xi} \tanh\left(\frac{(\zeta^*)^2\hat{\xi}}{2ct}\right) \cosh\left(\frac{(\zeta^*)^2(\zeta/\zeta^*)\hat{\xi}}{2ct}\right) \exp\left(\frac{-(\zeta^*)^2\hat{\xi}^2}{4ct}\right) d\hat{\xi} \qquad (12.119)$$

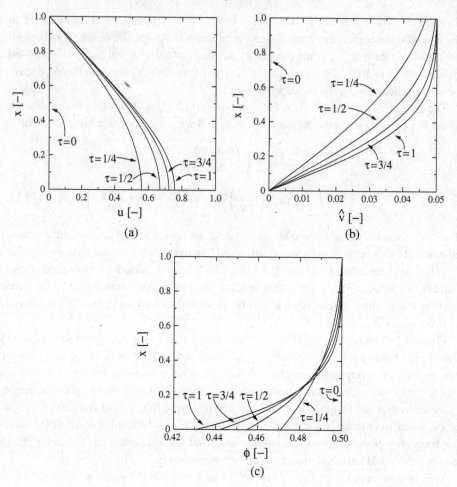

Figure 12.13. *The Gibson solution for $N_g = 0.1$ and $e_0 = 1$: (a) dimensionless overpressure, (b) dimensionless Darcy flux and (c) void ratio.*

where ωt is replaced by ζ^*. The integration over the variable ξ (with unit meter) is replaced by the dimensionless integration variable $\hat{\xi}$ that is related to ξ by $\xi = \zeta^* \hat{\xi}$. The ζ-coordinate is written as ζ/ζ^* to make it straightforward to replace it with the vertical unit coordinate x. We see that the factor $(\zeta^*)^2/(ct)$ appears repeatedly. This factor can be written in terms of the gravity number by replacing c with the definition (12.81) and scaling t by characteristic time t_0 (12.94). We then have

$$\frac{(\zeta^*)^2}{ct} = \frac{\mu\omega}{k_0 \Delta\varrho\, g}(1 + e_0)\tau = \frac{(1 + e_0)\tau}{N_g} \tag{12.120}$$

and we have the dimensionless Gibson solution (12.118).

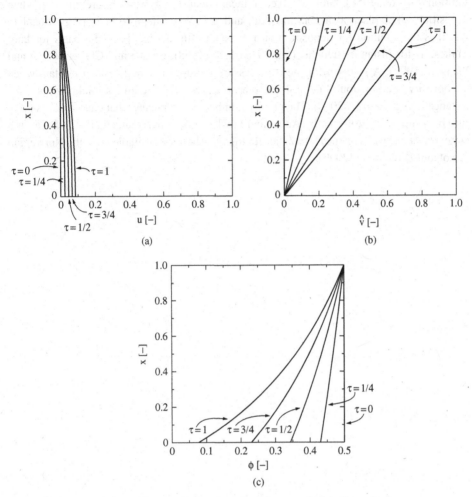

Figure 12.14. *The Gibson solution for* $N_g = 10$ *and* $e_0 = 1$: *(a) dimensionless overpressure, (b) dimensionless Darcy flux and (c) void ratio.*

Exercise 12.13 What is the surface permeability k_0 for the gravity numbers $N_g = 0.1$, 1 and 10, when 3000 m net sediment is deposited during 10 Ma? (Use that $\Delta\varrho = 1000$ kg m^{-3}, $\mu = 10^{-3}$ Pa s and $g = 10$ m s^{-2}.)

12.10 Further reading

There are not many exact solutions for the overpressure build-up in a sedimentary basin during deposition of sediments. It turns out to be very difficult to obtain the overpressure when compaction is mechanical. One of the few solutions that may exist is Gibson's solution (Gibson, 1958). Gibson solved a linear equation for overpressure build-up when deposition takes place at a constant rate, and also when deposition is proportional to $\sqrt{t}$. It is the solution for deposition at a constant rate that has been the basis for later studies, as for instance Bredehoeft and Hanshaw (1968), Bethke and Corbet (1988) and Wangen (1992). A solution for the overpressure build-up is much easier to obtain when the porosity is a function of depth (or depth measured as net sediment thickness), see for instance Wangen (1997, 2000). The gravity number as a parameter that characterizes pressure build-up was introduced by Audet and Fowler (1992) and Wangen (1992). Data for a large number of wells on the Norwegian continental shelf are available from the Norwegian Petroleum Directorate (2009).

13

Fluid flow: basic equations

There are several different types of fluid flow in a sedimentary basin, like for instance fluid flow from overpressure decay, overpressure build-up, meteoric fluid flow and thermal convection. The fluid flow is found by solving pressure equations. A rather general pressure equation is derived from mass conservations of fluid and solid, which serves as a common starting point for the study of different types of fluid flow. Conservation laws alone are not sufficient to make a pressure equation. They have to be supplied with properties for the fluid and the solid in order to form a closed system (a system with the same number of equations as unknowns). This chapter therefore presents some relationships for densities and the porosities as a function of pressure and temperature. The pressure equations are expressed with the fluid flow potential as the main unknown. An introduction to the fluid flow potential is therefore carried out before simple pressure equations are derived and solved. A part of the chapter is concerned with different models for overpressure build-up, and it turns out that various overpressure-generating processes differ mainly by a source term in the pressure equation. Much can therefore be said about overpressure-generating processes by only considering the source terms for fluid expulsion.

13.1 Conservation of solid

The (solid) rock matrix is now a composition of N minerals. Mass conservation of each mineral phase i is expressed using the continuity equation (3.161)

$$\frac{\partial (\phi_i \varrho_i)}{\partial t} + \nabla \cdot (\phi_i \varrho_i \mathbf{v}_s) = q_i, \tag{13.1}$$

where ϕ_i is the bulk volume fraction of mineral i, ϱ_i is the density of mineral i, $\mathbf{v}_s$ is the velocity of the solid phase and q_i is a source term for mineral i. The source term gives the rate at which the mineral is produced or consumed in units of mass per bulk volume and per time. The velocity $\mathbf{v}_s$ of the solid is the same for all mineral phases i. Equation (13.1) can be rearranged as

$$\phi_i \nabla \cdot \mathbf{v}_s = \frac{q_i}{\varrho_i} - \frac{D\phi_i}{dt} - \frac{\phi_i}{\varrho_i} \frac{D\varrho_i}{dt} \tag{13.2}$$

where the operator D/dt is the material differentiation $D/dt = \partial/\partial t + \mathbf{v}_s \cdot \nabla$ with respect to the movement of the solid (the sediment matrix). Equation (13.2) gives an expression for the divergence of the solid velocity by a summation over each component

$$\nabla \cdot \mathbf{v}_s = \frac{1}{(1-\phi)} \sum_{i=1}^{N} \left(\frac{q_i}{\varrho_i} - \frac{\phi_i}{\varrho_i} \frac{D\varrho_i}{dt} \right) + \frac{1}{(1-\phi)} \frac{D\phi}{dt} \tag{13.3}$$

where it is used that the volume fractions of each mineral phase and the porosity (ϕ) add up to one:

$$\phi + \sum_{i=1}^{N} \phi_i = 1. \tag{13.4}$$

We notice that the divergence of $\mathbf{v}_s$ gives the rate of change of porosity

$$\frac{D\phi}{dt} = (1-\phi)\nabla \cdot \mathbf{v}_s - \sum_{i=1}^{N} \left(\frac{q_i}{\varrho_i} - \frac{\phi_i}{\varrho_i} \frac{D\varrho_i}{dt} \right). \tag{13.5}$$

The two equivalent equations (13.3) and (13.5) seem a little unmotivated at this stage, but in the next section we will see that they are needed in the derivation of a pressure equation from mass conservation of fluid. These relations show that the porosity is closely connected to the divergence of the solid velocity. This is clearly seen when the solid densities ϱ_i are constant and when the source terms q_i are zero. We then have

$$\frac{1}{(1-\phi)} \frac{D\phi}{dt} = \nabla \cdot \mathbf{v}_s. \tag{13.6}$$

This relationship tells us that where there is a rate of change of porosity there is a non-constant solid velocity. For example, in a 1D vertical model where there is porosity loss with increasing depth there will also be an increase in the vertical solid velocity (see Exercise 13.1). It is also worth recalling from Section 12.2 that it is possible to integrate equation (13.6).

Exercise 13.1 This exercise looks at relationship (13.6) between $D\phi/dt$ and $\nabla \mathbf{v}_s$ in the Lagrangian ζ-coordinate system. We assume that the porosity decreases exponentially with depth from the basin surface as

$$\phi(\zeta) = \phi_0 \, e^{-(\zeta^* - \zeta)/\zeta_0} \tag{13.7}$$

where $\zeta^* - \zeta$ is the depth measured as net (porosity-free) rock. The height of the basin surface from the basement is ζ^* and ζ_0 is a depth that characterizes the compaction. (This porosity function is discussed in Section 5.1.) Furthermore, we let the basin be in a state of constant deposition with $\zeta^* = \omega t$ where ω is the deposition rate. Deposition starts at $t = 0$. The porosity at a ζ-position is then the following function of time:

$$\phi(\zeta, t) = \phi_0 \, e^{-(\omega t - \zeta)/\zeta_0} \tag{13.8}$$

for $\zeta < \omega t$. The porosity at the base of the basin (at $\zeta = 0$) is

$$\phi_{bot}(t) = \phi_0 \, e^{-\omega t/\zeta_0} \tag{13.9}$$

as a function of time.

(a) Use relationship (13.6) to show that the gradient of the rock velocity v_s gives

$$\frac{\partial v_s}{\partial \zeta} = -\frac{1}{t_0} \frac{\phi}{(1-\phi)^2} \tag{13.10}$$

where the characteristic time $t_0 = \zeta_0/\omega$ is the time needed for deposition to build a basin with height ζ_0.

(b) Assume that $\phi \ll 1$ (or alternatively that $\omega t - \zeta \gg \zeta_0$) which allows us to make the approximation $\partial v_s/\partial \zeta \approx -\phi/t_0$. Show that the velocity of the rock at position ζ is

$$v_s \approx -\omega(\phi - \phi_{bot}) \tag{13.11}$$

where ϕ_{bot} is the porosity at the base of the basin (at $\zeta = 0$).

(c) Show that the z-position of the ζ-coordinate at time t sinks a distance

$$s = \zeta_0 \left(e^{\zeta/\zeta_0} - 1 \right) e^{-\omega t/\zeta_0} \tag{13.12}$$

relative to the base $\zeta = 0$ after infinite time.

(d) Calculate the height of the sedimentary column from the basement ($\zeta = 0$) to ζ at time t using the approximation

$$h(t) = \int_0^\zeta \frac{d\zeta}{(1-\phi)} \, d\zeta \approx \int_0^\zeta (1+\phi) \, d\zeta \tag{13.13}$$

which applies when $\phi \ll 1$. The difference $h(t) - \zeta$ is the extra height of the column due to the porosity. This column is reduced in height from $h(t)$ to ζ when it compacts completely. Check that the difference $h(t) - \zeta$ is the same as the distance in (c).

Solution: (a) The left-hand side of expression (13.6) is

$$\frac{1}{(1-\phi)} \frac{D\phi}{dt} = \frac{1}{(1-\phi)} \frac{\partial \phi}{\partial t} = -\left(\frac{\omega}{\zeta_0}\right) \frac{\phi}{(1-\phi)} \tag{13.14}$$

where the first equality holds because ζ is a Lagrange coordinate. The right-hand side is

$$\frac{\partial v_s}{\partial z} = (1-\phi) \frac{\partial v_s}{\partial \zeta} \tag{13.15}$$

and the left- and right-hand sides together give (a).

(b) It follows from the gradient (13.10) that the velocity of a ζ-position is

$$v_s = -\frac{1}{t_0} \int_0^\zeta \phi(\zeta', t) \, d\zeta' = -\omega \, \phi_0 \, e^{-\omega t/\zeta_0} (e^{\zeta/\zeta_0} - 1) \tag{13.16}$$

when the solid velocity is $v_s = 0$ at the base $\zeta = 0$. It is straightforward to rewrite this velocity in terms of the porosities ϕ and ϕ_{bot}.

(c) The distance the ζ-position moves from time t until infinity is

$$s = \int_t^\infty v_s(\zeta, t')\, dt' = -\omega\, \phi_0\, (e^{\zeta/\zeta_0} - 1) \int_t^\infty e^{-\omega t'/\zeta_0}\, dt' \tag{13.17}$$

which is the wanted answer (13.12).

(d) The height of the column at time t is

$$h(t) = \zeta + \phi_0\, e^{-\omega t/\zeta_0} \int_0^\zeta e^{\zeta'/\zeta_0} d\zeta' \tag{13.18}$$

which becomes $\zeta + s$ when the integration is carried out.

13.2 Conservation of fluid

A fluid pressure equation is now derived from mass conservation of both solid and fluid. Mass conservation of fluid is

$$\frac{\partial(\phi\varrho_f)}{\partial t} + \nabla \cdot \left(\phi\varrho_f \mathbf{v}_f\right) = q_f, \tag{13.19}$$

as shown in Section 3.18, where ϕ is the porosity, ϱ_f is the fluid density, $\mathbf{v}_f$ is the fluid velocity and q_f is a source term for either production or consumption of fluid in units of mass per unit time and per bulk volume. The difference between the fluid velocity and the solid velocity is proportional to the Darcy flux $\mathbf{v}_D$

$$\mathbf{v}_D = \phi(\mathbf{v}_f - \mathbf{v}_s) = -\frac{\mathbf{K}}{\mu}\left(\nabla p_f - \varrho_f g \mathbf{n}_z\right) \tag{13.20}$$

where $\mathbf{K}$ is the permeability tensor and μ is the viscosity of the fluid. Darcy's law (13.20) is introduced in Section 2.4. The fluid pressure is denoted by p_f and $\mathbf{n}_z$ is the unit normal in the vertical direction (z-direction). Conservation of fluid mass (13.19) can be rearranged in a way similar to mass conservation of the minerals using the material derivative with respect to the solid velocity, which gives

$$\phi\frac{D\varrho_f}{dt} + \varrho_f\frac{D\phi}{dt} + \nabla \cdot (\varrho_f \mathbf{v}_D) + \phi\varrho_f \nabla \cdot \mathbf{v}_s = q_f. \tag{13.21}$$

It is at this point that we need mass conservation of solid, formulated by equation (13.3) or the equivalent equation (13.5), in order to go further. However, we are faced with two alternatives – we can either use equation (13.3) to eliminate $\nabla \cdot \mathbf{v}_s$ or we can use equation (13.5) to eliminate $D\phi/dt$. These two alternatives are quite different. In the last alternative, where $D\phi/dt$ is eliminated, we have to supply the model with equations that allow the displacement field of the solid phase to be calculated, and thereby the velocity $\mathbf{v}_s$. In the following we will continue with the first alternative, where we eliminate $\nabla \cdot \mathbf{v}_s$. The other alternative, that involves the displacement field, is discussed later (see Note 13.6). Replacing $\nabla \cdot \mathbf{v}_s$ in equation (13.19) using expression (13.3) leads to the following expression for conservation of fluid mass:

$$\frac{\phi}{\varrho_f} \frac{D\varrho_f}{dt} - \frac{\phi}{\varrho_s} \frac{D\varrho_s}{dt} + \frac{1}{\varrho_f} \nabla \cdot (\varrho_f \mathbf{v}_D)$$

$$= -\frac{1}{(1-\phi)} \frac{D\phi}{dt} + \frac{q_f}{\varrho_f} - \frac{\phi}{(1-\phi)} \sum_i \frac{q_i}{\varrho_i}. \tag{13.22}$$

The solid density for each component has now been replaced with the (average) density of the solid phase, ϱ_s (see Note 13.1). This equation is the starting point for making pressure equations. Density functions, Darcy's law, and various source terms have to be provided. (The simplest assumption about source terms is to assume that they are all zero.) The next section does this for poroelastic rocks.

Note 13.1 The solid density of each component is ϱ_i and it occupies a volume fraction ϕ_i of the (bulk) volume. The density of each component is then replaced by an average density ϱ_s by letting

$$\frac{1}{1-\phi} \sum_i \frac{\phi_i}{\varrho_i} \frac{D\varrho_i}{dt} = \frac{1}{\varrho_s} \frac{D\varrho_s}{dt} \tag{13.23}$$

which turns out to be a simple linear averaging of the solid compressibilities of the components.

13.3 Poroelastic pressure equation

Conservation of fluid (13.22) forms the basis for various pressure equations and the pressure equation for a poroelastic material is presented first. The source terms for fluid and solid are both for simplicity assumed zero ($q_f = 0$ and $q_i = 0$). The fluid pressure enters the conservation law (13.22) by means of compressibilities of the fluid phase, the solid density and the porosity (see Note 13.2). The following poroelastic pressure equation then appears:

$$\phi(\alpha_f + \alpha_{pp}) \frac{Dp_f}{dt} - \frac{1}{\varrho_f} \nabla \cdot \left(\frac{\varrho_f \mathbf{K}}{\mu} \left(\nabla p_f + \varrho_f g \mathbf{n}_z \right) \right) = \phi \alpha_{pc} \frac{Dp_b}{dt} \tag{13.24}$$

where α_f is the compressibility of the fluid, α_{pp} is the pore volume expansion coefficient and α_{pc} is the drained pore compressibility (see Section 4.1). The poroelastic pressure equation (13.24) applies, for example, for rocks that undergo burial. Material derivatives are needed when the rock compacts, and the time derivative of the bulk pressure is non-zero because of burial.

The factor $\phi(\alpha_f + \alpha_{pp})$ in front of the time derivative Dp_f/dt is called the *unconstrained specific storage*. This storage coefficient follows directly from the conservation laws for fluid and solid, and the definitions of the rock compressibilities. The unconstrained specific storage is also derived for a simpler pressure equation in Section 13.4

and in Exercise 14.3. It is possible to define other specific storage coefficients, under various assumptions for how the pore volume responds to the fluid pressure, as done in Section 13.4. Notice that the unconstrained specific storage becomes dominated by the fluid compressibility when it is much larger than the pore volume expansion coefficient.

Note 13.2 The compressibilities of the pore space and the solid volume of the porous rock are taken from Section 4.3. (See equation (4.47) for $\Delta\phi/\phi$ and equation (4.48) for $\Delta\varrho_s/\varrho_s$.) The time derivatives of the solid density and the porosity in the conservation law (13.22) then become

$$
-\frac{\phi}{\varrho_s}\frac{D\varrho_s}{dt} + \frac{1}{(1-\phi)}\frac{D\phi}{dt}
$$

$$
= \frac{\phi}{1-\phi}\left\{(\phi\alpha_{pc} - \alpha_{bc})\frac{Dp_b}{dt} + (\alpha_{bp} - \phi\alpha_{pp})\frac{Dp_f}{dt}\right\}
$$

$$
+ \frac{\phi}{1-\phi}\left\{(\alpha_{bc} - \alpha_{pc})\frac{Dp_b}{dt} + (\alpha_{pp} - \alpha_{bp})\frac{Dp_f}{dt}\right\}
$$

$$
= \phi\alpha_{pp}\frac{Dp_f}{dt} - \phi\alpha_{pc}\frac{Dp_b}{dt}. \tag{13.25}
$$

Recall that $\Delta V_s/V_s = -\Delta\varrho_s/\varrho_s$. Adding together these contributions, and using the definition of fluid compressibility $\alpha_f = (1/\varrho_s)(d\varrho_f/dp_f)_T$, gives the poroelastic pressure equation (13.24).

13.4 Storage coefficients

The previous section introduced the unconstrained specific storage coefficient in a pressure equation for a compacting porous medium. This storage coefficient is now obtained more directly from conservation of fluid mass in a test volume, as for example the box shown in Figure 3.18. Mass conservation of fluid in the box is

$$
\Delta(\varrho_f V_p) = \Big((\varrho_f v_D)(x) - (\varrho_f v_D)(x + \Delta x)\Big)A\Delta t + q_f\Delta V\Delta t \tag{13.26}
$$

where ϱ_f is the fluid density, and q_f is the rate of fluid production inside the box in units of mass per bulk volume and time. The box has a (bulk) volume $\Delta V = A\Delta x$, where A is the cross-section of the box, and $V_p = \phi\Delta V$ is the pore volume. The left-hand side of the mass conservation (13.26) is the increase in the fluid mass inside the box during a time step Δt. This increase in the mass is equal to the net mass that flows into the box plus the mass of fluid created inside the box, which are the two terms on the right-hand side, respectively. The Darcy velocity v_D, which is a fluid flux in units of fluid volume per cross-section and time, gives the mass flux $\varrho_f v_D$. Conservation of fluid mass (13.26) then becomes

$$
\frac{\phi}{\varrho_f V_p}\frac{\partial(\varrho_f V_p)}{\partial t} + \frac{\partial v_D}{\partial x} = \frac{q_f}{\varrho_f} \tag{13.27}
$$

in the limit where both the time step and the length of the box approach zero. (It is for simplicity assumed that the density does not depend on the position x, see Note 13.3.) Expression (13.27) is rewritten as

$$\phi\left(\frac{1}{\varrho_f}\frac{\partial \varrho_f}{\partial t} + \frac{1}{V_p}\frac{\partial V_p}{\partial t}\right) + \frac{\partial v_D}{\partial x} = \frac{q_f}{\varrho_f} \tag{13.28}$$

by carrying out the differentiation in time. The rate of change of the pore volume and the fluid density are

$$\frac{\partial \varrho_f}{\partial t} = \alpha_f \varrho_f \frac{\partial p_f}{\partial t} \tag{13.29}$$

$$\frac{\partial V_p}{\partial t} = \alpha_{pp} V_p \frac{\partial p_f}{\partial t} \tag{13.30}$$

when expressed by the compressibilities of the fluid and the pore volume (4.5). It is assumed that the (confining) bulk pressure is constant. The expression (13.27) now becomes the pressure equation,

$$\phi(\alpha_f + \alpha_{pp})\frac{\partial p_f}{\partial t} + \frac{\partial v_D}{\partial x} = \frac{q_f}{\varrho_f} \tag{13.31}$$

with the *unconstrained specific storage coefficient* $S_\sigma = \phi(\alpha_f + \alpha_{pp})$. Mass conservation gives that the unconstrained storage coefficient is defined as

$$S_\sigma = \phi\left(\frac{1}{\varrho_f}\frac{\partial \varrho_f}{\partial p_f} + \frac{1}{V_p}\frac{\partial V_p}{\partial p_f}\right) \tag{13.32}$$

when there is no constraint on the pore volume. The same expression also defines other storage coefficients, which apply under different conditions for how the pore volume changes for a changing fluid pressure. An example is when the bulk volume remains constant for changes in the fluid pressure. A constant bulk volume requires that the bulk pressure and the fluid pressure change in such a way that

$$\frac{\Delta p_b}{\Delta p_f} = \frac{\alpha_{bp}}{\alpha_{bc}} \tag{13.33}$$

which follows from the compressibilities of the bulk volume (4.5). The relative change of pore volume with respect to the fluid pressure becomes, when the bulk volume is constant,

$$\frac{1}{V_p}\left(\frac{\partial V_p}{\partial p_f}\right)_{V_b} = \alpha_{pp} - \frac{\alpha_{bp}\alpha_{pc}}{\alpha_{bc}} \tag{13.34}$$

and the *constrained specific storage coefficient* is therefore

$$S_\epsilon = \phi\left(\alpha_f + \alpha_{pp} - \frac{\alpha_{bp}\alpha_{pc}}{\alpha_{bc}}\right). \tag{13.35}$$

Another possibility is the *unjacketed specific storage coefficient*, which applies when the fluid pressure and bulk pressure are equal. The relative change of pore volume with respect to the fluid pressure for this case is

$$\frac{1}{V_p}\left(\frac{\partial V_p}{\partial p_f}\right)_{p_s} = \alpha_{pp} - \alpha_{pc} \tag{13.36}$$

and the unjacketed storage coefficient is

$$S_\gamma = \phi(\alpha_f + \alpha_{pp} - \alpha_{pc}). \tag{13.37}$$

Notice that the pressure difference $p_s = p_b - p_f$ is constant when $p_b = p_f$. At this point it is natural to introduce the quantity

$$\Delta\zeta = \phi\left(\frac{\Delta\varrho_f}{\varrho_f} + \frac{\Delta V_p}{V_p}\right) \tag{13.38}$$

which is called the *increment in fluid content*. The specific storage coefficients are often defined in terms of the increment in fluid content, since

$$S_\sigma = \frac{\partial\zeta}{\partial p_f}, \quad S_\epsilon = \left(\frac{\partial\zeta}{\partial p_f}\right)_{V_b} \quad \text{and} \quad S_\gamma = \left(\frac{\partial\zeta}{\partial p_f}\right)_{p_s}. \tag{13.39}$$

The increment in fluid content is simply the same as

$$\Delta\zeta = \frac{\Delta V_p - \Delta V_f}{V_b} \tag{13.40}$$

where ΔV_p is the increase in the pore volume, and where ΔV_f is the volume of compressed fluid in the pore space. Both volume changes are caused by the same increase in the (pore) fluid pressure. The increment in fluid content is therefore the net increase in fluid volume per bulk volume of rock, caused by an increase in the fluid pressure.

Note 13.3 The divergence of mass flux into the test volume has been approximated the following way in equation (13.27)

$$\frac{1}{\varrho_f}\frac{\partial(\varrho_f v_D)}{\partial x} \approx \frac{\partial v_D}{\partial x} \tag{13.41}$$

assuming that the density does not depend on the x-position.

Exercise 13.2
(a) Show that the unconstrained specific storage coefficient can be written as

$$S_\sigma = \left(\frac{1}{K} - \frac{1}{K_s'}\right) + \phi\left(\frac{1}{K_f} - \frac{1}{K_\phi}\right). \tag{13.42}$$

(b) Show that the unjacketed specific storage coefficient can be written as

$$S_\gamma = \phi\left(\frac{1}{K_f} - \frac{1}{K_\phi}\right). \tag{13.43}$$

Hint: use relation (4.19) and the compressibilities defined in Section 4.2.

Exercise 13.3 Show that the constrained specific storage coefficient can be written

$$S_\epsilon = S_\sigma - \frac{\gamma_b^2}{K}. \tag{13.44}$$

Hint: use that $\phi\alpha_{pc} = \alpha_{bc} - \alpha_s$.

Exercise 13.4 Show that the increment of fluid volume (13.38) is the same as the alternative expression (13.40).

13.5 Stress, strain and poroelasticity

Section 3.15 anticipated how the fluid pressure changes the relationship between stress and strain in a porous medium, like for instance a sedimentary rock. The next encounter with fluid pressure and strain was in Chapter 4, where the compressibilities of a fluid saturated porous medium were introduced. Equation (4.59) for relative bulk volume changes is simply an equation for bulk volume strain (3.13), which gives

$$\epsilon = \epsilon_{kk} = -\alpha_{bc}(\sigma - \gamma_b p_f) \tag{13.45}$$

where σ is the average confining stress, which is the same as the bulk pressure p_b. The symbol σ is used instead of p_b because bulk pressure will soon be related to the stress tensor σ_{ij}. Expression (13.45) can be compared with the definition of the bulk modulus (3.3), and we have

$$\alpha_{bc} = \frac{1}{K}. \tag{13.46}$$

The bulk modulus for a zero-porosity material is the drained bulk compressibility of the porous medium. Bulk volume strain (13.45) is rewritten as

$$-\sigma = K\epsilon - \gamma_b p_f \tag{13.47}$$

which allows for a generalization of the Lamé equations (3.98). The introduction of an isotropic fluid pressure in the pores does not change the state of shear stress, and we have as before that

$$-\sigma_{ij} = 2G\epsilon_{ij} \quad \text{for} \quad i \neq j. \tag{13.48}$$

The cases for normal stress ($i = j$) become

$$-\sigma_{xx} = 2G\epsilon_{xx} + \lambda\epsilon_{kk} - \gamma_b p_f \tag{13.49}$$
$$-\sigma_{yy} = 2G\epsilon_{yy} + \lambda\epsilon_{kk} - \gamma_b p_f \tag{13.50}$$
$$-\sigma_{zz} = 2G\epsilon_{zz} + \lambda\epsilon_{kk} - \gamma_b p_f \tag{13.51}$$

because these three equation give that

$$-\sigma = -\frac{1}{3}\sigma_{kk} = K\epsilon_{kk} - \gamma_b p_f \tag{13.52}$$

which is expression (13.47) for the compressibilities. Recall that the bulk modulus is $K = (2G/3) + \lambda$. The Lamé equations for a fluid saturated porous medium are therefore

$$- \sigma_{ij} = 2G\epsilon_{ij} + \lambda\epsilon_{kk}\delta_{ij} - \gamma_b p_f \delta_{ij}. \tag{13.53}$$

The bulk volume strain (13.45) therefore gives how the fluid pressure should be incorporated into the Lamé equations.

The equilibrium equations (3.176) (for a z-axis pointing downwards)

$$\sigma_{ij,j} = \varrho g \delta_{iz} \tag{13.54}$$

are three equations for the displacements u_i. Inserting the Lamé equations (13.53) into the equilibrium equations (13.54) gives the following three equations for the displacement field u_i:

$$Gu_{i,jj} + (G + \lambda)u_{k,ki} - \gamma_b p_{f,i} = \varrho g \delta_{iz} \tag{13.55}$$

when using the summation convention. These equations are more readable when written out as one equation for each component of the displacement field. Indices are avoided by letting $u = u_1$, $v = u_2$ and $w = u_3$, and equations (13.55) become

$$G\nabla^2 u + (G + \lambda)\left(\frac{\partial^2 u}{\partial x^2} + \frac{\partial^2 v}{\partial x \partial y} + \frac{\partial^2 w}{\partial x \partial z}\right) = \gamma_b \frac{\partial p_f}{\partial x} \tag{13.56}$$

$$G\nabla^2 v + (G + \lambda)\left(\frac{\partial^2 u}{\partial y \partial x} + \frac{\partial^2 v}{\partial y^2} + \frac{\partial^2 w}{\partial y \partial z}\right) = \gamma_b \frac{\partial p_f}{\partial x} \tag{13.57}$$

$$G\nabla^2 w + (G + \lambda)\left(\frac{\partial^2 u}{\partial z \partial x} + \frac{\partial^2 v}{\partial z \partial y} + \frac{\partial^2 w}{\partial z^2}\right) = \gamma_b \frac{\partial p_f}{\partial x} - \varrho g. \tag{13.58}$$

The equations for the displacement (13.55) and the poroelastic pressure equation (13.24) are then four equations for four unknowns (the displacement field u, v, w and the fluid pressure p_f).

The poroelastic pressure equation (13.24) is not straightforward to couple with the equations for the displacement, because of the time derivative of the bulk pressure p_b. This problem is fixed by replacing the bulk pressure p_b with bulk strain using the compressibility relationship (13.47). The poroelastic pressure equation (13.24) then becomes (see Note 13.4 for details)

$$S_\epsilon \frac{Dp_f}{dt} + \gamma_b \frac{D\epsilon}{dt} + \nabla \cdot \mathbf{v}_D = 0 \tag{13.59}$$

where

$$\epsilon = \frac{\partial u}{\partial x} + \frac{\partial v}{\partial y} + \frac{\partial w}{\partial z} \tag{13.60}$$

is the volume strain when it is written without indices. The use of indices is shown with equation (13.45). The poroelastic pressure equation (13.59) is the version that couples to the equations (13.56)–(13.58) for the displacement field. These four equations go under the name Biot equations after Biot who first introduced them (Biot, 1941). The Biot equations become simpler when they are reduced to the vertical direction, and assuming that there is zero displacement in the horizontal plane ($u = 0$ and $v = 0$). The two remaining equations are then

$$S_\epsilon \frac{\partial p_f}{\partial t} - \frac{k}{\mu} \frac{\partial^2 p_f}{\partial z^2} = -\gamma_b \frac{\partial^2 w}{\partial t \partial z} \tag{13.61}$$

$$(2G + \lambda)\frac{\partial^2 w}{\partial z^2} - \gamma_b \frac{\partial p_f}{\partial z} = -\varrho g \tag{13.62}$$

since there is no x or y dependence. The permeability k and the viscosity μ are both constant, and it is assumed that the deformations are so small that the material derivative can be replaced by the partial derivative. The most simple solution to these two coupled equations is for a hydrostatic fluid. The fluid pressure is then $p_f = \varrho_f g z$. The equation for the vertical displacement w is

$$(2G + \lambda)\frac{d^2 w}{dz^2} = -(\varrho - \gamma_b \varrho_f)g. \tag{13.63}$$

We have already solved this equation in Exercise 3.28 for an unsaturated rock, and for a saturated rock in Exercise 3.30. Notice that the effect of gravity becomes zero when bulk density is as low as $\varrho = \gamma_b \varrho_f$, in which buoyancy from the pore fluid carries the entire rock matrix.

Note 13.4 The derivation of the poroelastic pressure equation makes use of the average pressure (13.47). This stress–strain relationship is not compatible with strain derived from conservation of solid with a source term. The reason is that the stress–strain relationship (13.47) does not account for changes in strain caused by a source term, for example strain caused by dissolution of the solid.

Note 13.5 The pressure equation (13.24) becomes

$$S_\sigma \frac{D p_f}{dt} - \phi\alpha_{pc}\left(-\frac{1}{\alpha_{bc}}\frac{D\epsilon}{dt} + \gamma_b \frac{D p_f}{dt}\right) + \nabla \cdot \mathbf{v}_D = 0 \tag{13.64}$$

when the confining pressure (13.47) replaces the bulk pressure p_b. Furthermore, we have that

$$\frac{\phi\alpha_{pc}}{\alpha_{bc}} = \frac{\alpha_{bc} - \alpha_s}{\alpha_{bc}} = 1 - \frac{\alpha_s}{\alpha_{bc}} = \gamma_b \tag{13.65}$$

and that

$$\phi\alpha_{pc}\gamma_b = \gamma_b^2 \alpha_{bc} = \frac{\gamma_b^2}{K}. \tag{13.66}$$

Using that $S_\epsilon = S_\sigma - \gamma_b^2/K$ (see Exercise 13.3) gives the pressure equation (13.59).

Note 13.6 There is an alternative route to the poroelastic pressure equation (13.59), which begins with mass conservation of fluid and rock. Conservation of fluid (13.21) is

$$\frac{\phi}{\varrho_f}\frac{D\varrho_f}{dt} + \frac{D\phi}{dt} + \frac{1}{\varrho_f}\nabla(\varrho_f \cdot \mathbf{v}_D) + \phi\nabla \cdot \mathbf{v}_s = 0 \tag{13.67}$$

and the gradient of the rock velocity $\nabla \cdot \mathbf{v}_s$ is

$$\nabla \cdot \mathbf{v}_s = \frac{\partial v_{s,k}}{\partial x_k} = \frac{\partial}{\partial x_k}\left(\frac{\partial u_k}{\partial t}\right) = \frac{\partial}{\partial t}\left(\frac{\partial u_k}{\partial x_k}\right) = \frac{\partial \epsilon}{\partial t} \tag{13.68}$$

where u_k is the displacement field in a coordinate system that follows the rock deformations. Conservation of rock mass was used in Section 13.2 to replace $\nabla \cdot \mathbf{v}_s$ with an expression with $D\phi/dt$. The opposite is now done; $D\phi/dt$ is replaced with expression (13.5) that contains $\nabla \cdot \mathbf{v}_s$, and we arrive at

$$\frac{(1-\phi)}{\varrho_s}\frac{D\varrho_s}{dt} + \frac{\phi}{\varrho_f}\frac{D\varrho_f}{dt} + \frac{D\epsilon}{dt} + \frac{1}{\varrho_f}\nabla(\varrho_f \cdot \mathbf{v}_D) = 0. \tag{13.69}$$

The time differentiation of the densities becomes time differentiation of the fluid pressure and the bulk (confining) pressure by means of the compressibilities. Expression (13.47) for bulk pressure replaces the bulk strain, and we have pressure equation (13.59). In the case of constant fluid and rock densities, conservation of fluid mass (13.69) gives

$$\nabla \cdot \mathbf{v}_D = -\frac{D\epsilon}{dt} \tag{13.70}$$

which shows how expulsion of fluid is linked to compaction in terms of the time differentiation of bulk strain.

13.6 Stress caused by overpressure

Rock behaves as a poroelastic material on a human time scale, but on a geological time scale the deformations may be dominated by flow. The state of the stress into the Earth is therefore often isotropic. We will in this section look at the stress change caused by a change in overpressure, for instance by fluid injection (or production) in an aquifer by a well. The initial stress state is isotropic before the overpressure is changed, and the initial fluid pressure is for simplicity taken to be hydrostatic. The change in the overpressure takes place on a human time scale, while the initial isotropic stress state is the result of creep over millions of years.

The Lamé equations (13.53) show that it is the effective stress

$$\sigma'_{ij} = \sigma_{ij} - \gamma_b \, p_f \delta_{ij} \tag{13.71}$$

which is the cause for strain in a fluid saturated porous medium. The equilibrium equations (13.54) become

$$\sigma_{ij,j} = \sigma'_{ij,j} + \gamma_b \, p_{f,i} = \varrho g \delta_{iz} \tag{13.72}$$

in terms of the effective stress (13.71), where g is the bulk divisity. We can subtract the hydrostatic fluid pressure from both sides of this equation, which gives

$$\sigma'_{ij,j} + \gamma_b \, p_{e,i} = (\varrho - \gamma_b \varrho_f)g \, \delta_{iz} \tag{13.73}$$

where p_e is the overpressure. The overpressure is the difference between the fluid pressure and the hydrostatic fluid pressure (p_h)

$$p_e = p_f - p_h \tag{13.74}$$

where the hydrostatic pressure is the weight of the water column,

$$p_h = \int_0^z \varrho_f g \, dz. \tag{13.75}$$

We now introduce the isotropic reference state

$$\sigma_{ij}^{(0)} = \int_0^z (\varrho - \gamma_b \varrho_f) g \, dz \, \delta_{ij} \tag{13.76}$$

for effective stress. It is the solution of the equilibrium equations (13.73), for zero overpressure, since

$$\sigma_{ij,j}^{(0)} = (\varrho - \gamma_b \varrho_f) g \, \delta_{iz}. \tag{13.77}$$

We are interested in the difference in effective stress from the reference state $\sigma_{ij}^{(0)}$, when the difference is caused by a change in the fluid pressure. The effective stress is therefore written as the sum

$$\sigma_{ij}' = \sigma_{ij}^{(0)} + \sigma_{ij}^{(1)} \tag{13.78}$$

where $\sigma_{ij}^{(1)}$ is the difference due to a change in overpressure. It is the effective stress that creates displacements and we are interested in the displacements and strain caused by the overpressure. We therefore have that the effective stress difference from the reference state is

$$-\sigma_{ij}^{(1)} = G\varepsilon_{ij} + \lambda \, \varepsilon_{kk} \delta_{ij} \tag{13.79}$$

and that the full effective stress is

$$-\sigma_{ij}' = -\sigma_{ij,j}^{(0)} - \sigma_{ij}^{(1)} = -\sigma_{ij,j}^{(0)} + G\varepsilon_{ij} + \lambda \, \varepsilon_{kk} \delta_{ij}. \tag{13.80}$$

The Lamé equations for the difference in the effective stress become

$$\sigma_{ij,j}^{(1)} = -\gamma_b p_{e,i} \tag{13.81}$$

or

$$G\varepsilon_{ij,j} + \lambda \, \varepsilon_{kk,i} = \gamma_b p_{e,i} \tag{13.82}$$

which shows the coupling between the overpressure and the strain field. The gradient of the overpressure acts like a body force on the saturated porous rock.

These equations are considerably simplified when the overpressure is only dependent on the depth, $p_e = p_e(z)$. The Biot coefficient is assumed to be one $(\gamma_b = 1)$ and the bulk density ϱ is constant. It is also assumed that there is no x- or y-dependence in the effective stress. The equilibrium equations (13.81) then become

$$\sigma_{zz,z}^{(1)} = -p_{e,z} \tag{13.83}$$

because $\sigma_{zx,x}^{(1)} = \sigma_{zy,y}^{(1)} = 0$, which is integrated to

$$\sigma_{zz}^{(1)} = -p_e. \tag{13.84}$$

The effective stress difference in the vertical difference is simply the negative overpressure. Equation (13.79) gives that

$$-\sigma_{zz}^{(1)} = (2G + \lambda)\varepsilon_{zz} \tag{13.85}$$

$$-\sigma_{xx}^{(1)} = \lambda\varepsilon_{zz} \tag{13.86}$$

when there is no strain in the xy-plane ($\varepsilon_{xx} = \varepsilon_{zz} = 0$). We therefore have

$$\sigma_{xx}^{(1)} = \frac{\lambda}{(2G + \lambda)}\sigma_{zz}^{(1)} = \frac{\nu}{(1 - \nu)}\sigma_{zz}^{(1)} = -\frac{\nu}{(1 - \nu)}p_e. \tag{13.87}$$

(This is the same reasoning as in Exercise 3.13.) The excess lithostatic stress is

$$\sigma_{ij}^{(0)} = (\varrho - \varrho_f)gz\,\delta_{ij} \tag{13.88}$$

and the effective stress (13.78) becomes

$$\sigma_{xx}' = (\varrho - \varrho_f)gz - \frac{\nu}{(1 - \nu)}p_e \tag{13.89}$$

$$\sigma_{zz}' = (\varrho - \varrho_f)gz - p_e. \tag{13.90}$$

The solution for effective stress shows that anisotropy develops with increasing overpressure. The effective stress state is isotropic for $p_e = 0$ and it increases until $\sigma_{zz}' = 0$, in which case

$$\sigma_{xx}' = \left(\frac{1 - 2\nu}{1 - \nu}\right)(\varrho - \varrho_f)gz. \tag{13.91}$$

The stress is the effective stress added to the fluid pressure and it is

$$\sigma_{xx} = \varrho gz + \left(\frac{1 - 2\nu}{1 - \nu}\right)p_e \tag{13.92}$$

$$\sigma_{zz} = \varrho gz. \tag{13.93}$$

The stress in the vertical direction is the weight of the sedimentary column regardless of overpressure, but the stress in the horizontal plane increases with increasing overpressure.

We have from Section 8.6 that a sufficiently large anisotropy in the effective stress leads to hydrofracturing. The fracture criterion is in terms of a Coulomb fracture envelope, where fracture takes place when the Mohr's circle for the state of effective stress touches the fracture envelope. The effective stress that gives hydrofracture is then, according to equation (8.35),

$$a\sigma_1' - b\sigma_3' = 2S_0 \tag{13.94}$$

where $a = \sqrt{1 + \mu^2} - \mu$ and $a = \sqrt{1 + \mu^2} + \mu$. (The coefficient of internal friction is $\mu = \tan\phi$, where ϕ is the angle of internal friction. See Section 8.6.) Inserting σ_{xx}' and σ_{zz}' as the largest and least principal stresses gives that the overpressure necessary for fracture is

$$p_e = \left(\frac{b-a}{b-fa}\right)(\varrho - \varrho_f)gz + \frac{2S_0}{(b-fa)} \tag{13.95}$$

where f is the coefficient $f = \nu/(1-\nu)$. The factor $(b-a)/(b-fa)$ increases from ~ 0.8 at $\phi = 30°$ to 1 at $\phi = 90°$. The overpressure necessary for hydrofracturing is $\sim 90\%$ of the excess lithostatic pressure at depths where $(\varrho - \varrho_f)gz \gg 2S_0$.

The same procedure can be applied to any initial state (or reference state) for the effective stress that solves the equilibrium equations, when we are interested in the poroelastic displacement and strain caused by a pressure difference.

13.7 The rate of change of porosity

The porosity is changed by a series of processes, like mechanical compaction of shallow unlithified sediments, dehydration of clay, oil generation, cementation of pore space, and thermal expansion of fluids and solids. These different processes will later be studied in turn. The porosity changing processes are now simplified and divided into three groups. The first is poroelastic porosity changes from variations in the pore fluid pressure and the (confining) bulk pressure. Poroelasticity is linear mechanical compaction, where porosity changes are linearly related to pressure changes. The next group is nonlinear mechanical compaction, where the porosity is a function of the effective (vertical) stress, as for example compaction by the normal consolidation line. The third group is temperature-controlled porosity changes, for instance, cementation of pore space and oil generation. This group also includes the thermal expansion of the rock. The pressure equation (13.24), which already covers poroelastic porosity changes, is now extended to cover thermal porosity changes as well. Poroelasticity applies for consolidated and lithified rocks, while nonlinear mechanical compaction takes place in unlithified sediments like sand and clay. Thermal compaction can operate in both lithified and unlithified sediments. These three groups are summarized in this table.

Group	Process	Variable
1	*Linear Mechanical Compaction:*	p_f and p_b
	Poroelasticity	p_f and p_b
2	*Nonlinear Mechanical Compaction:*	$p_b - p_f$
	Normal Consolidation Line	$p_b - p_f$
3	*Thermal Compaction:*	T
	Thermal expansion	T
	Cementation of the pore space	T
	Dehydration	T
	Oil generation	T

The rate of change of porosity is written as the sum of poroelastic porosity changes and temperature controlled porosity changes

$$\frac{D\phi}{dt} = \frac{D\phi}{dt}_{\text{poro}} + \frac{D\phi}{dt}_{\text{thermal,1}} + \frac{D\phi}{dt}_{\text{thermal,2}} + \cdots . \tag{13.96}$$

All the thermal processes that change the porosity could in principle operate at the same time, although it is not likely. An important difference between the different thermal compaction processes is that thermal expansion is linearly related to temperature, while the other thermal compaction processes are controlled by Arrhenius kinetics. Thermal expansion is proportional to temperature changes, which implies that

$$\frac{D\phi}{dt}_{\text{thermal}} = \beta \frac{DT}{dt} \tag{13.97}$$

where β is the thermal expansibility of the rock. It is shown in Exercise 13.5 that the porosity has a thermal expansibility

$$\beta = (1 - \phi)(\beta_b - \beta_s) \tag{13.98}$$

where β_b and β_s are bulk and average solid thermal expansibilities, respectively. The other thermally controlled porosity changes are simply given by a rate that is a function of the temperature

$$\frac{D\phi}{dt}_{\text{thermal}} = f(T) \tag{13.99}$$

where the function $f(T)$ will normally be an Arrhenius relationship. An example of a function f is the rate of porosity loss from quartz cementation (11.8). This and other examples of thermally controlled porosity changes are studied in later sections.

Exercise 13.5 Show that the porosity responds thermally as given by equation (13.98). Explain why this relationship is reasonable for the special cases $\beta_b = \beta_s$ and $\phi \to 1$.
Solution: The rate of change of porosity $d\phi/dT$ becomes

$$\frac{d\phi}{dT} = \frac{d}{dT}\left(\frac{V_p}{V_b}\right) = \frac{d}{dT}\left(\frac{V_b - V_s}{V_b}\right)$$
$$= -\frac{1}{V_b}\frac{dV_s}{dT} + \frac{V_s}{V_b^2}\frac{dV_b}{dT} \tag{13.100}$$

when expressed by the bulk volume V_b, solid volume V_s and the pore volume $V_p = V_b - V_s$ of a rock sample. We then get equation (13.98) using the definitions

$$\beta_b = \frac{1}{V_b}\frac{dV_b}{dT} \quad \text{and} \quad \beta_s = \frac{1}{V_s}\frac{dV_s}{dT} \tag{13.101}$$

of the thermal expansibilities of the bulk and the solid, respectively.

13.8 A general pressure equation

We are now ready to write down a general pressure equation for porous rocks that accounts for poroelasticity, thermal expansibility and thermal compaction processes. The equation for the (pore) fluid pressure is then (as shown in Note 13.7)

$$S_\sigma \frac{D p_f}{dt} - \frac{1}{\varrho_f} \nabla \cdot \left(\frac{\varrho_f \mathbf{K}}{\mu} \left(\nabla p_f + \varrho_f g \mathbf{n}_z \right) \right) = Q_f \qquad (13.102)$$

where

$$S_\sigma = \phi (\alpha_f + \alpha_{pp}) \qquad (13.103)$$

is the unconstrained storage coefficient, and where the source term is

$$
\begin{aligned}
Q_f = & \phi \alpha_{pc} \frac{D p_b}{dt} \\
& - \frac{1}{(1 - \phi)} \frac{D \phi}{dt}_{\text{thermal}} \\
& + \left(\phi \beta_f + (1 - \phi) \beta_s - \beta_b \right) \frac{DT}{dt} \\
& + \frac{q_f}{\varrho_f} - \frac{\phi}{(1 - \phi)} \sum_i \frac{q_i}{\varrho_i}.
\end{aligned}
\qquad (13.104)
$$

This is still a poroelastic pressure equation. The only difference between this equation and the poroelastic equation (13.24) is the source term Q_f, which now accounts for several more compaction processes.

A pressure equation is an expression for conservation of fluid mass, and the interpretation of the terms in the pressure equation (13.102) follows from the basic formulation of fluid mass conservation in a small test volume (see Section 13.4). The first term is a storage term, the divergence term represents the net flow of fluid that enters a test volume, and on the right-hand side there is a source term.

All contributions to the source (or sink) term Q_f give fluid production (or consumption) in units of volume fluid per bulk volume rock and per time. The first contribution (13.104) to the source term is the rate of fluid expulsion caused by the increasing bulk pressure. The second term (13.104) is the rate of fluid expulsion from thermal compaction, and the third term (13.104) is the rate of fluid expulsion from thermal expansion. The last term (13.104) is the rate of fluid expulsion from chemical reactions.

Note 13.7 All parts of the source term in the pressure equation (13.102) follow from conservation of fluid mass (13.22) and the poroelastic pressure equation (13.24), except part (13.104) for thermal expansion. Thermal expansion applies for the fluid, the solid and the pore space, and is

$$- \frac{\phi}{\varrho_f} \frac{\partial \varrho_f}{\partial T} + \frac{\phi}{\varrho_s} \frac{\partial \varrho_s}{\partial T} - \frac{1}{(1 - \phi)} \frac{d\phi}{dT} = \phi \beta_f + (1 - \phi) \beta_s - \beta_b. \qquad (13.105)$$

The thermal expansibilities of the pore space, the solid (β_s) and the bulk rock matrix (β_b) are taken from Exercise 13.5. The thermal expansibility of the fluid is $\beta_f = -(1/\varrho_f)\partial\varrho_f/\partial T$.

13.9 Potential flow

The fluid flow potential is now revisited as an alternative to the fluid pressure as the main variable (see Section 2.5). The advantage of using a potential as the main variable is that fluid flow becomes potential flow, which is flow that is proportional to the gradient of a potential. It is often simpler mathematically to deal with potential flow compared to the alternatives, which involve the gravity term in Darcy's law. The potential (Φ) is the fluid pressure p_f less the hydrostatic pressure relative to a (constant) reference level ($p_{h,0}$):

$$\Phi = p_f - p_{h,0}. \tag{13.106}$$

The reference level is $z = 0$ for convenience, and the hydrostatic pressure relative to the reference level is

$$p_{h,0} = \int_0^z \varrho_f g\, dz = \varrho_f g z \tag{13.107}$$

when the fluid density ϱ_f is constant and g is the gravitational acceleration. Darcy's law becomes simplified to potential flow

$$v_D = -\frac{K}{\mu}\nabla\Phi \tag{13.108}$$

when the fluid pressure is replaced by the potential in Darcy's law (13.20) where $p_f = \Phi + \varrho g z$. A constant fluid density is the condition for the gravity term in Darcy's law (13.20) to disappear, and the fluid density can in most cases safely be assumed constant. Darcy's law is often expressed as

$$v_D = -\frac{K g \varrho_f}{\mu}\nabla h \tag{13.109}$$

using the *hydraulic head*

$$h = z + \frac{p_f}{\varrho g} \tag{13.110}$$

as an alternative version of the potential. It turns out that the hydraulic head allows for a generalization that accounts for pressure-dependent density (see Note 13.8). The potential Φ is energy per unit volume (when the fluid density is constant), and the hydraulic head is energy measured per unit mass (and g). The hydraulic head measures an energy difference as an equivalent vertical difference in the gravitational field. The existence of a potential is a property of a *conservative field*, see Section 10.3.

The density is no longer constant when it depends on temperature or pressure. This is the case for thermal convection, where thermal density differences are responsible for the fluid flow. In this case Darcy's law (13.20) becomes

$$\mathbf{v}_D = -\frac{\mathbf{K}}{\mu}\left(\nabla\Phi + \frac{\partial p_{h,0}}{\partial x}\mathbf{n}_x + \frac{\partial p_{h,0}}{\partial y}\mathbf{n}_y\right) \tag{13.111}$$

as a function of Φ, because the reference hydrostatic pressure $p_{h,0}$ may have lateral variations when the fluid density depends on temperature. This is not potential flow any more and Φ is now called a *pseudo-potential*. The vectors $\mathbf{n}_x$ and $\mathbf{n}_y$ are the unit vectors in the x- and y-directions respectively (see Exercise 13.6.)

The potential along the basin surface is often needed as a boundary condition. The basin surface below sea level has a zero potential, because $p_f = p_{h,0}$ along the seafloor when sea level is the reference level $z = 0$. The surface above sea level has a potential defined by the height of the water table above the reference level ($z = 0$). If the water table is $z = h(x, y)$, the potential along the surface is

$$\Phi_{surf} = \int_0^{h(x,y)} \varrho_f g\, dz = \varrho_f g\, h(x, y). \tag{13.112}$$

The surface-potential Φ_{surf} is greater than zero for all parts of the basin surface above sea level, and it is zero for all parts that are below sea level. We will later see how (a non-zero) surface potential is the cause for meteoric (ground) water flow.

The potential can be converted to the fluid pressure by use of relation (13.106). The potential can also be converted to the excess pressure, p_e, by subtracting the fluid potential at the surface

$$p_e = \Phi - \Phi_{surf}. \tag{13.113}$$

Equation (13.113) shows that the excess pressure is equal to the fluid potential for those parts of a basin which are below sea level, because $\Phi_{surf} = 0$ there. The hydrostatic pressure can also be written as the sum of the reference hydrostatic pressure and the pressure of the weight of the fluid column above the reference level

$$p_h = p_{h,0} + \int_0^{h(x,y)} \varrho_f g\, dz = p_{h,0} + \Phi_{surf}. \tag{13.114}$$

As a consistency check, it is seen that the excess pressure (13.113) added to the hydrostatic pressure (13.114) yields the fluid pressure (13.106). The *effective pressure*, defined as the overburden minus the fluid pressure, becomes

$$\sigma' = p_b - p_{h,0} - \Phi \tag{13.115}$$

when it is expressed with the potential Φ.

Note 13.8 *The Hubbert potential.* The condition for potential flow has so far been that the fluid density is constant. The fluid flow is then driven by the gradient of the hydraulic head, which is

$$\nabla h = \mathbf{n}_z + \frac{1}{\varrho_f g} \nabla p_f. \tag{13.116}$$

This gradient remains unchanged when the hydraulic head (13.110) is generalized to

$$h = z + \int_0^{p_f} \frac{dp}{\varrho_f(p)\, g} \tag{13.117}$$

in the case of the pressure-dependent density. The generalized head therefore gives potential flow too. This extension was introduced by Hubbert (1940), but it only allows for the density as a function of pressure. It is straightforward to verify that the gradient of the potential (13.117) has the same form as the gradient (13.116).

Note 13.9 *The potential and overpressure.* The potential Φ is defined in a similar way as the *excess pressure* p_e, which is the fluid pressure minus the hydrostatic pressure p_h:

$$p_e = p_f - p_h, \tag{13.118}$$

where the hydrostatic pressure is the pressure from the weight of the entire water column above a point. The hydrostatic pressure is

$$p_h = \int_z^{h(x,y)} \varrho_f g \, dz \tag{13.119}$$

where $h = h(x, y)$ is the topography above the reference level. A natural choice for a reference level ($z = 0$) is sea level or a water surface, where $h = 0$ along the sea surface, see Figure 13.1. The potential Φ and the excess pressure p_e are therefore identical for those parts of a sedimentary basin that are covered by water, because the hydrostatic pressure and the reference hydrostatic pressure are the same there. These two hydrostatic

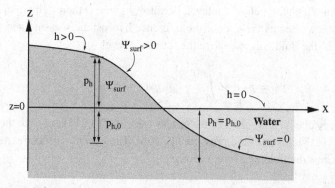

Figure 13.1. *The reference level is defined by the water surface, and it has the vertical position $z = 0$. The topography is given by the function $z = h(x, y)$ above sea level, but it is $h = 0$ along the water surface.*

pressures are not the same where the basin surface is above the water (the reference level), where their difference is

$$p_h - p_{h,0} = \int_0^{h(x,y)} \varrho_f g \, dz \qquad (13.120)$$

which is the weight of the water column above the reference level.

Exercise 13.6 Show Darcy's law (13.111) in terms of the pseudo-potential Φ.

13.10 A general equation for the fluid flow potential

A pressure equation in Φ can now be written with basis in the general poroelastic pressure equation (13.102). The fluid pressure $p_f = \Phi + p_{h,0}$ is replaced by the potential Φ added to the reference hydrostatic pressure $p_{h,0}$. The general poroelastic equation for the potential is then

$$S_\sigma \frac{D\Phi}{dt} - \frac{1}{\varrho_f} \nabla \left[\frac{\varrho_f \mathbf{K}}{\mu} \left(\nabla \Phi + \frac{\partial p_{h,0}}{\partial x} \mathbf{n}_x + \frac{\partial p_{h,0}}{\partial y} \mathbf{n}_y \right) \right] = Q_{\text{pot}} \qquad (13.121)$$

where S_σ is

$$S_\sigma = \phi(\alpha_f + \alpha_{pp}) \qquad (13.122)$$

and where the right-hand side is

$$\begin{aligned}
Q_{\text{pot}} = & -S_\sigma \frac{D p_{h,0}}{dt} + \phi \alpha_{pc} \frac{D p_b}{dt} \\
& - \frac{1}{(1-\phi)} \frac{D\phi}{dt}\bigg|_{\text{thermal}} \\
& + \left(\phi \beta_f + (1-\phi)\beta_s - \beta_b \right) \frac{DT}{dt} \\
& + \frac{q_f}{\varrho_f} - \frac{\phi}{(1-\phi)} \sum_i \frac{q_i}{\varrho_i}.
\end{aligned} \qquad (13.123)$$

The equation for the potential is a straightforward step from the same equation for fluid pressure. It is written out with all the details because it is this equation we will build on in the sections that follow.

13.11 Simple pressure equations

The general equation (13.121) can be simplified considerably by assuming that both the rock density and the fluid density are constant. If we furthermore assume that there are no source terms that allow for mass exchange between the solid and the fluid, and that the porosity and the permeability are constant, we simply get

$$\nabla^2 \Phi = 0. \qquad (13.124)$$

Fluid flow is then given by a Laplace equation for the potential. Recall that the Laplace equation is an expression for mass conservation of an incompressible fluid, as shown in Section 3.22. The Laplace equation is without any parameters, and the solution for the potential is therefore controlled by boundary conditions.

Another simplification is obtained by restricting equation (13.121) to the vertical direction. Such a 1D equation, where all densities are constant, and source terms are zero, except for the porosity, is

$$\frac{\partial}{\partial z}\left(\frac{k}{\mu}\frac{\partial \Phi}{\partial z}\right) = \frac{1}{(1-\phi)}\frac{D\phi}{dt}. \tag{13.125}$$

This is precisely equation (12.11), which was obtained directly from fluid conservation in Section 12.1 (see Exercise 13.7). We have already seen that the pressure equation can be formulated in the Lagrange coordinate denoted ζ. The ζ-coordinate measures a vertical position as the height from the basement as fully compacted (zero-porosity) sediments. The pressure equation is then

$$\frac{1}{(1-\phi)^2}\frac{\partial \phi}{\partial t} = \frac{\partial}{\partial \zeta}\left((1-\phi)\frac{k(\phi)}{\mu}\frac{\partial \Phi}{\partial \zeta}\right) \tag{13.126}$$

where the material derivative is replaced by a partial derivative. The permeability's dependence on the porosity is denoted explicitly. A more general version of the equation (13.125) is

$$\frac{\partial}{\partial z}\left(\frac{k}{\mu}\frac{\partial \Phi}{\partial z}\right) = Q_{\text{pot}} \tag{13.127}$$

where the rate of change of porosity is included in the source term Q_{pot}. This equation shows that the condition for pressure build-up is a source term, unless there are boundary conditions that create overpressure. Another simple version of the general pressure equation (13.121) is

$$S_\sigma \frac{D\Phi}{dt} - \frac{\partial}{\partial z}\left(\frac{k}{\mu}\frac{\partial \Phi}{\partial z}\right) = 0 \tag{13.128}$$

which does not have any source term. It cannot model pressure build-up, but it is the equation for pressure decay. It has already been anticipated that S_σ controls the decay of the overpressure, which is studied in more detail in the next chapter.

Exercise 13.7 Show that the pressure equation (12.11) in the Lagrangian coordinate ζ is the same as the pressure equation (13.125) in the Euler coordinate z.
Hint: use that the void ratio is $e = \phi/(1 - \phi)$ and that $\partial p/\partial z = (1 - \phi)\, \partial p/\partial \zeta$.

Exercise 13.8 Derive the following expression for fluid conservation:

$$\frac{\phi}{\varrho_f}\frac{D\varrho_f}{dt} + \frac{1}{(1-\phi)}\frac{D\phi}{dt} = -\frac{\partial v_D}{\partial z} \tag{13.129}$$

based on expression (12.1) for fluid conservation (take a look at Figure 12.2). Notice that the fluid density is not assumed constant.

13.12 Further reading

The book *Dynamics of Fluids in Porous Media* by Bear (1972) has become a standard reference for most aspects of fluid flow in porous media. Basic equations are derived, and important assumptions are discussed in great detail. A large number of analytical results are obtained, and numerical methods are introduced.

Wang (2000) *Theory of Linear Poroelasticity* is a comprehensive treatment of poroelasticity, with a large number of analytical results, and an introduction to numerical solutions with the finite-element method.

14

Fluid flow: basic equations

14.1 Unconfined flow

Fluid flow with a free surface is called *unconfined* flow. Figure 14.1 shows unconfined subsurface flow from one water down to another water over an impermeable base. It is assumed that the rock is saturated with groundwater below the free surface, and that there is no groundwater above the free surface. The free surface is named the *water table* or the *phreatic surface*. The height of the water table at position x is $h(x)$, and the fluid pressure at the point (x, z) is taken to be hydrostatic:

$$p(x, z) = \varrho g(h(x) - z). \tag{14.1}$$

The assumption of hydrostatic fluid pressure gives the horizontal pressure gradient

$$\frac{dp}{dx} = \varrho g \frac{dh}{dx} \tag{14.2}$$

which is the *Dupuit approximation*. The horizontal component of the Darcy flux is then

$$u = -\frac{k}{\mu} \frac{dp}{dx} = -\frac{k\varrho g}{\mu} \frac{dh}{dx} \tag{14.3}$$

which says that the flow is proportional to the steepness of the water table. Conservation of mass requires that the rate of water transport is constant along the profile, when the flow is steady state (time-independent). The volume flux (volume per length and time),

$$Q_0 = u(x) h(x) = -\frac{k\varrho g}{\mu} h \frac{dh}{dx} \tag{14.4}$$

is therefore constant. This expression can be rewritten as an equation for the water table $h(x)$,

$$h \, dh = -\frac{\mu Q_0}{k\varrho g} dx \tag{14.5}$$

which is integrated to

$$h(x) = \left(h_0^2 - \frac{2\mu Q_0}{k\varrho g} x \right)^{1/2} \tag{14.6}$$

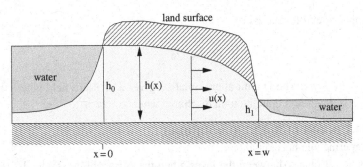

Figure 14.1. *The left water is flowing through the subsurface to the right water.*

where h_0 is the height of the surface of the water to the left (see Figure 14.1). The equation for the water table can be used to estimate the flow rate Q_0 from the upper water to the lower water, when the heights h_0 and h_1 of the water surfaces are known. The total discharge from the upper water is then

$$Q_0 = \frac{k\varrho g}{2\mu w}\left(h_0^2 - h_1^2\right) \tag{14.7}$$

where w is the lateral difference between the waters. Expression (14.7) is called the *Dupuit–Fucheimer* discharge formula.

Note 14.1 *The accuracy of the Dupuit approximation.* Notice that lateral Darcy flux was not obtained by solving a pressure equation. Instead, the Dupuit approximation was used to find equation (14.5) for the water table. The Dupuit approximation assumes that the fluid pressure is hydrostatic, which means that there is no vertical flow.

However, the flow field $\mathbf{u} = (u, 0)$ must fulfill the continuity equation $\nabla \cdot \mathbf{u} = 0$ to be mass conservative, where u is the horizontal Darcy flux (14.3). However the divergence of the flow field gives the non-zero source term

$$S = -\frac{k\varrho g}{\mu}\frac{d^2h}{dx^2} \tag{14.8}$$

which can be used to measure the accuracy of the Dupuit approximation.

Recall that the source term gives the fluid production as volume fluid per bulk rock and time. The total amount of fluid gained or lost per time is found by integrating the source term over the area of flow from ($x = 0$ to $x = w$)

$$Q_m = \int_0^w\int_0^{h(x)} S\,dz\,dx < \int_0^w\int_0^{h_0} S\,dz\,dx = -\frac{k\varrho g}{\mu}h_0\left(\frac{dh}{dx}(w) - \frac{dh}{dx}(0)\right). \tag{14.9}$$

The rate Q_m is constrained by integrating up to the maximum height h_0, and rewriting equation (14.5) as a slope of the water table as

$$\frac{dh}{dx} = -\frac{\mu Q_0}{k\varrho g}\frac{1}{h} \tag{14.10}$$

gives the upper bound for the rate

$$Q_m < Q_0 \frac{\Delta h}{h_1} \qquad (14.11)$$

where $\Delta h = h_0 - h_1$. The Dupuit approximation gives a pressure field where the rate of fluid leakage is much less than the total discharge when $\Delta h \ll h_1$.

Exercise 14.1 A dam of width 100 m separates two waters, which are 50 m and 20 m above an impermeable horizontal base.
(a) What is the discharge between the lakes when the permeability is $k = 1 \cdot 10^{-15}$ m^2? The water viscosity is $\mu = 1 \cdot 10^{-3}$ Pa s and the water density is $\varrho = 1000$ kg m^{-3}.
(b) How much time is needed for 1 m^3 of water per meter dam to leak out of the upper water?

Exercise 14.2 Show that the Dupuit–Fucheimer discharge formula (14.7) can be written as

$$Q_0 = \frac{k\varrho g}{\mu} \bar{h} \frac{(h_0 - h_1)}{w} \qquad (14.12)$$

where $\bar{h}$ is the average $(h_0 + h_1)/2$. The discharge is obtained using the average height $\bar{h}$ and the gradient $(h_0 - h_1)/w$.

14.2 Meteoric fluid flow

Section 3.22 gives an example of meteoric fluid flow driven by a sine-shaped water table. This section gives a similar example, where the fluid flow is driven by a water table shaped as a sine function superposed on a linear slope

$$h_s(x) = cx + a \sin(bx). \qquad (14.13)$$

The starting point for making a pressure equation is mass conservation of fluid, and conservation of an incompressible fluid is $\nabla \cdot \mathbf{u} = 0$, where $\mathbf{u} = 0$ is the Darcy flux. The gradient of the fluid flow potential Φ gives the flux $\mathbf{u} = -(k/\mu)\nabla\Phi$. The potential Φ is the natural choice for a main variable, when the permeability over viscosity is constant, because the equation for the potential becomes a Laplace equation

$$\nabla^2 \Phi = 0. \qquad (14.14)$$

(See also Section 13.11.) The potential is defined as the fluid pressure minus the hydrostatic pressure relative to a reference level as shown in Section 13.9. The reference level is taken to be $z = 0$. It is now assumed that the fluid pressure along the line $z = 0$ is $p_f(x) = \varrho_f g\, h_s(x)$, which is the weight of the fluid above the line $z = 0$. This implies that the potential also becomes $\Phi(x) = \varrho_f g h_s(x)$ along $z = 0$, because the reference hydrostatic pressure is zero for $z = 0$. The Laplace equation is solved for a rectangular domain of width l and depth h. The boundaries of the domain are closed for fluid flow, except the line $z = 0$, where fluid is entering or leaving the domain. The boundary conditions for the Laplace equation are therefore

$$\Phi(x, z=0) = \varrho_f g h_s(x) \tag{14.15}$$

$$\frac{\partial \Phi}{\partial z}(x, z=h) = 0 \tag{14.16}$$

$$\frac{\partial \Phi}{\partial x}(x=0, z) = 0 \tag{14.17}$$

$$\frac{\partial \Phi}{\partial x}(x=l, z) = 0 \tag{14.18}$$

where boundary conditions (14.17) and (14.18) give zero horizontal fluid flow along the vertical boundaries $x = 0$ and $x = l$. The boundary condition (14.16) gives zero vertical fluid flow along the base of the domain, $z = h$. The solution of the Laplace equation for these boundary conditions are

$$\Phi(x, z) = a_0 + \sum_{n=1}^{\infty} a_n \cos(\lambda_n x) \cos(\lambda_n(z - h)) \tag{14.19}$$

with the Fourier coefficients

$$a_0 = \frac{1}{2}\varrho_f g c l + \frac{\varrho_f g a}{lb}\left(1 - \cos(bl)\right) \tag{14.20}$$

and

$$a_n = \frac{2\varrho_f g}{l \cosh(\lambda_n h)}\left(\frac{c\left((-1)^n - 1\right)}{\lambda_n^2} + \frac{ab\left(1 - \cos(\lambda_n l)\cos(bl)\right)}{b^2 - \lambda_n^2}\right) \tag{14.21}$$

and where $\lambda_n = n\pi/l$ (see Note 14.2). (This solution is simply a superposition of the solutions shown in Section 3.22.) The Darcy velocities follow directly from the potential by differentiation with respect to x and z:

$$u_x = -\frac{k}{\mu}\frac{\partial \Phi}{\partial x} = \frac{k}{\mu}\sum_{n=1}^{\infty}\lambda_n a_n \sin(\lambda_n x)\cosh(\lambda_n(z - h)) \tag{14.22}$$

$$u_z = -\frac{k}{\mu}\frac{\partial \Phi}{\partial x} = -\frac{k}{\mu}\sum_{n=1}^{\infty}\lambda_n a_n \cos(\lambda_n x)\sinh(\lambda_n(z - h)). \tag{14.23}$$

In 2D the Darcy flux is related to a stream function Ψ as

$$u_x = \frac{\partial \Psi}{\partial z} \quad \text{and} \quad u_z = -\frac{\partial \Psi}{\partial x} \tag{14.24}$$

which always gives mass conservation for stationary fluid flow (see Section 3.22). The stream function therefore follows from the Darcy flux field by integration:

$$\Psi(x, z) = \int u_x \, dz = \Psi_0 + \frac{k}{\mu}\sum_{n=1}^{\infty}\lambda_n a_n \sin(\lambda_n x)\sinh(\lambda_n(z - h)). \tag{14.25}$$

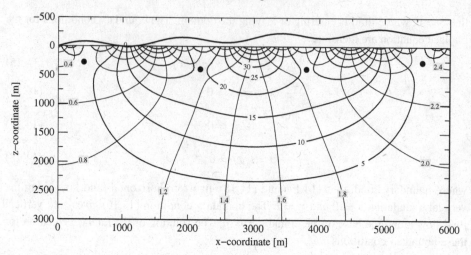

Figure 14.2. *Fluid flow potential (thin lines, MPa) and streamlines (thick curves, $10^6 \times m^2 \, years^{-1}$) for meteoric fluid flow driven by the water table shown above $z = 0$. The black bullets show the stagnation points.*

The stream function becomes zero along the closed boundaries by setting in the integration constant $\Psi_0 = 0$. Recall that it is only differences in the stream function that are interesting, not its absolute value. The difference between two stream functions is the volumetric flow in the steam tube made by the two corresponding streamlines.

Figure 14.2 shows the potential and the stream function for the case with length $l = 6000$ m, domain height $h = 3000$ m, amplitude $a = 61$ m, wave number $b = 8\pi/l$ and steepness $c = 0.05$. The wave number b says that there are 4 wavelengths from $x = 0$ to $x = l$. This case was originally presented by Tóth (1963) (see his Figure 2i). The flow pattern in Figure 14.2 is more complex than in Figure 3.21. The plot shows that there are local and shallow flow fields created by the sine-shape of the topography, and in addition there is a regional flow field that is created by the slope. The case also shows stagnation points, which are points of zero fluid flow. These points therefore have no streamlines going through them.

The fluid flow in a shallow basin is shown in Figures 14.3a and 14.3b, where the flow is dominated by the linear slope in Figure 14.3a and by the sine-shaped topography in Figure 14.3b.

Note 14.2 The solution of the Laplace equation with the boundary conditions (14.15)–(14.18) is done by separation of variables. The potential is written as a product of two functions, $\Phi(x, z) = U(x)V(z)$, where $U(x)$ is only a function of x and $V(z)$ is only a function of z. Inserting the product UV into the Laplace equation leads to

$$-\frac{U''}{U} = \frac{V''}{V} = \lambda^2. \tag{14.26}$$

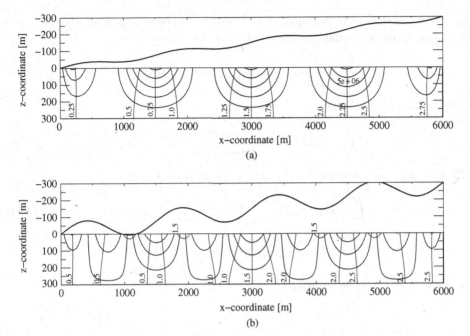

Figure 14.3. *Fluid flow potential (thin lines, MPa) and streamlines (thick curves) for meteoric fluid flow driven by the water table shown above $z = 0$. Notice that the (near vertical) contours of the potential correspond to the height of the topography. The difference between two streamlines is $0.166 \times 10^6 \times m^2\,years^{-1}$. Panels (a) and (b) correspond to Figures 2c and 2d in Tóth (1963), respectively.*

The term U''/U is only a function of x and term V''/V is only a function of z. Since both these terms are equal they must therefore be equal to a constant, which is written as λ^2. The two equations (14.26) are

$$U'' + \lambda^2 U = 0 \quad \text{and} \quad V'' - \lambda^2 V = 0 \tag{14.27}$$

where a solution for $U(x)$ is a linear combination of $\sin(\lambda x)$ and $\cos(\lambda x)$, and where a solution for $V(z)$ is a linear combination of $\exp(\lambda z)$ and $\exp(-\lambda z)$. The boundary conditions (14.17) and (14.18) are fulfilled by $U(x) = \cos(\lambda x)$, because $U'(x) = -\lambda \sin(\lambda x)$ is zero for $x = 0$ and also for $x = l$ when λ is such that $\lambda l = n\pi$, where n is an integer. There is therefore not just one λ, but $\lambda_n = n\pi/l$ for every integer n. The function $V(z) = c_1 e^{\lambda z} + c_2 e^{-\lambda z}$ fulfills boundary condition (14.16), when $V'(z = h) = 0$. That happens for $c_1 = e^{-\lambda h}$ and $c_2 = e^{\lambda h}$. We therefore get that $V(z) = \cosh(\lambda_n(z - h))$. Any function $\Phi(x, z) = \sum_n a_n \cos(\lambda_n x)\cosh(\lambda_n z)$ solves the Laplace equation, and also fulfills the boundary conditions (14.16)–(14.18). The last boundary condition (14.15) requires that

$$\Phi(x, h) = \sum_n a_n \cos(\lambda_n x)\cosh(\lambda_n h) = \varrho_f g\,(cx + a \sin(bx)). \tag{14.28}$$

Multiplying both sides of this equation by $\cos(\lambda_m x)$ and integrating from $x = 0$ to $x = l$ and using the (orthogonality) property

$$\int_0^l \cos(\lambda_n x) \cos(\lambda_m x)\, dx = \frac{1}{2} l \delta_{nm} \tag{14.29}$$

gives the Fourier coefficient

$$a_n = \frac{2 \int_0^l \varrho_f g \, (cx + a \sin(bx)) \cos(\lambda_n x)\, dx}{l \cosh(\lambda_n h)}. \tag{14.30}$$

The remaining work with the coefficient a_n gives the two integrals

$$\int_0^l x \cos(\lambda_n x)\, dx = \frac{1}{\lambda_n^2}\left((-1)^n - 1\right) \tag{14.31}$$

$$\int_0^l \sin(bx) \cos(\lambda_n x)\, dx = \frac{b(1 - \cos(\lambda_n l) \cos(bl))}{b^2 - \lambda_n^2} \tag{14.32}$$

which are both solved by integration by parts.

14.3 Decay of overpressure and pressure seals

Chapter 12 shows that rapid deposition of low-permeability sediments creates overpressure. The question now is how fast (or slow) the overpressure decays after overpressure generation has stopped? The answer is given as a solution of pressure equation (13.128) for the overpressured sedimentary column shown in Figure 14.4a, where the initial overpressure increases linearly with depth (at time $t = 0$). The top of the column has zero fluid pressure, and the fluid flow potential (Φ) and the overpressure (p_e) are therefore identical

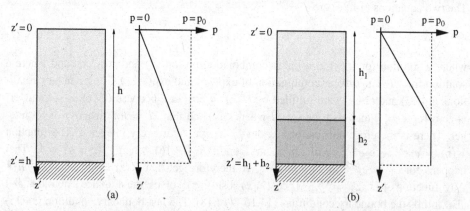

Figure 14.4. *(a) A homogeneous sedimentary column where the overpressure increases linearly with depth. (b) A sedimentary column with two layers, a permeable one at the base with a sealing one above. The initial pressure is constant in the permeable layer and it increases linearly with depth through the seal.*

quantities. The base of the column is impermeable and it has an initial overpressure p_0. The pressure equation is first made dimensionless before it is solved, using the dimensionless z-coordinate $\hat{z} = z/h$, where h is the height of the column, and the dimensionless overpressure $\hat{p} = p_e/p_0$. The pressure equation (13.128) is

$$\left(\frac{\mu h^2 S_\sigma}{k} \right) \frac{\partial \hat{p}}{\partial t} - \frac{\partial^2 \hat{p}}{\partial \hat{z}^2} = 0 \tag{14.33}$$

assuming that the (mobility) ratio k/μ does not depend on the z-coordinate. The characteristic time is identified as

$$t_0 = \frac{\mu h^2 S_\sigma}{k}, \tag{14.34}$$

and the dimensionless pressure equation is

$$\frac{\partial \hat{p}}{\partial \hat{t}} - \frac{\partial^2 \hat{p}}{\partial \hat{z}^2} = 0 \tag{14.35}$$

where dimensionless time is $\hat{t} = t/t_0$. We notice that there are no parameters in the dimensionless equation (14.35) for pressure decay. Dimensionless solutions of pressure equation (14.35) are therefore universal for pressure decay regardless of the thickness of the sedimentary column or the value for the permeability of the sediments. The dimensionless overpressure in the case of the initial condition and the boundary conditions shown in Figure 14.4a is

$$\hat{p}(\hat{z}, \hat{t}) = \sum_{n=0}^{\infty} a_n e^{-\lambda_n^2 \hat{t}} \sin(\lambda_n \hat{z}) \tag{14.36}$$

as shown in Note 14.3, where

$$a_n = \frac{2(-1)^n}{\lambda_n^2} \quad \text{and} \quad \lambda_n = \frac{\pi}{2} + n\pi. \tag{14.37}$$

The overpressure with units follows directly from the scaling, which is simply $p(z, t) = p_0 \hat{p}(t/t_0, z/h)$. All terms in series (14.36) decay to zero with increasing $\hat{t}$, which shows that the overpressure decays to zero with increasing time. Term $n = 0$ has the slowest decay, and the decay becomes faster as n increases. The first term can be used to estimate the half-life of the pressure decay, which is $\hat{t}_{1/2} = \ln 2/\pi^2 \approx 0.07$. The half-life estimate with units is

$$t_{1/2} = \frac{\ln 2}{\pi^2} t_0 \approx \frac{\ln 2 \, \mu h^2 S_\sigma}{\pi^2 k}. \tag{14.38}$$

Figure 14.5 shows the solution (14.36) for decay of overpressure and that the half-life estimate $\hat{t}_{1/2} \approx 0.07$ is roughly a factor 2 less than the more accurate half-life $\hat{t}_{1/2} \approx 0.2$. Nevertheless, the characteristic time t_0 tells us much about the time span of overpressure decay. There are two parameters in t_0 that may be difficult to assess: firstly, the permeability, which is often difficult to bound within an order of magnitude; secondly, the storage coefficient, which is approximately $S_\sigma \approx \phi \alpha_f$, when the pore fluid is much

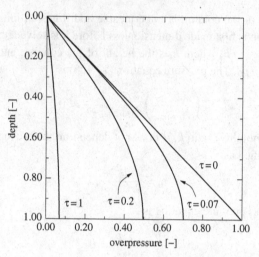

Figure 14.5. *The dimensionless pressure solution (14.36) at times* $\tau = 0$, $\tau = 0.07$, $\tau = 0.2$ *and* $\tau = 1$.

more compressible than the pore space. If the sedimentary column is $h = 1000$ m thick, $\alpha_f = 5 \cdot 10^{-10}$ Pa^{-1}, $\phi = 0.1$, $k = 1 \cdot 10^{-19}$ m^2, the characteristic time constant is $t_0 = 16\,000$ years. A 1 km thick layer of shale therefore needs to have a permeability as low as $k = 1 \cdot 10^{-21}$ m^2 for overpressure to be preserved for several million years, unless the storage coefficient for some reason is much larger than $\phi \alpha_f$.

It is often difficult to solve a pressure equation that involves two lithologies. This situation is shown in Figure 14.4b, where an overpressured fluid reservoir is leaking through a low permeable and low porosity seal. The pressure decay of such a sealed reservoir is studied in Exercise 14.3.

Note 14.3 The dimensionless pressure equation (14.35), with boundary conditions $\hat{p} = 0$ at $\hat{z} = 0$ and $\partial \hat{p}/\partial \hat{z} = 0$ at $\hat{z} = 1$, and initial condition $\hat{p} = \hat{z}$ is solved using separation of variables. The pressure is then written as the product $\hat{p}(\hat{z}, \hat{t}) = U(\hat{t}) \cdot V(\hat{z})$, where one factor U is a function of only $\hat{t}$ and the other factor V is a function of only $\hat{z}$. When the product $\hat{p} = U \cdot V$ is inserted into the pressure equation we get

$$\frac{U'}{U} = \frac{V''}{V} = -\lambda^2 \tag{14.39}$$

where $-\lambda^2$ has to be a constant. (A prime denotes derivation.) The ratio U'/U is only a function of $\hat{t}$ and the ratio V''/V is only a function of $\hat{z}$, and since they are equal they have to be equal to a constant. The solution of $U'/U = -\lambda^2$ is $U(\hat{t}) = a \exp(-\lambda^2 \hat{t})$, where a is the value of U at $\hat{t} = 0$. (The constant $-\lambda^2$ has to be a negative number, because $U(\hat{t})$ would otherwise grow exponentially with time, which is impossible.) The solution of equation $V'' + \lambda^2 V = 0$ for V is any linear combination of $\sin(\lambda \hat{z})$ and $\cos(\lambda \hat{z})$. The boundary condition $\hat{p} = 0$ at $\hat{z} = 0$ is fulfilled for $\sin(\lambda \hat{z})$, but not for $\cos(\lambda \hat{z})$ regardless of λ, and we have that $V(\hat{z}) = \sin(\lambda \hat{z})$. The other boundary condition is also fulfilled for

$\sin(\lambda \hat{z})$ by choosing $\lambda_n = \pi/2 + n\pi$. There is therefore not only one solution for just one λ, but solutions for all λ_n, $n = 0, 1, \ldots$. The pressure can be written as a sum of all these solutions:

$$\hat{p}(\hat{z}, \hat{t}) = \sum_{n=0}^{\infty} a_n \exp(-\lambda_n^2 \hat{t}) \sin(\lambda_n \hat{z}) \tag{14.40}$$

because the pressure equation is linear. The unknown coefficients a_n are obtained from the initial condition $\hat{p} = \hat{z}$ at $\hat{t} = 0$, which is $\sum_{n=0}^{\infty} a_n \sin(\lambda_n \hat{z}) = \hat{z}$. Both sides of this equality are multiplied by $\sin(\lambda_m \hat{z})$, and then integrated from 0 to 1. We used

$$\int_0^1 \sin(\lambda_n \hat{z}) \sin(\lambda_m \hat{z}) dz = \begin{cases} 0 & n = m \\ \frac{1}{2} & n \neq m \end{cases} \tag{14.41}$$

which implies that only the term m survives, and we get

$$\frac{1}{2} a_m = \int_0^1 \hat{z} \sin(\lambda_m \hat{z}) = \frac{(-1)^m}{\lambda_m^2} \tag{14.42}$$

which is the same coefficient as in the solution (14.36).

Exercise 14.3 An overpressured reservoir unit is placed on an impermeable base underneath a sealing layer, and fluid from the reservoir leaks through the seal, see Figure 14.4b. Let the reservoir be one (high-permeable) unit with an initial pressure p_e, and assume that the seal has zero compressibility. The reservoir has the porosity ϕ and the thickness h_2. The seal has the permeability k and the thickness h_1. Let A be the surface area of the column and μ the viscosity of the fluid.

(a) Derive an equation for the overpressure in the reservoir based on the mass balance for the reservoir fluid. Assume that only the fluid is compressible.

(b) Make the overpressure equation dimensionless. What is the characteristic time for overpressure decay?

(c) Obtain the reservoir pressure as a function of time.

(d) Derive the same equation as in (a), but account for the compressibility of both the pore space and the solid density as well as the fluid.

(e) How could pressure equation (13.128) be used as a starting point for the solution in (c)?

(f) Find the mass of the fluid that has leaked out as a function of time.

Solution:

(a) Mass balance is when the mass of fluid expelled (by decompression) from the reservoir is equal to the mass of fluid that leaks through the seal. The Darcy flux through the seal is $v_D = (k/\mu)(p_e/h_1)$, where p_e is the overpressure in the reservoir. (The overpressure gradient through the seal is p_e/h_1.) The mass flux is then

$$r_1 = -A\varrho_f \frac{k}{\mu} \cdot \frac{p_e}{h_1}. \tag{14.43}$$

A minus sign is added because the rate r_1 is a negative number when fluid is leaking out. The rate of expulsion of fluid from the reservoir due to decompression is

$$r_2 = \frac{d}{dt}(\phi V_b \varrho_f) = \phi V_b \frac{d\varrho_f}{dt} = \phi V_b \alpha_f \varrho_f \frac{dp_f}{dt} \tag{14.44}$$

where the mass of fluid in the reservoir is $\phi V_b \rho_f$ and the bulk volume is $V_b = Ah_2$. The definition of the fluid compressibility, $\Delta\varrho_f/\varrho_f = \alpha_f \Delta p_f$, is used in the last equality in equation (14.44). Furthermore, $dp_f/dt = dp_e/dt$ because the fluid pressure is the hydrostatic pressure (p_h) added to the overpressure, $p_f = p_h + p_e$, and the hydrostatic pressure does not change through time. The rate of expulsion (r_2) is equal to the rate of leakage, $r_2 = r_1$, which gives the following equation for the overpressure:

$$\phi \alpha_f h_2 \frac{dp_e}{dt} = -\frac{k}{\mu} \frac{p_e}{h_1}. \tag{14.45}$$

The volume of the reservoir is replaced by $V_b = h_2 A$, and both the area A and the density drop out.

(b) Equation (14.45) can be rewritten as

$$\left(\frac{\phi \mu \alpha_f h_1 h_2}{k}\right) \frac{dp_e}{dt} = -p_e \tag{14.46}$$

where the characteristic time is identified as

$$t_0 = \frac{\phi \mu \alpha_f h_1 h_2}{k}. \tag{14.47}$$

This characteristic time is different from the characteristic time (14.34) for decay of over-pressure in a homogeneous column. It is proportional to both the reservoir thickness and the seal thicknesses $h_1 h_2$, while the characteristic time (14.34) is proportional to h^2 (the thickness of the column to the power of 2). The dimensionless equation is simply $d\hat{p}/d\hat{t} = -\hat{p}$, where dimensionless time is $\hat{t} = t/t_0$ and dimensionless pressure is $\hat{p} = p_e/p_0$.

(c) The solution of equation (14.46) for the overpressure is

$$p_e(t) = p_0 \exp(-t/t_0). \tag{14.48}$$

The pressure decays exponentially with time, and the half-life is $t_{1/2} = \ln 2\, t_0$.

(d) The rate of expulsion from the reservoir due to decompression is

$$r_2 = \frac{d}{dt}\left(V_p \rho_f\right) = \frac{dV_p}{dt}\varrho_f + V_p \frac{d\varrho_f}{dt} \tag{14.49}$$

where V_p is the pore volume. The time derivatives of the pore volume (13.30) and the fluid density (13.29) are then inserted. The rate of change of the bulk pressure is zero ($dp_b/dt = 0$), because the weight of the basin remains the same during pressure decay. The fluid pressure is the overpressure added to the hydrostatic pressure, $p_f = p_e + p_h$, which implies that $dp_f/dt = dp_e/dt$. (The hydrostatic pressure, just like the bulk pressure, does not change with pressure reduction.) Putting these together gives

$$r_2 = \phi(\alpha_{pp} + \alpha_f)V_b\varrho_f\frac{dp_e}{dt}$$

$$= S_\sigma V_b\varrho_f\frac{dp_e}{dt} \tag{14.50}$$

using that $V_p = \phi V_b$. The unconstrained storage coefficient (13.122) is derived one more time, but now in a simpler setting. Mass conservation requires that $r_1 = r_2$, or when written in terms of the overpressure

$$S_\sigma h_2\frac{dp_e}{dt} = \frac{k}{\mu} \cdot \frac{p_e}{h_1} \tag{14.51}$$

which is similar to the pressure equation (14.45) with the only difference that the compressibility term ($\phi\alpha_f$) is replaced with the unconstrained storage coefficient S_σ.

(e) Subtasks (a) to (d) give an equation for decay of reservoir overpressure found directly from the mass balance of the reservoir fluid. The same equation can also be obtained by starting with pressure equation (13.128). The z-dependence can be integrated out as follows:

$$\int_0^h S_\sigma\frac{\partial p_e}{\partial t}dz = \int_0^h \frac{\partial}{\partial z}\left(\frac{k}{\mu}\frac{\partial p_e}{\partial z}\right)dz \tag{14.52}$$

where equation (13.128) is integrated over z from 0 to h. The left-hand side can be approximated by $S_\sigma(dp_e/dt)h_2$, because the compressibility of the seal is zero and the pressure is almost the same everywhere in the reservoir. The right-hand side is

$$\int_0^h \frac{\partial}{\partial z}\left(\frac{k}{\mu}\frac{\partial p_e}{\partial z}\right)dz = \frac{k}{\mu}\frac{\partial p_e}{\partial z}\bigg|_h - \frac{k}{\mu}\frac{\partial p_e}{\partial z}\bigg|_0 \tag{14.53}$$

where the gradient at the base ($z = h$) is zero ($\partial p_e/\partial z = 0$), because the base is impermeable. The gradient at the top ($z = 0$) is approximated by $\partial p_e/\partial z \approx p_e/h_1$, which gives equation (14.45) for the decay of reservoir overpressure.

(f) The rate r_1 gives that the mass of fluid leaking out through a horizontal area A during a time step dt is

$$dm = \varrho_f A\frac{k}{\mu}\frac{p_e(t)}{h_1}dt \quad \text{where} \quad p_e(t) = p_0 e^{-t/t_0} \tag{14.54}$$

After integration we get that

$$m(t) = m_0(1 - e^{-t/t_0}) \quad \text{and} \quad m_0 = \phi\alpha_f\varrho_f p_0 h_2 A \tag{14.55}$$

where m_0 is the initial mass of fluid compressed by the initial pressure p_0. The integration constant in this case is found from $m(t=0) = 0$. We see that 63% of the mass m_0 has leaked out after a time t_0.

14.4 Overpressure decay in clay

Clay is a soft porous material that may compact substantially during decreasing overpressure. Reduction of overpressure in a thick layer of clay is a nonlinear problem, because the permeability depends on the porosity, and the porosity depends on the overpressure. Nevertheless, it is possible to make an estimate for the time scale of pressure decay. The overpressure decay can be modeled with equation (13.126) when it is combined with a constitutive equation. We have from Section 4.5 that the normal compaction line gives the void ratio of clay as a function of effective vertical stress

$$e = e_0 - C_c \ln(\sigma'/\sigma_0') \tag{14.56}$$

where σ' is the effective (vertical) stress (in 1D), and where e_0 is the void ratio at the effective stress σ_0'. (Stresses are assumed to be isotropic in 1D, and the term pressure can therefore be used instead of stress.) The parameter C_c is the compression index. A formulation in the Lagrange ζ-coordinate rather than the real z-coordinate simplifies the pressure equation because the material derivative becomes a partial derivative. The effective vertical stress σ' is related to the overpressure p_e by

$$\sigma' = (\zeta^* - \zeta)\Delta\varrho\, g - p_e \tag{14.57}$$

where ζ^* is the ζ-position of the basin surface (see Sections 3.17, 4.6 and 12.7). The difference $\Delta\varrho = \varrho_s - \varrho_f$ is between the solid (matrix) density (ϱ_s) and the density of the fluid (ϱ_f). Notice that there is no porosity dependence in the relationship between σ' and p_e when the ζ-coordinate is used. Both overpressure build-up and overpressure decay are strongly dependent on the permeability, which is normally taken to be a function of the porosity, $k = k(\phi)$. Such permeability functions can also be written as $k(e) = k_1 \hat{k}(e)$, where k_1 is the permeability at the reference void ratio e_1. (The dimensionless function $\hat{k}(e)$ is therefore $\hat{k} = 1$ for $e = e_1$.) Equation (13.126) then becomes

$$\frac{\partial p_e}{\partial t} - \frac{k_1}{\mu C_c}\left((\zeta^* - \zeta)\Delta\varrho\, g - p_e\right)\frac{\partial}{\partial \zeta}\left(\frac{\hat{k}(e)}{1+e}\frac{\partial p_e}{\partial \zeta}\right) = 0 \tag{14.58}$$

as shown in Note 14.4. It is assumed that there is water above the basin surface and the overpressure is therefore equal to the fluid potential. The permeability function is now chosen to be

$$k = k_1\hat{k}(e) \quad\text{where}\quad \hat{k}(e) = \left(\frac{1+e}{1+e_1}\right)\left(\frac{e}{e_1}\right)^n \tag{14.59}$$

which behaves similarly to the Kozeny–Carman permeability function. Boundary conditions are zero overpressure at the basin surface and zero fluid flux into the base of the sedimentary column. There is no simple exact solution of this equation, and it is therefore studied numerically. It should be noted that the equation is not well-defined at the basin surface, because the normal consolidation line does not have a void ratio at zero effective stress. (This is handled numerically using a constant void ratio close to the basin surface.) Numerical solutions of the equation are shown in Figure 14.6 for the parameters found in Table 14.1. Figure 14.6a shows the overpressure at the time steps $t = 0$ Ma, 1 Ma, 3 Ma

Table 14.1. *The data used in the case shown in Figure 14.6.*

Parameter	Value	Units
e_0	1.5	[–]
σ_0'	18316	[Pa]
C_c	0.2	[–]
ζ^*	500	[m]
μ	0.001	[Pa s]
n	3	[–]
e_1	0.43	[–]
k_1	$4.5 \cdot 10^{-20}$	[m^2]
e_2	1.5	[–]
k_2	$5 \cdot 10^{-18}$	[m^2]

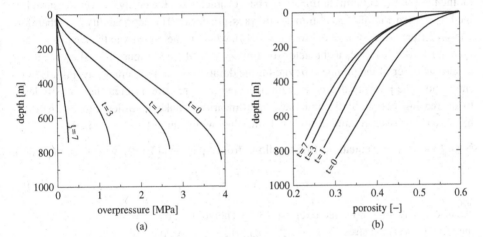

Figure 14.6. *Pressure decay in a column of 500 m net (porosity-free) sediments. Time is in unit Ma, and the initial pressure is at $t = 0$ Ma. Notice that the column is compacting as the overpressure is decreasing.*

and 7 Ma. (The initial state is at $t = 0$ Ma.) It is possible to make a dimensionless version of the pressure equation (14.58) by introducing the dimensionless vertical unit coordinate $x = \zeta/\zeta^*$, dimensionless pressure $u = p/\Delta\varrho\, g\zeta^*$, and dimensionless time $\tau = t/t_0$ where

$$t_0 = \frac{\mu_0\zeta^*}{C_c k_1 \Delta\varrho g}. \qquad (14.60)$$

(The dimensionless vertical coordinate x should not be confused with the x-coordinate in the horizontal plane.) The dimensionless pressure equation is then

$$\frac{\partial u}{\partial \tau} - (1 - x - u)\frac{\partial}{\partial x}\left(\hat{k}(e)\frac{\partial u}{\partial x}\right) = 0 \qquad (14.61)$$

with boundary conditions being zero pressure at the surface ($u = 0$ at $x = 1$) and zero flux into the base of the column ($\partial u / \partial x = 0$ at $x = 0$). The dimensionless pressure equation is without explicit parameters, just like the previous equation (14.35) for pressure decay. The void ratio is related to the overpressure as

$$e = e_0 - C_c \ln\left(N_c(1 - x - u)\right) \tag{14.62}$$

where the number is $N_c = \Delta \varrho \, g \zeta^* / \sigma_0'$. There are then four dimensionless numbers implicit in the dimensionless formulation, and they are e_0, C_c and N_c in the normal consolidation line, and n in the permeability. The four parameters appear in the permeability function when expressed in terms of the overpressure. All sedimentary columns with the same dimensionless parameters will have the same pressure decay. The question is now how the permeability k_1 should be chosen. The permeability decreases by more than two orders of magnitude from the surface to the base of the basin, and the choice of k_1 is therefore important for the characteristic time t_0. It turns out that the lowest permeability (at the base of the column) at time $t = 0$ best characterizes the decay of overpressure. It is the lowest permeability that dominates the pressure decay. The permeability k_1 is therefore chosen for the porosity $\phi = 0.3$ (or $e_1 = 0.43$), which is the porosity at the base of the column at $t = 0$. That gives a characteristic time $t_0 = 3$ Ma, and Figure 14.6 shows that it is at least an order of magnitude estimate for the duration of the pressure decay. Figure 14.6b shows how the porosity decreases with decaying overpressure. Decreasing porosity leads to decreasing permeability, and these non-linearities make it difficult to estimate the half-life of the overpressure decay any better than within an order of magnitude.

Note 14.4 Pressure equation (14.58) follows from equation (13.126), which can be written

$$\frac{\partial e}{\partial t} = \frac{\partial}{\partial \zeta}\left(\frac{k(e)}{(1 + e)}\frac{\partial p_e}{\partial \zeta}\right) \tag{14.63}$$

because $e = \phi/(1 - \phi)$. (See Exercise 13.7.) The void ratio is $e = e_0 - C_c \ln(\sigma'/\sigma_0')$ and the effective vertical stress is $\sigma' = (\zeta^* - \zeta)\Delta \varrho \, g - p_e$. We then get

$$\frac{\partial e}{\partial t} = -\frac{C_c}{\sigma'}\frac{\partial \sigma'}{\partial t} = \frac{C_c}{\sigma'}\frac{\partial p_e}{\partial t} \tag{14.64}$$

which gives equation (14.58) for overpressure decay in clay when it is inserted into equation (14.63).

14.5 Overpressure build-up in clay

Deposition of clay leads to overpressure build-up. We have already seen that the gravity number can be used to determine when deposition is sufficiently rapid (or the permeability is sufficiently low) for strong overpressure build-up to take place. The same use of the gravity number is now shown for sediments that compact according to a normal compaction line. The starting point is the same as for overpressure decay, pressure equation (13.126) combined with the normal consolidation line (14.56). The Lagrangian

ζ-coordinate is once more used, because the material derivative then becomes a partial derivative. The other advantage of using the ζ-coordinate is that the effective vertical stress relationship (14.57) becomes

$$\sigma' = (\zeta^* - \zeta)\Delta\varrho g - p_e \tag{14.65}$$

which does not include the porosity. The model is simplified by studying overpressure build-up during constant deposition of sediments. The height of the sedimentary column is then $\zeta^* = \omega t$, where ω is the deposition rate as net (porosity-free) sediments. The permeability is written as the product $k(e) = k_2 \hat{k}(e)$ where k_2 is the permeability at the reference void ratio e_2. The reference void ratio is now a void ratio close to the basin surface, and k_2 is the permeability of the surface sediments. The equation for overpressure build-up becomes

$$\frac{\partial p_e}{\partial t} - \frac{k_2}{\mu C_c}\left((\zeta^* - \zeta)\Delta\varrho\, g - p_e\right)\frac{\partial}{\partial \zeta}\left(\frac{\hat{k}(e)}{1+e}\frac{\partial p_e}{\partial \zeta}\right) = \omega\Delta\varrho g. \tag{14.66}$$

It differs from equation (14.58) for overpressure decay only by the right-hand side, which is the source term responsible for overpressure generation. We assume that there is water at the surface of the sediments, and the fluid potential and the overpressure are therefore identical quantities. Boundary conditions are the usual ones – zero overpressure at the surface and zero fluid flux into the base of the sedimentary column. There are no simple analytical solutions of equation (14.66), and it is therefore studied numerically. It should be noted once more that the normal consolidation line is undefined for zero effective vertical stress. This problem is solved numerically using 0.6 as a maximum porosity. A numerical solution of pressure equation (14.66) is shown in Figure 14.7 using the data from Table 14.1. The overpressure and the porosity are plotted after deposition of 500 m net (porosity-free) sediments during 2 Ma, 10 Ma and 50 Ma. (The deposition rates are 250 m/Ma, 50 m/Ma and 10 Ma/m, respectively.) Figure 14.7a shows that the porosity

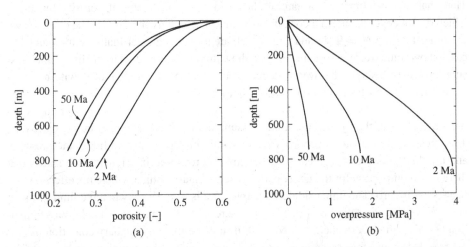

Figure 14.7. *Overpressure build-up (a) and porosity (b) after deposition during 2 Ma, 10 Ma and 50 Ma.*

increases with increasing overpressure, and Figure 14.7b shows that the overpressure increases with increasing deposition rate.

We have already seen in Sections 12.4 and 12.8 that the gravity number tells us whether there will be strong overpressure generation or not. The gravity number appears when pressure equation (14.66) is made dimensionless using the dimensionless overpressure $u = p_e/\Delta\varrho g\omega t$, the dimensionless vertical coordinate $x = \zeta/\omega t$ and dimensionless time $\tau = t/t_0$, where t_0 is a characteristic time. The dimensionless version of the pressure equation (14.66) then becomes

$$\tau\frac{\partial u}{\partial \tau} + u - x\frac{\partial u}{\partial x} - N(1 - x - u)\frac{\partial}{\partial x}\left(\hat{k}(e)\frac{\partial u}{\partial x}\right) = 1 \qquad (14.67)$$

where N is the gravity number

$$N = \frac{N_g}{C_c}, \qquad (14.68)$$

N_g is the familiar gravity number

$$N_g = \frac{k_2\Delta\varrho\,g}{\mu\omega} \qquad (14.69)$$

and where equation (14.62) is the dimensionless version of the normal consolidation line. The gravity number N is different from the gravity number N_g by the compression index C_c in the denominator. Another thing to notice is that the dimensionless pressure equation (14.67) does not depend on a specific choice for a characteristic time. Different choices for t_0 give the same dimensionless equation and the same gravity number. One possible characteristic time is $t_0 = \zeta^*_{max}/\omega$, which is the time needed to deposit a column of the wanted height ζ^*_{max}. This choice differs from the characteristic time used in Section 12.8, which is when enough sediments have been deposited for compaction to be noticeable. That characteristic time was a natural choice for a special group of porosity functions, which do not cover the normal consolidation line. The gravity number for the cases shown in Figure 14.7 are $N = 9$, 43 and 213, which are all much greater than 1. Although these cases show some overpressure they do not show fluid pressure equal to the lithostatic pressure. Figure 14.8 shows the overpressure and the porosity in the case of gravity numbers $N = 0.1$, 1, 10 and 100. The figure shows that $N_g \gg 1$ is the regime of low overpressure, while $N_g \ll 1$ is the regime of high overpressure. Low and high overpressure are fluid pressures close to the hydrostatic fluid pressure and the lithostatic pressure, respectively. The effective vertical stress becomes zero for fluid pressure equal to lithostatic pressure, and the fluid pressure is then carrying the entire porous matrix. Figure 14.8b shows that there is no compaction for the upper part of a sedimentary column where the effective vertical stress is zero. The dimensionless overpressure is $u = 1 - x$ when the fluid pressure is equal to the lithostatic pressure. The pressure $u = 1 - x$ is also a solution of equation (14.67), which corresponds to $N_g = 0$. It fulfills only one boundary condition ($u = 0$ for $x = 1$) because the pressure equation becomes a first-order equation for $N_g = 0$. The solution $u = 1 - x$ is nevertheless a good approximation of the dimensionless overpressure for the regime $N_g < 1$, although the boundary condition at $x = 0$ is not fulfilled.

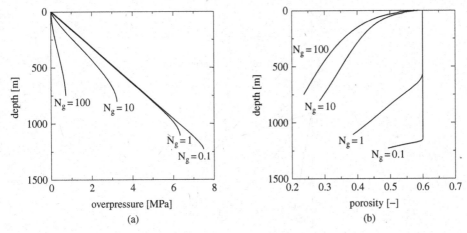

Figure 14.8. *Overpressure build-up (a) and porosity (b) for gravity number N = 0.1, 1, 10 and* 100.

Note 14.5 Pressure equation (14.66) follows from equation (13.126) which can be written as equation (14.63). The void ratio is $e = e_0 - C_c \ln(\sigma'/\sigma_0')$ and the effective vertical stress is $\sigma' = (\omega t - \zeta)\Delta\varrho\, g - p_e$. We then get that

$$\frac{\partial e}{\partial t} = -\frac{C_c}{\sigma'}\frac{\partial \sigma'}{\partial t} = \frac{C_c}{\sigma'}\left(\frac{\partial p_e}{\partial t} - \omega\Delta\varrho\, g\right) \tag{14.70}$$

which inserted into (14.63) gives equation (14.66) for the overpressure build-up in clay.

Exercise 14.4
(a) Verify that $u = 1 - x$ is a solution of the dimensionless pressure equation (14.67).
(b) Show that $u = 1 - x$ corresponds to the overpressure $p_e = (\omega t - \zeta)\Delta\varrho g$.
(c) Verify that $p_e = (\omega t - \zeta)\Delta\varrho g$ is a solution of equation (14.66).

14.6 The gravity number

We have already seen that the gravity number N_g tells us whether there is high, intermediate or low overpressure build-up depending on whether $N_g \ll 1$, $N_g \sim 1$ or $N_g \gg 1$, respectively. The gravity number at a given depth is now defined as

$$N_g = \frac{k\,\Delta\varrho\, g}{\mu v_D} \tag{14.71}$$

where k is the permeability, $\Delta\varrho = \varrho_s - \varrho_f$ is the difference between the sediment matrix density and the fluid density, g is the gravitational constant, μ is the viscosity and v_D is Darcy's flux in the vertical direction. The gravity number is simply an expression for the overpressure gradient in terms of the Darcy flux. The fluid pressure is normally bounded by the lithostatic pressure, and the overpressure is bounded by the excess lithostatic pressure, which is the lithostatic pressure minus the hydrostatic fluid pressure. The overpressure gradient is therefore compared with the gradient of the excess lithostatic pressure,

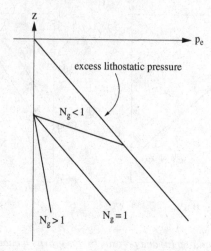

Figure 14.9. *Low, intermediate and high overpressure gradients.*

$(\varrho_b - \varrho_f)g$, where $\varrho_b - \varrho_f$ is the difference between the bulk rock density and the fluid density. The bulk rock density is $\varrho_b = \phi\varrho_f + (1-\phi)\varrho_s$, and the excess lithostatic gradient is $(1-\phi)(\varrho_s - \varrho_f)g$, as the gradient of the ζ-coordinate is $(\varrho_s - \varrho_f)g$. The overpressure gradient is compared with the gradient $\Delta\varrho\, g$, and low, intermediate and high overpressure gradients are characterized by

$$\frac{\partial p_e}{\partial z} \ll \Delta\varrho\, g, \qquad \frac{\partial p_e}{\partial z} = \Delta\varrho\, g \quad \text{and} \quad \frac{\partial p_e}{\partial z} \gg \Delta\varrho\, g. \qquad (14.72)$$

The overpressure gradient can be replaced by the Darcy flux using Darcy's law $\partial p_e/\partial z = v_D\mu/k$. The overpressure regimes (14.72) are then the same as

$$N_g \gg 1 \qquad N_g = 1 \quad \text{and} \quad N_g \ll 1 \qquad (14.73)$$

when expressed in terms of the gravity number. Figure 14.9 illustrates the overpressure for these three regimes. The Darcy flux, which is still unknown, can be estimated from the stationary pressure equation (13.121), which is

$$\frac{dv_D}{dz} = Q_{\text{pot}} \qquad \text{or} \qquad \frac{dv_D}{d\zeta} = (1+e)Q_{\text{pot}} \qquad (14.74)$$

in the vertical direction, where Q_{pot} is a source term. The Darcy velocity is therefore

$$v_D = \int_{z_0}^{z} Q_{\text{pot}}\, dz \qquad \text{or} \qquad v_D = \int_{0}^{\zeta} (1+e)Q_{\text{pot}}\, d\zeta \qquad (14.75)$$

where z_0 is the base of the basin. The integrals show that the vertical Darcy velocity increases in every depth interval where the source term is positive. Section 12.2 shows how the Darcy flux is obtained from such an integral when the source term

$$Q_{\text{pot}} = -\frac{1}{(1-\phi)}\frac{\partial \phi}{\partial t} \tag{14.76}$$

is integrated over ζ. An important result from Section 12.2 is that the Darcy flux is

$$v_D \approx e\omega \tag{14.77}$$

where ω is the net burial rate, when the porosity depends only on the depth from the basin surface measured as net (porosity-free) rock. The overpressure build-up is then characterized by the (familiar) gravity number where $v_D \approx \omega$ (assuming that $e \sim 1$).

The gravity number as a measure for overpressure build-up was introduced by Audet and Fowler (1992) and Wangen (1992).

Exercise 14.5 Show that

$$v_D = \int_0^\zeta (1+e)Q_{\text{pot}}\,d\zeta = -\int_0^\zeta \frac{\partial e}{\partial t} \tag{14.78}$$

for the source term (14.76).

14.7 Overpressure from thermal expansion

Thermal overpressure generation happens when thermal expansion of the pore fluid leads to fluid expulsion during burial. There are several ways to address the potential of thermal overpressure build-up. The first considers the thermal expansion as the only process that generates overpressure. A second approach adds fluid and rock compressibilities to the system, and a third approach is to compare the source term from thermal expansion with the source terms from other overpressure generating processes.

The first approach leads to the stationary overpressure equation

$$-\frac{k}{\mu}\frac{d^2 p_e}{dz^2} = Q_{\text{pot}} = \beta_{\text{eff}}\frac{\partial T}{\partial t} \tag{14.79}$$

where the only non-zero parts in the source term are the thermal expansibilities. All the expansibilities are collected in the effective thermal expansibility

$$\beta_{\text{eff}} = \phi(\beta_f - \beta_s) - (\beta_b - \beta_s) \tag{14.80}$$

where β_f, β_s and β_b are the thermal expansibilities of fluid, (pure) solid and the bulk, respectively. The stationary pressure equation is applied to a layer buried at a constant rate ω along a constant geothermal gradient dT/dz. The layer has thickness l, porosity ϕ and permeability k. There is no compaction of the layer. Boundary conditions for the layer are zero overpressure at the surface and zero fluid flux into the base. Integration of pressure equation (14.79) using the boundary conditions gives the overpressure

$$p(z) = p_{\max}\frac{z}{l}\left(2 - \frac{z}{l}\right) \tag{14.81}$$

where p_{max} is the maximum overpressure found at the base of the layer:

$$p_{max} = \frac{\mu Q_{pot} l^2}{2k} = \frac{\mu \beta_{eff} \omega \, (dT/dz) \, l^2}{2k}. \tag{14.82}$$

The top of the layer is at $z = 0$ and the base of the layer is at $z = l$. Notice that the origin ($z = 0$) is at the surface of the layer, and therefore follows it during burial. The following parameters give an example of what the overpressure could be in a layer with a thickness of $l = 1000$ m, thermal gradient $dT/dz = 0.035°C/m$, permeability $k = 1 \cdot 10^{-20}$ m^2, porosity $\phi = 0.1$, viscosity $\mu = 10^{-3}$ Pa s and thermal expansibility of the fluid $\beta_f = 4 \cdot 10^{-4}$ K^{-1}. The burial rate is taken to be $\omega = 315$ m/Ma (or $\omega = 10^{-11}$ m/s), and the effective thermal expansibility is approximated as $\beta_{eff} \approx \phi \beta_f$, assuming that $\beta_f \gg \beta_s$ and that $\beta_s \approx \beta_b$. The maximum overpressure is then $p_{max} = 0.7$ MPa at the base of the layer. A similar exercise can also be made for a layer at hydrostatic pressure at both the base and the top (see Exercise 14.6).

The pressure gradient decreases from its largest value $2p_{max}/l$ at the top of the layer to zero at the base of the layer. In other words, the maximum overpressure gradient is two times the linear gradient p_{max}/l. The maximum gradient $2p_{max}/l$ is now compared with the gradient of the excess lithostatic pressure $\Delta \varrho g$ in order to decide whether the overpressure gradient is small, intermediate or large. This ratio turns out to be the gravity number

$$N_g = \frac{\Delta \varrho g}{2 p_{max}/l} = \frac{k \Delta \varrho g}{\mu \upsilon_D} \tag{14.83}$$

where υ_D is the Darcy velocity out of the surface of layer

$$\upsilon_D = \int_0^l Q_{pot} \, dz = \int_0^l \beta_{eff} \frac{dT}{dt} dz = \beta_{eff} \frac{dT}{dz} \omega l. \tag{14.84}$$

Using the excess lithostatic gradient $\Delta \varrho g = 1500$ Pa/m and the parameters above the gravity number becomes $N_g = 1.1$, which is in the intermediate regime.

The next step is to add compressibilities to the model above and the overpressure equation becomes time-dependent

$$S_\sigma \frac{\partial p_e}{\partial t} - \frac{k}{\mu} \frac{\partial^2 p_e}{\partial z^2} = Q_{pot} \tag{14.85}$$

where equation (13.122) gives the poroelastic storage coefficient S_σ as

$$S_\sigma = \phi(\alpha_f + \alpha_{pp}) \tag{14.86}$$

The source term (13.123) gives

$$Q_{pot} = \beta_{eff} \frac{\partial T}{\partial z} \omega - S_\sigma \varrho_f g \omega + \phi \alpha_{pc} \varrho_b g \omega \tag{14.87}$$

and where ϱ_f and ϱ_b are the fluid density and the bulk rock density, respectively. The rates of change of the fluid pressure and the bulk pressure are $\partial p_h/\partial t = \varrho_f g \omega$ and $\partial p_b/\partial t = \varrho_b g \omega$, respectively, because the burial is a constant rate. Note that it is possible for the compressibilities in the source term to add up to zero, which happens when the compression of the fluid matches the compressibility of the pore space (see Exercise 14.7). Compression reduces the overall source term Q_{pot} if the fluid becomes compressed more than the pore space. The opposite situation leads to an increased source term, when the compression of the fluid is less than the compression of the pore space. The stationary solution for overpressure is the same as for thermal expansion alone, except that the source term in the maximum overpressure (14.82) is now the expanded source term (14.87).

A layer that has hydrostatic fluid pressure at the start of the burial does not immediately attain the stationary fluid pressure. A solution of the overpressure equation (14.85) tells us how fast the overpressure increases from zero towards the stationary overpressure. Before the equation for overpressure is solved it is rewritten in dimensionless form as

$$\frac{\partial \hat{p}}{\partial \hat{t}} - \frac{\partial^2 \hat{p}}{\partial \hat{z}^2} = 2 \tag{14.88}$$

where the variables are the dimensionless overpressure $\hat{p} = p_e / p_{max}$, the dimensionless z-coordinate $\hat{z} = z/l$ and the dimensionless time $\hat{t} = t/t_0$. The characteristic time for overpressure build-up in a layer of given thickness is

$$t_0 = \frac{S_\sigma \mu l^2}{k} \tag{14.89}$$

and it is the same as the characteristic time (14.34) for overpressure decay (see Note 14.6). The boundary conditions for the layer are unchanged: zero overpressure at the surface and zero fluid flux into the base. The solution of the dimensionless overpressure equation (14.88) is

$$\hat{p}(\hat{z}, \hat{t}) = \hat{z}(2 - \hat{z}) + \sum_{n=1}^{\infty} a_n \sin(\lambda_n \hat{z}) e^{-\lambda_n^2 \hat{t}} \tag{14.90}$$

where

$$a_n = -\frac{4}{\lambda_n^3} \quad \text{and} \quad \lambda_n = \frac{\pi}{2} + n\pi \tag{14.91}$$

which is the stationary overpressure solution added to a transient overpressure term that dies out on a time scale of length t_0 (see Note 14.7). The dimensionless solution (14.90) for overpressure is plotted in Figure 14.10, which shows that the overpressure is close to the stationary overpressure after a time $\hat{t} = 1$ (or $t = t_0$). The characteristic time is $t_0 = 0.2$ Ma using the parameters above in addition to $\alpha_f = 5 \cdot 10^{-10}$ Pa^{-1} and $\alpha_{pp} = \alpha_{pc} = 1 \cdot 10^{-10}$ Pa^{-1}. These numbers serve as an order of magnitude estimate for t_0 in this case. The maximum overpressure is $p_{max} = 0.4$ MPa, which is nearly one-half of the overpressure estimated when compressibilities are ignored. Both the maximum overpressure and the characteristic time have the factor $\mu l^2 / k$ in common. That means that decreasing the permeability of the layer by one order of magnitude increases both the maximum overpressure and the characteristic time by one order of magnitude. The maximum overpressure is therefore related to the characteristic time as

$$\frac{p_{max}}{t_0} = \frac{Q_{pot}}{2S_\sigma}. \tag{14.92}$$

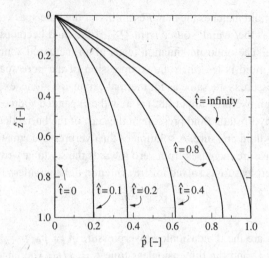

Figure 14.10. *Overpressure build-up from zero towards stationary overpressure.*

The parameters above give that this ratio is $p_{max}/t_0 = 2.1$ MPa/Ma. It should be noted that the ratio is proportional to the burial rate. Thermal overpressure generation is much weaker than processes that reduce the pore space (see Exercise 14.8), although it may be an important process for overpressure generation in shales, a rock that usually has a sufficiently low permeability for overpressure build-up to be strong. Note 14.8 shows how the Gibson solution can be applied to overpressure generation from thermal expansion under constant deposition of sediments.

Note 14.6 The overpressure could be scaled with overpressure $p_0 = \mu l^2 Q_{pot}/k$ rather than $p_{max} = p_0/2$. The dimensionless source term is then 1 rather than 2, but the maximum dimensionless overpressure becomes $\hat{p}_{max} = 1/2$.

Note 14.7 The solution $\hat{p}(\hat{z}, \hat{t})$ of the dimensionless overpressure equation (14.88) is written as the sum

$$\hat{p}(\hat{z}, \hat{t}) = \hat{p}_1(\hat{z}) + \hat{p}_2(\hat{z}, \hat{t}) \tag{14.93}$$

where $\hat{p}_1(\hat{z}) = \hat{z}(2 - \hat{z})$ is the stationary solution that takes care of the source term, and where the transient solution $\hat{p}_2(\hat{z}, \hat{t})$ solves the (homogeneous) equation

$$\frac{\partial \hat{p}}{\partial \hat{t}} - \frac{\partial^2 \hat{p}}{\partial \hat{z}^2} = 0 \tag{14.94}$$

with the initial condition $\hat{p}_2 = -\hat{z}(2 - \hat{z})$ (at $\hat{t} = 0$). This initial condition ensures that $\hat{p} = \hat{p}_1 + \hat{p}_2 = 0$ at $\hat{t} = 0$. Equation (14.94) is the same as the dimensionless pressure equation in Section 14.3, and it is solved in almost the same way. The only difference is that the Fourier coefficients are

$$\frac{1}{2}a_n = -\int_0^1 \hat{z}(2 - \hat{z}) \sin(\lambda_n \hat{z}) d\hat{z} = -\frac{2}{\lambda_n^3}. \tag{14.95}$$

Note 14.8 Thermal overpressure build-up can also be treated analytically in a sedimentary column that is increasing in height due to deposition at a constant rate. The overpressure equation (14.85) is then written in the same form as equation (12.57), which allows for Gibson's solution for overpressure generation. Gibson's solution can be applied to thermal overpressure build-up because the source term is constant.

Exercise 14.6 (a) Show that the thermal overpressure in a layer that is hydrostatic both at the surface and the base is

$$p(z) = 4 \, p_{\max} \frac{z}{l} \left(1 - \frac{z}{l}\right) \tag{14.96}$$

where

$$p_{\max} = \frac{\mu \beta_{\text{eff}} \omega \, (dT/dz) \, l^2}{8k} \tag{14.97}$$

is the maximum overpressure at the center of the layer (at $z = l/2$).
(b) What is $p_{\max}$ using the parameters from the text?

Exercise 14.7 Show that the parts involving compressibilities add up to zero in the source term (13.123) when the drained bulk compressibility α_{pc} is

$$\alpha_{pc} = (\alpha_f + \alpha_{pp}) \frac{\varrho_f}{\varrho_b} \tag{14.98}$$

where α_f is the fluid compressibility and α_{pp} is the pore volume expansion coefficient.

Exercise 14.8 Compare the Darcy velocity (14.84) with the Darcy velocity from porosity reduction $v_D = \omega(e - e_{\text{bot}})$, where $e - e_{\text{bot}}$ is the difference in void ratio between the base and the top of the layer. Use the numbers from the text and find how much porosity reduction corresponds to thermal expansion.
Solution: $d\phi/(1 - \phi)^2 = e - e_{\text{bot}} = \beta_{\text{eff}} l \, (dT/dz)$.

14.8 Special cases of fluid expulsion and mineral reactions

The source term in the pressure equation (13.123) is what causes overpressure build-up. There are two parts in the source term that have not been discussed so far – the term for fluid generation q_f/ϱ_f and the term for mineral production/consumption $\sum_i q_i/\varrho_i$. These terms are important when mineral reactions produce fluids and change the volume of the solid part of the porous matrix. Recall that q_f is the rate of fluid generation in units of mass per bulk volume of rock and time, and q_f/ϱ_f is therefore the rate of fluid generation in units of volume fluid per bulk rock volume and per time. The same applies for q_i/ϱ_i, which gives the rate of production/consumption of each mineral phase i in units of volume per bulk volume and per time. We will look at the three different situations with respect to reactions of the solid phase and generation of pore fluid. The first situation has conserved solid volume. The next situation has conserved bulk volume, which implies that solid volume lost in generation of fluids becomes pore space. The final situation, that allows both the solid volume and the pore volume to change, is treated in Section 14.11. It is assumed that reactions take place locally, which means that there is no "long" distance transport of mineral species. There is therefore no need for transport modeling of dissolved species.

The first case has conserved solid volume of the rock, and the term for production/consumption of minerals is therefore zero ($\sum_i q_i/\varrho_i = 0$). Quartz cementation of the pore space is an example of such a process, where the solid volume is conserved and where there is no fluid generation. The full source term for quartz cementation is therefore

$$Q_{cem} = -\frac{1}{(1-\phi)}\frac{D\phi}{dt} \tag{14.99}$$

when the contributions from compressibilities and thermal expansibilities are ignored.

The next case is when mineral reactions do not alter the bulk volume of the rock. This means that there is no compaction, although both the porosity and the volume of solid are allowed to change. Vertical intervals without compaction have a constant velocity $\mathbf{v}_s$ and therefore $\nabla \cdot \mathbf{v}_s = 0$. The conservation law for solids (13.2) then gives

$$\sum_i \frac{q_i}{\varrho_i} = \sum_i \frac{D\phi_i}{dt} = -\frac{D\phi}{dt} \tag{14.100}$$

when the density for each mineral phase ϱ_i is constant. The last equality in (14.100) follows from the sum of the volume fractions, which is $\phi + \sum_i \phi_i = 1$. Relationship (14.100) shows that the lost volume fraction of solid becomes porosity when the bulk volume is conserved. The full source term for conserved bulk volume is then

$$Q_b = -\frac{1}{(1-\phi)}\frac{D\phi}{dt} + \frac{q_f}{\varrho_f} - \frac{\phi}{(1-\phi)}\sum_i \frac{q_i}{\varrho_i}$$

$$= -\frac{D\phi}{dt} + \frac{q_f}{\varrho_f} \tag{14.101}$$

when terms with compressibilities and thermal expansibilities are ignored. Such a source term could for instance model the melting of ice in the pore space, dehydration of rock or oil generation from kerogen (organic matter). An example is oil generation from kerogen where the volume of kerogen transformed to oil becomes pore space. The source term (14.101) is then

$$Q_b = -\left(\frac{\varrho_k}{\varrho_o} - 1\right)\frac{Dc}{dt} \tag{14.102}$$

where c is the volume fraction of reactive kerogen that can be completely transformed to oil if sufficiently heated. The densities of kerogen and oil are ϱ_k and ϱ_o, respectively. (The derivation of expression (14.102) is Exercise 14.9.) The source term (14.102) is zero when $\varrho_o = \varrho_k$, which means that the oil volume generated replaces exactly the volume of kerogen lost, and oil generation does not contribute to overpressure generation. The condition for oil generation to contribute to overpressure build-up is that $\varrho_o < \varrho_k$.

Exercise 14.9 Assume that the bulk volume remains constant when kerogen is transformed to oil, and that the volume fraction of kerogen converted to oil becomes pore space. Then show that

$$-\frac{D\phi}{dt} + \frac{q_f}{\varrho_f} = -\left(\frac{\varrho_k}{\varrho_o} - 1\right)\frac{Dc}{dt} \qquad (14.103)$$

where the kerogen content is the volume fraction c of the bulk volume, and where ϱ_k and ϱ_o are the densities of kerogen and oil, respectively.

Solution: The volume of kerogen ΔV_k converted to oil generates a volume ΔV_o of oil, where mass conversion requires that $\varrho_k \Delta V_k + \varrho_o \Delta V_o = 0$. The volume ΔV_k of kerogen is related to the bulk volume by $\Delta V_k = \Delta c V_b$. The rate of porosity generation is then

$$\frac{D\phi}{dt} = \frac{1}{V_b}\frac{DV_k}{dt} = -\frac{Dc}{dt} \qquad (14.104)$$

and the volume rate of oil generation per bulk volume of rock is

$$\frac{q_f}{\varrho_f} = \frac{1}{V_b}\frac{DV_o}{dt} = -\frac{1}{V_b}\frac{\varrho_k}{\varrho_o}\frac{DV_k}{dt} = -\frac{\varrho_k}{\varrho_o}\frac{Dc}{dt}. \qquad (14.105)$$

The two terms (14.104) and (14.105) inserted into the term (14.101) give the wanted source term (14.103).

- 14.9 Overpressure from quartz cementation

Quartz cementation is an example of a mineral reaction where the volume of the solid phase is conserved, but where the porosity decreases. There is no fluid generation, but the quartz dissolved along stylolites or quartz/mica contacts precipitates in the pore space. The decreasing porosity is therefore the main contribution to the source term (13.123), which is

$$Q_{cem} = -\frac{1}{(1-\phi)}\frac{D\phi}{dt} = \frac{v c k_f(T) S(\phi)}{1-\phi} \qquad (14.106)$$

when all other terms are ignored. The rate of change of porosity (11.8) is used, where k_f is the dissolution rate in units of mole $m^{-2}\,s^{-1}$, $S(\phi)$ is the specific surface as a function of the porosity, c is the degree of supersaturation of silica in the pore space and v is the molar volume of quartz.

14.10 Overpressure from cementation of pore space

Cementation of pore space is in the following studied as the main cause of overpressure generation during burial. The cementation process now applies to silic-clastic sediments in general, and not only for quartzose sandstones. This compaction model has in common with the quartz cementation model in Section 11.1 that the source for the cement is local dissolution, and it serves as a simple model of chemical compaction. We consider a simple model where the precipitation rate is exponential in temperature

$$r(T) = c k_f(T) = b e^{aT} \tag{14.107}$$

which is the same exponential precipitation rate as suggested by Walderhaug (1994b, a) for quartz cementation. The unit for temperature in the rate law is °C. The overpressure is found by going through the same steps as in Section 12.4, and using the same technique, where the overpressure was obtained from a known function of the void ratio. The porosity is found first, then the Darcy flux and finally the overpressure. The rate of change of porosity is

$$\frac{d\phi}{dt} = -v\, S(\phi) r(T) \tag{14.108}$$

where v is the molar volume of the pore-filling cement and $S(\phi)$ is the specific surface of the pore space. (The unit for v is $m^3\,mole^{-1}$, the unit for S is m^2/m^3 and the unit for r is mole $m^{-2}\,s^{-1}$.) The specific surface decreases as mineral cement fills the pore space, and the reduction is assumed to be a function of the porosity

$$S(\phi) = S_0 \left(\frac{\phi - \phi_c}{\phi_o - \phi_c} \right)^n \tag{14.109}$$

where S_0 is the initial specific surface at the initial porosity ϕ_0. The porosity ϕ_c is the minimum porosity where the pore space becomes disconnected. An integration of the rate-law (14.108) is simplified when heating takes place at constant rate. The porosity is then the function

$$\phi(T) = \phi_c + (\phi_0 - \phi_c) \exp\left(- N_b (e^{aT} - e^{aT_0}) \right) \tag{14.110}$$

of the temperature for a linear specific surface function ($n = 1$). The initial temperature at the beginning of burial is T_0, and the (dimensionless) number N_b is

$$N_b = \frac{b\, v\, S_0}{(\phi_0 - \phi_c) a\, (dT/dt)} \tag{14.111}$$

where the heating rate dT/dt is constant. An example of the porosity–temperature function (14.110) is shown in Figure 14.11 for the data given in Table 14.2. The next step is to find the Darcy flux generated by the porosity reduction (14.110) with increasing temperature. We already have the general relationship (12.15) between porosity and Darcy flux, which applies for burial at a constant rate, when the porosity is only a function of the distance from the basin surface measured as net (porosity-free) sediments. In order to use the Darcy flux (12.15) we assume that the temperature is a function of the net amount of rock from the basin surface,

$$T(\zeta) = d\, (\zeta^* - \zeta) + T_0 \tag{14.112}$$

Table 14.2. *The parameters used in the overpressure case shown in Figures 14.11 and 14.12.*

Parameter	Value	Units
S_0	$5 \cdot 10^5$	$[m^2/m^3]$
v	$2.4 \cdot 10^{-5}$	$[m^3 mole^{-1}]$
a	0.051	$[1/{}^\circ C]$
b	$2 \cdot 10^{-18}$	$[mole\, m^{-2}s^{-1}]$
ϕ_0	0.3	$[-]$
ϕ_c	0.03	$[-]$
n	1	$[-]$
m	3	$[-]$
ω	31.5	$[m\,Ma^{-1}]$
d	0.045	$[{}^\circ C/m]$
μ	10^{-3}	$[Pa\,s]$
ϱ_s	2200	$[kg\,m^{-3}]$
ϱ_w	1000	$[kg\,m^{-3}]$

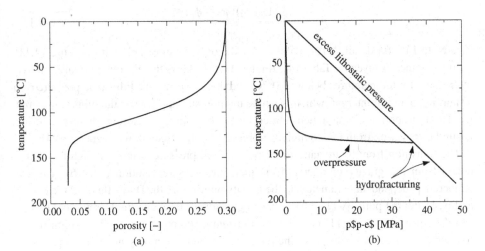

Figure 14.11. *(a) Porosity (14.110) as a function of temperature. (b) Overpressure (14.116) as a function of temperature. The overpressure corresponds to the porosity in (a).*

using a constant gradient $dT/d\zeta = -d$, where ζ is the position in the basin measured as net (porosity-free) rock from the basement. The basin surface is at the ζ-position $\zeta^* = \omega t$, when deposition takes place at the constant rate ω. (The ζ-position is now linearly related to the Lagrange coordinate ζ.) The Darcy flux then becomes

$$v_D = \omega\,(e - e_{bot}) \qquad (14.113)$$

where e_{bot} is the void ratio at the base of the basin ($\zeta = 0$). The final step is to obtain the overpressure by integrating Darcy's law

$$v_D = \frac{k(\phi)}{(1+e)\mu} \frac{\partial p_e}{\partial \zeta} \tag{14.114}$$

where the permeability is the following function of the porosity:

$$k(\phi) = k_0 \frac{(1+e)}{(1+e_0)} \left(\frac{\phi - \phi_c}{\phi_0 - \phi_c} \right)^m . \tag{14.115}$$

The permeability decreases with decreasing porosity to the power of the exponent m, and k_0 is the initial permeability at the initial porosity ϕ_0. The scaling factor $(1+e)/(1+e_0)$ is included to simplify the calculation of the overpressure, but its importance is negligible compared to the porosity to the power of m. The approximation $v_D \approx \omega (\phi - \phi_c)$, which is quite accurate for small porosities, simplifies the calculation too. The overpressure as a function of the temperature then becomes

$$p_e(T) = p_0 \int_{u_0}^{u(T)} \frac{e^x}{x} dx \tag{14.116}$$

where $u(T) = (m-1)N_b e^{aT}$ and $u_0 = u(T_0)$, with the pressure coefficient

$$p_0 = \frac{\mu\omega(1+e_0)(\phi_0 - \phi_c)e^{-u_0}}{k_0 a d} . \tag{14.117}$$

(See Note 14.9 for details.) This solution for the overpressure is plotted in Figure 14.11b for the parameters given by Table 14.2. Figure 14.11b shows that the overpressure is nearly hydrostatic for temperatures below 100 °C and that it reaches the lithostatic pressure over a short temperature interval, which is where the porosity approaches the minimum porosity. The fluid pressure is assumed limited by the lithostatic pressure by hydrofracturing. A similar transition from low to high overpressure is often seen in sedimentary basins.

The transition from hydrostatic to lithostatic fluid pressure is more straightforward to obtain from the gravity number (14.71) than the pressure solution (14.116). The two temperature-dependent parameters in the gravity number are the Darcy flux and the permeability. Using the porosity (14.110) it is straightforward to obtain the Darcy flux (14.113) and the permeability (14.115) as functions of temperature. The remaining parameters in the gravity number are constants. The permeability and the gravity number are plotted in Figure 14.12 for the case shown in Figure 14.11. The figure shows that the hydrostatic region is the temperature interval where $N_g \gg 1$. The temperature interval where the overpressure goes from (nearly) hydrostatic to lithostatic pressure corresponds to the depth interval where the gravity number goes from the region $N_g \gg 1$ to $N_g \ll 1$. The usefulness of the gravity number is that it is often much easier to obtain than the full pressure solution, especially when the simple expression (12.15) gives the Darcy flux for burial at a constant rate.

Figure 14.12a shows that the permeability decreases steeply towards zero when the porosity approaches ϕ_c. The Darcy velocity also decreases with decreasing porosity, but

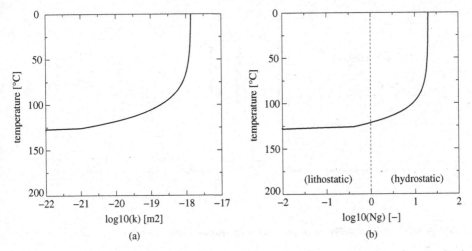

Figure 14.12. *The permeability (a) and the gravity number (b) correspond to the porosity in Figure 14.11a.*

the permeability decreases more rapidly when $m > 1$. That is why high overpressure develops at depths where there is not much porosity left. This is seen directly from the gravity number as a function of the porosity, when it is approximated as

$$N_g(\phi) = \frac{k(\phi)\,\Delta\varrho g}{\mu\,v_D(\phi)}$$

$$\approx N_{g,0}\,\frac{(\phi - \phi_c)^{m-1}}{(\phi_0 - \phi_c)^m}. \tag{14.118}$$

The first factor $N_{g,0} = k_0\Delta\varrho g/\mu\omega$ is the surface gravity number, and the second factor represents how the permeability decreases relative to the Darcy flux. (The Darcy flux is approximated as $v_D = \omega\,(\phi - \phi_c)$ for ϕ close to ϕ_c.) The surface gravity number tells us if there is high or low overpressure at shallow depths. The upper part of the sediments are hydrostatic if $N_{g,0}$ is much larger than 1. As seen in Section 12.4 the permeability exponent has to be $m > 1$ for overpressure to develop when $\phi \to \phi_c$. When m is as high as 3 there is a sharp transition from (almost) hydrostatic fluid pressure ($N_g \gg 1$) to lithostatic pressure ($N_g \ll 1$), when the porosity decreases towards ϕ_c.

Note 14.9 The following steps leads to the overpressure (14.116) as a function of temperature. The Darcy law (14.114), when combined with the Darcy flux (14.113), gives the overpressure gradient

$$\frac{\partial p_e}{\partial \zeta} = \frac{\mu\omega\,(1+e)(e - e_{bot})}{k(\phi)}. \tag{14.119}$$

The next steps are the approximation of the factor $(e - e_{bot})$ by $(\phi - \phi_c)$, and the insertion of the permeability function (14.115), and the overpressure gradient becomes

$$\frac{\partial p_e}{\partial \zeta} = \frac{1}{k_0}\mu\omega\,(1 + e_0)(\phi_0 - \phi_c)\left(\frac{\phi - \phi_c}{\phi_0 - \phi_c}\right)^{1-m}. \tag{14.120}$$

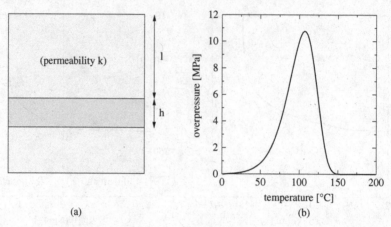

(a)

Figure 14.13. *(a) A layer being buried. (b) The overpressure in the layer as a function of temperature.*

The porosity is then replaced by expression (14.110) in T, and the integration of the gradient is carried out by changing the integration variable from ζ to $u = u(T)$, using relationship (14.112) between ζ and T.

Exercise 14.10 Let a sedimentary layer be buried at a constant rate (ω), when the compaction of the layer is controlled by the rate (14.108). Figure 14.13a shows a sketch of the layer. Assume that the overburden rock does not compact, and that it has the permeability $k = k_0/(1 - \phi)$ during the burial, when ϕ is the porosity of the layer being buried, and where k_0 is a constant permeability.

(a) What is the overpressure in the layer during burial?
(b) At what depth, time and temperature is the overpressure at a maximum?
(c) What is the maximum overpressure?
Solution:
(a) The Darcy flux

$$v_D = \frac{k}{\mu}\frac{p_e}{z} \tag{14.121}$$

gives the leakage of the fluid from the layer through the overburden, when p_e is the overpressure in the layer, z is the thickness of the overburden (which is the same as the depth of burial of the layer), and μ is the viscosity. The Darcy flux is equal to the rate of fluid expulsion per surface area of the layer

$$v_D = -h\frac{d\phi}{dt} \tag{14.122}$$

where h is the thickness of the layer. The thickness h decreases during burial because of compaction, and it is therefore related to the net thickness of the rock h_0 by $h = h_0/(1-\phi)$ (which is independent of the porosity). The two fluxes (14.121) and (14.122), which are equal, give the overpressure

$$p_e = -\frac{\mu z h}{k}\frac{d\phi}{dt} = -\frac{\mu z h_0}{k_0}\frac{d\phi}{dt}. \tag{14.123}$$

We see that the factor $1/(1-\phi)$ in the permeability is just a tactical assumption to cancel the same factor in the thickness h. (The minus sign is needed because $d\phi/dt$ is a negative number.) The overpressure as a function of temperature becomes

$$p_e(T) = \frac{\mu v h S_0 T}{kd}\exp\left(-nN_b(e^{aT}-e^{aT_0})\right)b\,e^{aT} \tag{14.124}$$

when the porosity (14.110) is inserted into the specific surface function (14.109). The burial depth is related to temperature through the constant thermal gradient d by $z = T/d$. The overpressure in a layer with thickness $h_0 = 200$ m and an overburden with the permeability $k_0 = 10^{-20}$ m^2 is shown in Figure 14.13b, with all the other parameters taken from Table 14.2.

(b) The equation $dp_e/dT = 0$ gives the temperature for maximum overpressure, but this equation is difficult to solve exactly. It turns out that the temperature of maximum overpressure is controlled mainly by

$$\frac{d\phi}{dt} = -v\,S_0\exp\left(-nN_b(e^{aT}-e^{aT_0})\right)b\,e^{aT} \tag{14.125}$$

when porosity (14.110) is inserted into the rate (14.108). The temperature at the maximum rate of porosity loss is easier to obtain from the function $f(T) = \ln(-d\phi/dt)$, rather than from $d\phi/dt$ itself. The maximum is given by $f'(T_{max}) = 0$, which has the solution

$$T_{max} = -\frac{\ln(nN_b)}{a}. \tag{14.126}$$

The burial depth for T_{max} is $z_{max} = T_{max}/d$, and the time is $t_{max} = z_{max}/\omega$. The data in Table 14.2 gives that $T_{max} = 104\,°C$.

(c) Inserting the temperature (14.126) into the porosity rate (14.125) gives the maximum rate of compaction

$$\left(\frac{d\phi}{dt}\right)_{max} = \frac{S_0 bv\exp(-1)}{nN_b} = a(\phi_0-\phi_c)\frac{dT}{dz}\omega\frac{\exp(-1)}{n} \tag{14.127}$$

and thereby the maximum overpressure

$$p_{e,max} = \frac{h_0\,\mu\,\omega}{nk_0}\ln(nN_b)(\phi_0-\phi_c)\exp(-1). \tag{14.128}$$

An alternative formulation of maximum overpressure is to divide it by the excess lithostatic pressure at the same depth,

$$\hat{p}_{e,max} = \frac{p_{e,max}}{(\varrho_s-\varrho_w)gl} = \frac{a\Delta T}{nN_g}(\phi_0-\phi_c)\exp(-1) \tag{14.129}$$

where $\Delta T = hd$ is the temperature difference across the layer, and N_g is the gravity number for the overburden. We have seen before, with the overpressure solution (12.32), that the overpressure is inversely proportional to the gravity number, when the porosity is a known function of depth. In addition the overpressure is proportional to the temperature difference across the layer.

14.11 Fluid expulsion and mineral reactions

This section deals with mineral reactions, and how they may be represented by the source term (13.123) in the equation for overpressure, when both the solid volume and the bulk volume are allowed to vary. Mineral reactions control the volume of solid – the porosity together with the volume of solid control the bulk rock volume. How the porosity changes does not need to be given by mineral reactions alone. Texture and special processes like, for example, stylolites may be important.

We will now see what the terms q_f/ϱ_f and $\sum_i q_i/\varrho_i$ might look like in the source term (13.123); they represent fluid generation and mineral reactions, respectively. The following example assumes that a fraction of the initial solid volume (slowly) reacts to produce pore fluid together with a product that is also solid. Mass conservation requires that

$$\varrho_r \Delta V_r = \varrho_f \Delta V_f + \varrho_p \Delta V_p \tag{14.130}$$

when a volume ΔV_r of reactant reacts to produce a volume ΔV_f of fluid and a volume ΔV_p of product, where ϱ_r, ϱ_f and ϱ_p are the densities of the reactant, fluid and product, respectively. Let the mass of fluid generated be given as a fraction w_f of the mass of reactant, and the mass of product be the mass fraction w_p of the reactant,

$$\varrho_f \Delta V_f = w_f \varrho_r \Delta V_r \qquad \text{and} \qquad \varrho_p \Delta V_p = w_p \varrho_r \Delta V_r \tag{14.131}$$

where mass conservation implies that $w_f + w_p = 1$. A volume ΔV_r of reactant becomes the following volumes of fluid and product when it has reacted:

$$\Delta V_f = a_f \Delta V_r \quad \text{where} \quad a_f = w_f \frac{\varrho_r}{\varrho_f} \tag{14.132}$$

$$\Delta V_p = a_p \Delta V_r \quad \text{where} \quad a_p = w_p \frac{\varrho_r}{\varrho_p}. \tag{14.133}$$

The rate of net volume reduction (or production) of solid is then

$$r_s = \frac{DV_p}{dt} - \frac{DV_r}{dt} = -(a_p - 1)V_0 \frac{Dc}{dt} \tag{14.134}$$

where V_0 is the initial volume of solid, and the reactant is given as a fraction c of V_0. Similar reasoning gives that the rate of fluid volume generation is

$$r_f = \frac{DV_f}{dt} = -a_f V_0 \frac{Dc}{dt}. \tag{14.135}$$

Notice that $Dc/dt < 0$ because the volume fraction of reactant is decreasing. The volume rate of fluid generation is therefore positive, and the volume rate of the solid is negative when $a_p < 1$. The parts q_f/ϱ_f and $\sum_i q_i/\varrho_i$ in the source term give the time rate of volume production/reduction per bulk volume. We therefore need the current bulk volume when the initial solid volume V_0 is known. An intermediate step towards the current bulk volume is the current solid volume, which is the sum of the inert part and the current volumes of reactant and product,

$$V_s = (1 - c_0)V_0 + cV_0 + a_p(c_0 - c)V_0$$
$$= \left(1 + (a_p - 1)(c_0 - c)\right) V_0. \tag{14.136}$$

The porosity relates the current bulk volume to the current solid volume:

$$V_b = \frac{V_s}{1 - \phi}. \tag{14.137}$$

The wanted parts in the source term then become

$$\frac{q_f}{\varrho_f} = \frac{r_f}{V_b} \quad \text{and} \quad \sum_i \frac{q_i}{\varrho_i} = \frac{r_s}{V_b} \tag{14.138}$$

and putting together equations (14.134) and (14.138) we get the source term

$$\frac{q_f}{\varrho_f} - \frac{\phi}{(1 - \phi)} \sum_i \frac{q_i}{\varrho_i} = -\frac{(1 - \phi)\,a_f - \phi\,(a_p - 1)}{1 + (a_p - 1)(c_0 - c)}\frac{Dc}{dt}. \tag{14.139}$$

The solid volume is conserved when $\varrho_p = w_p\varrho_r$, and the source term then becomes that of fluid generation. The denominator can be approximated by 1 in cases where $\varrho_p \approx w_p\varrho_r$ and the initial fraction c_0 is small compared to 1. The source term (14.139) is the basis of the discussion of overpressure generation from dehydration of clay in Section (14.12).

A Lagrange coordinate like ζ cannot be based on the solid volume when it is not a conserved property. Instead the inert solid volume can be used, which is the inert fraction of the initial solid volume $V_i = (1 - c_0)V_0$. A small ζ-step is then $\Delta\zeta = V_i/A$, where A is a unit area of the base (and assuming that the volume V_i has vertical sides and a planar top and base). The bulk volume then gives the corresponding (real) vertical step $dz = V_b/A$, and volume relations (14.136) and (14.137) give that the (real) z-coordinate is related to the ζ-coordinate by

$$z = \int_\zeta^{\zeta^*} \frac{1 + (a_p - 1)(c_0 - c)}{(1 - \phi)(1 - c_0)} \, d\zeta \tag{14.140}$$

where ζ^* is the ζ-coordinate for the basin surface. (The basin surface defines $z = 0$.) The relationship (14.140) between the ζ-coordinate and the z-coordinate can be approximated by the familiar integral $\int d\zeta/(1 - \phi)$ when the fraction c_0 is much smaller than 1 and $a_p \approx 1$.

14.12 Overpressure from dehydration of clay

Dehydration of clay may be a process that generates overpressure. The source terms of the last section are now used as a simple means to model Darcy flux when hydrate is converted to dehydrate and fluid. (We keep the notation of the previous section, by letting the reactant be hydrate and the product be dehydrate.) Two scenarios are considered, where the first

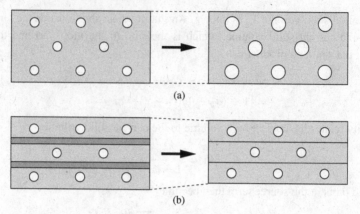

Figure 14.14. *Mineral reactions for two types of texture: (a) increasing porosity, but unchanged bulk volume; (b) constant porosity, but compaction by dissolution of (hydrate) bands.*

assumes that the porosity remains constant under the dehydration process, and the second where the loss in solid volume becomes porosity. Figure 14.14 schematically illustrates the rock texture for the two different cases.

The first scenario is represented by the source term (14.139), which assumes a constant porosity. In order to address the problem of overpressure generation we could solve the pressure equation. That is not straightforward in the case of the source term (14.139). A much easier approach is first to calculate the Darcy flux from the dehydration process, and then the gravity number N_g. This is the same approach as for quartz cementation in Section 14.10.

The source term (14.139) in the case of constant porosity is

$$Q_1 = \frac{a_f - \phi\,(a_f + a_r - 1)}{1 + (a_p - 1)(c_0 - c)}\frac{Dc}{dt} \qquad (14.141)$$

and the Darcy flux from the source term at a ζ-position is approximated by the integral (14.75)

$$v_{D1} = \int_0^\zeta Q_1 d\zeta. \qquad (14.142)$$

The hydrate conversion is now needed before we can proceed with the Darcy flux. A simple description of hydrate conversion to dehydrate is by means of a one-step Arrhenius equation,

$$\frac{\partial c}{\partial t} = -k_h c \quad \text{where} \quad k = A_h e^{-E_h/RT}. \qquad (14.143)$$

The initial value of hydrate (at burial) is c_0 and it decreases with increasing temperature during burial. The hydrate concentration is obtained as a function of temperature by assuming burial at a constant (net) rate ω, and that the temperature is proportional to the net amount of rock from the basin surface, as given by equation (14.112). We then have

$$c = c_0 \exp\left(-a\,(x^2 e^{-1/x} - x_0^2 e^{-1/x_0})\right) \qquad (14.144)$$

Table 14.3. *The data used in the case shown in Figure 14.6. Arrhenius kinetics for hydrate and the parameter c_0 are taken from Audet (1995). For the parameters a_f and a_p see Audet (1995) and Wangen (2001).*

Parameter	Value	Units
E_h	60	$[\text{kJ mole}^{-1}]$
A_h	$4.2 \cdot 10^{-5}$	$[\text{mole m}^{-2}\text{s}^{-1}]$
a_f	0.14	$[-]$
a_p	0.96	$[-]$
ω	315	$[\text{m Ma}^{-1}]$
R	8.314	$[\text{J K}^{-1}\text{mole}^{-1}]$
d	0.04	$[^\circ\text{C/m}]$
c_0	0.2	$[-]$
ϕ	0.2	$[-]$

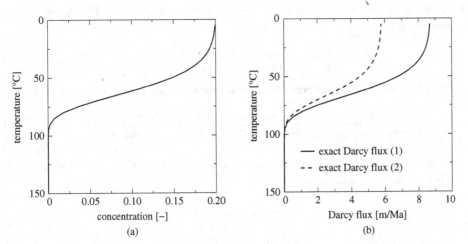

Figure 14.15. *(a) The hydrate concentration as a function of temperature. (b) The Darcy flux from the dehydration of clay as a function of temperature.*

where $x = T/T_h$, $x_0 = T_0/T_h$, $T_h = E_h/R$ and $a = A_h T_h/(\omega d)$. The temperatures are in kelvin, and T_0 is the temperature at the basin surface. (The same integration as for the porosity during quartz cementation gives the concentration (14.144). See Section 11.6 for details.) Figure 14.15a shows the hydrate concentration for the data from Table 14.3. The data set gives a rather shallow conversion of hydrate to dehydrate.

The Darcy velocity is then approximated by the integration (12.15), using the hydrate concentration instead of the porosity, which gives

$$v_{D1} = b_1 \omega c \tag{14.145}$$

where $b_1 = a_f - \phi (a_f + a_p - 1)$. (The integration assumes that the hydrate concentration is a function of the ζ-depth during the burial. See Section 12.2.) The (exact) Darcy flux is plotted in Figure 14.15b for the same data as for the hydrate concentration in Figure 14.15a. The approximate expression (14.145) gives $v_{D1} = 7.6 \text{ m Ma}^{-1}$ at the surface, which is roughly 10% of the exact Darcy flux $v_{D1} = 8.7 \text{ m Ma}^{-1}$. The approximate expression for the Darcy flux is sufficient for simple assessments of overpressure build-up. The gravity number, which tells us if there is strong or weak overpressure generation, is

$$N_g = \frac{k \Delta \varrho g}{\mu \, v_{D1}} = \frac{1}{b_1 c_0} \frac{k \Delta \varrho g}{\mu \omega}. \tag{14.146}$$

The gravity number is the usual gravity number $k \Delta \varrho g / \mu \omega$ for overpressure generation (by porosity reduction when $\phi \sim 0.5$) divided by the product $b_1 c_0$. The data in Table 14.3, which is an upper bound for the dehydration process, gives that $b_1 c_0 \approx 0.024$, and the permeability must therefore be at least two orders of magnitude less than for the case of the porosity reduction to achieve the same overpressure build-up.

The estimation of the Darcy flux above assumes that the porosity is constant. An alternative for the porosity is that the solid volume lost becomes porosity. The rate of change of porosity is then

$$\frac{D\phi}{dt} = -\frac{r_s}{V_b} = \frac{(1-\phi)(a_p - 1)}{V_s} \frac{DV_r}{dt}. \tag{14.147}$$

Adding this rate of change of porosity to the source terms for mineral reactions gives

$$-\frac{1}{(1-\phi)} \frac{D\phi}{dt} + \frac{q_f}{\varrho_f} - \frac{\phi}{(1-\phi)} \sum_i \frac{q_i}{\varrho_i} = -\frac{(1-\phi)(a_f + a_p - 1)}{1 + (a_p - 1)(c_0 - c)} \frac{Dc}{dt}. \tag{14.148}$$

The Darcy flux is less in this case because the fluid that is generated occupies the porosity that is formed. The gravity number is

$$N_g = \frac{1}{b_2 c_0} \frac{k \Delta \varrho g}{\mu \omega} \tag{14.149}$$

where $b_2 = (1 - \phi)(a_f + a_p - 1)$. Figure 14.15 shows the Darcy flux in this case plotted as a dashed line.

The model above for dehydration can also be applied to oil generation and expulsion in source rocks. Organic matter in the form of kerogen is then converted to a fluid oil and a solid coke by one or more Arrhenius steps.

Exercise 14.11 Show that $a_f + a_p - 1$ is the relative volume increase when the reactant is converted to the product.

Solution: A volume V_r of reactant becomes a volume $V_f = a_f V_r$ of fluid and a volume $V_p = a_p V_r$ of product after it has reacted. The relative increase in the (total) volume is

$$\frac{V_f + V_p - V_r}{V_r} = a_f + a_p - 1. \tag{14.150}$$

Exercise 14.12 Let V_s be the volume of solid and ϕ the porosity of the test volume that undergoes dehydration. The solid volume is divided into an inert part V_i and a reactive part V_r (and we have $V_s = V_i + V_r$). Show that the total amount of fluid expelled from the test volume is

$$\Delta V_f = \frac{1}{(1-\phi)}\left(a_f - \phi\,(a_f + a_p - 1)\right)V_r \qquad (14.151)$$

where the parameters a_f and a_p are the same as in the text above.
Solution: The volume of solid becomes $V_s' = a_p V_r + V_i$ after the hydrate has reacted and produced a volume $\Delta V_1 = a_f V_r$ of fluid. The shrinking of the solid volume leads to expulsion of the fluid volume

$$\Delta V_2 = \frac{\phi}{(1-\phi)}(V_s - V_s') = \frac{\phi}{(1-\phi)}(a_p - 1)V_r. \qquad (14.152)$$

(Recall that the fluid volume over the solid volume is the void ratio.) Adding the contributions ΔV_1 and ΔV_2 gives the total (14.151).

14.13 Weak (non-Rayleigh) thermal convection

Density variations may be the cause of fluid flow. One example is a type of weak fluid flow driven by lateral variations of the temperature field. The density is linearly dependent on the temperature by means of the thermal expansibility β

$$\varrho = \varrho_0(1 - \beta(T - T_0)) \qquad (14.153)$$

where ϱ_0 is a reference density at the reference temperature T_0. The linearization of the density is an approximation that holds for quite large temperature intervals. The weak fluid flow is not sufficiently strong to have any convective feedback on the temperature, and the temperature field is therefore conduction-dominated. This type of fluid flow is sometimes called non-Rayleigh convection to distinguish it from thermal convection where the fluid flow and the temperature are strongly coupled. Correspondingly, the latter type of fluid flow is sometimes called Rayleigh convection.

This section deals with non-Rayleigh convection, where we will see that fluid flow in a sealed aquifer is possible when the temperature field has non-horizontal isotherms ($\partial T/\partial x \neq 0$). An impervious boundary has

$$\mathbf{v}_D \cdot \mathbf{n} = 0 \qquad (14.154)$$

where $\mathbf{v}_D$ is the Darcy flux and $\mathbf{n}$ is the outward unit-vector along the boundary. Darcy's law is according to equation (13.111)

$$\mathbf{v}_D = -\frac{k}{\mu}\left(\nabla\Phi + \frac{\partial p_{h,0}}{\partial x}\mathbf{n}_x\right) \qquad (14.155)$$

where $p_{h,0}$ is the reference hydrostatic pressure

$$p_{h,0} = \int_0^z \varrho_f g \, dz. \tag{14.156}$$

The x-derivative of $p_{h,0}$ is

$$\frac{\partial p_{h,0}}{\partial x} = \int_0^z \frac{\partial \varrho}{\partial x} g \, dz = -\varrho_0 \, g \, \beta \int_0^z \frac{\partial T}{\partial x} \, dz. \tag{14.157}$$

Mass conservation assuming a constant density is

$$\nabla \cdot \mathbf{v}_D = 0 \tag{14.158}$$

which gives

$$\nabla^2 \Phi = \varrho_0 \, g \, \beta \int_0^z \frac{\partial^2 T}{\partial x^2} \, dz. \tag{14.159}$$

This is an example of the Boussinesq approximation, which states that density differences in buoyancy-driven flow are sufficiently small to be neglected, except where they appear in terms multiplied by gravity. Although it might be possible to approximate the right-hand side with zero, for instance in the case of almost linear (but not horizontal) isotherms, this equation is difficult to solve due to the boundary condition (14.154). A streamline approach is simpler. The definition (3.203) of the Darcy flux using the stream function Ψ gives

$$v_x = \frac{\partial \Psi}{\partial z} = -\frac{k}{\mu} \left(\frac{\partial \Phi}{\partial x} - \varrho_0 \, g \, \beta \int_0^z \frac{\partial T}{\partial x} \, dz \right) \tag{14.160}$$

$$v_z = -\frac{\partial \Psi}{\partial x} = -\frac{k}{\mu} \frac{\partial \Phi}{\partial x}. \tag{14.161}$$

The term $\partial v_x / \partial z - \partial v_z / \partial x$ eliminates the pressure and leads to Poisson's equation

$$\nabla^2 \Psi = \frac{k}{\mu} \varrho_0 \, g \, \beta \frac{\partial T}{\partial x} \tag{14.162}$$

where it is the source term (the right-hand side) involving the temperature that is responsible for the flow. A change $d\Psi$ in the stream function for a step (dx, dz) is

$$d\Psi = \frac{\partial \Psi}{\partial x} dx + \frac{\partial \Psi}{\partial z} dz = \mathbf{v}_D \cdot \mathbf{n} \tag{14.163}$$

where $\mathbf{v}_D = (\partial \Psi / \partial z, -\partial \Psi / \partial x)$ is the Darcy flux and $\mathbf{n} = (dz, -dx)$ is the normal vector to the step. (A step vector of unit length gives a normal vector that also has unit length.) Zero flow through the boundary implies that the stream function is constant along the boundary. We are free to choose any reference value for the stream function, since it is only a difference or gradient of the stream function that has a physical meaning. The stream function is therefore chosen to be zero along the boundary of the aquifer. The temperature field that covers the aquifer is assumed to be linear, $T = T_1 + ax + bz$, and the equation for the stream function is

$$\nabla^2 \Psi = \frac{k}{\mu} \varrho_0 \, g \, \beta \, a \qquad (14.164)$$

with $\Psi = 0$ along the boundary. The solution to this problem in dimensionless coordinates is

$$\Psi(\hat{x}, \hat{z}) = \Psi_0 \sum_{n=1}^{\infty} a_n \sin(n\pi\hat{x}) \, F_n(\hat{z}) \qquad (14.165)$$

where $\hat{x} = x/L, \hat{z} = z/H, \epsilon = H/L$ (the aspect ratio) and

$$\Psi_0 = \frac{k}{\mu} \varrho_0 \, g \, \beta \, a \, H^2 \qquad (14.166)$$

$$a_n = \frac{2(1 - (-1)^n)}{n\pi} \qquad (14.167)$$

$$F_n(\hat{z}) = \frac{\sinh(n\pi\epsilon\hat{z}) - \sinh(n\pi\epsilon(\hat{z} - 1)) - \sinh(n\pi\epsilon)}{(n\pi\epsilon)^2 \sinh(n\pi\epsilon)} \qquad (14.168)$$

and where L is the length and H is the height of the aquifer. Note 14.10 gives the details of how the solution (14.168) is obtained. The stream function gives the Darcy flux (14.160)–(14.161), which is

$$v_x = v_0 \, f_x(\hat{x}, \hat{z}), \quad f_x(\hat{x}, \hat{z}) = \sum_{n=1}^{\infty} a_n \sin(n\pi\hat{x}) \frac{d F_n}{d\hat{z}}(\hat{z}) \qquad (14.169)$$

$$v_z = v_0 \, f_z(\hat{x}, \hat{z}), \quad f_z(\hat{x}, \hat{z}) = -\sum_{n=1}^{\infty} n\pi\epsilon \, a_n \cos(n\pi\hat{x}) \, F_n(\hat{z}) \qquad (14.170)$$

where

$$v_0 = \frac{k}{\mu} \varrho_0 \, g \, \beta \, a H. \qquad (14.171)$$

Figure 14.16 shows an example of the stream function solution (14.165) and the corresponding Darcy flux (14.169)–(14.170). The Darcy flux is largest along the boundaries. The x-component of the Darcy flux is largest along the horizontal boundaries, and the z-component is largest along the vertical boundaries. The dimensionless functions f_x and f_z are plotted in Figure 14.17 in order to find out how the Darcy flux depends on the aspect ratio. The f_x-component is plotted at $\hat{x}$–$\hat{z}$-position $(1/2, 0)$ and the f_z-component at $\hat{x}$–$\hat{z}$-position $(0, 1/2)$. The f_x-component is roughly $v_0/2$, independent of the aspect ratio, while the f_z-component is proportional to the aspect ratio for aspect ratios less than 10^{-2}.

The Darcy flux (14.171) characterizes non-Rayleigh convection inside a sealed aquifer. The flux can be split into the parts

$$v_0 = \frac{k}{\mu} g \Delta\varrho, \qquad \Delta\varrho = \varrho_0\beta\Delta T \qquad \text{and} \qquad \Delta T = aH \qquad (14.172)$$

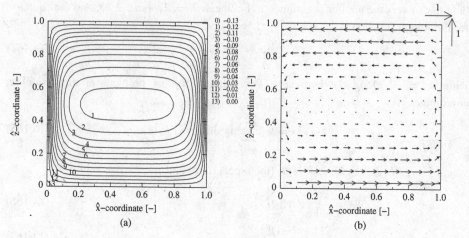

Figure 14.16. *Non-Rayleigh convection in a closed rectangular aquifer with aspect ratio $\epsilon = 0.2$. (a) The streamline solution (14.165) for $\Psi_0 = 1$. (b) The Darcy flow field (14.169)-(14.170) for $v_0 = 1$.*

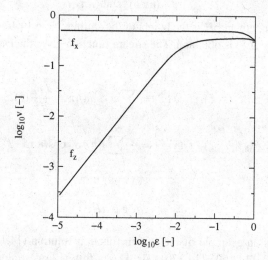

Figure 14.17. *The flux components v_x/v_0 and v_z/v_0 at positions $(1/2, 0)$ and $(0, 1/2)$, respectively, as a function of the aspect ratio ϵ.*

which shows that the fluid flow is driven by the density difference $\Delta\varrho = \varrho_0\beta\Delta T$ caused by the temperature difference $\Delta T = aH$. The characteristic flux is $v_0 = 9450$ m/Ma for a layer that is $H = 100$ m thick, has a permeability $k = 10^{-12}$ m^2 and gradient $a = 0.01°$C/m. The last parameter says that the temperature increases by 1°C over a horizontal distance of 100 m. The other parameters are $\mu = 10^{-3}$ Pa s, $g = 10$ m s^{-2} and $\varrho_0 = 1000$ kg m^{-3}. Although the Darcy flux is low in this example, it is often much larger than the flux from compaction fluid flow. Aquifers that drain compaction fluid may therefore be dominated by this weak type of thermal convection flow. It is possible to model the combined effect of (weak) convection fluid flow and compaction fluid flow by superposing the two respective solutions from the same linear and stationary pressure equation.

Note 14.10 We want to solve the Poisson equation $\nabla^2 \Psi = 1$ for a rectangular domain with length L and height H with $\Psi = 0$ along the boundaries. The equation is solved by expressing the right-hand side as a Fourier series

$$\sum_{n=1}^{\infty} a_n \sin(\lambda_n x) = 1 \tag{14.173}$$

and where the stream function is expressed by a similar Fourier series

$$\Psi(x, z) = \sum_{n=1}^{\infty} a_n \sin(\lambda_n x) F_n(z). \tag{14.174}$$

The Fourier coefficient then have to be

$$a_n = \frac{2}{n\pi}\left(1 - (-1)^n\right) \tag{14.175}$$

in order for the right-hand side (14.173) to be fulfilled. The stream function (14.174) fulfills the boundary condition $\Psi = 0$ along the vertical boundaries when $\lambda_n = n\pi/L$. Inserting the stream function (14.174) into the Poisson equation leads to the left-hand side

$$\nabla^2 \Psi = \sum_{n=1}^{\infty} a_n\left(-\lambda_n^2 F_n(z) + F_n''(z)\right) \sin(\lambda_n z) \tag{14.176}$$

which gives

$$F_n'' - \lambda_n^2 F_n = 1 \tag{14.177}$$

when it is equal to the right-hand side (14.173). The solution of this equation is

$$F_n(z) = c_1 e^{\lambda_n z} + c_2 e^{-\lambda_n z} - \frac{1}{\lambda_n^2} \tag{14.178}$$

where c_1 and c_2 are two coefficients that are obtained from the boundary conditions $F_n(0) = 0$ and $F_n(H) = 0$. These coefficients are

$$c_1 = \frac{(1 - e^{-\lambda_n H})}{2\lambda_n^2 \sinh(\lambda_n H)} \quad \text{and} \quad c_2 = -\frac{(1 - e^{\lambda_n H})}{2\lambda_n^2 \sinh(\lambda_n H)} \tag{14.179}$$

which gives expression (14.168) for function F_n.

Note 14.11 A layer is often rotated, and the rectangular geometry of a layer is simplified in a coordinate system that is rotated with the layer. It is straightforward to work in the rotated coordinate system since the Poisson equation (14.164) remains unchanged when going from the unrotated to the rotated coordinate system. The reason is that the Laplace operator is a scalar (rotation invariant) as shown in Exercise 2.15.

14.14 Thermal convection

The fluid density (14.153) decreases when the temperature increases. If the temperature increases with depth we get a situation where a relatively dense fluid is lying on top of a less dense fluid. This situation may become unstable, in which case we get thermal convection. Convection is in terms of circular rolls or cells where the fluid is heated at the base of the cell. The "hot" and less dense fluid ascends, then cools and becomes more dense, and finally descends to complete the cycle. The fluid flow has a strong feedback on the temperature field, which becomes substantially perturbed from the conduction-dominated stationary state. Figure 14.18 shows an example of convection cells.

This section looks at when thermal convection is possible in a sealed horizontal layer heated from below. The technique used is linear stability analysis, which gives a condition for when an infinitesimal perturbation of the stationary temperature field starts to grow. The basis for the problem is a temperature equation that is coupled to the Darcy flow. Equation (14.162) for the stream function

$$\nabla^2 \Psi = \frac{k}{\mu} \varrho_0 \, g \, \beta \frac{\partial T}{\partial x} \tag{14.180}$$

shows how the temperature controls the stream function and thereby the flow field. The derivatives (14.160)–(14.161) of the stream function give the Darcy flux, and when the Darcy flux components are inserted into the temperature equation (6.15) we get

$$\varrho_b c_b \frac{\partial T}{\partial t} + \varrho_f c_f \left(\frac{\partial \Psi}{\partial z} \frac{\partial T}{\partial x} - \frac{\partial \Psi}{\partial x} \frac{\partial T}{\partial z} \right) - \lambda \nabla^2 T = 0. \tag{14.181}$$

Equations (14.180) and (14.181) are a pair of coupled equations for the stream function and the temperature. These equations are also non-linear because of the convection term. Boundary conditions for a sealed layer are $\Psi = 0$ along the boundaries (see equation (14.163)). The temperature is T_1 along the top surface and T_2 along the base. The fluid in the layer is initially at rest and the initial stream function is therefore $\Psi_I = 0$. The heat flow is initially by conduction and the temperature increases linearly with depth,

$$T_I(z) = (T_2 - T_1) \left(\frac{z - z_1}{z_2 - z_1} \right) + T_1 \tag{14.182}$$

where z_1 and z_2 are the depth of the top and the base of the layer, respectively.

The next step is now to make a dimensionless version of the coupled equations (14.180) and (14.181). Time is scaled with the characteristic time for heat conduction $t_0 = H^2/\kappa$ as $\hat{t} = t/t_0$, where $\kappa = \lambda/\varrho_b c_b$ is the thermal diffusivity. The length is scaled as $\hat{x} = x/H$ and $\hat{z} = z/H$ with the thickness of the layer H. The Darcy flux is scaled with the characteristic velocity $v_0 = H/t_0$, and the stream function is therefore scaled with characteristic stream value $\Psi_0 = \kappa$. The dimensionless stream function and temperature are

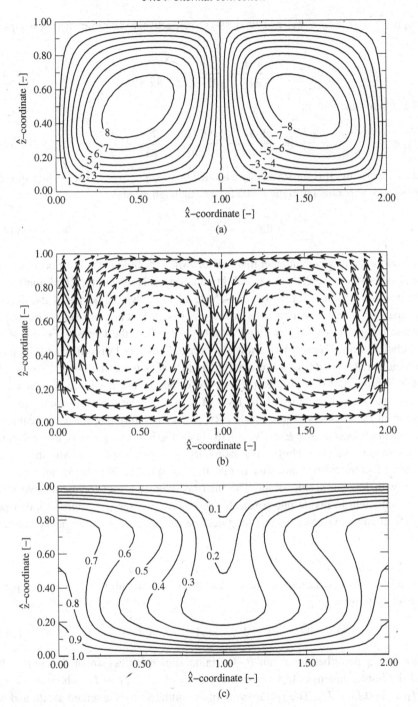

Figure 14.18. *Numerical solution for Rayleigh convection with* Ra $= 200$ *in a closed rectangular aquifer with aspect ratio* $\epsilon = 0.5$. *(a) The streamlines. (b) Darcy flux. (c) The temperature field.*

therefore $\hat{T} = T/(T_2 - T_1)$ and $\hat{\Psi} = \Psi/\kappa$, respectively. The dimensionless equations are

$$\frac{\partial^2 \hat{\Psi}}{\partial \hat{x}^2} + \frac{\partial^2 \hat{\Psi}}{\partial \hat{z}^2} = \text{Ra} \frac{\partial \hat{T}}{\partial \hat{x}} \tag{14.183}$$

for the stream function and

$$\frac{\partial \hat{T}}{\partial t} + \frac{\partial \hat{\Psi}}{\partial \hat{z}} \frac{\partial \hat{T}}{\partial \hat{x}} - \frac{\partial \Psi}{\partial \hat{x}} \frac{\partial \hat{T}}{\partial \hat{z}} = \frac{\partial^2 \hat{T}}{\partial \hat{x}^2} + \frac{\partial^2 \hat{T}}{\partial \hat{z}^2} \tag{14.184}$$

for the temperature. It is for simplicity assumed that $\varrho_b c_b = \varrho_f c_f$. The scaling shows that there is just one parameter in the problem, the Rayleigh number

$$\text{Ra} = \frac{k\beta \varrho_0 g (T_2 - T_1) H}{\kappa \mu}. \tag{14.185}$$

It is therefore the Rayleigh number that controls when the fluid in the layer becomes unstable and starts to circulate. A linear stability analysis tells us that the condition for onset of thermal convection is $\text{Ra} > 4\pi^2$, and Note 14.12 shows the details of the linear stability analysis. The convection cells at the onset of convection are stationary and have the same length as height. For "high" Rayleigh numbers the thermal convection turns into turbulent flow. Figure 14.18 shows the streamlines, the Darcy flux and the temperature field for $\text{Ra} = 200$ in a box with a length that is twice the height. The Rayleigh number is proportional to the temperature difference across the layer, the thickness of the layer and the permeability of the layer. It is in particular these three parameters that differ for different layers. The condition $\text{Ra} > 4\pi^2$ is rarely met for a permeable layer in sedimentary basins, see Exercise 14.13. One possible exception is thermal convection that might take place above magmatic intrusions. One reason for why the condition for onset of convection is not met is that thick permeable formations are often interbedded with low-permeable strata. The inter-bedding of a thin low-permeable stratum may be sufficient to divide one permeable layer into two separate units. Another possibility for thermal convection is that the thickness of the layer is large, maybe several km thick.

Note 14.12 *Linear stability analysis.* We are seeking a condition for when a small perturbation of the initial (stationary) state starts to grow. The stream function and temperature are therefore written as

$$\hat{\Psi} = \hat{\Psi}_I + \psi \qquad \text{and} \qquad \hat{T} = \hat{T}_I + \theta \tag{14.186}$$

in terms of the perturbations ψ and θ. The initial dimensionless stream function is $\hat{\Psi}_I = 0$ and the initial dimensionless temperature is $\hat{T}_I = \hat{z} - \hat{z}_1 + \hat{T}_1$, where $\hat{z}_1 = z_1/H$ and $\hat{T}_1 = T_1/(T_2 - T_1)$. The perturbation of the initial state is assumed small, and when the temperature and the stream function (14.186) are inserted into equations (14.183) and (14.184) we get that the non-linear convection term is, to the first order,

$$\frac{\partial \hat{\Psi}}{\partial \hat{z}}\frac{\partial \hat{T}}{\partial \hat{x}} - \frac{\partial \hat{\Psi}}{\partial \hat{x}}\frac{\partial \hat{T}}{\partial \hat{z}} = \frac{\partial \psi}{\partial \hat{z}}\frac{\partial \theta}{\partial \hat{x}} - \frac{\partial \psi}{\partial \hat{x}}\left(1 + \frac{\partial \theta}{\partial \hat{z}}\right) \approx -\frac{\partial \psi}{\partial \hat{x}}. \tag{14.187}$$

The coupled equations to the first order in the perturbations are therefore

$$\frac{\partial^2 \psi}{\partial \hat{x}^2} + \frac{\partial^2 \psi}{\partial \hat{z}^2} = \text{Ra}\frac{\partial \theta}{\partial \hat{x}} \tag{14.188}$$

and

$$\frac{\partial \theta}{\partial \hat{t}} - \frac{\partial \psi}{\partial \hat{x}} = \frac{\partial^2 \theta}{\partial \hat{x}^2} + \frac{\partial^2 \theta}{\partial \hat{z}^2}. \tag{14.189}$$

The linearized equations for the perturbation can be solved by separation of variables, and we are seeking solutions of the form

$$\psi = \cos(k\hat{x})f(\hat{z})e^{\lambda \hat{t}} \quad \text{and} \quad \theta = \cos(k\hat{x})g(\hat{z})e^{\lambda \hat{t}}. \tag{14.190}$$

When the choices for ψ and θ are inserted into equations (14.188) and (14.189) we get the following equations for the functions f and g:

$$f'' - k^2 f = \text{Ra}\,kg \tag{14.191}$$

$$\lambda g + kf = g'' - k^2 g. \tag{14.192}$$

These equations are solved by letting $f = A\sin(\pi\hat{x})$ and $g = B\sin(\pi\hat{x})$, where the choices for f and g fulfill the boundary conditions $f = 0$ and $g = 0$ for $\hat{z} = 0, 1$. Both the perturbation of the stream function and the temperature are zero along the top surface and the base of the layer. Inserting f and g into (14.191) gives

$$A = -\frac{\text{Ra}\,k}{\pi^2 + k^2}B \tag{14.193}$$

and similarly when inserted into (14.192):

$$A = -\frac{\pi^2 + k^2 + \lambda}{k}B. \tag{14.194}$$

The coefficients A and B can be eliminated to give the following equation for the parameter λ:

$$\pi^2 + k^2 + \lambda = \frac{\text{Ra}\,k^2}{\pi^2 + k^2}. \tag{14.195}$$

A perturbation of the form (14.190) begins to grow exponentially when $\lambda > 0$ and it dies out when $\lambda < 0$. The onset of convection takes place for $\lambda = 0$, which is for the Ra-number

$$\text{Ra} = \frac{(\pi^2 + k^2)^2}{k^2} \tag{14.196}$$

and shows that the Ra-number for onset of convection is different for different wave numbers. An arbitrary perturbation contains a spectrum of wave numbers and the wave number that begins to grow at a lowest possible Ra-number is $k = \pi$, see Figure 14.19. The critical

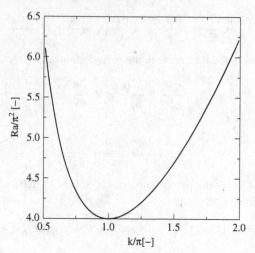

Figure 14.19. Ra/π^2 *as a function of k/π.*

Ra-number for $k = \pi$ is $4\pi^2$. (See Exercise 14.15.) The wave number $k = \pi$ corresponds to cells with the same width as height.

Exercise 14.13

(a) What is the Ra-number of a layer of thickness $H = 100$ m, permeability $k = 10^{-13}$ m^2, and a temperature difference $T_2 - T_1 = 4\,°$C? The other parameters are $\beta = 2.5 \cdot 10^{-5}$ K^{-1}, $\varrho_0 = 1000$ kg m^{-3}, $\kappa = 10^{-6}$ m^2 s^{-1}, $\mu = 10^{-3}$ Pa s and $g = 10$ m s^{-2}.
(b) Is the Ra-number sufficient for thermal convection?

Exercise 14.14 Show that the layer thickness at the onset of thermal convection is

$$H = 2\pi \left(\frac{\kappa \mu}{k\beta\varrho_0 g \, dT/dz} \right)^{1/2} \tag{14.197}$$

when the thermal gradient is dT/dz.

Exercise 14.15 Show that the Ra-number (14.196) has a minimum for $k = \pi$, which is $Ra = 4\pi^2$.

14.15 Further reading

The paper *Hydrodynamics of sedimentary basins* by Bredehoeft (2002) gives a non-mathematical view of fluid flow in sedimentary basins, while the paper *Basin-scale hydrogeological model* by Person *et al.* (1996) is a more technical review of hydrodynamics on a basin scale. The book *Dynamics of Fluids in Porous Media* by Bear (1972) treats analytically a large number of examples of fluid flow in porous media. A more applied book is *Physical and Chemical Hydrogeology* by Domenico and Schwartz (1998), which covers a large range of topics related to fluid flow in porous media. *Origin and Prediction of Abnormal Formation Pressure* by Chilingar *et al.* (2002) deals with over-pressured formations both with respect to evaluation, drilling, completing and hydrocarbon production. *Convection in Porous Media* by Nield and Bejan (1998) accounts for the theory behind convection in general (not only thermal), and treats a variety of applications. Wangen (2001) discusses overpressure build-up in terms of source terms for fluid expulsion and the gravity number.

15

Wells

Much of what we know about fluid flow in the sub-surface is obtained from pressure measurements in wells. This chapter looks at pressure and fluid flow in aquifers caused by fluid production and injection by wells.

15.1 Stationary pressure from a well

It is possible to make a simple model of stationary fluid flow around a well by assuming a horizontal aquifer of constant thickness and a constant permeability, see Figure 15.1. The stationary pressure then becomes the solution of a pressure equation in cylinder coordinates. Mass conservation with cylinder symmetry is

$$\frac{v_r}{r} + \frac{\partial v_r}{\partial r} = 0, \tag{15.1}$$

where v_r is the radial Darcy flux at the radius r. This equation is derived in Exercise 15.1. The radial Darcy flux is $v_r = -(k/\mu)\partial p/\partial r$ where k is a constant permeability in the horizontal plane of the aquifer, μ is the viscosity, and p is the fluid pressure. Inserting the Darcy flux into equation (15.27) for conservation of fluid yields the pressure equation

$$\frac{1}{r}\frac{\partial p}{\partial r} + \frac{\partial^2 p}{\partial r^2} = 0, \tag{15.2}$$

assuming that both the permeability and viscosity are constant. This pressure equation can be rewritten as

$$\frac{1}{r}\frac{\partial}{\partial r}\left(r\frac{\partial p}{\partial r}\right) = 0, \tag{15.3}$$

which is seen by carrying out the differentiation. We see that the factor $1/r$ does not matter, and we therefore get

$$r\frac{\partial p}{\partial r} = a, \tag{15.4}$$

where a is an integration constant. The pressure $p(r)$ is found by the next integration, which gives

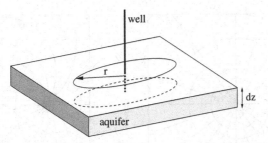

Figure 15.1. *A horizontal aquifer of thickness dz with a vertical well. The aquifer is assumed to be sealed above and below. Fluid is injected or produced at the same rate along the well inside the aquifer, and the fluid pressure is assumed to be only a function of the radius and the time.*

$$p(r) = a \ln(r) + b. \tag{15.5}$$

The integration constants a and b can be obtained from two pressure measurements, p_1 and p_2 at the radii r_1 and r_2, respectively. The stationary pressure then becomes

$$p(r) = (p_2 - p_1) \frac{\ln(r/r_1)}{\ln(r_2/r_1)} + p_1 \tag{15.6}$$

which shows that the stationary pressure surrounding a well depends on the radius as $\ln r$.

The integration constants can also be obtained by using the flow rate at the well instead of the pressure. The total injection (or production) rate is then (in units of volume per unit time)

$$Q = -2\pi r \, dz \frac{k}{\mu} \frac{\partial p}{\partial r} = -2\pi \, dz \frac{k}{\mu} a \quad \text{or} \quad a = -\frac{\mu Q}{2\pi \, dz \, k}. \tag{15.7}$$

Q is obtained by multiplication of the Darcy flux v_r with the area of the cylinder, $2\pi r \, dz$. (Recall that the Darcy flux is the volume of fluid per unit area and unit time.) Q is therefore the volume of fluid passing through the walls of the cylinder per time. The solution for the stationary pressure is

$$p(r) = -\frac{\mu Q}{2\pi \, dz \, k} \ln\left(\frac{r}{r_1}\right) + p_1 \tag{15.8}$$

where p_1 is the pressure at the radius r_1. Notice that $Q > 0$ for fluid injection into the aquifer, because the fluid pressure then decreases away from the well, and $Q < 0$ for fluid production, because the fluid pressure increases away from the well. The pressure (15.8) gives the permeability as

$$k = \frac{\mu Q \ln(r_1/r_2)}{2\pi \, dz \, (p_2 - p_1)} \tag{15.9}$$

from two pressure measurements p_1 and p_2 at the radii r_1 and r_2, respectively. Equation (15.9) makes it possible to estimate the permeability of the aquifer using two pressure measurements.

The pressure (15.8) around a single well, where $r = (x^2 + y^2)^{1/2}$, is also a solution of the Laplace equation. (See Exercise 15.2.) Recall that the Laplace equation is an expression for

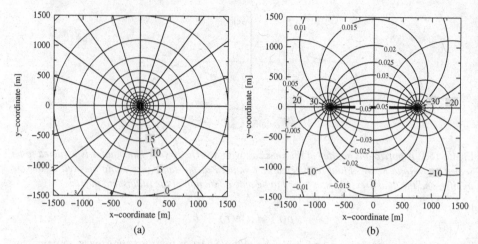

Figure 15.2. *(a) The isobars (MPa) and the streamlines ($m^2\,s^{-1}$) around a single (injector) well.*
(b) The isobars (MPa) and the streamlines ($m^2\,s^{-1}$) around two wells, an injector and a producer.

mass conservation of an incompressible fluid in a homogeneous aquifer, see Section 3.22.
The Laplace equation is linear, which means that any linear combination of solutions is
also a solution. The pressure (15.8) around a single well can therefore be generalized to a
solution for the pressure field from several wells as follows:

$$p(x, y) = -\frac{\mu}{2\pi\,dz\,k} \sum_{i=1}^{N} Q_i \ln\left(\frac{r_i(x, y)}{r_i(x_0, y_0)}\right) + p_0 \qquad (15.10)$$

where

$$r_i(x, y) = \left((x - x_i)^2 + (y - y_i)^2\right)^{1/2} \qquad (15.11)$$

is the distance from position (x, y) to well number i, which is found at position (x_i, y_i).
The total number of wells is N. The flow rate in well i is Q_i, and the pressure at position
(x_0, y_0) is p_0. It is straightforward to verify that each term in the pressure (15.10) is a
solution of the Laplace equation and that the pressure at position (x_0, y_0) is p_0.

Figure 15.2a shows the circular pressure contours of solution (15.8) around a single
injection well. The injection rate is $Q = 1\ m^3\,s^{-1}$, the aquifer has a thickness $dz = 10$ m,
the aquifer permeability is $k = 10^{-12}\ m^2$ and the water viscosity is $\mu = 10^{-3}$ Pa s.
The pressure is fixed at $p_1 = 0$ at the radius $r_1 = 1500$ m, and the pressure is
$p = \mu Q/(2\pi\,dz\,k) = 15.9$ MPa at the radius $r = \exp(-1)\,r_1 = 552$ m (as seen from
Figure 15.2a).

Exercise 15.1

(a) Derive equation (15.1) for fluid conservation with cylinder symmetry. (Hint: look at
Figure 6.13 and rewrite equation (6.116) for a circle rather than a sphere.)
(b) Assume that the fluid is compressible with the compressibility α and that the aquifer has
a porosity ϕ. What is the equation for fluid conservation? (The answer is equation (15.27).)

Exercise 15.2 Show that the pressure (15.8), where $r = (x^2 + y^2)^{1/2}$ is a solution of the Laplace equation $\nabla^2 p = 0$.

15.2 Wells and streamlines

Equation (15.10) in the previous section gives the pressure field from any number of wells in an homogeneous aquifer (constant permeability and thickness). Darcy's law then gives the corresponding flow field:

$$u_x = -\frac{k}{\mu}\frac{\partial p}{\partial x} = \frac{1}{2\pi \, dz}\sum_{i=1}^{N} Q_i \frac{(x - x_i)}{r_i^2} \tag{15.12}$$

$$u_y = -\frac{k}{\mu}\frac{\partial p}{\partial y} = \frac{1}{2\pi \, dz}\sum_{i=1}^{N} Q_i \frac{(y - y_i)}{r_i^2}. \tag{15.13}$$

The flow is very well visualized by streamlines, and a stream function can be obtained from the Darcy flux as shown in Section 3.22, since the flow field is stationary in 2D. The stream function becomes

$$\Psi = \int u_x \, dy$$

$$= \frac{1}{2\pi \, dz}\sum_{i=1}^{N} Q_i \int \frac{(x - x_i)}{(x - x_i)^2 + (y - y_i)^2} \, dy$$

$$= \frac{1}{2\pi \, dz}\sum_{i=1}^{N} Q_i \tan^{-1}\left(\frac{y - y_i}{x - x_i}\right) \tag{15.14}$$

using the integral in Note 15.1. Figure 15.2a shows the streamlines around a single injection well. The difference in value between the stream function for two streamlines is $\Delta\Psi = 0.005$ m^2 s^{-1}. There are 20 streamtubes around the well in Figure 15.2a, which give that the total flow into the well is $Q = 20\Delta\Psi \, dz = 1$ m^3 s^{-1}, as expected.

Figure 15.2b shows a production well and an injection well where production and injection take place with the same volume rate. The pressure solution (15.10) then becomes

$$p(x, y) = \frac{Q_0\mu}{4\pi \, dz \, k}\ln\left(\frac{(x + d)^2 + y^2}{(x - d)^2 + y^2}\right) \tag{15.15}$$

when the pressure is zero at $(0, 0)$, $Q_1 = -Q_0$, $Q_2 = Q_0$, $x_1 = -d$, $x_2 = d$ and $y_1 = y_2 = 0$. Figure 15.2b shows that the isobars seems to be circles. Note 15.2 verifies that the isobar of a pressure (p_c) in this case is a circle, and that its radius is $r = 2d\sqrt{c}/(1 - c)$ and its center is at $(-d(1+c)/(1-c), 0)$, where $\ln c = p_c \, 4\pi \, dz \, k/(\mu Q)$. The parameters in Figure 15.2b are $Q_0 = 1$ m^3 s^{-1}, $dz = 10$ m, $k = 10^{-12}$ m^2, $\mu = 10^{-3}$ Pa s and $d = 750$ m. The c-parameter that defines the contours for the pressure $p_c = 20$ MPa is

$c = 12.3$, and the radius of the circle is $r = 465$ m and the center along the x-axis is at $x = -882$ m. A look at the figure verifies this contour.

The stream function (15.14), for this case, is

$$\Psi = \frac{Q_0}{2\pi\, dz} \left(\tan^{-1}\left(\frac{y}{x-d}\right) - \tan^{-1}\left(\frac{y}{x+d}\right) \right). \tag{15.16}$$

These streamlines are also circular and they are orthogonal to the isobars, as shown in Section 3.22 for 2D stationary flow. Note 15.3 show that a streamline with stream function Ψ_c is a circle with radius $r = d\sqrt{1+1/b}$ and center at position $(0, -d/b)$, where $b = -\tan(2\pi\, dz\, \Psi_c/Q_0)$. The streamlines in Figure 15.2b can be checked in the same way as the isobars. The b-parameter for the stream function $\Psi_c = -0.02$ m^2 s^{-1} is $b = 3.1$, which gives a radius $r = 863$ m of the (circular) streamline and a center along the y-axis at $y = -244$ m.

Note 15.1

$$\int \frac{du}{a^2 + u^2} = \frac{1}{a}\tan^{-1}\left(\frac{u}{a}\right). \tag{15.17}$$

Note 15.2 The pressure (15.15) is rewritten as

$$\frac{(x+d)^2 + y^2}{(x-d)^2 + y^2} = c \tag{15.18}$$

where $c = \exp(4\pi\, dz\, k\, p_c/\mu Q)$. A little more work leads to

$$\left(x + \left(\frac{1+c}{1-c}\right)d\right)^2 + y^2 = \frac{4c}{(1-c)^2}d^2 \tag{15.19}$$

which is the equation for a circle with radius $r = 2d\sqrt{c}/(1-c)$ and center on the x-axis at $x = -d\,(1+c)/(1-c)$.

Note 15.3 The stream function (15.16) is rewritten as

$$\frac{\dfrac{y}{x-d} - \dfrac{y}{x+d}}{1 + \dfrac{y}{(x-d)}\dfrac{y}{(x+d)}} = \tan\left(\frac{\Psi_c\, 2\pi\, dz}{Q_0}\right) = -b \tag{15.20}$$

using that $\tan(\alpha - \beta) = (\tan\alpha - \tan\beta)/(1 + \tan\alpha\,\tan\beta)$. A few more steps lead to $2dy/(x^2 - d^2 + y^2) = -b$, which is rewritten as

$$x^2 + \left(y + \frac{d}{b}\right)^2 = d^2\left(1 + \frac{1}{c}\right). \tag{15.21}$$

This is the equation for a circle with radius $r = d\sqrt{1+1/b^2}$ and center on the y-axis at position $y = -d/b$.

15.3 The skin factor

The skin factor is a dimensionless number that tells us about inhomogeneities in the rock around a well. It is based on a simple model where the rock in a cylindrical area around the well has a different permeability than in the aquifer further away, see Figure 15.3. The rock between the well radius r_w and the radius r_f has the permeability k_f, and the rock outside radius r_f has the permeability k. Let p_w be the well pressure, and p_f and p the pressure at the radii r_f and r, respectively, where the radius r is greater than r_f. We have from equation (15.8) that the stationary pressures at these radii are

$$p - p_f = -\frac{a}{k}\ln\left(\frac{r}{r_f}\right) \tag{15.22}$$

$$p_f - p_w = -\frac{a}{k_f}\ln\left(\frac{r_f}{r_w}\right) \tag{15.23}$$

where $a = \mu\, Q/(2\pi\, dz)$. We can use these two relations to eliminate pressure p_f, and we get

$$p - p_w = -\frac{a}{k}\left(\ln\left(\frac{r}{r_f}\right) + \frac{k}{k_f}\ln\left(\frac{r_f}{r_w}\right)\right). \tag{15.24}$$

This expression can be rewritten in the familiar form of a stationary pressure solution

$$p - p_w = -\frac{a}{k}\left(\ln\left(\frac{r}{r_w}\right) + S\right) \tag{15.25}$$

where the offset S is the skin factor

$$S = \left(\frac{k}{k_f} - 1\right)\ln\left(\frac{r_f}{r_w}\right). \tag{15.26}$$

We used that $\ln(r/r_f) = \ln(r/r_w) - \ln(r_f/r_w)$ when going from equation (15.24) to (15.25). We see that the skin factor is a negative number if $k_f > k$ and that it is a positive number when $k_f < k$, and the skin factor is zero when $k_f = k$. Figure 15.4 shows how the skin factor can be obtained from pressure observations at different radii. A straight line is fitted to pressure observations plotted as $\ln(r/r_w)$. The position where this line intersects the horizontal $p = p_w$ is the negative of the skin factor. The skin factor is often a negative

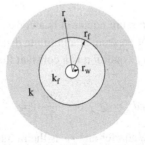

Figure 15.3. *A well with a different permeability close to the well.*

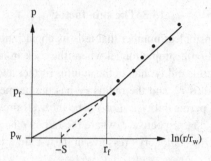

Figure 15.4. *The figure shows pressure observations at different radii from a well, plotted as* $\ln(r/r_w)$, *where r_w is the well radius. The solid line shows the real pressure, and the dotted line is an extrapolation of the pressure observations. The skin factor is minus the value of $\ln(r/r_w)$ where the extrapolated pressure is p_w (the well pressure).*

number because damage to the rocks from the drilling increases the permeability close to the well.

15.4 Transient pressure from a well

The pressure will be time-dependent when a compressible fluid is produced or injected into an aquifer of compressible rock. Fluid conservation in cylindrical coordinates is then

$$-\phi\alpha\frac{\partial p}{\partial t} = \frac{v_r}{r} + \frac{\partial v_r}{\partial r} \tag{15.27}$$

as shown in Exercise 15.1, where v_r is the radial Darcy flux at the radius r, p is the fluid pressure, ϕ is the porosity and α is the total compressibility of fluid and rock. (See Section 13.3 for the storage coefficient $\phi\alpha$.) Figure 15.1 shows a sketch of a horizontal aquifer with a vertical well. It also shows a cylinder with radius r from the center of the well. The radial Darcy flux is $v_r = -(k/\mu)\partial p/\partial r$ where k is a constant isotropic permeability of the aquifer and μ is the viscosity. Inserting the Darcy flux into equation (15.27) for fluid conservation yields the pressure equation

$$\frac{\phi\alpha\mu}{k}\frac{\partial p}{\partial t} = \frac{1}{r}\frac{\partial p}{\partial r} + \frac{\partial^2 p}{\partial r^2}. \tag{15.28}$$

The initial condition for this pressure equation is a constant aquifer pressure p_0 and one boundary condition is a constant pressure at an infinite distance away from the well, $p(r = \infty, t) = p_0$. The other boundary condition is a constant flow rate Q into the aquifer from the well:

$$-\frac{k}{\mu}\frac{\partial p}{\partial r}(r{=}r_w) \cdot 2\pi r_w\, dz = Q \quad \text{or} \quad \frac{\partial p}{\partial r}(r{=}r_w) = -\frac{\mu Q}{2\pi r_w k\, dz} \tag{15.29}$$

where dz is the thickness of the aquifer and r_w is the radius of the well. Boundary condition (15.29) shows that the rate Q has units m^3 s^{-1}. Notice that the volume rate is $Q > 0$

for fluid injection and that it is $Q < 0$ for fluid production. ($Q > 0$ implies that $\partial p/\partial r < 0$ and $v_r > 0$, which says that fluid is flowing into the aquifer.) The solution to the pressure equation (15.28) is

$$p(r, t) = \frac{\mu Q}{4\pi k\, dz} \operatorname{E_1}\left(\frac{\phi \alpha \mu r^2}{4kt}\right) + p_0 \tag{15.30}$$

as shown in Note 15.4, where $\operatorname{E_1}(x)$ is the exponential integral function defined as

$$\operatorname{E_1}(x) = \int_x^\infty \frac{e^{-u}}{u}\, du. \tag{15.31}$$

Time t in the solution (15.30) is the time since the rate Q was turned on. The pressure solution (15.30) is often referred to as the Theis equation, because it first originated in the work of Theis (1935). It is often seen in the literature that the pressure solution (15.30) is written using a related exponential integral function denoted Ei, which yields $\operatorname{E_1}$ by the relationship, $\operatorname{E_1}(x) = -\operatorname{Ei}(-x)$.

It is useful to study a dimensionless version of the solution (15.30) for the pressure as a function of time at a fixed radius. We see that the factor

$$p_c = \frac{\mu Q}{4\pi k\, dz} \tag{15.32}$$

has units of pressure and it can therefore be used to define the dimensionless pressure $\hat{p} = (p - p_0)/p_c$. The dimensionless pressure is the pressure deviation from the initial constant pressure p_0 measured in units p_c. We can also define the characteristic time

$$t_r = \frac{\phi \alpha \mu r^2}{4\,k} \tag{15.33}$$

associated with the radius r. The pressure evolution at the fixed radius r is then

$$\hat{p}_r(\hat{t}) = \operatorname{E_1}(1/\hat{t}). \tag{15.34}$$

Figure 15.5 shows the dimensionless pressure as a function of dimensionless time. The pressure is $\hat{p}_r(\hat{t}) \approx 1$ for $\hat{t} = 4$ because $\operatorname{E_1}(1/4) = 1.04$, which means that $p \approx p_c$ for $t = t_r/4$.

When $t \gg t_r$ or when $x = 1/\hat{t} \ll 1$ the function $\operatorname{E_1}(x)$ can be approximated as

$$\operatorname{E_1}(x) \approx -0.577 - \ln(x) = -\ln(1.78\,x) = -2.3\,\log_{10}(1.78\,x) \tag{15.35}$$

(see Exercise 15.4). The solution for the pressure can therefore be approximated as

$$p(r, t) - p_0 \approx -\frac{\mu Q}{4\pi k\, dz} \ln\left(1.78 \cdot \frac{\phi \alpha \mu r^2}{4kt}\right) \tag{15.36}$$

or as dimensionless pressure at the fixed radius r as

$$\hat{p}_r(\hat{t}) \approx -0.577 + \ln(\hat{t}). \tag{15.37}$$

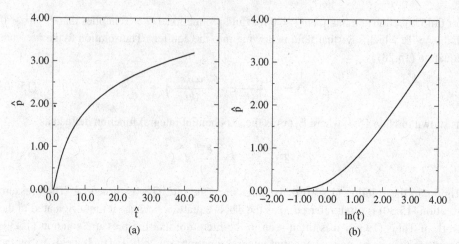

Figure 15.5. *(a) The dimensionless pressure solution (15.34) plotted as a function of $\hat{t}$. (b) The dimensionless pressure solution (15.34) plotted as a function of $\ln \hat{t}$. It is seen that pressure is almost a linear function of $\ln \hat{t}$ for $\ln \hat{t} > 2$, $(\hat{t} > e^2)$.*

The condition $t \gg t_r$ will first apply around the well, because the radius is at its minimum (r_w) there. The approximation (15.36) can normally be assumed in the vicinity of the well too. If we have two pressure measurements p_1 and p_2 at the same radius taken at times t_1 and t_2, respectively, we then get that the pressure difference is

$$\Delta p = p_2 - p_1 = p_c \ln\left(\frac{t_1}{t_2}\right) = -p_c \ln\left(\frac{t_2}{t_1}\right). \tag{15.38}$$

A series of pressure measurements p_i at times t_i should therefore become a line when plotted as a function of $\ln(t_i/t_1)$. It is seen that two pressure measurements give the characteristic pressure as

$$p_c = \frac{p_2 - p_1}{\ln(t_1/t_2)}. \tag{15.39}$$

From the definition (15.32) of the characteristic pressure we have that two pressure measurements also give the permeability, when we know the production/injection rate Q, the viscosity μ and the thickness of the aquifer. The permeability is then

$$k = \frac{\mu \, Q \ln(t_1/t_2)}{4\pi \, dz \, (p_2 - p_1)}. \tag{15.40}$$

This equation is the basis for well testing, where the permeability can be obtained from a series of pressure measurements forming a line when plotted as the logarithm of time. Notice that the permeability estimate (15.40) is similar to estimate (15.9) for stationary flow – the only difference is that the radii r_1 and r_2 have been replaced by the times t_1 and t_2.

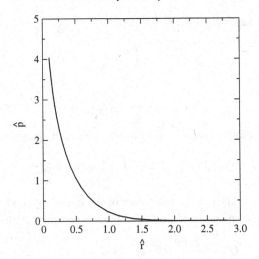

Figure 15.6. *The pressure into the aquifer as a function of $\hat{r}$ in the case of fluid injection. Notice that the fluid pressure is almost zero for $\hat{r} > 2$.*

The pressure (15.30) as a function of r at a fixed time is in dimensionless form

$$\hat{p}(\hat{r}) = \mathrm{E}_1\left(\frac{r^2}{r_t^2}\right) \tag{15.41}$$

where the characteristic radius at the time t is

$$r_t = \sqrt{\frac{4kt}{\phi\alpha\mu}}. \tag{15.42}$$

The pressure as a function of the radius is plotted in Figure (15.6) for an injection well, which shows that the pressure has decreased to almost zero for $\hat{r} > 2$.

Note 15.4 The pressure written as $p = \Theta(\eta)$ in terms of the similarity variable

$$\eta = \frac{r^2}{4\kappa t} \tag{15.43}$$

where $\kappa = k/(\phi\alpha\mu)$, solves the pressure equation (15.28) in cylindrical coordinates. Differentiation yields that

$$\frac{\partial p}{\partial t} = -\Theta' \cdot \frac{r^2}{4\kappa t^2} \tag{15.44}$$

$$\frac{\partial p}{\partial r} = \Theta' \cdot \frac{2r}{4\kappa t} \tag{15.45}$$

$$\frac{\partial^2 p}{\partial r^2} = \Theta'' \cdot \left(\frac{2r}{4\kappa t}\right)^2 + \Theta' \cdot \frac{2}{4\kappa t}. \tag{15.46}$$

Inserting these partial derivatives into the pressure equation (15.28) gives

$$(\eta + \eta^2)\Theta' + \eta^2\Theta'' = 0. \tag{15.47}$$

This equation can be rewritten with $f = \Theta'$ as

$$\frac{df}{f} = -\left(1 + \frac{1}{\eta}\right) d\eta \tag{15.48}$$

and integrated to

$$f = f_0 \exp(-\eta - \ln \eta) \tag{15.49}$$

where f_0 is an integration constant. Integration of f gives the pressure as a function of η

$$p = \Theta(\eta) = -f_0 \int_\eta^\infty \frac{e^{-u}}{u} du + p_0, \tag{15.50}$$

where the integration limit $\eta = \infty$ takes care of the boundary condition $p \to p_0$ when $r \to \infty$. The second boundary condition (15.29) gives the integration constant f_0:

$$\begin{aligned}
Q &= -\frac{k}{\mu} \frac{\partial p}{\partial r}(r=r_w) \cdot 2\pi r_w \, dz \\
&= -\frac{k}{\mu} f_0 \exp(-\eta - \ln \eta) \frac{2r_w}{4\kappa t} 2\pi r_w \, dz \\
&= -\frac{k}{\mu} 4\pi \, dz \, f_0 \exp(-\eta) \tag{15.51}
\end{aligned}$$

where η is evaluated for $r = r_w$. Assuming now that $t \gg r_w^2/(4\kappa)$ we get that $\exp(-\eta) \approx 1$, and that

$$f_0 = -\frac{\mu Q}{4\pi k \, dz} \tag{15.52}$$

and we have the pressure solution (15.30). We could also have approximated the well radius with $r_w = 0$ which implies that $\exp(-\eta) = 1$. Notice that the solution (15.30) is also approximate and does not apply unless $t \gg r_w^2/(4\kappa)$ or $r_w = 0$.

Exercise 15.3 The following table shows the pressure increase in an aquifer close to an injection well during several years of water injection. Use the pressure measurements to obtain the permeability in the aquifer when $\mu = 0.001$ Pa s, $Q = 1$ m^3 s^{-1} and $dz = 10$ m.

time (years)	well pressure (MPa)
2.8	33.0
4.1	36.2
5.5	38.4
6.9	40.2
8.3	41.6
9.6	42.8

Hint: use equation (15.40).

Exercise 15.4 Show that $E_1(x) \approx -0.57 - \ln(x) + x$ for $x \ll 1$.

Solution: A simple and approximate way is to write $E_1(x)$ as

$$E_1(x) = \int_x^{0.01} \frac{e^{-u}}{u} du + \int_{0.01}^{\infty} \frac{e^{-u}}{u} du \qquad (15.53)$$

where the integration limit 0.01 is any number much less than 1, but larger than x. The first integral can be approximated as

$$\int_x^{0.01} \frac{e^{-u}}{u} du \approx \int_x^{0.01} \frac{1-u}{u} du = \ln(0.01) - 0.01 - \ln(x) + x \qquad (15.54)$$

and the last integral is the constant $c = E_1(0.01) = 4.04$. By adding the two terms in equation (15.53) we get that $E_1(x) \approx -0.57 - \ln(x) + x$.

15.5 Well testing

Well testing involves methods for obtaining the permeability and the storativity of aquifers by measuring the pressure transients in production and injection wells. We have already seen from equation (15.40) how two pressure measurements at two different times give the permeability. It should be noted that the permeability is not measured directly, but it is obtained indirectly through a model for pressure transients in the aquifer. The model behind equation (15.40) assumes that the aquifer is non-leaky, that it has a well-defined thickness, an infinite extent and a constant isotropic permeability. The estimated permeability is not only a valuable estimate, but it is often the best possible estimate available, although the assumptions behind the model are idealizations of a real aquifer.

We have from equation (15.36) that the pressure in the aquifer at a given radius is approximated as

$$p(t) - p_0 = -p_c \ln\left(1.78 \cdot \frac{t_r}{t}\right) \qquad (15.55)$$

where the characteristic pressure p_c and the characteristic time t_r for the radius r are given by equations (15.32) and (15.33), respectively. Equation (15.39) shows how the characteristic pressure p_c is obtained from two pressure measurements at two different times. Once we have p_c we can also obtain the characteristic time t_r for the radius r. From equation (15.55) we get

$$t_r = 0.56 \cdot \exp\left(-\frac{p_1 - p_0}{p_c}\right) t_1 \qquad (15.56)$$

using the pressure measurement p_1 at time t_1. We could also have used any other pressure measurement p_i at time t_i. Recall that equation (15.56) assumes that $t_i \gg t_r$, and once t_r is estimated we can check that this condition is fulfilled. If the measurements are made in an observation well at a large distance from the production/injection well, then this condition may not apply. In that case one has to carry out curve matching against the

equation (15.28) is linear. The solution of the pressure equation for the recovery (after the well has been closed) is

$$p(r, t) - p_0 = p_c \, \mathrm{E}_1 \left(\frac{t_r}{t} \right) - p_c \, \mathrm{E}_1 \left(\frac{t_r}{t - t_s} \right), \quad t > t_s \qquad (15.59)$$

where t_s is the time when pumping is stopped. Both terms (on the right-hand side) solve the pressure equation (15.28) and they satisfy the boundary condition $p(r=\infty, t) = 0$. The first term has as boundary condition the flow rate Q, and the second term has as boundary condition the flow rate $-Q$. We get a solution with zero flow in the well by adding these terms together. Figure 15.7 shows the dimensionless version of solution (15.59). The pressure drop since production started is called the *draw down*, a term that has a literal meaning when the pressure drop is measured as a hydraulic head (in meters). The pressure solution (15.59) can also be approximated as

$$p(r, t) \approx p_c \ln \left(\frac{t}{t - t_s} \right) \qquad (15.60)$$

at a given radius using approximation (15.35). This equation is the basis for a well test called the *recovery test* or the *pressure build-up test*. In petroleum engineering it is also called the *drill stem test*.

Appendix

Fourier series, the discrete Fourier transform and the fast Fourier transform

A.1 Fourier series

There have been several occasions where we have needed to approximate a function with a Fourier series, for instance with heat flow, flexure, gravity and fluid flow. One particular example is the deflection of the lithosphere under a load, where the deflection can be obtained directly from the Fourier series of the load (see Section 9.4). Other examples are time-dependent pressure and temperature solutions where the Fourier series for initial conditions give the Fourier coefficients in the solution. A function $f(x)$ is then approximated as

$$f(x) = a_0 + \sum_{j}^{m} \left(a_j \cos(2\pi jx/L) + b_j \sin(2\pi jx/L) \right), \tag{A.1}$$

over the x-interval from 0 to L, where the Fourier coefficients are (see Note A.1)

$$a_0 = \frac{1}{L} \int_0^L f(x) \, dx \tag{A.2}$$

$$a_j = \frac{2}{L} \int_0^L f(x) \, \cos(2\pi jx/L) \, dx \tag{A.3}$$

$$b_j = \frac{2}{L} \int_0^L f(x) \, \sin(2\pi jx/L) \, dx. \tag{A.4}$$

These expressions for the Fourier coefficients follow from the orthogonality property of the cosine and sine functions. The approximation of a function with a Fourier series assumes that the function is periodic, $f(x + L) = f(x)$, and it repeats itself after a distance L along the x-axis. In a discussion of Fourier series it is useful to introduce the notion of *even* and *odd* functions. An even function has the property that $f(-x) = -f(x)$ and an odd function fulfills $f(-x) = f(x)$. We notice that sine and cosine are odd and even functions, respectively. An odd function $f(x)$ therefore is represented by a sin-series – the coefficients a_j in the Fourier series (A.1) are all zero. Similarly, an even function is a pure cos-series (where the coefficients a_j are all zero). Most functions encountered in examples are either even and odd, and we then know that the Fourier series is either a pure sine-series

502

or a pure cosine-series. Any function $f(x)$ can also be written as a sum of an even and an odd function, since

$$f(x) = \frac{1}{2}\big(f(x) + f(-x)\big) + \frac{1}{2}\big(f(x) - f(-x)\big) \tag{A.5}$$

where the first part is an even function and the second is odd. We therefore need both sine and cosine to approximate any function.

Note A.1 *Orthogonality*. The sine and cosine functions of Fourier series are mutually orthogonal functions, because we have the following relations:

$$\int_0^L \sin(2\pi mx/L)\,\cos(2\pi nx/L)\,dx = 0 \quad \text{for all } m \text{ and } n \tag{A.6}$$

$$\int_0^L \cos(2\pi mx/L)\,\cos(2\pi nx/L)\,dx = \begin{cases} L & \text{for } m = n = 0, \\ \frac{1}{2}L & \text{for } m = n > 0, \\ 0 & \text{for } m \neq n \end{cases} \tag{A.7}$$

$$\int_0^L \sin(2\pi mx/L)\,\sin(2\pi nx/L)\,dx = \begin{cases} 0 & \text{for } m = n = 0, \\ \frac{1}{2}L & \text{for } m = n > 0, \\ 0 & \text{for } m \neq n. \end{cases} \tag{A.8}$$

The functions $\cos(2\pi nx/L)$ and $\sin(2\pi mx/L)$ can therefore be regarded as an orthogonal basis with respect to the inner product

$$(f, g) = \int_0^L f(x)\,g(x)\,dx. \tag{A.9}$$

Multiplication of both sides of equation (A.1) with either $\cos(2\pi nx/L)$ or $\sin(2\pi mx/L)$ and then integration from $x = 0$ to $x = L$ gives the Fourier coefficients (A.2)–(A.4).

A.2 Interpolation with Fourier series

When working with real data we normally do not have a function – we have only a stream of data points (x_k, y_k) for $k = 0, \ldots, n$. The question we want to answer is how can we approximate such a data set like a curve with a Fourier series. In other words, how can we obtain the Fourier coefficients for a curve (or a function) represented by discrete data points? The answer is the fast Fourier transform (FFT), which is computationally very efficient. We will see that the approximation of a data set actually becomes interpolation, because the approximation goes through the points on the curve. Figure A.1 shows an example of Fourier interpolation of a simple piecewise linear curve with 16 and 32 points. A related question is the inverse problem – how to obtain the interpolated points of a curve from the Fourier coefficients. It is assumed that the point 0 is equal to point n of the curve, and that the curve can be viewed as one period of a periodic signal. The length of a period is then $L = x_n - x_0$, and the x-coordinates are $x_k = x_0 + k\,dx$ with $dx = L/n$ when the points are equally spaced along the x-axis. For simplicity we let

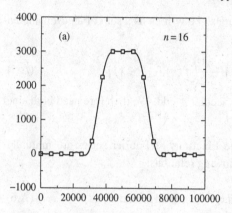
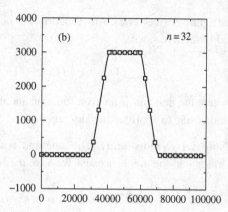

Figure A.1. *A piecewise linear curve is interpolated by a Fourier series with 16 points in (a) and 32 points in (b).*

$x_0 = 0$, and we are then looking for a Fourier series that interpolates the points (x_k, y_k) such that

$$y_k = a_0 + \sum_j^m \left(a_j \cos(2\pi j x_k/L) + b_j \sin(2\pi j x_k/L) \right), \quad k = 0, \ldots, n-1, \quad (A.10)$$

where $n = 2m + 1$. The case for an even n, which is similar, is treated below. We notice that equations (A.10) for the Fourier coefficients can be written

$$y_k = a_0 + \sum_j^m \left(a_j \cos(2\pi j k/n) + b_j \sin(2\pi j k/n) \right), \quad k = 0, \ldots, n-1 \quad (A.11)$$

which means that we don't have to know the x-coordinates or the length L of one period to obtain the coefficients. It is also seen from equation (A.10) that the number of Fourier coefficients a_j and b_j is the same as the number of points in the curve. (There are $m + 1$ coefficients a_j, m coefficients b_j and $2m + 1$ points along the curve.) There is therefore the same number of equations as there are coefficients (unknowns). A direct way to obtain the Fourier coefficients would simply be to solve the linear equation system (A.11). But such an approach becomes very computationally demanding when the number of points is large. It can be shown that the number of operations is $O(n^3)$ for this approach, where the capital O denotes the order of the number of operations. Only multiplication and division are counted as operations, because these two operations are much more time consuming than addition and subtraction. A better approach is to compute the coefficients a_j and b_j using the discrete Fourier transform.

A.3 The discrete Fourier transform

Although we want the Fourier coefficients in a series that interpolates the real values y_k of a curve, it turns out to be simpler to deal with the same problem assuming complex values y_k and complex Fourier coefficients c_j. The Fourier series (A.11) is written as

$$y_k = \sum_{j=0}^{n-1} c_j e^{i2\pi jk/n}, \quad k = 0, \ldots, n-1 \tag{A.12}$$

where $i = \sqrt{-1}$. The relationship between the exponential of a complex number and trigonometric functions is Euler's formula $e^{ix} = \cos x + i \sin x$. An important reason for dealing with a Fourier series in terms of complex numbers is the following property of exponentials:

$$\sum_{j=0}^{n-1} e^{i2\pi j(k-l)/n} = \begin{cases} n, & k = l \\ 0, & \text{otherwise} \end{cases} \tag{A.13}$$

which is shown in Exercise A.2. The property (A.13) can be used to obtain a series for the Fourier coefficients c_j in terms of the values y_k,

$$c_j = \frac{1}{n} \sum_{k=0}^{n-1} y_k e^{-i2\pi jk/n}, \quad l = 0, \ldots, n-1. \tag{A.14}$$

This expression for the Fourier coefficients is the *discrete Fourier transform* (DFT), which is the discrete version of the continuous Fourier transform. Exercise (A.3) shows how the DFT follows from the complex Fourier series (A.12) and the property (A.13). It is also seen that the computation of all n Fourier coefficients using the series (A.14) requires $O(n^2)$ operations, which is much better than $O(n^3)$ for the solution of the linear equation system. Once we have the Fourier coefficients c_j we can recover the interpolation points y_k using the series (A.12). The Fourier series (A.12) is therefore the *inverse* discrete Fourier transform. The DFT and the inverse DFT are similar series – only the factor $1/n$ and the sign in front of i in the exponent are different.

A.4 The fast Fourier transform

There is a really fast way to carry out the summation (A.14) in the DFT, which is the fast Fourier transform (FFT). The FFT needs only $O(n \log n)$ operations, but it assumes that $n = 2^p$ for a positive integer p. We will now see how we can utilize the FFT, where the Fourier coefficients are complex numbers, to obtain the real Fourier coefficients (a_j and b_j) for a curve represented by real values y_k. The inverse FFT gives the same series as the inverse DFT – the factor $1/n$ and the sign in front of i in the exponent are different. The complex Fourier coefficients c_l are not independent of each other when the values y_k are real numbers. We then have that $\bar{c}_{n-l} = c_l$ (which is shown in Exercise A.4). A bar above a symbol denotes complex conjugation. The relation $\bar{c}_{n-l} = c_l$ can be used to obtain the Fourier coefficients in the more familiar (real) series (A.11). The case with an odd number of points, $n = 2m + 1$, is treated first, where the complex Fourier series (A.12) becomes the real Fourier series

Table A.1. *The relation between the complex Fourier coefficients* c_j *and the real Fourier coefficients* a_i *and* b_i.

Even, $n = 2m$	Odd, $n = 2m + 1$
$a_0 = \text{Re}\, c_0$	$a_0 = \text{Re}\, c_0$
$b_0 = 0$	$b_0 = 0$
$a_j = 2\text{Re}\, c_j, \quad j = 1, \ldots, m-1$	$a_j = 2\text{Re}\, c_j, \quad j = 1, \ldots, m$
$b_j = -2\text{Im}\, c_j, \quad j = 1, \ldots, m-1$	$b_j = -2\text{Im}\, c_j, \quad j = 1, \ldots, m$
$a_m = \text{Re}\, c_m$	
$b_m = 0$	

$$y_k = a_0 + \sum_{j=1}^{m} \left(a_j \cos(2\pi jk/n) + b_j \sin(2\pi jk/n) \right) \tag{A.15}$$

where the Fourier coefficients are $a_0 = \text{Re}\, c_0$, $a_j = 2\text{Re}\, c_j$ and $b_j = -2\text{Im}\, c_j$, for $j = 1, \ldots, m$ (as shown in Exercise A.5). In the even case, when $n = 2m$, we have that

$$y_k = a_0 + \sum_{j=1}^{m-1} \left(a_j \cos(2\pi jk/n) + b_j \sin(2\pi jk/n) \right) + a_m \cos(2\pi mk/n) \tag{A.16}$$

where the Fourier coefficients are $a_0 = \text{Re}\, c_0$, $a_m = \text{Re}\, c_m$, $a_j = 2\text{Re}\, c_j$ and $b_j = -2\text{Im}\, c_j$, for $j = 1, \ldots, m-1$ (see Exercise A.6, which also shows that c_m is a real number). We notice that we have the same number of Fourier coefficients a_j and b_j as values y_k in both the odd and the even cases. These expressions for a_j and b_j in terms of c_j are easily inverted to expressions for c_j in terms of a_j and b_j. The relationship between the complex Fourier coefficients c_j and the real Fourier coefficients a_i and b_i are summarized in Table A.1.

Note A.2 *Octave.* The program Octave is an open source implementation of the basic features of Matlab, and it is available for all platforms (Octave, 2009). Octave and Matlab are very handy tools for numerical modeling, like for instance Fourier analysis. The following function returns the Fourier coefficients for a real input array y. It shows how the FFT-function stores the Fourier coefficients. We recognize the relation between the complex and the real Fourier coefficients given by table A.1, except for the factor $1/N$. The next function is an application that tests the Fourier coefficients by reproducing the input data with a wanted number of points. (These functions have only been tested under Octave.)

```
function [a0, an, bn] = make_fourier_coefs(y)
    c = fft(y);
    N = rows(y);
    M = floor(N / 2);
    an = zeros(M, 1);
```

```
      bn = zeros(M, 1);
      a0 = real(c(1)) / N;
      for i = 1:M
            an(i) =   2 * real(c(i+1)) / N;
            bn(i) = - 2 * imag(c(i+1)) / N;
      endfor
      if (rem(N, 2) == 0)
            an(M) = real(c(M+1)) / N;
            bn(M) = 0;
      endif
endfunction

function y2 = test_fourier_interpolation(y, N)
      [a0, an, bn] =  make_fourier_coefs(y);
      Ncoefs = rows(an);
      y2 = zeros(N, 1);
      for i = 1:N
            x = (i - 1) / (N - 1);
            y2(i) = a0;
            for n = 1:Ncoefs
                  knx = 2 * pi * n * x;
                  y2(i) += an(n) * cos(knx) + bn(n) * sin(knx);
            endfor
      endfor
endfunction
```

Exercise A.1 Carry out the following test of the function `make_fourier_coefs`. Does the output make sense?

```
N = 16;
t = [0:N-1]'/N;
y = cos(2*pi*1*t) + sin(2*pi*3*t)
[a0, an, bn] = make_fourier_coefs(y)
```

Exercise A.2 Show that

$$\sum_{j=0}^{n-1} e^{i2\pi(l-k)j/n} = n\delta_{k,l} \tag{A.17}$$

for $k = 0, \ldots, n - 1$.

Solution: (a) The case $l = k$ is easy. (b) When $l \neq k$ the series is the summation of a "star" of unit vectors that simply adds up to zero. See Figure A.2 for examples of "stars" of unit vectors when n is even and odd.

Comment 1: The Fourier series (A.12) can be written as a matrix-vector-product $\mathbf{Ac} = \mathbf{y}$, where $\mathbf{A}$ is the symmetric matrix with elements $A_{k,j} = e^{i2\pi jk/n}$. The property (A.13) can then be written as $\bar{\mathbf{A}}^T \mathbf{A} = n\mathbf{I}$, where $\mathbf{I}$ is the identity matrix. The complex conjugate and transpose of the matrix $\mathbf{A}$ is its inverse, except for the factor n. The rows (or columns) of $\mathbf{A}$ are therefore orthogonal vectors, and property (A.13) is referred to as an orthogonality property.

Appendix

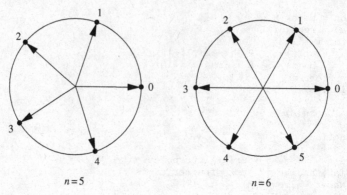

$$n = 5 \qquad\qquad n = 6$$

Figure A.2. *The stops through the unit circle for $n = 5$ (odd) and $n = 6$ (even).*

Comment 2: The discrete Fourier transform is defined with $e^{-i2\pi jk/n}$ and the inverse transform with $e^{i2\pi jk/n}$. The orthogonality relationship remains the same even if opposite signs are used. The discrete Fourier transform is therefore sometimes defined with $e^{i2\pi jk/n}$ while the inverse transform is defined with $e^{-i2\pi jk/n}$.

Exercise A.3 Show that

$$c_l = \frac{1}{n} \sum_{k=0}^{n-1} y_k e^{-i2\pi lk/n}, \quad l = 0, \ldots, n-1 \tag{A.18}$$

follows from the property (A.13).
Solution: We have

$$\sum_{k=0}^{n-1} y_k e^{-i2\pi lk/n} = \sum_{k=0}^{n-1} \left(\sum_{j=0}^{n-1} c_j e^{i2\pi jk/n} \right) e^{-i2\pi lk/n}$$

$$= \sum_{j=0}^{n-1} c_j \left(\sum_{k=0}^{n-1} e^{i2\pi (j-l)k/n} \right)$$

$$= \sum_{j=0}^{n-1} c_j n\, \delta_{j,l}$$

$$= nc_l. \tag{A.19}$$

The Fourier series (A.12) is substituted for y_k in equation (A.19), the summation of j and k is interchanged in equation (A.19), and the (orthogonality) property (A.13) is used in equation (A.19).

Exercise A.4 Show that $\bar{c}_{n-l} = c_l$ when y_k are real numbers.
Solution: From $y_k = \bar{y}_k$ it follows that

$$\bar{y}_k = \sum_{j=0}^{n-1} \bar{c}_j e^{-i2\pi jk/n}$$

$$= \sum_{j=0}^{n-1} \bar{c}_j e^{i2\pi(n-j)k/n}$$

$$= \sum_{l=0}^{n-1} \bar{c}_{n-l} e^{i2\pi lk/n}. \tag{A.20}$$

Equation (A.20) is first multiplied with $e^{i2\pi n/n} = 1$, before the index in equation (A.20) is changed from j to $l = n - j$ (or $j = n - l$). When series (A.20) is compared with series (A.12) for y_k we get that $\bar{c}_{n-l} = c_l$. Notice that $\bar{c}_n = c_0$.

Exercise A.5 Let the interpolation values y_k be real numbers. Use relation $\bar{c}_{n-j} = c_j$, which then applies, to show that the Fourier coefficients in series (A.15) are $a_0 = \text{Re } c_0$, $a_j = 2\text{Re } c_j$ and $b_j = -2\text{Im } c_j$ for $j = 1, \ldots, m$ when n is odd, $n = 2m + 1$.
Solution: We first split the series (A.12) into three parts when $n = 2m + 1$:

$$y_k = c_0 + \sum_{j=1}^{m} c_j e^{i2\pi jk/n} + \sum_{j=m+1}^{2m} c_j e^{i2\pi jk/n}. \tag{A.21}$$

The last summation can be rewritten as follows:

$$\sum_{j=m+1}^{2m} c_j e^{i2\pi jk/n} = \sum_{j=m+1}^{2m} c_j e^{i2\pi(j-n)k/n}$$

$$= \sum_{j=m+1}^{2m} \bar{c}_{n-j} e^{i2\pi(j-n)k/n}$$

$$= \sum_{l=1}^{m} \bar{c}_l e^{-i2\pi lk/n} \tag{A.22}$$

and the series for y_k is

$$y_k = c_0 + \sum_{j=1}^{m} \left(c_j e^{i2\pi jk/n} + \bar{c}_j e^{-i2\pi jk/n} \right)$$

$$= c_0 + \sum_{j=1}^{m} \left((c_j + \bar{c}_j) \cos(2\pi jk/n) + i(c_j - \bar{c}_j) \sin(2\pi jk/n) \right) \tag{A.23}$$

where Euler's formula $e^{ix} = \cos x + i \sin x$ is used to replace the exponentiations. The Fourier coefficients are then identified as

$$a_j = (c_j + \bar{c}_j) = 2\text{Re}\, c_j \qquad (A.24)$$

$$b_j = i(c_j - \bar{c}_j) = -2\text{Im}\, c_j \qquad (A.25)$$

for $j = 1, \ldots, m$, and $a_0 = \text{Re}\, c_0$.

Exercise A.6 Let us assume that the interpolation values y_k are real numbers. Use relation $\bar{c}_{n-j} = c_j$, which then applies, to show that the Fourier coefficients in the series (A.16) for even $n = 2m$ are $a_0 = \text{Re}\, c_0$, $a_m = \text{Re}\, c_m$, $a_j = 2\text{Re}\, c_j$ and $b_j = -2\text{Im}\, c_j$ for $j = 1, \ldots, m - 1$.
Solution: The solution to the even case is similar to the odd case with the exception that the sum over j is split as follows:

$$y_k = c_0 + \sum_{j=1}^{m-1} \bar{c}_j e^{i2\pi jk/n} + c_m e^{i\pi k} + \sum_{j=m+1}^{2m-1} c_j e^{i2\pi jk/n}. \qquad (A.26)$$

Notice that $\bar{c}_m = c_m$ follows from relation $\bar{c}_{n-j} = c_j$ when $n = 2m$, which means that c_m is a real number.

Exercise A.7 Show that the symmetry $y_k = y_{n-k}$ for real values y_k implies that $c_j = \bar{c}_j$, which means that c_j are real numbers.
Solution: We have

$$y_k = y_{n-k} = \sum_{j=0}^{n-1} c_j e^{i2\pi j(n-k)/n} = \sum_{j=0}^{n-1} c_j e^{-i2\pi jk/n} \qquad (A.27)$$

and

$$y_k = \bar{y}_k = \sum_{j=0}^{n-1} \bar{c}_j e^{-i2\pi jk/n}, \qquad (A.28)$$

and it then follows that $c_j = \bar{c}_j$.

Note A.3 *The FFT algorithm.* The summation in the discrete Fourier transform (A.14) leads to $O(n^2)$ operations for the computations of all Fourier coefficients. The FFT is an algorithm that does the summations faster, using only $O(n \log n)$ operations. The FFT carries out the summations in a recursive way that works only when $n = 2^p$ for a positive integer p. How recursion can be used to evaluate the series is seen by rewriting the discrete Fourier transform in two parts, one for the even array elements and another for the odd elements,

$$nc_j = \sum_{k=0}^{n-1} y_k e^{-i2\pi jk/n}$$

$$= \sum_{k=0}^{n/2-1} y_{2k} e^{-i2\pi j2k/n} + \sum_{k=0}^{n/2-1} y_{2k+1} e^{-i2\pi j(2k+1)/n}$$

$$= \sum_{k=0}^{m-1} f_k e^{-i2\pi jk/m} + e^{-i2\pi j/n} \sum_{k=0}^{m-1} g_k e^{-i2\pi jk/m}. \tag{A.29}$$

The even array elements are $f_k = y_{2k}$ and the odd elements are $g_k = y_{2k+1}$, and these two arrays have half the length of the initial array, with $k = 0, \ldots, m - 1$ where $m = n/2$. The original discrete Fourier transform is now written as the sum of two discrete Fourier transforms, where both are only half as long as the initial one. Each of the two discrete Fourier transforms can recursively be split further into new discrete Fourier transforms, until we end up with a series that is only one element long. The discrete Fourier transform of just one element is the same as multiplication by 1. It is the successive division by 2 that requires the initial length of the series to be 2^p for a positive integer p. How the recursion goes is illustrated for the case $n = 8$, where the Fourier coefficients are

$$8c_j = \big((y_0 + w^{4j}y_4) + w^{2j}(y_2 + w^{4j}y_6)\big)$$
$$+ w^j\big((y_1 + w^{4j}y_5) + w^{2j}(y_3 + w^{4j}y_7)\big) \tag{A.30}$$

with $w = e^{-i2\pi/8}$. The computation of the Fourier coefficients is therefore carried out as shown in Figure A.4, after the array has been reorganized as shown in Figure A.3. This explains why the FFT only needs $O(np) = O(n\log_2 n)$ operations. The splitting of the

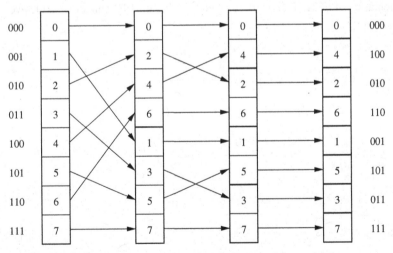

Figure A.3. *The recursive regrouping of an array into even and odd parts leads to a reorganization of the array, which is obtained directly by the reversal of the bit patterns of the array index.*

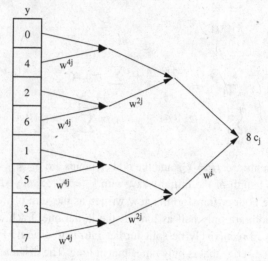

Figure A.4. *The computation of a complex Fourier coefficient using the reorganized array from Figure A.3. See also equation (A.30).*

discrete Fourier transform shown by equation (A.29) applies equally well to the inverse transform (A.12), because it is only the sign of the complex exponent that is different.

Note A.4 This Octave code is an example of how a FFT function can be programmed using recursion. The if-branch is an implementation of equation (A.29), which shows that the FFT of an array can be reduced to two FFT-calls on two half-as-long sub-arrays. The output from myFFT can be compared with the output from Octave's built-in fft-function. Although the example seems simple it not a trivial matter to make an efficient and robust implementation of FFT. Much work is saved by using the FFT-functions in Octave, Matlab or similar tools.

```
function c = myFFT(x)
    x = x(:);
    N = length(x);
    omega = exp(-i * 2 * pi / N);
    if (rem(N, 2) == 0)
        k = (0:N/2-1)';
        w = omega .^ k;
        u = myFFT(x(1:2:N-1));
        v = w .* myFFT(x(2:2:N));
        c = [u+v; u-v];
    elseif (N == 1)
        c = x;
    else
        printf("Error: N != 2^p for integer p\n");
        exit(1);
    endif
endfunction
```

References

Aase, N., P. Bjørkum, and P. Nadeau (1996). The effect of grain coating micro-quartz on preservation of reservoir porosity. *AAPG Bulletin* **80**(10), 1654–1673.

Allen, P. and J. Allen (1990). *Basin Analysis – Principles and Applications*. Oxford: Blackwell.

Anderson, M. (2005). Heat as a ground water tracer. *Ground Water* **43**(6), 951–968.

Angevine, C. and D. Turcotte (1993). Porosity reduction by pressure solution: a theoretical model for quartz arenites. *Bulletin of the Geological Society of America* **94**, 1129–1134.

Aris, R. (1962). *Vectors, Tensors and the Basic Equations of Fluid Mechanics*. New York: Dover.

Athy, L. (1930). Density, porosity, and compaction of sedimentary rocks. *AAPG Bulletin* **14**, 1–24.

Audet, D. M. and A. C. Fowler (1992). A mathematical model for compaction in sedimentary basins. *Geophysical Journal International* **110**, 577–590.

Audet, D. M. (1995). Mathematical-modelling of gravitational compaction and clay dehydration in thick sedimentary layers. *Geophysical Journal International* **122**, 283–298.

Balling, N. (1995). Heat flow and thermal structure of the lithosphere across the Baltic Shield and northern Tornquist Zone. *Tectonophysics* **244**, 13–50.

Bear, J. (1972). *Dynamics of Fluids in Porous Media*. Elsevier, New York.

Bellingham, P., and N. White (2000). A general inverse method for modelling extensional sedimentary basins. *Basin Research*, **12**, 2190–226.

Bellingham, P. and N. White (2002). A two-dimensional inverse model for extensional sedimentary basins: 2 Application. *Journal of Geophysical Research* **107**(B10), 1–18.

Beltrami, H. (2001). Surface heat flux histories from geothermal data: Inferences from inversion. *Geophysical Research Letters* **28**(4), 655–658.

Beltrami, H., L. Cheng, and J.-C. Mareschal (1997). Simultaneous inversion of borehole temperature data for determination of ground surface temperature history. *Geophysical Journal International* **129**, 311–318.

Beltrami, H. and J.-C. Mareschal (1992). Ground temperature histories for central and eastern Canada from geothermal measurements. *Geophysical Research Letters* **19**, 689–692.

Beltrami, H. and J.-C. Mareschal (1995). Resolution of ground temperature histories inverted from borehole temperature data. *Global and Planetary Change* **11**, 57–70.

Benfield, A. E. (1949). The effect of uplift and denudation on underground temperatures. *Journal of Applied Physics* **220**, 66–70.

Berryman, J. (1987). Relationship between specific surface area and spatial correlation functions for anisotropic porous media. *Journal of Mathematical Physics* **28**(1), 244–245.

Bethke, C. and T. Corbet (1988). Linear and nonlinear solutions for one-dimensional compaction flow in sedimentary basins. *Water Resources Research* **24**(3), 461–467.

Bickle, M. and D. McKenzie (1987). The transport of heat and matter by fluids during metamorphism. *Contributions to Mineralogy and Petrology* **95**, 384–392.

Biot, M. (1941). General theory of three dimensional consolidation. *Journal of Applied Physics* **12**, 155–164.

Birchwood, R. and D. Turcotte (1994). A unified approach to geopressuring, low-permeability zone formation, and secondary porosity generation in sedimentary basins. *Journal of Geophysical Research* **99**(B10), 20.051–20.058.

Bjørkum, P. (1994). How important is pressure in causing dissolution of quartz in sandstones? *AAPG Annual Meeting Program with Abstracts* **3**, 105.

Bjørkum, P. (1996). How important is pressure in causing dissolution of quartz in sandstones? *Journal of Sedimentary Research* **66**(1), 147–154.

Bjørkum, P., E. Oelkers, P. Nadeau, O. Walderhaug, and W. Murphy (1998). Porosity prediction in quartzose sandstones as a function of time, temperature, depth stylolite frequency, and hydrocarbon saturation. *AAPG Bulletin* **82**(4), 637–648.

Bjørlykke, K. (1993). Fluid flow in sedimentary basins. *Sedimentary Geology* **86**, 137–138.

Blackwell, D. and J. Steele (1989). *Thermal Conductivity of Sedimentary Rocks: Measurement and Significance*, pp. 13–36. New York: Springer.

Blakely, R. J. (1996). *Potential Theory in Gravity and Magnetic Applications*. Cambridge: Cambridge University Press.

Bloch, S., R. Lander, and L. Bonnell (2002). Anomalously high porosity and permeability in deeply buried sandstone reservoirs: origin and predictability. *American Association of Petroleum Geologists Bulletin* **86**(2), 301–328.

Bloomer, J. (1981). Thermal conductivities of mudrocks in the United Kingdom. *Quarterly Journal of Engineering Geology* **14**, 357–362.

Bourbie, T. and B. Zinszner (1985). Hydraulic and acoustic properties as a function of porosity in Fontainebleau sandstone. *Journal of Geophysical Research* **90**(B13), 11,524–11,532.

Bracewell, R. (1965). *The Fourier Transform and its Applications*. New York: McGraw-Hill.

Bredehoeft, J. D. (2002). Hydrodynamics of sedimentary basins. *Encyclopedia of Physical Science and Technology* **7**, 471–488.

Bredehoeft, J. and B. Hanshaw (1968). On the maintenance of anomalous fluid pressures: I. thick sedimentary sequences. *Geological Society of America Bulletin* **79**, 1097–1106.

Bredehoeft, J. and I. Papadopolous (1965). Rates of vertical groundwater movement estimated from the Earth's thermal profile. *Water Resources Research* **1**, 325–328.

Burnham, A. and J. Sweeney (1989). A chemical kinetic model of vitrinite maturation and reflectance. *Geochimica et Cosmochimica Acta* **53**(10), 2649–2657.

Byerlee, J. (1977). Friction of rocks. In J. Evernden (ed.), *Experimental Studies of Rock Friction with Application to Earthquake Prediction*, pp. 55–77. Menlo Park, CA: US Geological Survey.

Byerlee, J. (1978). Friction of rocks. *Pure and Applied Geophysics* **116**, 615–626.

Carman, P. (1937). Fluid through a granular bed. *Transactions of the Institute of Chemical Engineering London* **15**, 150–156.

Carslaw, H. and J. Jaeger (1959). *Conduction in Solids,* 2nd edn. Oxford: Clarendon Press.

Chilingar, G., V. Serebryakov, and J. Robertson (2002). *Origin and Prediction of Abnormal Formation Pressures.* Amsterdam: Elsevier.

Chouinard, C. and J.-C. Mareschal (2007). Selection of borehole temperature depth profiles for regional climate reconstructions. *Climate of the Past* **3**(2), 297–313. See www.clim-past.net/3/297/2007/.

Clauser, C. and E. Huenges (1995). *Thermal Conductivity of Rocks and Minerals,* pp. 105–126. Washington DC: American Geophysical Union.

Crosby, A. and N. White and G. R. Edwards (2010). Self-consistent strain rate and heat flow modelling of lithospheric extension: application to Newfoundland-Iberia conjugate margins. *Petroleum Geoscience* **16**, 247–256.

Davis, G. and S. Reynolds (1996). *Structural Geology of Rock and Regions.* Wiley, New York.

Dewers, T. and P. Ortoleva (1990). A coupled reaction–transport-mechanical model for intergranular pressure solution, stylolites and differential compaction and cementation in clean sandstones. *Geochimica et Cosmochimica Acta* **36**, 1337–1358.

Domenico, P. and V. Palciauskas (1973). Theoretical analysis of forced convective heat transfer in regional ground-water flow. *Geological Society of America Bulletin* **84**, 3803–3813.

Domenico, P. and F. Schwartz (1998). *Physical and Chemical Hydrogeology,* 2nd edn. Wiley, New York.

Davis, R. O. and A. P. S. Selvadurai (1996). *Elasticity and geomechanics.* Cambridge University Press.

Davis, R.O. and A.P.S. Selvadurai (2002). *Plasticity and geomechanics.* Cambridge University Press.

Doré, A., E. R. Lundin, L. N. Jensen, Ø. Birkeland, P. E. Eliassen, and C. Fichler (1999). Principal tectonic events in the evolution of the northwest European Atlantic margin. In A. J. Fleet and S. A. R. Boldy (eds.), *Petroleum Geology of Northwest Europe: Proceedings of the Fifth Conference,* pp. 41–61. London: Geological Society.

Dullien, F. (1979). *Porous Media, Fluid Transport and Pore Structure.* New York: Academic Press.

Dutton, S., W. Flanders, and M. Barton (2003). Reservoir characterization of a Permian deep-water sandstone, East Ford field, Delaware basin, Texas. *AAPG Bulletin* **87**(4), 609–627.

Ferguson, G., A. Woodbury, and G. Matile (2003). Estimating deep recharge rates beneath an interlobate moraine using temperature logs. *Ground Water* **41**(5), 640–646.

Forsyth, D. (1985). Subsurface loading and estimates of the flexural rigidity of continental lithosphere. *Journal of Geophysical Research* **90**(B14), 12,623–12,632.

Fung, Y. (1965). *Foundations of Solid Mechanics.* Englewood Cliffs, NJ: Prentice Hall.

Ge, S. (1998). Estimation of groundwater velocity in localised fracture zones from well temperature. *Journal of Volcanology and Geothermal Research* **84**, 93–101.

Gemmer, L., S. Ings, S. Medvedev, and C. Beaumont (2004). Salt tectonics driven by differential sediment loading: Stability analysis and finite element experiments. *Basin Research* **16**, 199–218.

Gernigon, L., F. Lucazeau, F. Brigaud *et al.* (2006). A moderate melting model for the Vøring Margin (Norway) based on structural observations and thermo-kinematical modeling: Implication for the meaning of the lower crustal body. *Tectonophysics* **412**, 255–278.

Gernigon, L., J. Ringenbach, and S. Planke (2003). Extension, crustal structure and magmatism at the outer Vøring Basin, North Atlantic margin, Norway. *Journal of the Geological Society London* **160**, 197–208.

Gibson, R. (1958). The progress of consolidation in a clay layer increasing in thickness with time. *Geotechnique* **8**, 171–182.

Gonzalez-Rouco, J. F., H. Beltrami, E. Zorita, and M. B. Stevens (2008). Borehole climatology: a discussion based on contributions from climate modeling. *Climate of the Past Discussions* **4**, 1–80. See www.clim-past-discuss.net/4/1/2008/.

Goto, S. and O. Matsubayashi (2008). Inversion of needle-probe data for sediment thermal properties of the eastern flank of the Juan de Fuca Ridge. *Journal of Geophysical Research* **113**(B08105), 1–17.

Guéguen, Y. and V. Palciaukas (1994). *Introduction to the physics of rocks*. Princeton University Press, Princeton, NJ.

Guillou, L., J. Mareschal, C. Jaupart *et al.* (1994). Heat flow and gravity structure of the Abitibi belt, Superior province. *Earth and Planetary Science Letters* **122**, 447–1460.

Gunnarsson, I. and S. Arnórsson (2000). Amorphous silica solubility and the thermodynamic properties of H_4SiO_2 in the range of 0°C to 350°C. *Geochimica et Cosmochimica Acta* **64**(13), 2295–2307.

Halvorsen, G. (2003). Personal communication.

Haq, B., J. Hardenbol, and P. Vail (1987). Chronology of fluctuating sea levels since the Triassic. *Science* **235**, 1156–1167.

Haug, E. and H. Langtangen (1997). Basic equations in eulerian continuum mechanics. In M. Daehlem and A. Tveito (eds.), *Numerical Models and Software Tools in Industrial Mathematics*, pp. 1–36. Basel: Birkäuser.

Heald, M. (1955). Stylolites in sandstones. *Journal of Geology* **63**, 101–114.

Helland-Hansen, W., C. Kendall, I. Lerche, and K. Nakayama (1988). A simulation of continental basin margin sedimentation in response to crustal movements, eustatic sea level change, and sediment accumulation rates. *Mathematical Geology* **20**, 777–802.

Hermanrud, C. and H. M. N. Bolås (2002). *Leakage from Overpressured Hydrocarbon Reservoirs at Haltenbanken and in the Northern North Sea*, pp. 221–231. Amsterdam: Elsevier.

Hofmeister, A. (1999). Mantle values of thermal conductivity geotherm from phonon lifetimes. *Science* **283**, 1699–1709.

Huang, S., H. Pollack, and P. Shen (2000). Temperature trends over the past five centuries reconstructed from borehole temperatures. *Nature* **403**, 756–758.

Hubbert, M. (1940). The theory of groundwater motion. *Journal of Geology* **48**, 785–944.

Hubbert, M. (1948). A line-integral method of computing the gravimetric effects of two-dimensional masses. *Geophysics* **30**, 215–225.

Hudec, M. and M. Jackson (2007). Understanding salt tectonics. *Earth-Science Reviews* **82**, 1–28.

Huismans, R. S. and C. Beaumont (2008). Complex rifted continental margins explained by dynamical models of depth-dependent lithospheric extension. *Geology* **36**(2), 63–166.

Hutchison, I. (1985). The effect of sedimentation and compaction on oceanic heat flow. *Geophysical Journal of the Royal Astronomical Society* **82**, 439–459.

International Association of Geodesy (2009). See site: http://Earth-info.nga.mil/GandG/ wgs84/agp/index.html.

International Heat Flow Commission (2008). See site: http://www.geo.lsa.umich.edu/ IHFC/.

Isaksen, K. and J. Sollid (2002). Permafrosten tiner. *Tidsskrift fra CICERO* **11**(4), 4–6. See site: www.cicero.uio.no.

Issler, D. (1984). Calculation of organic maturation levels for offshore eastern Canada – implications for general application of Lopatin's method. *Canadian Journal of Earth Sciences* **21**, 477–488.

Jaeger, J. (1964). Thermal effects of intrusions. *Reviews of Geophysics* **2**(3), 443–466.

Jaeger, J. (1968). Cooling and solidification of igneous rocks. In H. Hess and A. Poldervaat (eds.), *Basalts*, Volume 2, pp. 503–536. New York: Wiley.

Jaeger, J. and N. Cook (1969). *Fundamentals of Rock Mechanics*. Chapman and Hall, London.

Jaeger, J., N. Cook and R. Zimmerman. (2007). *Fundamentals of rock mechanics* (4th ed.), Oxford: Blackwell.

Japsen, P. and J. Chalmers (2000). Neogen uplift and tectonics around the North Atlantic: overview. *Global and Planetary Change* **24**, 165–173.

Jarvis, G. and D. McKenzie (1980). Sedimentary basin formation with finite extension rates. *Earth and Planetary Science Letters* **48**, 42–52.

Jaupart, C., S. Labrosse, and J.-C. Mareschal (2007). Temperatures, heat and energy in the mantle of the Earth. In Bercovici and Schubert (eds.), *Mantle Dynamics*, Treatise on Geophysics, pp. 253–303. Amsterdam: Elsevier.

Jones, S. M. and N. White and P. Faulkner and P. Bellingham (2004). Animated models of extensional basins and passive margins. *Geochemistry Geophysics Geosystems* **5**, 1–38, 10.1029/2003GC000658.

Judd, A. and M. Hovland (2007). *Sea Bed Fluid Flow*. Cambridge: Cambridge University Press.

Karner, G. (1991). Sediment blanketing and the flexural strength of the extended continental lithosphere. *Basin Research* **3**, 177–185.

Kaus, B., J. Connolly, Y. Podladchikov, and S. Schmalholz (2005). The effect of mineral phase changes on sedimentary basin subsidence and uplift. *Earth and Planetary Science Letters* **233**, 213–228.

Kozeny, J. (1927). Über kapillare leitung des wassers im boden. *Sitzungsber Akademie der Wissensehaften Wien* **136**, 271–306.

Kreyszig, E. (2006). *Advanced engineering mathematics* (9th ed.). New York: Wiley.

Kümpel, H.-J. (1991). Poroelasticity: parameters reviewed. *Geophysical Journal International* **105**, 783–799.

Lemée, C. and Y. Guéguen (1996). Modeling of porosity loss during compaction and cementation of sandstones. *Geology* **24**(10), 875–878.

Lerche, I. (1990a). *Basin analysis. Quantitative Methods*, Volume 1. New York: Academic Press.

Lerche, I. (1990b). *Basin analysis. Quantitative Methods*, Volume 2. New York: Academic Press.

Li, X. and H.-J. Götze (2001). Ellipsoid, geoid, gravity and geophysics. *Geophysics* **66**(6), 1660–1668.

Lopatin, N. (1971). Temperature and geological time as a factor in coalification. *Izvestiya Akademii Nauk SSSR, Seriia Geologicheskie* **3**, 95–106.

Lubis, R. and Y. Sakura (2008). Groundwater recharge and discharge processes in the Jakarta groundwater basin, Indonesia. *Hydrogeology Journal* **16**, 927–938.

Lucazeau, F. and S. L. Douaran (1985). The blanketing effect of sediments from by extension: a numerical model. Application to the Gulf of Lion and Viking Graben. *Earth and Planetary Science Letters* **74**, 92–102.

Lundin, E. R. and A. G. Doré (1997). A tectonic model for the Norwegian passive margin with implications for the NE Atlantic Early Cretaceous to break-up. *Journal of the Geological Society London* **154**, 545–550.

Mallon, A. and R. Swarbrick (2008). How should permeability be measured in fine-grained lithologies? Evidence from the chalk. *Geofluids* **8**, 35–45.

Manning, C. and S. Ingebritsen (1999). Permeability of the continental crust: implications of geothermal data and metamorphic systems. *Reviews of Geophysics* **40**(1), 127–150.

Mareschal, J.-C. (2005). University of Michigan borehole database - CA-8907. See site: www.ncdc.noaa.gov/paleo/indexbore.html.

Mase, G. (1970). *Theory and Problems of Continuum Mechanics*. Schaum's Outline Series. New York: McGraw Hill.

McAdoo, D. and D. Sandwell (1989). On the source of cross-grain lineations in the central Pacific gravity field. *Journal of Geophysical Research* **94**(B7), 9341–9352.

McKenzie, D. (1978). Some remarks on the development of sedimentary basins. *Earth and Planetary Science Letters* **40**, 25–32.

McKenzie, D. (1984). The generation and compaction of partially molten rock. *Journal of Petrology* **25**, 713–765.

McKenzie, D. (2003). Estimating t_e in the presence of internal loads. *Journal of Geophysical Research* **108**(B9, 2438), 1–21.

McKenzie, D. and M. Bickle (1988). The volume and composition of melt generated by extension of the lithosphere. *Journal of Petrology* **29**, 625–679.

McKenzie, D. and C. Bowin (1976). The relationship between bathymetry and gravity in the Atlantic Ocean. *Journal of Geophysical Research* **81**, 1903–1915.

McKenzie, D. and D. Fairhead (1997). Estimates of the effective elastic thickness of the continental lithosphere from Bouguer and free-air gravity anomalies. *Journal of Geophysical Research* **102**, 27,523–27,552.

McKenzie, D. and K. P. J. Jackson (2005). Thermal structure of oceanic and continental lithosphere. *Earth and Planetary Science Letters* **233**, 337–349.

Mello, U., G. Karner, and R. Anderson (1995). Role of salt in restraining the maturation of subsalt source rocks. *Marine and Petroleum Geology* **12**(7), 697–716.

Midttømme, K., E. Roaldset, and P. Aagaard (1998). Thermal conductivity of selected claystones and mudstones from England. *Clay Minerals* **33**, 131–145.

Mjelde, R., S. Kodaira, H. Shimamura *et al.* (1997). Crustal structure of the central part of the Vøring basin, mid-Norway margin, from ocean bottom seismographs. *Tectonophysics* **277**, 235–257.

Monicard, R. (1980). *Properties of Reservoir Rocks: Core Analysis*. Editons Technip, Paris.

Morrow, D. and D. Issler (1993). Calculation of vitrinite reflectance from thermal histories: A comparison of some methods. *AAPG Bulletin* **77**(4), 610–624.

Nadai, A. (1963). *The Theory of Flow and Fracture of Solids*, Volume 2. New York: McGraw-Hill.

National Imagery and Mapping Agency (1984). World geodetic system 1984 – its definition and relationships with local geodetic systems. Report NIMA TR8350.2. See site: http://www1.nga.mil/ProductsServices/GeodesyGeophysics.

Nelson, P. (2004a). Permeability–porosity data sets for sandstones. *The Leading Edge* **23**(11), 1143–1144.

Nelson, P. (2004b). Permeability-porosity relationships in sedimentary rocks. *The Log Analyst* **35**(3), 38–62.

Nelson, P. (2005). Permeability, porosity, and pore-throat size – a three-dimensional perspective. *Petrophysics* **46**(6), 452–455.

Nelson, P. and J. Kibler (2003). A catalog of porosity and permeability from core plugs in siliciclastic rocks. *Technical Report*, U.S. Geological Survey. Open-file report 03-420: http://pubs.usgs.gov/of/2003/ofr-03-420/ofr-03-420.html.

Neuzil, C. (1994). How permeable are clays and shales? *Water Resources Research* **30**(2), 145–150.

Nield, D. and A. Bejan (1998). *Convection in Porous Media*, 2nd edition. New York: Springer.

Norden, B. and A. Förster (2006). Thermal conductivity and radiogenic heat production of sedimentary and magmatic rocks in the Northeast German Basin. *AAPG Bulletin* **90**(6), 939–962.

Norwegian Petroleum Directorate (2009). Data for a large number of wells are available from the site: http://www.npd.no/engelsk/cwi/pbl/en/wdss_index.htm.

Octave (2009). See site: http://www.gnu.org/software/octave.

Oelkers, E., P. Bjørkum, and W. Murphy (1992). The mechanism of porosity reduction, stylolite development and quartz cementation in North Sea sandstones. In Y. Kharaka and A. Maest (eds.), *Water–Rock Interaction*, Vol. 2, pp. 1183–1186. Rotterdam: Balkema.

Oelkers, E., P. Bjørkum, and W. Murphy (1993). Calculation of the rate and distribution of chemically driven quartz cementation on North Sea sandstones. In M. Cuney and M. Cathelineau (eds.), *Proceedings of the Fourth International Symposium on Hydrothermal Reactions*, pp. 169–172. Nancy, France: Institut Lorain des Geosciences.

Oelkers, E., P. Bjørkum, and W. Murphy (1996). A petrographic and computational investigation of quartz cementation and porosity reduction in North Sea sandstones. *American Journal of Science* **296**, 420–452.

Oelkers, E., P. Bjørkum, O. Walderhaug, P. Nadeau, and W. Murphy (2000). Making diagenesis obey thermodynamics and kinetics: the case of quartz cementation in sandstones from offshore mid-Norway. *Applied Geochemistry* **15**, 295–309.

Parker, R. (1972). The rapid calculation of potential anomalies. *Geophysical Journal of the Royal Astronomical Society* **31**, 447–455.

Parks, R. (1989). *Foundation of Structural Geology*. Glasgow: Blackie Academic and Professional.

Pavlis, N., S. Holmes, S. Kenyon, and J. Factor (2008). An Earth gravitational model to degree 2160. *Technical Report EGM2008*, 2008 Vienna: General Assembly of the European Geosciences Union, April 13–18, 2008. See site: http://Earth-info.nga.mil/GandG/wgs84/gravitymod/egm2008/anomalies_dov.html.

Perry, H., C. Jaupart, J.-M. Mareschal, and G. Bienfait (2006). Crustal heat production in the Superior province, Canadian shield, and North America inferred from heat flow data. *Journal of Geophysical Research 11, B04401*, 1–20.

Person, M., J. P. Raffensperger, S. Ge, and G. Garven (1996). Basin-scale hydrogeological modeling. *Reviews of Geophysics* **34**(1), 61–87.

Petersen, K. and I. Lerche (1995). Quantification of thermal anomalies in sedimentary basins. *Geothermics* **24**, 253–268.

Podladchikov, Y., A. Poliakov, and D. Yuen (1994). The effect of lithospheric phase transitions on subsidence of extending continental lithosphere. *Earth and Planetary Science Letters* **124**, 95–103.

Poplavskii, K., Y. Podladchikov, and A. Stephenson (2001). Two-dimensional inverse modeling of sedimentary basins subsidence. *Journal of Geophysical Research* **106**, 6657–6671.

Pribnow, D., M. Kinoshita, and C. Stein (2000). Thermal data collection and heat flow recalculations for ocean drilling program legs 101–180. *Technical Report*, Institut für Geowissenschaftliche Gemainschaftsaufgaben. See site: www-odp.tamu.edu.

Rajesh, R. and D. Mishra (2003). Admittance analysis and modelling of satellite gravity over Himalayas-Tibet and its seismogenic correlation. *Current Science* **84**(2), 224–230.

Ramm, M. (1992). Porosity-depth trends in reservoir sandstones: theoretical models related to Jurassic sandstones offshore Norway. *Marine and Petroleum Geology* **9**, 553–567.

Ranalli, G. (1995). *Rheology of the Earth*, 2nd edn. London: Chapman & Hall.

Reemst, P. and S. Cloetingh (2000). Polyphase rift evolution of the Vøring margin (mid-Norway): Constraints from forward tectonostatigraphic modeling. *Tectonophysics* **19**, 225–240.

Reiter, M. (2001). Using precision temperature logs to estimate horizontal and vertical groundwater flow components. *Water Resources Research* **37**(3), 663–674.

Ren, S., J. Faleide, O. Eldholm, J. Skogseid, and F. Gardstein (2003). Late Cretaceous–Paleocene tectonic development of the NW Vøring basin. *Marine and Petroleum Geology* **20**, 177–206.

Riley, K., M. Hobson, and S. Bence (1998). *Mathematical Methods for Physics and Engineering*. Cambridge: Cambridge University Press.

Roy, R., D. Blackwell, and F. Birch (1968). Heat generation of plutonic rocks and continental heat flow provinces. *Earth and Planetary Science Letters* **5**, 1–12.

Royden, L. and C. Keen (1980). Rifting process and thermal evolution of the continental margin of eastern Canada determined from subsidence curves. *Earth and Planetary Science Letters* **51**(2), 343–361.

Rudnick, R. and D. Fountain (1995). Nature and composition of the continental crust: A lower crustal perspective. *Reviews of Geophysics* **33**, 267–309.

Rüpke, L., S. Schmalholz, D. Schmid, and Y. Podladchikov (2008). Automated thermotectonostratigraphic basin reconstruction: Viking Graben case study. *AAPG Bulletin* **92**(3), 309–326.

Sandwell, D. and W. Smith (1997). Marine gravity anomaly from Geosat and ERS1 satellite altimetry. *Journal of Geophysical Research* **102**(B5), 10,039–10,054. See site: http://topex.ucsd.edu/cgi-bin/get_data.cgi.

Sandwell, D. T. and W. H. F. Smith (2009). Marine gravity from satellite altimetry, Scripps Institution of Oceanography, UC San Diego. See site: http://topex.ucsd.edu/marine_grav/mar_grav.html.

Sawyer, D. and D. Fackler (2007). Data report: automatic method for associating lithologic type with laboratory measurements taken aboard the JOIDES Resolution. In B. Tucholke, J.-C. Sibuet, and A. Klaus (eds.), *Proceedings of the Ocean Drilling Program, Scientific Results*, Volume 210, pp. 1–21. Ocean Drilling Program. See site: www-odp.tamu.edu.

Schneider, F., J. Potdevin, S. Wolf, and I. Faille (1996). Mechanical and chemical compaction model for sedimentary basin simulators. *Tectonophysics* **263**, 307–317.

Sclater, J. and P. Christie (1980). Continental stretching: An explanation for the post-mid-Cretaceous subsidence of the central North Sea basin. *Journal of Geophysical Research* **85**(B7), 3711–3739.

Seipold, U. (1995). The variation of thermal transport properties in the Earth's crust. *Journal of Geodynamics* **20**(2), 145–154.

Seipold, U. (1998). Temperature dependence of thermal transport properties of crystaline rocks – a general law. *Tectonophysics* **291**, 161–171.

Simon, N. and Y. Podladchikov (2008). The effect of mantle composition on density in the extending lithosphere. *Earth and Planetary Science Letters* **272**, 148–157.

Song, I. and J. Renner (2008). Hydromechanical properties of Fontainebleau sandstone: experimental determination and micromechanical modeling. *Journal of Geophysical Research* **113**(B09211), 1–16.

Søreng, J. (1989). Formulation and application of element methods in the numerical simulation of geological basins. *International Journal for Numerical and Analytical Methods in Geomechanics* **13**, 525–543.

Stallman, R. (1963). Computation of ground-water velocity from temperature data. In R. Bentall (ed.), *Methods of Collecting and Interpreting Ground-Water Data*, pp. 36–46. Geological Survey Water Supply paper 1544-H, Washington, DC: USGS.

Stüwe, K. (2002). *Geodynamics of the Lithosphere*. Springer, Berlin.

Swanberg, C., M. Chessman, G. Simmons, S. Smithson, G. Grøle, and K. Heier (1974). Heat flow and heat generation studies in Norway. *Tectonophysics* **23**, 31–48.

Sweeney, J. and A. Burnham (1990). Evaluation of a simple model of vitrinite reflectance based on chemical kinetics. *AAPG Bulletin* **74**, 1559–1570.

Talwani, M., J. Worzel, and M. Landisman (1959). Rapid computation for two-dimensional bodies with application to the Mendocino submarine fracture zone. *Journal of Geophysics Research* **64**, 49–59.

Terzaghi, K. (1925). *Erdbaumechanik auf Bodenphysikalischer Grundlage*. Leipzig: F. Deuticke.

Tester, J., G. Worley, B. Robinson, C. Grigsby, and J. Feerer (1994). Correlating quartz dissolution kinetics in pure water from 25 to 625°C. *Geochimica et Cosmochimica Acta* **58**(11), 2407–2420.

Theis, C. (1935). The relation between the lowering of the piezometric surface and the rate and duration of discharge of a well using groundwater storage. *Transactions of the American Geophysical Union* **2**, 519–524.

Tissot, B. and D. Welte (1978). *Petroleum Formation and Occurrence*. Berlin: Springer.

Tóth, J. (1963). A theoretical analysis of groundwater flow in small drainage basins. *Journal of Geophysical Research* **68**(16), 4795–4812.

Turcotte, D. and G. Schubert (1982). *Geodynamics*. New York: Wiley.

Turquato, S. (2002). *Random Heterogeneous Materials: Microstructure and Macroscopic Properties*. New York: Springer.

USGS (1999). See site: http://earthquake.usgs.gov/research/structure/crust/download.php.

Van Schmus, W. (1995). Natural radioactivity of the crust and mantle. In *A Handbook of Physical Constants*, AGU reference shelf 1, Global Earth Physics. American Geophysical Union.

Vasseur, G., P. Bernard, J. V. de Meulebrouck, Y. Kast, and J. Jolivet (1983). Holocene paleotemperatures deduced from geothermal measurements. *Paleogeography, Paleoclimatology, Paleoecology* **43**, 237–259.

Vasseur, G. and L. Demongodin (1995). Convective and conductive heat transfer in sedimentary basins. *Basin Research* **7**, 67–79.

Verdoya, M., V. Pasquale, and P. Chiozzi (2008). Inferring hydro-geothermal parameters from advectively perturbed thermal logs. *International Journal of Earth Sciences (Geologische Rundschau)* **97**, 333–344.

Walderhaug, O. (1994a). Precipitation rates for quartz cement in sandstones determined by fluid-inclusion microthermometry and temperature history modeling. *Journal of Sedimentary Research* **A64**(2), 324–333.

Walderhaug, O. (1994b). Temperatures of quartz cementation in Jurassic sandstones from the Norwegian continental shelf - evidence from fluid inclusions. *Journal of Sedimentary Research* **A64**(2), 311–323.

Walderhaug, O. (1996). Kinetic modeling of quartz cementation and porosity loss in deeply buried sandstone reservoirs. *AAPG Bulletin* **80**(5), 731–745.

Wang, H. (2000). *Theory of Linear Poroelasticity*. Princeton, NJ: Princeton University Press, Princeton.

Wang, K. (1992). Estimation of ground surface temperatures from borehole temperature data. *Journal of Geophysical Research* **97**, 2095–2106.

Wangen, M. (1992). Pressure and temperature evolution in sedimentary basins. *Geophysics Journal International* **110**, 601–613.

Wangen, M. (1993). A finite element formulation in Lagrangian coordinates for heat and fluid flow in compacting sedimentary basins. *International Journal for Numerical and Analytical Methods in Geomechanics* **15**, 705–733.

Wangen, M. (1995). The blanketing effect in sedimentary basins. *Basin Research* **7**, 283–298.

Wangen, M. (1997). Simple model of pressure build-up during burial. *Geophysical Journal International* **130**, 757–764.

Wangen, M. (1999). Modelling quartz cementation of quartzose sandstones. *Basin Research* **11**, 111–126.

Wangen, M. (2000). Generation of overpressure by cementation of pore space in sedimentary rocks. *Geophysical Journal International* **143**, 608–620.

Wangen, M. (2001). A quantitative comparison of some mechanisms generating overpressure in sedimentary basins. *Tectonophysics* **334**, 211–234.

Wangen, M. and J. Faleide (2008). Estimation of crustal thinning by accounting for stretching and thinning of the sedimentary basin – an example from the Vøring margin, NE Atlantic. *Tectonophysics* **457**, 224–238.

Wangen, M., W. Fjeldskaar, J. Faleide *et al.* (2007). Forward modeling of stretching episodes and paleo heat flow of the Vøring margin, NE Atlantic. *Journal of Geodynamics* **45**, 83–98.

Waples, D. (1980). Time and temperature in petroleum formation: application of Lopatin's method to petroleum exploration. *AAPG Bulletin* **64**, 916–926.

Watts, A. (2001). *Isostasy and Flexure of the Lithosphere*. Cambridge: Cambridge University Press.

Weissel, J. and G. Karner (1989). Flexural uplift of rift flanks due to mechanical unloading of the lithosphere during extension. *Journal of Geophysical Research* **94**(B10), 13,919–13,950.

Wernicke, B. (1985). Uniform-sense normal simple shear of the continental lithosphere. *Canadian Journal of Earth Sciences* **22**, 108–125.

White, N. (1994). An inverse method for determining lithospheric strain rate variation on geological timescales. *EPSL* **122**, 3–4, 351–371.

Xu, Y., T. Shankland, S. Linhardt, D. Rubie, F. Lagenhorst, and K. Klasinsk (2004). Thermal diffusivity and conductivity of olivine, wadsleyite and ringwoodite to 20 GPa and 1373 K. *Physics of the Earth and Planetary Interiors* **143–144**, 321–336.

Yamasaki, T. and M. Nakada (1997). The effect of the spinel-garnet phase transition on the formation of rifted sedimentary basins. *Geophysical Journal International* **130**, 681–692.

Zimmerman, R., W. Somerton, and M. King (1986). Compressibility of porous rocks. *Journal of Geophysical Research* **91**(B12), 12.765–12.777.

Index

Printed in the United States
By Bookmasters